Cytochrome P-450

Cytochrome P-450

**Structural and Functional Relationships
Biochemical and Physicochemical Aspects
of Mixed Function Oxidases**

Edited by

Klaus Ruckpaul
Horst Rein

with Contributions by

I. Blanck
R. Bernhardt
G. Butschak
J. Friedrich
G. Greschner
H. Honeck
G.-R. Jänig
C. Jung
M. Kühn
J. Lampe
S. Maricic
P. Mohr
H.-G. Müller
D. Pfeil
K. Pommerening
H. Rein
R. Renneberg
P. Riege
O. Ristau
K. Ruckpaul
W. Scheler
F. Scheller
F. Schubert
W.-H. Schunck
G. Smettan

with 104 Figures and 98 Tables

AKADEMIE-VERLAG · BERLIN
1984

Erschienen im Akademie-Verlag, DDR - 1086 Berlin, Leipziger Straße 3–4

Lizenznummer: 202 · 100/522/84
Printed in the German Democratic Republic
Gesamtherstellung: VEB Druckhaus „Maxim Gorki", 7400 Altenburg
Lektor: Christiane Grunow
Einbandgestaltung: Eckhardt Steiner
LSV 1315
Bestellnummer: 763 110 9 (6703)

08000

Preface

Several reasons induced the idea and continuously encouraged the authors to review current topics of the liver endoplasmic cytochrome P–450 system. The most obvious motive is the dynamic development of knowledge in the cytochrome P–450 field which originates from the functional implication of this enzyme system in problems of broad interest ranging from lipid and steroid metabolism, biotransformation of drugs and foreign compounds to related phenomena of carcinogenesis and mutagenesis. Its biological relevance has challenged pharmacologists, biochemists, toxicologists, biophysicists, physiologists, biologists and physicians thus determining the interdisciplinary character of research work in this field.

An increasing number of contributions has revealed more and more structural as well as functional details of this membrane bound enzyme system. Especially the last ten years are characterized by a remarkable progress culminating in the purification and characterization of the essential components and of identification of isozymes of P–450, in the analysis of molecular interactions between the components, and in the correlation of substrate induced spectral and magnetic changes to functional parameters. The aim of this review therefore is to compile these recent data in order to focus on important problems from which further developments and new ideas could be derived, and by this to actively contribute to the scientific progress in the field.

The second reason which the authors would like to point out is the devotion to a more comprehensive approach with respect to regulation and control, respectively, of the enzyme system. These topics are mainly concerned with the physicochemical aspects of the enzymatic mechanism in order to elucidate the functional adaptation of the system in the organism. This analysis includes data from other hemoproteins as well as from the *Pseudomonas putida* P–450, as being the best studied cytochrome.

Last but not least, the interdisciplinary character of research in the P–450 field represents the driving force to imply the necessary sight of application. It is now well acknowledged that almost all cells in the multicell organisms contain either mitochondrial and/or microsomal cytochrome P–450. The reason we have found decisive for focussing on the monooxygenatic system in the endoplasmic reticulum of the mammalian liver is its involvement not only in the metabolism of cholesterol, bile acids, fatty acids, bilirubin and heme, thyroxin, various indoles, methylated purines and sympathicomimetic amines, i.e. of endogenous substrates, but also in the biotransformation of exogeneous substances like drugs and xenobiotics and by this into detoxification as well as in toxification processes. The implication of the enzyme system in so many fields of more applied biological disciplines becomes more and more relevant and therefore efforts increase to contribute to problems of pollutional, medical and also of biotechnological importance. Therefore we are obliged to keep our professional vision wideangled to make this broad ranging research field useful for science and society.

Klaus Ruckpaul
Horst Rein

Contents

Chapter 1

Biochemical Aspects of the Monooxygenase System in the Endoplasmic Reticulum of Mammalian Liver

K. RUCKPAUL and R. BERNHARDT

Contents

Abbreviations

P-450	Cytochrome P-450 (EC 1.14.14.1)
P-450 LM2, LM4 ...	Electrophoretically homogeneous subfractions of rabbit liver P-450
P-450 reductase	NADPH-dependent cytochrome P-450 reductase (EC 1.6.2.4)
PC	sn-glycero-3-phosphorylcholine
PE	sn-glycero-3-phosphorylethanolamine
PS	sn-glycero-3-phosphorylserine
PA	sn-glycero-3-phosphatidic acid
Tc	Temperature of the phase transition from the crystalline (or gel state) to the liquid-crystalline phase

1. Introductory Remarks

The aim of the two introducing chapters is to provide the reader with data about structure/function relationships of the endoplasmic reticulum and with biochemical data about its components which are necessary for an understanding of the more biophysical aspects of the following chapters. Due to the dynamic development of the research field the authors have tried to include recent data and own experimental results to stress on such aspects of interaction processes of the hydroxylating microsomal hepatic enzyme system which are still under discussion.

Studies of AXELROD (1955) and BRODIE et al. (1955) in the early fifties revealed the existence of an enzymatic system in the endoplasmic reticulum of the liver capable of converting different drugs and aromatic compounds into more polar products by means of NADPH, Mg^{2+} and glucose-6-phosphate dehydrogenase. It was established by these authors that the reaction only proceeds in the presence of molecular oxygen. In the same year

1955 independently MASON and HAYAISHI provided evidence for the existence of enzymes being able to catalyze the transfer of oxygen into substrates. HAYAISHI et al. (1955) evidenced the mechanism of the pyrocatechase as dioxygenase whereas MASON et al. (1955) analysing the phenolase mechanism derived the stoichiometry of the overall reaction as

$$AH + NADPH + H^+ + {}^{18}O_2 \rightarrow$$
$$A^{18}OH + NADP^+ + H_2{}^{18}O$$

thus first describing a monooxygenatic reaction.

3 years later GARFINKEL (1958) and KLINGENBERG (1958) detected a CO binding pigment in liver microsomes of pigs and rats, respectively, which was shown to be reducible either by NADPH or dithionite. They were the first to show that this pigment as CO complex exhibits a typical absorption maximum at 450 nm. But because of lacking alpha-(560 nm) and beta-(525 nm) bands they were not able to identify the heme character of the pigment. Not before 1964 the hemoprotein nature of the CO binding pigment was proven by OMURA and SATO (1964a, b). The identification as a b-type cytochrome with an atypical Soret band of the CO-complex at 450 nm made the authors designate it cytochrome P-450. The hydroxylation was found to proceed in a multistep reaction. The first step in the reaction cycle was determined as formation of an enzyme substrate complex the spectral characteristics of which depend on the respective substrate (NARASIMHULU et al., 1965; REMMER et al., 1966; SCHENKMAN et al., 1967). Substrates which are supposed to bind to apolar amino acid residues of the active site and being hydroxylated induce a blue shift of the Soret band with absorption bands at 385–390 nm as maximum and 425 to 435 nm as minimum in the difference spectrum (type I). The other type of compounds which is assumed to be bound directly to the heme iron induce a red shift thus leading to an absorption maximum at 425–435 nm and a minimum at 390–410 nm in the difference spectrum which are denoted as type II spectra (SCHENKMAN et al., 1967; SCHENKMAN, 1970). Most of them act as inhibitors rather than as substrates. The modified type II (M II) spectral change is produced by alcohols, ketones and certain drugs (e.g. phenacetin) and is characterized by a peak at 420 nm and a trough at 385 nm, the apparent mirror image of the type I change. SCHENKMAN et al. (1972) ascribed these changes to interactions of M II compounds at both the type I site and another site apart from the type II site. These results supported the idea of different binding sites.

In 1963 it was experimentally substantiated that P-450 really is the terminal oxidase in the monooxygenatic system. By means of the photochemical action spectrum (CO-complex) ESTABROOK et al., (1963) and COOPER et al. (1965) succeeded in demonstrating that in the bovine mitochondrial adrenal system P-450 is indeed the terminal oxidase of the system. Somewhat later in 1969 this result was confirmed for the system in the endoplasmic reticulum of mammalian liver by DIEHL et al. (1969).

In 1968 a certain breakthrough was achieved insofar as it was simultaneously shown by the GUNSALUS (P-450 CAM) (KATAGIRI et al., 1968; TYSON et al., 1972) and the COON (P-450 LM) group (LU et al., 1968, 1969b) that these monooxygenase systems are composed of a certain number of essential components as was shown earlier for the mitochondrial 11-β-hydroxylase system of bovine adrenals (TOMKINS et al., 1958; KIMURA and SUZUKI, 1965; SUZUKI and KIMURA, 1965). All these systems work as electron transferring chains transporting the electron from the nicotinamide adenine dinucleotide (phosphate) (NADH or NADPH) dependent flavoprotein via a non-heme-iron protein as electron transmitter (trans-

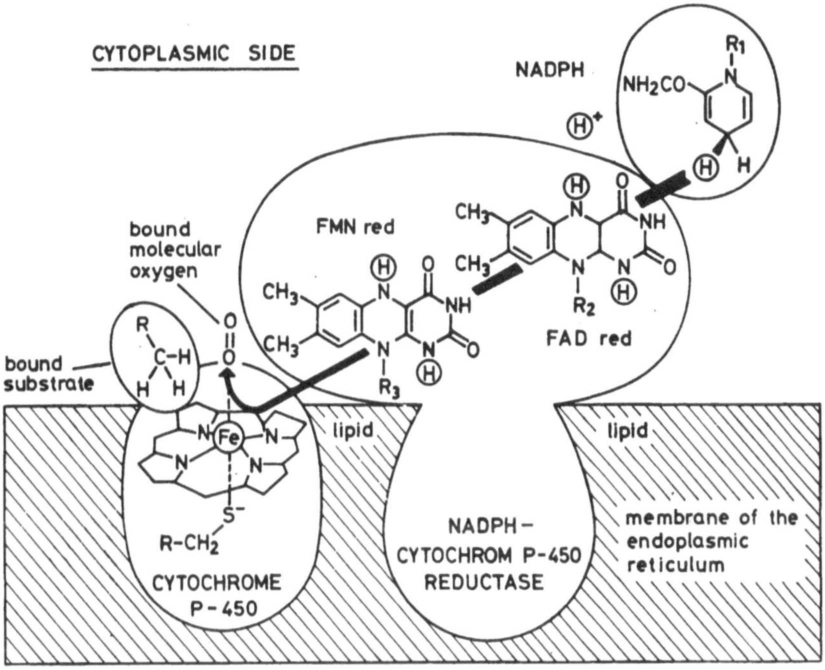

Fig. 1. Schematical drawing of the constituents of the liver endoplasmic monooxygenatic system

ferring the two electron transfer from NADPH or NADH into two one electron transfer processes) to cytochrome P–450 (EC 1.14.14.1) as terminal oxidase acting as electron acceptor. The liver microsomal system contains only a NADPH dependent FAD/FMN containing flavoprotein (EC 1.6.4.2) without the non-heme-iron protein as in mitochondria.

Comparing various thiol complexes of different hemoproteins Bayer et al. (1969) and Hill et al. (1970) found a great similarity of their EPR spectra with those of P–450 and deduced that one of the axial ligands in P–450 is a thiol residue, thus contributing essentially to knowledge of the electronic structure of the heme iron which enables the enzyme to perform its specific enzymatic function which has been supported by Ullrich et al. (1979) and recently experimentally evidenced by the detection of a Fe–S-bond in P–450 CAM using resonance Raman technique (Champion et al., 1982). The unusual red shift of the Soret band and the UV-band at 363 nm possessing the same polarization and integrated intensity have been explained in terms of allowed mercaptide sulfur-to-porphyrin charge transfer transitions extensively mixed with the Soret $\pi - \pi^*$ transition, thus shifting the transition energy to longer wavelengths (Hanson et al., 1977; Gunsalus and Sligar, 1978). The 6th ligand has not been identified. From EPR-studies on P–450 models the presence of a hydroxyl group (hard ligand) either from water (Griffin and Peterson, 1975) or a seryl or threonyl residue was suggested by Ullrich et al. (1979). Proton relaxation studies of Philson (1977) provide evidence against imidazole as 6th ligand and favour H_2O, ROH or $R-CONH_2$ as likely. Experiments by means of second derivative spectroscopy are compatible with tyrosine as 6th ligand (Ruckpaul et al., 1980a).

Ishimura et al. (1971) on P–450 CAM and Estabrook et al. (1971) on liver

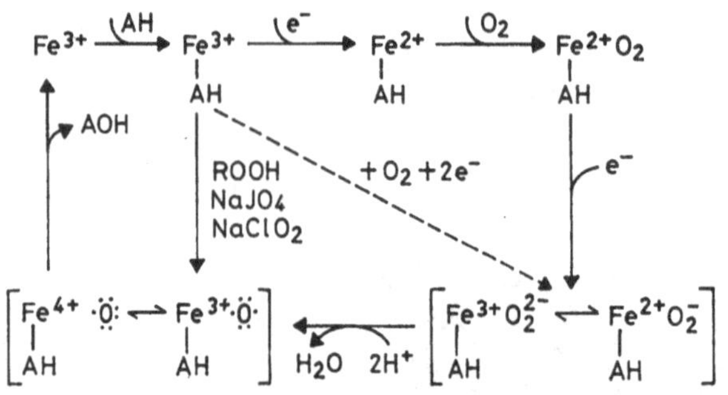

Fig. 2. Scheme of the reaction cycle of cytochrome P–450 (According to GUSTAFSSON et al. (1976))

microsomal P–450 simultaneously characterized spectrophotometrically an oxygen adduct of the reduced enzyme substrate complex the so called ternary complex. This result was an important step with respect to an understanding of the capability of the enzyme to activate molecular oxygen.

All investigations in the fifties and sixties, however, were performed with microsomes or with only partially purified preparations of P–450. In 1968 the group of GUNSALUS (KATAGIRI et al., 1968) succeeded in isolating a soluble P–450 from a bacterial mutant of *Pseudomonas putida* which hydroxylates camphor to 5-exohydroxycamphor. With this highly purified form the studies on P–450 could be continued on a highly sophisticated level and led to propose a reaction cycle which is shown in Fig. 2.

It is arranged in elementary steps. In the first reaction step the substrate is bound to a hydrophobic binding site in P–450 resulting in an enzyme substrate complex. The binding affinity is correlated with the partition coefficient of the respective substrate. The substrate binding is characterized by spin shift coupled spectral changes (REIN et al., 1977). The resulting ferric heme iron complex is assumed to exist as six-coordinated high spin complex which passing an as yet not characterized intermediate 5 coordinated ferric heme iron complex is reduced to the ferrous form by insertion of the first electron provided by the NADPH dependent FAD/FMN flavoprotein. This reaction is biphasic in character. The nature of this process is not understood in detail as yet. The next step is the attachment of an oxygen molecule to the ferrous heme iron resulting in the formation of the ternary complex which is followed by an intramolecular electron transfer leading to the oxidation of the ferrous heme iron and the insertion of the second electron into the oxygen molecule. We need to known more about the sources and the mechanism of transfer of this second electron.

An alternative pathway to hydroxylate substrates has been found by HRYCAY and O'BRIEN (1972) using the peroxidatic capability of P–450 in the presence of hydroperoxides. Because this pathway circumvents the electron inserting steps (reduction step and the subsequent insertion of the second electron) it has been called "shunt" mechanism. It should be mentioned, however, that in recent experiments CAPDEVILA et al. (1980) and RENNEBERG et al. (1981) have shown that the pattern of products formed by the peroxidatic pathway differs in dependence on the hydroperoxide used. H_2O_2 supported reactions are almost identical to the monooxygenatic pathway whereas alkylhydroperoxide supported reactions lead to another pattern of hydroxylated products.

The first 4 steps of the reaction cycle including the binding of oxygen are fairly well understood. The subsequent part of the cycle has not been experimentally proved in detail. The splitting of the molecular oxygen obviously is preceded by an acylation of the distal oxygen atom by which a transient acyl peroxide is formed that facilitates oxygen-oxygen bond cleavage (SLIGAR et al., 1980; WHITE et al., 1980). The remaining oxygen atom is assumed to exist as oxenoid peroxide (ULLRICH et al., 1979) which is inserted into the substrate thus forming a hydroxylated product and leaving the enzyme ready to enter into the next reaction cycle.

The existence of a protein bound peroxyacid as important intermediate during the activation of molecular oxygen is supported by the determination of benzyl alcohol in a P–450 promoted hydroxylation of different substrates by peroxyphenylacetic acid. The decarboxylation of peroxyphenylacetic acid has been shown to occur rather as a homolytic cleavage of the peroxide bond ($RO-OH \rightarrow RO\cdot + HO$) via the formation of a benzyl carbonium ion as quasi Fenton mechanism than as by a heterolytic decomposition ($RO - OH \rightarrow RO^- + HO^+$) with the resulting oxenium ion stabilized by coordination to the iron-sulfur center (compound I in the peroxidases (WHITE et al., 1980). This postulate is continuing and detailing ULLRICH's assumption that the hydroxylation proceeds via an oxenoid peroxide which is inserted into the substrate. But certainly the conclusive elucidation of the nature of the activated oxygen species is still an open question.

Strong efforts have been made in the seventies to understand the details of the action principles. In the course of the investigations 3 lines of experimental approach have been carried out:

- Biochemical studies have been performed with the aim of developing methods of isolation and purification for the membrane bound mitochondrial and microsomal systems. The characterization of structural and functional properties of the pure components including the determination of the amino acid sequence were directed to draw conclusions on the origin of the different P–450 fractions. A further aim of these studies was to characterize the molecular organization and the molecular interactions which determine the supramolecular structure.
- Biophysical studies with highly sophisticated methods have been performed to elucidate the electronic structure of the heme iron and the stereochemistry of the active centre in order to get insight into the functional linkage between electron transfer and oxygen activation and its relation to structural parameters.
- Biological studies have been undertaken to clarify the mechanism of induction and its inheritance and to understand which way epigenetic processes related to the biotransformation of drugs, steroids, mutagenic and carcinogenic substances may be regulated.

Milestones in the beginning of the eighties have already been set up. The amino acid sequence of P–450 CAM has been described (HANIU et al., 1982). Multiple forms have been identified as genetically determined products thus excluding an epigenetic origin of the different forms (COON et al., 1982). The molecular biology has just begun to disclose the dynamics of biosynthesis and the first coding nucleotide sequence of phenobarbital inducible cytochrome P–450 c DNA from rat liver has been reported (FUJIIKURIYAMA et al., 1982). But despite this encouraging progress the essential part of the reaction — the functional linkage between electron transfer and oxygen activation — is far from being fully understood. Many of open questions have remained unsolved or have been raised in the course of the investigations. The stereochemical para-

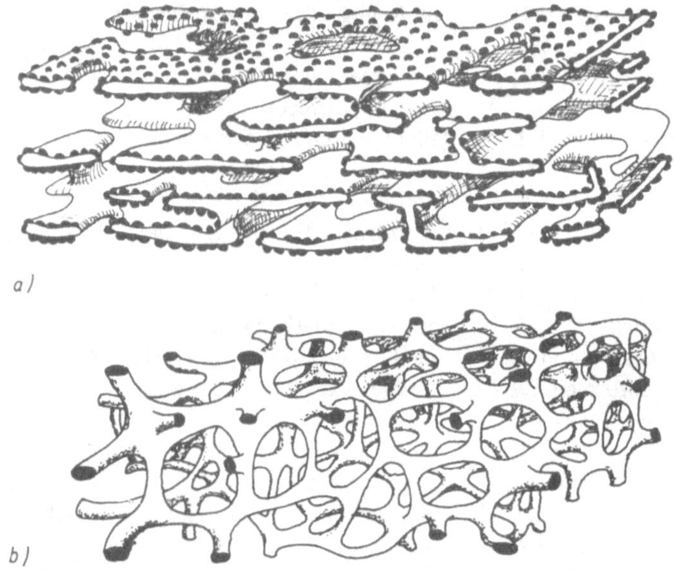

Fig. 3. (a) Threedimensional drawing of the granular endoplasmic reticulum. Ribosomes are shown only on the upper surface of the top cisternum
(b) Three dimensional drawing of the agranular endoplasmic reticulum
(From: The Ultrastructure of the humal cell by L.T. THREADGOLD (1969); reproduced with permission from Pergamon Press Ltd. Oxford)

meters for substrate and species specificity including the role of the substrate in the activation process have to be evaluated together with the determination of the axial ligands necessary for an understanding of the unique properties of the terminal oxidase. The definite conformation of the activated oxygen species is unsolved. The molecular organization is not understood and also the interactions between the essential components are puzzling. We are just in the beginning to understand more about the regulation of the induction mechanism and strong efforts are being made to elucidate correlationships between inducibility of P-450 activity and the occurrence of malignant tumors.

A lot of further questions could be derived which are challenging enough to intensify efforts for solving the problems connected with the reaction principles of this enzyme system and possibilities of its influencing.

2. Structure and Function of the Endoplasmic Reticulum

The discovery that the endoplasmic reticulum of liver cells contains a monooxygenating system consisting of several essential components has raised the question as to which way the components interact and how they are organized to function as a catalytically active complex. Therefore, discussing problems of the liver microsomal hydroxylating enzyme system, it seems reasonable first to look at the composition and structure of the corresponding matrix.

Most eukariotic cells are traversed by a membrane and tubular network which represents a basophilic component of the cytoplasm known as chromophilic substance or ergastoplasm. From its localization and form it has come to be referred to as endoplasmic reticulum by PORTER (1953). Subsequently it has been thoroughly electronmicroscopically characterized within different cells by PALADE

(1954) and shown to consist of rough and smooth membranes. The particles attached to the rough membranes (about 60% of the endoplasmic reticulum have been shown to be composed of ribonucleic acid (PORTER, 1953) and denoted as ribosomes by ROBERTS (1958). The endoplasmic reticulum passes the cytoplasm from the plasma membrane to the membrane of the nucleus separating a channel-like space from the cytoplasm. The cytoplasmic side of the barrier between two subcellular compartments provides sites for the attachment of ribosomes and accepts various substrates for synthesis, degradation and biochemical modification. At the luminal side the reaction products of membrane bound ribosomes and enzymes are transferred and segregated (SIEKEVITZ, 1972). That part of the endoplasmic reticulum which is involved in protein biosynthesis and occupied by ribosomes appears as rough membranes. In the peripheral part of the cell the ribosomes disappear and the endoplasmic reticulum morphologically appears as smooth membranes (SCHULZE and STAUDINGER, 1975).

After homogenization and differential centrifugation of cells a special fraction of small particles was first obtained by CLAUDE in 1938 which he denoted microsomes because of their small size. CLAUDE (1943) and later PALADE and SIEKEVITZ (1956) by means of electron microscopy showed that this microsomal fraction originates from the endoplasmic reticulum. Because of transversal rupture of the endoplasmic reticulum microsomes have the same inside outside orientation. Therefore they can be regarded as biochemical equivalent of the endoplasmic reticulum and despite of different morphology due to perturbation of the membrane structure during fragmentation it seems reasonable to assume that microsomes reflect functionally physiological conditions (DEPIERRE and DALLNER, 1975). Under usual experimental conditions particles with a mean diameter of 200 nm (150–300 nm) are obtained with bilayer membranes of 50–80 Å thickness (cf. DEPIERRE and DALLNER, 1975). HOLTZMANN et al. (1973) using inelastic laser light scattering determined a mean diameter for rat liver microsomes of 329 nm ± 5 nm. For morphometric and morphological details the reader is referred to review articles (DEPIERRE and DALLNER, 1975; DEPIERRE and ERNSTER, 1977).

Liver microsomes besides the endoplasmic reticulum contain also parts from other cell organelles. The total amount of components related to the protein content which does not emerge from the endoplasmic reticulum amounts to 20.1% and is composed of 7% plasma membranes, 3.9% Golgi-complex, 1% lysosomes, 4.3% mitochondria (without outer membrane) and 3.9% outer mitochondrial membranes (REMACLE et al., 1976). Microsomes have been subfractionated by isopycnic and differential centrifugation in density gradients and 27 components characterized which in dependence on their increasing median density have been divided into 4 groups (BEAUFAY et al., 1974). The components of the two median groups and RNA emerge from the endoplasmic reticulum whereas the other components can be attributed to other cell particles. The main constituents of the endoplasmic reticulum of hepatocytes — proteins and lipids — are contained in 1 g wet weight of liver in an amount of 40–50 mg protein and about 15 mg phospholipid meaning the membrane of the endoplasmic reticulum consists of 60–70% protein and 30–40% phospholipid by weight. Basing on a molecular weight of 800 for a typical phospholipid and 50.000 as the average molecular weight of a microsomal protein DEPIERRE and ERNSTER (1977) roughly calculated that are approximately 27–42 phospholipid molecules per protein molecule in the endoplasmic reticulum. Microsomal phospholipids have been found to consist

Table 1. Morphological and Biochemical Parameters of the Endoplasmic Reticulum (with modifications from DePierre (1975, 1977))

Lumen of the endoplasmic reticulum	200–300 Å (rough membrane) 300–600 Å (smooth membrane)
Thickness of membranes	50–80 Å
% of total cell volume	15.3
Surface of the endoplasmic reticulum per 1 g liver weight	7.3 – 11 m^2
Composition (related to 1 g wet liver weight)	40–50 mg protein 15 mg phospholipid
Cytochrome P-450 content (rabbit) non induced induced (phenobarbital	 2 nmole/mg protein 0.1 mg/mg protein 10% 4.2 nmole/mg protein 0.21 mg/mg protein 21%
Composition of the phospholipids (endoplasmic reticulum of the rat liver)	55% phosphatidylcholine 20–25% phosphatidylethanolamine 5–10% phosphatidylserine 5–10% phosphatidylinositol 4–7% sphingomyelin
Molar ratio phospholipid/protein P-450/reductase	 23 20–35

of about 55% phosphatidylcholine, 20 to 25% phosphatidylethanolamine, 5 to 10% phosphatidylserine, 5–10% phosphatidylinositol and 4–7% sphingomyelin (DePierre and Dallner, 1975). The fatty acid moieties of these phospholipids are mainly the 16:0, 16:1, 18:0, 18:1, 18:2, 20:4 and 22:6 species (Lee and Snyder, 1973).

The endoplasmic reticulum is not only matrix for the biosynthesis of numerous proteins but also location for a great number of enzymatic activities. Up to now about 50 different enzymatic activities have been assayed in liver microsomes (Bailey et al., 1974) which are distributed within at least 250 polypeptides of different molecular weights (Vlasuk and Walz, 1979). The endoplasmic reticulum besides biotransformation of xenobiotics is involved in phospholipid and cholesterol metabolism and in biosynthesis of lipoproteins, glycoproteins and glycolipids (Ernster and Orrenius, 1973).

The functional importance of all enzymatic activities can not be discussed in detail at this point but it seems reasonable to assume that they are integrated in the function of the whole cell. The structural arrangement of the endoplasmic reticulum as a tubular system indicates its implication in transportation processes meaning that the endoplasmic enzymes may be vectorial in nature. For glycoproteins it could be shown that they are synthesized at the cytoplasmic side of the membrane subsequently transported into the lumen and after passing the Golgi-apparatus are excreted by exocytose into the extracellular space (Schachter et al., 1970). The vectorial nature has been suggested also for the endoplasmic glucose-6-phosphatase thus actively transporting glucose from the interior of the cell via the tubular pathway of the endoplasmic reticulum to the blood (cf. Schulze and Staudinger, 1975). The same pathway is conceivable also for the transportation of drugs and xenobiotics after biotransformation by phase I and phase II enzymes.

Most hepatic enzymes which are involved in the metabolism of drugs and xenobiotics are embedded in the endoplasmic membrane. Besides the monooxygenase system with cytochrome P-450 as terminal oxidase the endoplasmic reticulum contains another electron transport system with cytochrome b_5 and a NADH dependent reductase which apparently is involved in some hydroxylating reactions. Shimataka et al. (1972) have shown cytochrom b_5 to be the intermediate electron carrier in the microsomal NADH-dependent stearyl-Co-A desaturase system. There is further contained an enzyme which catalyzes the oxidation of secondary and tertiary amines to N-oxides. This monooxygenatic reaction is dependent on O_2 and NADPH but is CO insensitive and independent of P-450. The enzyme involved is the flavine containing aminoxidase (Prough and Ziegler, 1977).

Functionally linked with the monooxygenase is the UDP-glucuronyltransferase (Bock and Lilienblum, 1979) and the epoxidehydratase (EC 4.2.1.63) which has been shown to form a stoichiometric 1:1 complex with the P-450 system (Seidegard et al., 1978). By this enzyme epoxides and areneepoxides which are formed by oxidation of aliphatic and aromatic double bonds are converted by addition of water into the respective diols. By this process the reactivity of epoxides and areneepoxides is destroyed (Oesch, 1977; Lu et al., 1977; Kapitulnik et al., 1977), but simultaneously in dependence on the respective compound ultimate carcinogens from polycyclic aromatic hydrocarbons are formed (Shen et al., 1980). Furthermore it contains conjugating enzymes involved in biotransformation of drugs and xenobiotics in addition to glucuronyltransferase and UDPGA-transferase (Ernster and Orrenius, 1973) which is discussed in more detail in chapter 6.

The endoplasmic reticulum of liver cells like other cell organelles is subjected to a continuous turnover which is relatively short. The half-lives vary specifically for each enzyme between 22 h (Greim, 1970) and 113 h (Omura et al., 1967). Proteins and phospholipids are turning over at different rates. Membrane lipids have 10—30% shorter half-lives indicating that the essential constituents (proteins and phospholipids) of the membrane are degraded and synthesized individually and not as a unit. The life time of the hepatocyte has been estimated to range from 160—400 days meaning that there is a significant turnover of important structural elements during the life of this cell (cf. Omura et al., 1967).

2.1. Physicochemical Properties of Phospholipids

Phospholipids are the matrix forming element of biomembranes. Despite different topology in the cell (e.g. plasmamembrane, mitochondrial and endoplasmic membrane) and different composition biomembranes are functionally characterized by a limited transport of metabolites and ions due to their selective permeability from which the compartmentalization emerges. The permeability of membranes significantly changes in dependence on the charge of the penetrating compound. Uncharged substances of a molecular weight up to at least 600 dalton freely penetrate the microsomal membrane whereas it becomes impermeable to charged substances at a molecular weight of 90 (Nilsson et al., 1973).

The knowledge of the physicochemical properties of phospholipids is a prerequisite for an understanding of membranes and their functional behaviour. From the amphiphilic character of the phospholipids the capability of self-organization emerges. The hydrophilic head groups are faced towards the aqueous phase and the hydrophobic fatty acid residues are directed towards each other, thus shielding themselves from water contact.

The lipid molecules exhibit high fluidity in biomembranes meaning that they rapidly move. With respect to its character intra- and intermolecular motions can be distinguished:

1) Rotational anisotropic reorientations and oscillations of individual functional segments around the symmetry axis of the molecule, with rotational correlation times for a highly fluid lamellar state from ^{13}C relaxation in the order of 10^{-10} s (LEE, 1974). Torsions perpendicular to this direction are strongly hindered.
2) Orientational fluctuations of the molecular symmetry axis.
3) Lateral diffusion of the molecules with diffusion coefficients ranging between 10^{-7} to 10^{-9} cm²/s (KUO and WADE, 1979).
4) Transversal motions between the two layers of the bilayer (flip-flop processes) with rates in the order of minutes, hours or even days (ROTHMAN and DAWIDOWICZ, 1975).
5) Exchanges of lipid molecules between different bilayers.

The lipids of eukaryotic cell membranes can be divided into 4 groups: phospholipids, glycolipids, sphingolipids and steroids (JONES, 1979). The most extensively studied group are the phospholipids which are diesters of sn-glycerol-3-phosphoric acid. The variety of the phospholipids originates from different head groups and different fatty acid residues. The various head groups (listed in Scheme 1) differ in their acidity (PC $<$ PS) and

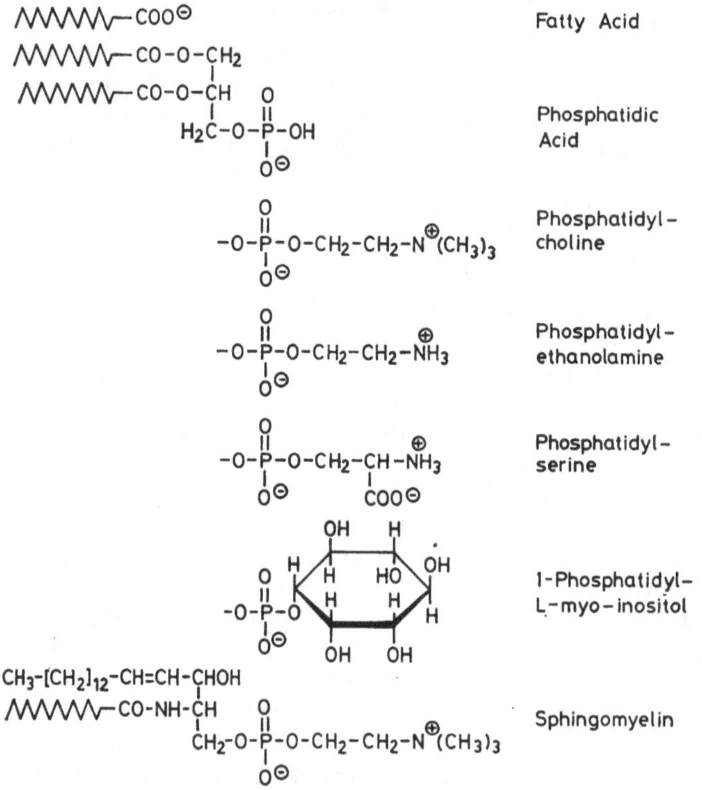

Scheme 1. Formulas of the head groups of biologically occurring phospholipids

Table 2. Mainly occurring fatty acid residues of phospholipids in biomembranes

R group	Configuration	Trivial name
$CH_3(CH_2)_{10}CO-$		lauric-
$CH_3(CH_2)_{12}CO-$		myristic-
$CH_3(CH_2)_{14}CO-$		palmitic-
$CH_3(CH_2)_{16}CO-$		stearic-
$CH_3(CH_2)_5CH{=}CH(CH_2)_7CO-$		palmitoleic-
$CH_3(CH_2)_7CH{=}CH(CH_2)_7CO-$	cis	oleic-
$CH_3(CH_2)_7CH{=}CH(CH_2)_7CO-$	trans	elaidic-
$CH_3(CH_2)_4CH{=}CHCH_2CH{=}CH(CH_2)_7CO-$	2 cis	linoleic-
$CH_3CH_2CH{=}CHCH_2CH{=}CHCH_2CH{=}$ $CH(CH_2)_7CO-$	3 cis	linolenic-
$CH_3(CH_2)_4CH{=}CHCH_2CH{=}CHCH_2$ $CH{=}CHCH_2CH{=}CH(CH_2)_3CO-$	4 cis	arachidonic-

their capability to form lamellar (PC) or non-lamellar (PE) structures which leads to asymmetric phospholipid distribution and the formation of protein-lipid domains.

On the other hand the chain length and the degree of saturation of the fatty acid residues of the respective phospholipid determine physical properties as fluidity and this way modulates the properties of membrane bound enzymes. The existence of two hydrocarbon chains per head group in phospholipids accounts for the formation of extended phospholipid bilayers whereas detergents with only one hydrocarbon chain form micelles. Lipids which contain only one acyl chain are designated with the prefix lyso. They are also unable to form bilayers even can destroy bilayer structures.

The acyl residues of fatty acids in biomembranes in general contain 16–22 carbon atoms (see Table 2). Saturated fatty acids may be in all-trans or due to the formation of so-called "kinks" in the gauche conformation (Fig. 4). The formation of rotational isomers in the hydrocarbon chain by "kinks" has consequences for the geometric arrangement of the fatty acids and their properties at phase transitions (see below). The fatty acids with double bonds may be cis-or trans-unsaturated which causes different geometric arrangements and leads to

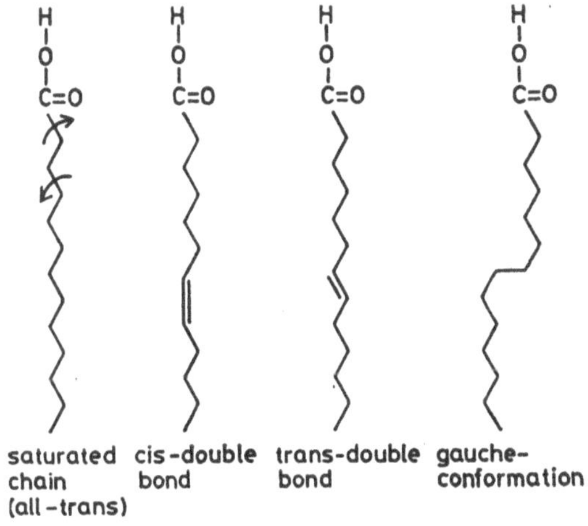

Fig. 4. Configuration of saturated and unsaturated fatty acids

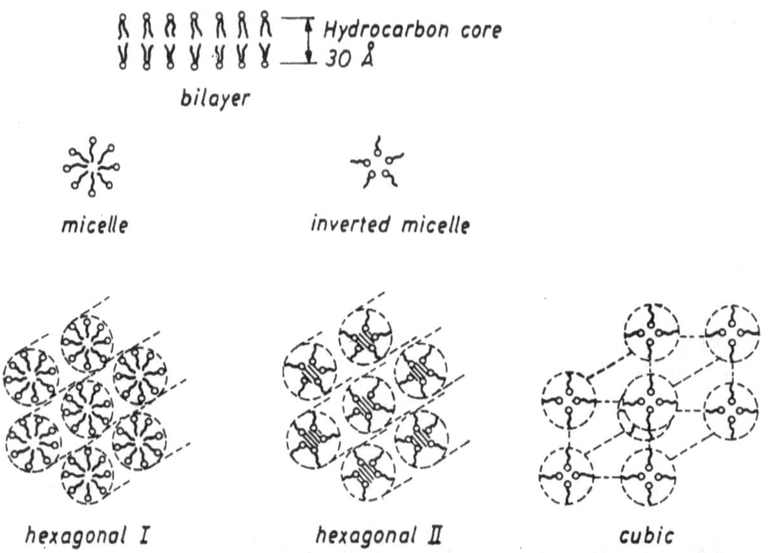

Fig. 5. Polymorphic phases of hydrated liquid crystalline phospholipids

consequences in the packing of the acyl chains within a bilayer or another lamellar lipid structure.

Besides physically isotropic liposomes which represent single or multilamellar concentric bilayers also other than lamellar structures have been described (Jones, 1979). Depending on the chemistry of lipids, their concentration, the composition of the medium and the temperature numerous structures of the lipids such as micelles, inverted micelles, bilayers, cubic, rhombic or hexagonal structures may be produced.

2.1.1. Model Membranes

Due to the heterogeneous composition and the complexity of biomembranes for biochemical and biophysical studies often model membranes are used which contain one or only few essential components. The aim of such studies on model membranes is to find general principles according to which membranes are organized to draw conclusions on biological membranes. Such kind of approach is commonly used. The basis for such studies is the thermodynamically determined capability of phospholipids to arrange themselves as multiwalled rather than as single-walled vesicles. Vigorous stirring by sonication and low initial concentration favour the formation of single-walled vesicles.

Many properties (e.g. permeability for ions and water, response to Ca^{2+}, incorporation of proteins) of bilayers resemble those of biomembranes. For biochemical and biophysical studies two types of bilayers can be used: (i) multibilayer dispersions, which are formed spontaneously whenever phospholipids are dispersed in an aqueous medium, (ii) homogeneous, small single-walled bilayer as vesicles, which are readily obtained under appropriate conditions as sonication. The formation of either multilayer or bilayer dispersions not only depends on the external physical conditions but also on the nature of the acyl residues. Electron microscopy confirmed that the majority of the particles present in sonicated dilauroyl-lecithin dispersions were aggregated and multilamellar. Attempts to produce single shelled vesicles failed (Hauser and Barratt, 1973). Contrary to dipalmityol lecithin particles those

Table 3. Transition temperature, transition enthalpy and transition entropy for the bilayer phase transition of phospholipids (data taken in part from BLUME (1979))

Parameters	DMPC	DPPC	DLPE	DMPE	DMPA (pH 5)
Tc (°C)	24.0	41.5	30.5	50.2	43.0
ΔH (kcal · mol^{-1})	6.3	8.3	4.0	6.1	5.6
ΔS (cal · mol^{-1} · deg^{-1})	21.1	26.4	13.0	18.9	17.7

DMPC, 1,2-dimyristoyl-PC
DPPC, 1,2-dipalmitoyl-PC
DLPE, 1,2-dilauroyl-PE
DMPE 1,2-dimyristoyl-PE
DMPA 1,2-dimyristoyl-PA

of dilauroyl lecithin are less stable (increase of turbidity at storage). This is consistent with results evidencing that in the liquid crystalline state the interaction energy of dimyristoyl lecithin and higher homologues is sufficient ($\geqq$6 kcal/mol $\geqq$10 kT) to guarantee a stable association of molecules in the bilayer (PHILLIPS and CHAPMAN, 1968; PHILIPPS et al., 1969).

In the solid state lipids generally exhibit more than one crystalline form, a phenomenon, which is known as polymorphism and depends on the symmetry relation between equivalent positions of one chain and its neighbours. In the liquid state lipids show both thermotropic and lyotropic mesomorphism, what means that on heating or by addition of water crystalline lipids do not pass directly to the liquid state, but form intermediate states, so-called mesomorphic or liquid-crystalline ones so that the following series of equilibria can be established (JONES, 1979):

$$\text{crystal} \rightleftharpoons \text{mesomorphic states} \rightleftharpoons \text{liquid} \rightleftharpoons \text{vapour}$$

The transition from the crystal to the liquid-crystalline state occurs at a well defined temperature known as the transition or chainmelting temperature Tc. Above Tc the lipid molecules undergo rotation about the axis formed by the C—C-bonds while the headgroups remain almost constant. The acyl chains exhibit no longer the fully extended all-trans configuration. The chainmelting temperature is possibly one of the most important parameters associated with the thermal behaviour of lipids. There are discontinuities in the entropy, volume and heat capacity of lipids when their temperature is varied through Tc. Some of these data determined by differential scanning calorimetry are summarized in Tab. 3. In addition, the Tc seems to be very important in studying some principal features of biological membranes although at physiological temperature (37 °C) the lipids should be in the liquid state due to the unsaturated nature of most naturally occurring lipids. As biological membranes are heterogeneous in lipid composition they do not exhibit sharp transition temperatures. The situation is further complicated by heterogeneous incorporation of proteins and cholesterol so that distinct lipid domains may exhibit rather different phase transition temperatures.

The phase transition from the crystalline (or gel state at a water content < 5%) to the liquid-crystalline phase has been studied especially by means of differential thermal analysis and differential scanning calorimetry, but also NMR-, ESR-, X-ray, infrared, laser-raman and optical measurements have been utilized (for review see OLDFIELD and CHAPMAN, 1972). From studies with synthetic lipid membranes it becomes evident that the temperature at which gel $\Rightarrow$ liquid-crystalline phase transition of lipids occurs is dependent on the nature of the respective head group as well as on the length and saturation level of the acyl side chain (see Tab. 4).

Table 4. Transition Temperatures of Saturated Diacylphosphatidylcholine, estimated by means of calorimetric investigations (data in part taken from MABREY and STURTEVANT (1976))

Lipid	Transition temperature (°C)	ΔH (kacl · mol^{-1})
DLPC	−1.8	1.7
DMPC	23.9	5.4
DPPC	41.4	8.7
DSPC	54.9	10.6

DLPC, 1,2-dilauroyl-PC
DMPC, 1,2-dimyristoyl-PC
DPPC, 1,2-dipalmitoyl-PC
DSPC, 1,2-distearoyl-PC

Due to its polarity the head groups of lipids are water soluble and able to interact with water, including the dielectric solvation of charged groups and the formation of hydrogen bonds between head groups and water. EIBL and WOOLEY (1979) could demonstrate that intrafacial hydrogen bonds can make a significant contribution to the energy of the membrane when lipids possess a suitable proton donor (N^+H, OH) and proton acceptor (N:, PO_4^-). Such bonding may play a role in the gel ⇌ liquid crystalline transition but also in determining distinct properties of biological membranes like diffusion or dissociation of single molecules. As demonstrated in Tab. 3 and 4 phosphatidylcholines with the same chain length and saturation level of the fatty acids have lower transition temperatures than phosphatidic acids and phosphatidyl-ethanolamines, being −1.8 °C and 24.0 °C for DLPC and DMPC, respectively and 30.5 °C and 50.2 °C for the respective phosphatidylethanolamines.

The Tc of a phospholipid is not only dependent on the head group, but also on the chemistry of the hydrocarbon chain. Lipids with longer fatty acid chains have a higher transition temperature than shorter ones (Tab. 3 and 4). Lipids with more unsaturated fatty acid residues but with the same head group and the same length of the acyl side chains have lower transition temperatures than those with more saturated ones. By means of differential scanning calorimetry it was shown that DSPC has a transition temperature of 54.9 °C (Tab. 4) whereas the DOPC already at −22 °C becomes liquid-crystalline (LADBROOKE et al., 1968). However, not only the head group, the acyl chain length and the saturation level determine Tc, but also, as was shown in studies of CHAPMAN et al. (cf. OLDFIELD and CHAPMAN, 1972) the conformation of unsaturated lipids. It was established by means of calorimetric methods that cis-unsaturated fatty acids have lower transition temperatures than trans-unsaturated ones.

In so far as the thermodynamic study of chainmelting in lipids relates to molecular motions of the system some remarks about the latter ones seem to be necessary. The molecular motion of the phosphorous atom and this way of the head group of the phospholipid has been studied by ^{31}P NMR. NMR studies have also been applied in addition to fluorescence probes and spin labels to study the molecular motion of the hydrocarbon chains. The molecular motion was found to be most rapid at the ends of the fatty acid chains. Below Tc the hydrocarbon chains of phospholipids are relatively immobile in the bilayer structure, especially in the head group region, whereas on heating to Tc the mobility of the whole molecule increases and the bilayer undergoes some lateral expansion (JONES, 1979). The increase of the volume of the bilayer above Tc could be explained by means

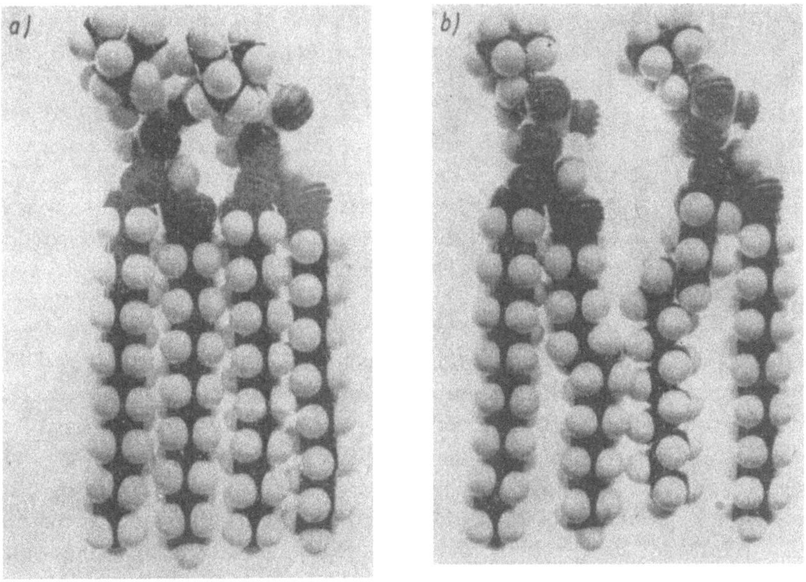

Fig. 6. Possible conformations for the head groups in dipalmitoyllecithin bilayers, (a) in the gel phase and (b) in the liquid crystalline phase (Fig. taken from LEE (1975); reproduced with permission from Pergamon Press Ltd., Oxford)

of a "kink-model". TRÄUBLE and HAYNES (1971) suggest that at Tc some C—C bonds change from the trans to the gauche conformation, allowing this way rotation around individual C—C bonds which is of lower energy than the rotation of the whole molecule and in addition shows a volume change in the range of the observed one as compared to a much larger increase in volume to be expected if the whole phospholipid molecule rotates at Tc. The most important rotational isomer could be observed when two trans conformations are transferred to gauche conformations.

The rapid transition between trans and gauche states of C—C single bonds in general is thought to be the molecular origin of hydrocarbon chain disorder. If cis double bonds are introduced which cannot exert such rotational jumps due to its specific chemical structure, then the bilayer must exhibit distinct conformational preferences. SEELIG and WAESPE-SARCEVIC (1978), studying 1-palmitoyl-2-oleyl-, 1-palmitoyl-2-elaidoyl-, and 1,2-dielaidoyl-GPC by means of deuterium NMR could show that only in a first approximation the hydrocarbon chains in this lipid bilayer above Tc are disordered and liquid-like, but there are distinct conformational constraints imposed on the fatty acyl chain.

A term, often used in connection with Tc values is the 'fluidity' of a lipid membrane. However, it has to be mentioned in this connection that this term until now is only rather poor defined and in addition to Tc values, there are spectral order parameters, relaxation times, partition coefficients, lateral and rotational diffusion coefficients involved.

2.1.2. Multicomponent Model Membranes and Biomembranes

Studies with model lipid membranes which usually consist of only 1—2 different lipids permit unambiguous results concerning the phase transition temperature derived from alterations of structural and functional properties of the vesicles (JACOBS et al., 1977). Studies with biological membrane systems, however, are more difficult to interpret because they are composed of a great variety of lipids differing in their head groups and constituent fatty acid residues (WUNDERLICH et al., 1978). A step nearer towards the in vivo membrane system is the study of multicomponent systems, containing water, cholesterol (which is of particular interest in the structure of cell membranes and function of membrane-bound enzymes) and phospholipids. It could be demonstrated that lateral phase separations in bilayers of two phospholipids can occur (cf. LEE, 1977). Furthermore phase diagrams were constructed for lipid-cholesterol and lipid-protein systems which correspond well to many experimental data (KLEEMANN and McCONNEL, 1976). The observation of the same authors that in phospholipid-protein and phospholipid-cholesterol-protein mixtures phase separation occurs led to the conclusion that the phospholipids in membranes are not homogeneously dispersed in the bilayer structure, so that in addition to the transversal asymmetry one has to take into account also an asymmetry within one layer of the bilayer. The transversal and lateral asymmetries of membrane lipid distribution implicate that different parts of the membrane may exhibit an identical lipid composition resulting in the same temperature of the phase transition.

The existence of phase transitions in biological systems was first demonstrated at the end of the 60ies by COHEN et al. (1968) and STEIM et al. (1969). The biological relevance of such transitions is evidenced by changed functional parameters after temperature change, change of ionic strength, incorporation of proteins etc. (TRÄUBLE and EIBL, 1974; VERKLEIJ et al., 1974; JACOBSON and PAPAHADJOPOULOS, 1975; HAUSER et al., 1977; VAN DIJCK et al., 1978).

Incorporation of proteins into a lipid bilayer was discussed to alter the phase state of the membrane lipids (METCALFE et al., 1972; MARSH et al., 1978; OWICKI and McCONNEL, 1979; PINK and CHAPMAN, 1979). The most frequently used model for protein-lipid interactions is based on the assumption that the lipids around the incorporated protein are immobilized (JOST et al., 1973a, b; MARSH et al., 1978) thus offering a cholesterol-like effect (KLEEMANN and McCONNEL, 1976). This assumption is conflicting with results indicating that there is not necessarily a rigid 'boundary layer' of lipids around the membrane proteins (RICE et al., 1979a, b; KANG et al., 1979a, b). It seems, however, reasonable to assume that not only the state of fluidity of the lipid phase should influence the rotational and lateral diffusion rates of the incorporated protein, and probably its ability to undergo conformational changes, but that also the proteins incorporated exert some influence on the adjacent lipid molecules.

Concerning the phospholipids and their distribution in the lateral plane of the endoplasmic reticulum experimentally supported ideas evidenced the existence of different lipid domains in the microsomal membrane (STIER and SACKMANN, 1973; MABREY et al., 1977). Further evidence for the existence of lipid domains in membranes has been derived from perturbations induced by different free fatty acids monitored by the emission polarization changes of two different fluorescent probes (1,6-diphenyl-1,3,5-hexatriene and 8-anilino-1-naphthalene sulfonate). The experiments revealed that cis-unsaturated fatty acids partition into fluid domains while the trans-unsaturated and saturated fatty acids preferentially partition into solid-like domains (KLAUSNER et al., 1980).

On the other hand, isolation of membrane-bound proteins including the removal of the lipid milieu can produce alterations in its activity and kinetic parameters analogously to alterations in the membrane fluidity (McMurchie and Raison, 1979; Chiang and Coon, 1979; French et al., 1980). About the molecular basis of protein-lipid interactions little is known as yet. In addition, one has to be very careful in discussing the influence of lipids and lipid specificity on proteins, because the degree of dependence on the physical parameters of the lipid may well differ for various proteins and may, for example, depend on the depth of penetration of the protein into the lipid phase.

Also there may be some experimental pitfalls, like difficulties to get aqueous dispersions of certain phospholipids, e.g. phosphatidylethanolamine, possible decomposition of lipids (dioleoyl-GPC), peroxidation of lipids containing highly unsaturated fatty acids and contamination with detergents (cf. Sandermann, 1978). To get more insight into this field systematic studies on lipids with varying head groups and acyl chains, but also with different free fatty acids incorporated in the bilayer interacting with phospholipids and with proteins have to be performed.

2.1.3. Current Models of Biological Membranes

Despite the large number of interdependencies of mutual interactions between membrane constituents some general principles of membrane organization and function have been developed which may help to understand the functional importance and its implications in fundamental living processes. Basing on the unit membrane model of Danielli and Davson (1935) which for a long time determined the knowledge of biochemists and biophysicists — the 'fluidmosaic' model of Singer and Nicolson (1972) and Singer (1974) postulates the membrane lipids to be in a fluid and relatively homogeneous phase into which proteins are inserted differently. The membrane contains proteins which (i) are only loosely associated with the hydrophobic surface of the bilayer, (ii) possess a hydrophobic anchor, penetrating into the apolar region of the lipid bilayer, whereas the more hydrophilic part of the protein is protruding the membrane, contacting its hydrophilic surface, (iii) span the bilayer, having two parts on both sides of the membrane and a hydrophobic one between surrounded by the apolar region of the bilayer and (iv) are completely embedded into the bilayer. An asymmetric transversal distribution of lipids and proteins as discussed in detail below was described for the erythrocyte membrane (Bretscher 1972, 1973), for the endoplasmic reticulum (Nilsson and Dallner, 1975, 1977a, b) and other biomembranes and has been discussed as a general phenomenon (Bergelson and Barsukov, 1977; Op den Kamp, 1979).

Basing on recent findings particularly by ^{31}P-NMR as well as by improved lipid isolation and synthesis Cullis and De Kruijff, (1978a, 1979) proposed a new model for biomembranes. They started from the characteristic behaviour of lipids to exhibit a great variety of phases as bilayer arrangement, hexagonal H_I and H_{II} configuration, micellar phase as well as those phases with cubic or rhombic structures (see Fig. 5).

Phosphatidylcholine and sphingomyelin seem to be responsible for the bilayer structure, whereas unsaturated phosphatidylethanolamines prefer to form the hexagonal (H_{II}) arrangement. In addition negatively charged phospholipids also can exert non-bilayer configuration particularly in the presence of divalent cations such as Ca^{2+}. Furthermore by means of ^{31}P-NMR and differential scanning calorimetry it was shown that addition of cholesterol to liquid-crystalline 18 : 1/18 : 1 and to 16 : 0/16 : 0 phosphatidylethanolamine stabilizes the bilayer whereas it destabilizes the bilayer when added to a mixture of 18 : 1/18 : 1

phosphatidylethanolamine and 18 : 1/18 : 1 phosphatidylcholine. Destabilization of the bilayer was assumed to occur by induction of the cubic or hexagonal H_{II} phase after addition of cholesterol (CULLIS et al., 1978b). This so-called "metamorphic mosaic" model of biological membranes from CULLIS and DEKRUIJFF (1979) substantially differs from the fluid mosaic model of SINGER and NICOLSON (1972). The authors stress on certain peculiarities of membrane lipids influencing the integral protein function: (i) lipids in the gelstate do not appear in most biological membranes (particularly those of eukariotic cells) as the unsaturated nature of most naturally occurring lipids results in phase transition temperatures well below the physiological one, (ii) integral membrane proteins obviously do not require specific lipids for their activity, and (iii) there seems to be a rapid exchange between annular and bulk lipids.

The occurrence of non-bilayer lipid structures in biological membranes has been suggested for the endoplasmic reticulum, sarcoplasmic reticulum and inner mitochondrial membrane (cf. CULLIS and DE KRUIJFF, 1979), where isotropic motion for some fractions of the endogeneous phospholipids at 37 °C has been observed (DEKRUIJFF et al., 1978). This would be consistent with the occurrence of inverted micellar or cylindrical H_{II} arrangements of lipids within the bilayer which are in rapid exchange with the bulk phospholipids (STIER and SACKMANN, 1973; STIER et al., 1978). These non-bilayer structures are described as being probably responsible for membrane fusion phenomena and transbilayer transport processes. The molecular mechanism of the influence of such non-bilayer structures on protein conformation and function is not fully understood as yet, but it seems possible that there may be a relationship between the alteration of the polymorphic phase state of lipids (in addition or instead of a hydrocarbon phase transition) and the function of integral protein.

In agreement with these results studies on lipid-protein interactions in different biological systems: *Escherichia coli* membrane (KANG et al., 1979b), lipid-gramicidin A interactions (RICE and OLDFIELD, 1979a) and lipid interaction with human brain lipophilin, beef brain myelin proteolipid apoprotein and sarcoplasmic ATPase (RICE et al., 1979a, b) led the respective authors to conclude that there is no evidence for the formation of a rigid boundary layer of lipid around the membrane proteins.

Extrapolating these results to physiological conditions one would ask for their relevance for the integrated situation in the liver cell and in the whole organ. In this connection a comparative ^{31}P-NMR study on rat liver microsomes, rat liver slices and intact perfused rat liver may be of importance (DEKRUIJFF et al., 1980). Contrary to microsomes where at 37°C the majority of phospholipids undergo isotropic motion the majority of phospholipids in liver slices shows spectra indicating bilayer structure and only 14% undergo isotropic motion. Preliminary experiments on perfused rat liver suggest a similar behaviour for the membrane phospholipids to that observed in the liver slices. If this observation can be ascertained it should have consequences for further experimental approaches.

On the other hand, studies on cytochrome oxidase-phospholipid interactions by means of NMR investigations evidenced that the methyl groups of 1-myristoyl-2-/14′, 14′, 14′-$^2H_3$/myristoyl-phosphatidylcholine are less ordered in the boundary lipid than in the free lipid bilayer thus giving evidence for boundary lipid (MARSH et al., 1978).

The discrepancies between the recently obtained NMR results and the earlier performed ESR measurements are explicable in part taking into account methodological differences, since ESR measurements reflect only static conditions whereas boundary lipids are in motion and disordered on the ^{2}H-NMR time scale (KANG et al., 1979a). The data available however are so controversial that it is impossible to draw decisive conclusions now on the structural organization of biomembranes and its functional implications.

2.2. Asymmetry of the Membrane Structure

2.2.1. Transversal Distribution of Lipids

Basing on the fluid mosaic model (SINGER and NICOLSON, 1972; SINGER, 1974) and the protein liquid crystal model (VANDERKOOI, 1974) of the membrane structure an asymmetric distribution of phospholipids between the two sides of the bilayer (transverse asymmetry) in different types of natural membranes has been well established. BRETSCHER (1972a, b, 1973) by means of labelling studies with the nonpenetrating marker ^{35}S-formylmethionylsulphone methylphosphate and ZWAAL et al. (1973) using phospholipases were the first to show that in the membranes of human erythrocytes phosphatidylcholine and sphingomyelin are predominantly located in the outer half of the lipid bilayer whereas phosphatidylserine and phosphatidylethanolamine are mainly present on the cytoplasmic side of the bilayer.

The existence of transverse lipid asymmetry has been further demonstrated in membranes of the endoplasmic reticulum (NILSSON and DALLNER, 1975, 1977a, b) as well as in the inner mitochondrial membrane (MARINETTI et al., 1976; KREBS et al., 1979). By use of phospholipase A_2 (Naja naja snake venom) NILSSON and DALLNER (1975, 1977a, b) have shown that most of phosphatidylethanolamine and phosphatidylserine are present on the outside (cytoplasmic side) of the microsomal membrane. Phosphatidylcholine has been found to be nearly symmetrically distributed over both sides. Phosphatidylinositol and sphingomyelin may be assigned to the inner side of the bilayer. The distribution of phospholipids is opposite to that found in erythrocyte membranes (ZWAAL et al., 1973) indicating that the cytoplasmic leaflets correspond to each other. It has to be proven if the distribution in the microsomes similarly corresponds to that in the hepatocyte plasma membrane. The results HIGGINS and DAWSON (1977) and BALLEN and HIGGINS (1980) obtained using phospholipase C of *Clostridium welchii* differ from those of NILSSON and DALLNER (1977a, b). The discrepancies between both groups of experimental results may be caused by the use of different phospholipases. The distribution of the phospholipids on the cytoplasmic side of the intact microsomal bilayer by the latter authors was calculated as 84% phosphatidylcholine, 8% phosphatidylethanolamine, 9% sphingomyelin and 4% phosphatidylserine. The inner leaflet revealed 28% phosphatidylcholine, 37% phosphatidylethanolamine, 6% phosphatidyl-serine and 5% sphingomyelin. Phosphatidylinositol was assigned to the inside (luminal) of the microsomal bilayer. The authors further found that smooth and rough membranes of the endoplasmic reticulum exhibit almost the same asymmetric phospholipid distribution with phosphatidylethanolamine and phosphatidylserine mainly distributed in the inner leaflet of the bilayer and phosphatidylcholine and sphingomyelin concentrated in the outer leaflet. Apart from lecithin which is shifted towards the outer leaflet in the smooth membranes the asymmetric distribution was conserved in rough and smooth membranes.

Specific hydrolysis of phosphatidylinositol provides evidence against its asymmetrical distribution in intracellular membranes. The results confirmed a symmetric distribution of this phospholipid across the microsomal membranes of rat liver (SUNDLER et al., 1977).

On the other hand the experimentally proved translocation of phospholipids ('flip-flop') that means the exchange between both sides of the membrane makes it difficult to understand by what mechanism asymmetry is maintained. Translocation of spin labelled dipalmitoylphosphatidylcholine at an artificial membrane of the same phosphatidylcholine was demonstrated to be slow with a half time of 6.5 h at 30 °C (KORNBERG and MCCONNEL, 1971). ROTHMAN and LENARD (1977) observed for the translocation of sphingomyelin and phosphatidylcholine in the membranes of influenza virus half times of 30 and 10 days, respectively with a strongly differing distribution between inside and outside oriented phospholipids as compared to that found in human erythrocytes.

Other natural membranes, however, such as those from electroplax of the electric eel *Electrophorus electricus* (MCNAMEE and MCCONNELL, 1973), of viable plasma membranes from *Acholeplasma laidlawii* (GRANT and MCCONNELL, 1973), erythrocytes (BLOJ and ZILVERSMIT, 1976) and microsomal membranes (ZILVERSMIT and HUGHES, 1977) gave half times in the order of minutes. As not only phosphatidylcholine but also other phospholipids participate in this 'flip-flop' mechanism, the asymmetric distribution of phospholipids observed may reflect an equilibrium distribution rather than being a direct consequence of biogenetic events. The affinity of asymmetrically inserted proteins to the polar head groups of specific phospholipids may contribute to the asymmetry.

As compared to the well documented asymmetric transversal distribution of phospholipids that of proteins has been less thoroughly studied. From the data, however, so far available it can be conclusively derived that also endoplasmic proteins exhibit an asymmetric distribution which is related (i) to the vectorial function of most endoplasmic enzymes and (ii) to different kinds of insertion (cf. p. 25). The existence of a transverse asymmetric distribution of proteins has been evidenced by utilizing hydrolytic enzymes (proteases, phospholipases) for a selective breakdown of the membrane and compared with the results obtained by detergents (see for details the following section).

A disadvantage of microsomes as compared to other membranes (inner mitochondrial membranes and erythrocyte membranes) is that it is not possible to achieve inside-out transformation of microsomes.

There are several possible explanations for protein and lipid asymmetry in microsomes. The availability of substrates and the final site of the deposition of the reaction products may be decisive for the localization. For example the NADPH-dependent P–450 reductase utilizes NADPH from the cytoplasmic pool and transfers the accepted electrons to the terminal oxidase which is embedded in the membrane. Therefore the reductase for being adapted to its function requires a hydrophilic acceptor site and a hydrophobic anchor. The information for this protein specific insertion into the bilayer membrane may be contained in its amino acid sequence determining its 3-dimensional structure, respectively. Individual phospholipids accumulate around asymmetrically distributed enzymes (STIER and SACKMANN, 1973). The distribution of lipid biosynthetic pathways suggests further explanation for the asymmetric lipid distribution in erythrocytes and possibly also in intracellular membranes.

2.2.2. Localization of the Electron Transfer Systems in the Endoplasmic Reticulum

Due to the membrane bound state of the proteins the localization of the liver endoplasmic monooxygenatic P–450 system should be discussed considering two aspects (i) how are the proteins embedded in the membrane with respect to their transversal orientation (ii) in which way are they distributed along the lateral plane of the membrane: heterogeneously or homogeneously. According to REMACLE et al. (1974) there are 2 possibilities for the enzymes to be arranged: the multiplicity model (each microsome contains enzymes of only 1 group) or the heterogeneity model (all microsomes contain components of all groups).

In order to localize the components of the electron transfering systems including the NADPH dependent cytochrome b_5 system three different kinds of approach have been utilized: differential centrifugation, proteolytic digestion and antibody labelling.

SCHULZE and STAUDINGER (1975) by means of differential centrifugation in a hyperosmotic gradient have demonstrated that irrespective of the character of the endoplasmic reticulum (rough or smooth membranes) lateral asymmetric distribution of enzymes in a macroscopic scale can be excluded. However analysis of very small microsomes revealed differing enzymatic activities indicating that cooperating enzymes are not randomly distributed but form protein domains (SCHULZE et al., 1972). These results have been supported by WINQUIST and DALLNER (1976) who demonstrated that adenosine monophosphatase, and inosine diphosphatase can be separated from the electron transporting enzymes as well as by STASIECKE (1979) who found epoxide hydratase, UDP glucuronyl-transferase and glutathione-S-transferase not to be enriched in the same fraction as the P–450 monooxygenase system. These data suggest the existence of a macrohomogeneity in the enzyme distribution in the lateral plane of the membrane at simultaneous spatial association of enzyme systems catalyzing successively the same substrate thus forming protein domains with a heterogeneous microdistribution in the lateral plane which supports the multiplicity model. WINQUIST and DALLNER (1976) moreover have shown that the NADPH- and NADH-dependent electron transport systems can be partially separated using detergent concentrations lower than those required to dissolve the membranes.

By trypsin treatment of intact microsomes it was found that NADH- and NADPH-dependent reductases as well as cytochrome b_5 were solubilized or inactivated whereas P–450 was only partially (40%) inactivated. P–450 is unique in its quite homogeneous distribution between both layers of the bilayer membrane (BAR-NUN et al., 1980). It is partially resistant towards proteolytic attack from both sides of the bilayer. Despite of certain objections due to limited accessibility of peptide bonds towards an enzymatic attack the data obtained (NILSSON and DALLNER, 1977b) are in good agreement with respective data obtained by other methods indicating that P–450 is embedded in the bilayer.

Labelling experiments support that P–450 has a transmembrane disposition because its accessibility to macromolecular labelling probes seems to be enhanced in the presence of detergents which increase the permeability of the membranes (RODRIGUEZ-BOULAN et al., 1978).

In contrast, the NADPH dependent cytochrome P–450 reductase is differently associated with the membrane. Comparing the molecular weights of detergent solubilized reductase (79.000) with that of a proteolytically solubilized one (71.000) WELTON et al. (1973) described a difference of about 8.000 which means that about 70 amino acid residues are embedded in the membrane bilayer acting as anchor for the whole molecule. A further example of membrane proteins which have a distinct amphipathic structure exhibits the cytochrome b_5 system. Cytochrome b_5 obtained by detergents solubilization has a molecular weight of 16.700 (SPATZ and

STRITTMATTER, 1971) and contains 40 amino acid residues more than that solubilized by trypsin or lipase extraction. By the same authors similar results have been obtained for the NADH dependent cytochrome b_5 reductase (SPATZ and STRITTMATTER, 1973). The detergent solubilized form has a molecular weight of 43.000 as compared to a molecular weight of 33.000 for reductase obtained by lysosomal extraction of microsomes. The difference is due to 99 amino acid residues forming a hydrophobic anchor for the molecule.

Cytochrome P–450 and cytochrome b_5 together with the respective flavoproteins were reacted with the respective proteins using ferritin labeled antibodies and the precipitates electromicroscopically visualized. The data revealed that P–450 is localized exclusively at the outer layer of the smooth as well as of the rough endoplasmic membrane and at the outer membrane of the nucleus (MATSUURA et al., 1978). Despite this controversial finding which may be due to specificities of the method used a transmembrane localization of P–450 is more likely.

The heterogeneous distribution of the ferritin particles on each microsomal vesicle indicates a cluster-like arrangement of P–450 (MATSUURA et al., 1978) which is supported by kinetic investigations (FRANKLIN and ESTABROOK, 1971; PETERSON et al., 1976). The results obtained by means of the ferritin labelling demonstrate a very low P–450 content in other cell organelles which may emerge from microsomal contaminations. Therefore its existence in other liver cell fractions except its experimentally proved occurrence in the membrane of the nuclei (SAGARA et al., 1978) and in the inner mitochondrial membrane (PEDERSEN, 1978) remains to be clarified.

Immunological investigations using ^{125}I-labelled antibodies against P–450 revealed that P–450 is exclusively synthesized by membrane bound ribosomes and is differently distributed between the outer surface of the rough and the smooth membranes (NEGISHI et al., 1976). Cytochrome b_5 in contrast is synthesized by free polysomes (RACHUBINSKI et al., 1980). The P–450 content has been shown also by immunological techniques to be higher in the smooth membranes and remarkably to increase after phenobarbital induction.

Similar to P–450 also the NADPH dependent P–450 reductase is synthesized by membrane bound ribosomes and without passing the vesicular cavity like albumin is directly inserted into the outer surface. This is evidenced by studies on rat liver microsomes which after incubation with ^{3}H-puromycin have been treated with a mixture of proteases (trypsin and chymotrypsin). Even the nascent P–450 reductase is accessible towards a proteolytic attack whereas albumin which is transported into the vesicular cavity is not digested (NEGISHI et al., 1975). By means of indirect immunoelectronmicroscopy it could be shown that 1—12 ferritin-conjugated antibodies against reductase are attached to 80% of the microsomes present in the liver slices. The morphological analysis revealed a homogeneous distribution over the outer surface of all microsomes but a heterogeneous distribution with respect to a single microsome (MORIMOTO et al., 1976).

Analogously to the P–450 system also cytochrome b_5 with its NADH dependent flavoprotein has been shown to be randomly distributed along the lateral plane (SCHULZE et al., 1972). The stoichiometry ranges similar to the P–450 system between an 10 to 20 fold excess of the hemoprotein over the reductase. Cytochrome b_5 has been assayed besides in microsomes also in the outer mitochondrial membrane and in small amounts also in peroxisomes. The Golgi-apparatus and the plasma membrane are free of the cytochrome b_5 system (FOWLER et al., 1976). The distribution is homogeneous over all vesicles and supports the heterogeneity model of REMACLE et al. (1976). Contrary to P-450 an intramicrosomal cluster-like association has not been shown with the cytochrome b_5 system (ROGERS and STRITTMATTER, 1974a).

Recent investigations indicate that NADH cytochrome b_5 reductase, cytochrome b_5 and NADPH-reductase despite differences in their molecular weights, their prosthetic groups and in their solubilization are characterized by similar properties. All 3 proteins are amphiphilic in nature, consist of a hydrophilic part which contains the prosthetic groups and a hydrophobic part which binds to the microsomal membrane. Moreover, there exists further structural similarities:

1) Treatment with carboxypeptidase split off 25—30 amino acid residues from the C-terminal leading to an inability to bind to the microsomal membrane (MIHARA et al., 1978).
2) There is no direct functional linkage between the 2 domains of the molecules which has been proved for cytochrome b_5 and the NADH cytochrome b_5 reductase.
3) The N-terminal amino acid residue is masked.

Briefly summarized the membrane can be regarded as vectorial structure maintained by a dynamic equilibrium. The lipids are present at each side of the membrane but in different composition. Its components — proteins and carbohydrates — are asymmetrically distributed between both surfaces. The protein specific insertion which for membrane bound proteins determines their asymmetric distribution in the membrane is assumed to emerge from protein phospholipid interactions the information of which is contained in the protein structure. The functional implications of the phospholipid asymmetry are not fully understood as yet. From different lipid composition, however, different fluidity of both monolayers results which may be of functional importance.

3. Functional Importance of the Endoplasmic Membrane

Discussing problems related to the membrane bound character of P-450 means an attempt to explain the uniqueness of action of the P-450 system when associated with the endoplasmic reticulum. With regard to the multicomponent character of the system, the multiplicity of the P-450 isozymes and the phospholipids taking account of different head group and fatty acid composition it is impossible within this chapter to discuss all problems involved. Therefore necessary simplifications have to be made. Disregarding the multiple character of P-450, problems concerned with liposomes and effectuating interactions with other protein components we will mainly stress on (i) what are the structural implications in the functional role of phospholipids (ii) what are the structural adaptions to almost definite stoichiometries with respect to the organization of the system within the membran, (iii) what mechanisms of regulating the system are conceivable.

3.1. *Mechanisms of Interations between Phospholipids and the Essential Components of the P-450 Dependent Monooxygenase*

Phospholipids appear to function in at least 3 different ways:

- phospholipids mediate interactions between P-450 and its corresponding reductase thus constituting a catalytically active arrangement,
- phospholipids modulate the electron transfer,
- phospholipids stabilize and/or induce a functionally active conformation of P-450.

In the following it will be discussed how far these assumptions are supported by experimental data implying that the different functional aspects do not exclude each other but more likely will overlap.

Since Lu et al. (1969b) succeeded in the solubilization of the rabbit liver microsomal mixed-function oxidase it is well established that lipids are essential for its hydroxylating activity. The reconstituted system, consisting of the terminal oxidase, the NADPH P–450 reductase and a heat-stable, organic solvent extractable factor, which later could be identified as phospholipid (Strobel et al., 1970) was found to catalyze the fatty acid ω-hydroxylation (Lu et al., 1969b) but also the hydroxylation of hydrocarbons and the demethylation of drugs (Lu et al., 1969a). Neutral fats proved ineffective for reconstituting an active system but synthetic phosphatidylcholine was found as active as the crude extract. Extending these results Strobel et al. (1970) observed that the extent of reduced P–450 raised at addition of phospholipids to the assay mixture and that there exists a correlation between the extent of reduction and the lipid concentration. This relation has been quantitated with respect to hydroxylating activity by Ingelman-Sundberg (1977).

All these data suggested as most relevant function of phospholipids with respect to the substrate metabolism their implication in electron transfer processes that means a relatively specific role. A lot of experimental data supported this working concept without solving the question for the mechanism of interaction. So Ingelman-Sundberg (1977) demonstrated that phospholipids facilitate the NADPH-dependent, but also the sodium periodate supported 6β-hydroxylation of androstenedione by a stabilization and/or induction of an active conformation of P–450. He found that in sodium periodate-supported 6β-hydroxylation of androstenedion renaturable P–450 was saturated with at least 10 nmoles of phosphatidylcholine per nmol P–450, whereas in NADPH-supported reactions saturation only occurs when a phosphatidylcholine/P–450 ratio of 1000 was reached thus indicating different requirements for phospholipids in dependence on the type of reaction (peroxidatic or monooxygenatic).

A similar requirement for phospholipids was also found for other P–450 dependent activities. Williamson and O'Donnell (1969) demonstrated that phospholipids can stabilize P–450, obtained from an aceton powder of bovine mitochondria, thereby preventing decomposition to the inactive P–420 and enhancing the rate of deoxycorticosteron 11β-hydroxylation. Hydroxylation of testosterone in 6β, 7-α or 16-position was also shown to depend highly on the presence of phospholipids in the assay system as does the 3,4-benzpyrenhydroxylation (Lu and Levin, 1974).

A very important step in the efforts to understand the role of phospholipids was the finding that certain non-ionic detergents can substitute phospholipids maintaining enzymatic activity. This finding suggests that the hydrocarbon chains of the phospholipids are more important than their characteristic properties to form bilayer structures (Lu et al., 1974). The relative unspecificity of phospholipid/protein interactions is further supported by the well known fact that in reconstitution experiments cross-reacting substitutions including phospholipids as well as reductase are possible over a wide phylogenetic range (yeast — mammalian) (Duppel et al., 1973). These results provide evidence for the functional importance of phospholipids without differentiating whether the phospholipids interact as molecular or supramolecular entities.

A role of the phospholipid component in substrate binding and control of substrate specificity was suggested by Lu et al. (1969a, b). The authors reported on the influence of phospholipids on the hydroxylation of benzphetamine and laurate but could not detect any influence on spectral binding constants of these two substrates. Phosphatidylcholine was found neither to serve as a substrate for the reconstituted system nor to cause any spectral changes in the absorption of CO difference spectra (Lu et al., 1969b). Later it was shown that partially purified P-450, which contained less than 1 nmol phospholipid per nmol P-450 still gives type I and type II binding spectra with a variety of substrates. Therefore obviously P-450 does not need phospholipid for substrate binding (Levin et al., 1974) but phospholipids increase the affinity of P-450 to bind substrates (Ristau et al., 1979), meaning that a conformational change is induced by phospholipids and/or the substrate partitioning in the non-aqueous phase is improved. The idea that phospholipids influence drug metabolism by partition effects was already suggested by Cho and Miwa (1973) studying the effect of lipids on substrate partition between microsomes and aqueous solutions. They found that in a mixture of buffer and heptan uncharged N,N-dimethylamine derivatives are preferably dissolved in the non-aqueous phase whereas charged compounds are distributed in the aqueous phase. An estimation of the partition coefficient of substrates in dependence on lipid concentration revealed a direct influence of lipids on substrate binding to P-450 (Ebel et al., 1978).

Chaplin and Mannering (1970) concluded from their experiments that the type I binding site is associated with the membrane lipid. After phospholipase C treatment they found a destruction of the type I binding site. An explanation of this finding was given by DePierre and Ernster (1975), who demonstrated that laurate and unsaturated fatty acids are able to compete with type I substrates for the binding site, thereby inhibiting the metabolism of type I compounds. No such effect was observed with aniline as a type II substrate. The free fatty acids could be produced in situ by diglyceryl lipase activity, which is contained in many membranes and gets its substrates from phospholipase C.

Delipidation of P-450 by phospholipases and organic solvent extractions removed 80—85% of phospholipids and essentially all neutral lipids, but as a consequence the measurable amount of P-450, determined by CO difference spectra, represented only 20—25% that of the starting material. The formation of P-420 probably is due to the effect of phospholipase-produced lysolecithins, which were shown to induce the conversion of P-450 to P-420 (Dyatlovitskaya et al., 1978a). Renaturation was most effective with a mixture of phosphatidylcholine and phosphatidylinositol of 10 : 1 (w/w), which corresponds to the proportion found in liver microsomes (De Pierre and Ernster, 1977), giving a 3 times higher P-450 content as in delipidated microsomes. Phosphatidylcholine alone was also very effective (50—62.5 %), whereas phosphatidylethanolamine (40 to 50%) and phosphatidylserine (26—32.5%) were less active in renaturing the P-450. The following combinations of different phospholipids were not able to increase the P-450 concentration to more than 200% of the content in delipidated microsomes, phosphatidylcholine : phosphatidylethanolamine = 2 : 1; phosphatidylethanolamine : phosphatidylserine = 2 : 1 and phosphatidylcholine : phosphatidylserine = 2 : 1 (w/w) (Ingelman-Sundberg, 1977).

Instead of modifying the membrane lipids by phospholipases, detergents or organic solvents Dyatlovitskaya et al. (1978b) therefore used a more physiological approach to study phospholipid/protein interactions altering the phospholipid composition of the membrane by the insertion of the respective lipids into the microsomes using the phospholipid incorporating protein. This way it was possible to show that P–450, inserted into the microsomal membrane needs phosphatidylcholine for its function, whereas phosphatidylethanolamine alone is not able to replace it. Thus, in a more physiological manner, without isolation of the respective proteins from the membrane and extraction of the concomitant lipids the significance of phosphatidylcholine for the function of P–450 was demonstrated.

After the first pioneering contributions to the essential role of phospholipids investigations more systematically were carried out aimed to prove due to the amphiphilic nature of phospholipids either the importance of the head group or that of the fatty acid chains by means of reconstitution experiments into differently composed liposomes. Already STROBEL et al. (1970) could demonstrate that addition of phosphatidylcholine can reactivate the delipidated P–450, whereas phosphatidylethanolamine or a mixture of both can reconstitute only a final activity in benzphetamine demethylation of about 17 and 46%, respectively.

On the other hand BÖSTERLING et al. (1979) demonstrated that phosphatidylethanolamine is essential to reconstitute P–450 and reductase into vesicles at high protein : phospholipid ratios. Previous studies of this group (STIER et al., 1978) revealed that the spectra of resuspended microsomal lipids in which the phosphatidylethanolamine : lecithin ratio is 1 : 2 resemble a typical bilayer, whereas at phosphatidylethanolamine : lecithin ratio = 1 : 1 in addition to the lamellar a non-lamellar phase appears, not clearly hexagonal, which allows motional averaging. Their ^{31}P-NMR studies show that part of the lipids exists in an inverted micellar state which may be indispensible for the structural integration of P–450 into the bilayer membrane. The spectra obtained from mixtures phosphatidylethanolamine : lecithin 1 : 1 are nearly identical in shape to those of microsomes at higher temperatures thus indicating ellipsoidal inverted micelles consisting predominantly of phosphatidylethanolamine within the bilayer. Only liposomes with a ratio of phosphatidylcholine : phosphatidylethanolamine : dipalmitoyl phosphatidic acid = 2 : 1 : 0.06 (w/w) were capable to have proteins being incorporated like microsomes thus representing a good model for studying rapid lateral and rotational diffusion as well as functional properties of the inserted P–450 reductase complex (BÖSTERLING et al., 1979). The role of phosphatidic acid thereby consists of preventing aggregation of the liposomes by introducing a surface charge.

Studying the influence of the head groups and hydrocarbon chains on the spin state transition after temperature-jump YANG and TSONG (1980) found that phosphatidylethanolamine was less effective than the respective phosphatidylcholine and they further demonstrated that in phosphatidylcholines the effectiveness in facilitating rapid conformational changes decreases in the following sequence: dilauroyl < dimyristoyl, dioleoyl < dipalmitoyl. By this they supported the finding of STROBEL et al. (1970), that already one double bond is sufficient (dioleoyl-lecithin) to form an active complex in functional studies. Since Triton X-100 and Emulgen 911 were also quite effective in accelerating the relaxation after T-jump the authors concluded that the hydrocarbon chain appears to play a more important role in affecting the rate of conformational changes than the head group. On the other hand the functional significance of negatively charged vesicles is evidenced for instance by HAAPARANTA et al. (1980), who demonstrated that acidic phospholipids are up to 6 times more active than neutral ones in P–450 LM2-catalyzed 0-demethylation of 7-ethoxycoumarin and hydroxylation of benzo(α)pyrene which will be discussed in more detail later.

Due to its amphipathic character the interaction of phospholipids with membrane bound enzymes can be altered also by different acyl chains. Reconstituted enzyme systems basing on the isolation and purification of the respective essential monooxygenatic components offer an appropriate means for studies of this kind. STROBEL et al. (1970) studied the influence of lecithins with various lengths of the fatty acid residues on the hydroxylation of drugs (ethylmorphine, benzphetamine), fatty acid (laurate) and hydrocarbons (hexane, octane). At low phospholipid concentration (about 35 nmoles per nmole P–450) dioleoyl-phosphatidylcholine was the most effective, whereas at higher concentrations of the phospholipid (160—190 nmoles per nmol P–450) mono- and dilauroyl-phosphatidylcholine exhibited the highest activities. Distearoyl- and dipalmitoyl-lecithin, in contrast, were ineffective.

Dilauroyl-lecithin often has been used in reconstituted systems because it has been found to give nearly maximum catalytic rates, is resistant towards lipid peroxidation and predominantly forms multilamellar aggregates. It is active not only in the drug-metabolizing enzyme system, but also in reconstituting the aryl hydrocarbon hydroxylase (LU et al., 1974; CHIANG et al., 1979) and the bile acid metabolism (BJÖRKHEM, 1974). The activity of the reconstituted system, however, is dependent on the respective reaction being catalyzed. So INGELMAN-SUNDBERG et al. (1979) demonstrated that the LM3-catalyzed 6β-hydroxylation of androstenedione proceeds 10 times faster in eggyolk phosphatidylcholine vesicles than in dilauroyl-lecithin particles, whereas the LM2-catalyzed 16-hydroxylation is inhibited in the same particles. More detailed investigations (INGELMAN-SUNDBERG and JOHANSSON, 1980) revealed that LM4-supported androstenedione- and 7-ethoxycoumarin-metabolism is equal in egg yolk lecithin vesicles and in the dilauroyl-lecithin system, whereas the LM3-supported hydroxylase activities for various substrates is influenced to a different amount. The LM3-catalyzed androstenedione and ethylmorphine metabolism is higher in egg yolk lecithin vesicles than in the dilauroyl-lecithin system, contrary to the LM3-catalyzed aminopyrine, 7-ethoxycoumarin and benzphetamine metabolism which is higher in the dilauroyl-lecithin system. In general LM2-dependent hydroxylase activities are inhibited in egg yolk lecithin vesicles as compared to the dilauroyl-lecithin system. On the other hand androstenedione, ethylmorphine, aminopyrine and 7-ethoxycoumarin hydroxylation is greatly enhanced in the presence of microsomal vesicular phospholipids thus evidencing the importance of certain membrane phospholipids which are not contained in dilauroyl-lecithin and egg yolk lecithin for the reconstitution of LM2-dependent activities and the dependence of activity of certain P–450 isozymes on vesicular or non vesicular phospholipids.

Up to this point of the discussion it has become evident that there are numerous experimental data indicating that the addition of phospholipids is not only essential to reconstitute an active enzyme system but moreover exerts a considerable stimulating effect on the rate of electron transfer and of substrate hydroxylation from 5—20-fold depending on the experimental conditions and the reaction under study not only in experimental animals (STROBEL et al., 1970; van der HOEVEN and COON, 1974) but also in the human liver system (BÖSTERLING and TRUDELL, 1980). But which way the mechanism of activation is functioning is still unclear. Taking into account the complex character of the system an activation may originate either from conformational changes of the components or by improved interactions between them or from both.

That in general phospholipids may alter the conformation is supported by the appearance of conformational changes of delipidated enzymes after reconstitution with phospholipids which has been described already previously for other enzymes, e.g. pyruvate oxidase, adenylatcyclase (cf. Burlakova et al., 1977; Sandermann, 1978). That conformational changes are involved indeed in phospholipid effects on P-450 was proved by CD-measurements (Chiang and Coon, 1979). In the presence of dilauroyl-lecithin the ellipticity of P-450 LM2 is enhanced indicating that the α-helix-content increases by about 10% whereas an increase of about 25% has been calculated for P-450 LM4. The same authors observed also an influence of dilauroyl-lecithin on the spin equilibrium of P-450 LM2 which is shifted to the high spin state. Utilizing these findings French et al. (1980) by titration experiments determined the equilibrium between P-450 LM2 and DLPC. Two apparent dissociation constants were obtained. The constant with high affinity was determined with 3–6 μM and for the second constant with lower affinity a value of 50–70 μM was found which was eliminated upon addition of reductase. The authors established that the phospholipid dependent interactions observed are either specific or the result of exclusion of water from nonpolar segments of the molecules involved therein.

By derivative spectroscopy in the median UV-region it has been evidenced that at interaction between P-450 and phospholipid the spectral properties of tyrosine are altered suggesting that tyrosine residues are involved therein (Ruckpaul et al., 1980a). Similar effects have not been observed by phospholipid interactions with reductase.

With regard to the phospholipid concentration within the membrane and considering further the critical micelle concentration of phospholipids which is in the μM range conformational changes are more likely to be induced by interactions of the proteins with phospholipids in the vesicular than in the molecular form. Dimyristoyl-lecithin, which in general incorporates LM2 very slowly (may be as a consequence of higher phase transition temperature) but in dependence on the lipid to protein ratio and on the temperature was shown to interact not directly with the substrate binding site of rat liver P-450 (Tsong and Yang, 1978). The authors could show that although dimyristoyl-lecithin did not alter the binding of benzphetamine, at 25 °C it significantly increased the rate of substrate-induced conformational changes of the protein and suggested that this phospholipid rather provides a proper environment to facilitate the conformational transitions of the enzyme.

These observations evidence that interactions between phospholipids and cytochrome P-450 induce conformational changes of the hemoprotein. The results, however, do not allow to conclude which way phospholipids exert their essential functional role. To better understand the mechanism of interactions between phospholipids and P-450 the influence of different head groups was systematically studied by an analysis of the spectral properties, of the spin equilibrium and the reduction reaction of P-450 LM2 (Ruckpaul et al., 1980, 1982).

The results prove that the effectivity of phospholipids to enhance the high spin portion of P-450 does not depend on any ligand function on a molecular level of interaction. The physiologically occurring phospholipids form vesicles at concentrations far below the K_s values of the enzyme substrate complex and the presence of phospholipids favours the binding of substrates rather than to compete with them. On the other hand detergents are capable to reconstitute a reducible system despite their ineffectivity to shift the spin equilibrium. Consequently rather a supramolecular structure (micelles or phospholipid vesicles) than molecular interactions seem to be necessary to reconstitute enzymatic activity independent of the capability to shift the spin equilibrium.

The importance of the acidity of the phospholipid head group to influence the activity of the P–450 system has been shown by several authors (INGELMAN-SUNDBERG, 1977; KISSEL et al., 1979; HAAPARANTA et al., 1980). The sequence of the effectivity with respect to the reduction reaction in dependence on varying head groups of the phospholipid which increases with increasing acidity correlates well with the shift of the spin equilibrium of P–450 (RUCKPAUL et al., 1980). Detergents and alkanes on the other hand do not fit this correlation. The functional integrity of the system, therefore, is not strictly correlated with the spin shift.

Extending these results INGELMAN-SUNDBERG et al. (1982) found a linear correlation between the negative charge of the vesicles and the reduction rate in various reconstituted P–450 LM2 systems. Furthermore, a comparison between the fraction of P–450 LM2 reduced in the fast phase and the mobilities of the vesicles indicate a linear relationship. Thus a more negative charge of the vesicles results in a more rapid reduction of P–450 LM2. Primarily the effect is indicated to be associated with a greater fraction of the protein reduced in the fast phase, whereas the rate constants for the fast and slow phases are not significantly affected by the membrane charge. These results are in line with data from studies of LU et al. (1980) on rat liver microsomal P–450. The authors demonstrated that dilauroyl-lecithin increases V_{max} for benzphetamine while the K_m value was not affected and concluded that the phospholipid facilitates the formation of a catalytically active binary complex between P–450 and reductase by lowering its dissociation constant.

Due to their highly ordered structure phospholipids make the proteins to orient themselves in a manner that parallels the axis of the proteins with the axis of the acyl residues of the phospholipids thus mediating a functional linkage between the essential proteins. This obviously is not only valid for proteins with a hydrophobic anchor but also for integral proteins (see for details section 3.3.). Detergent micelles can produce this orientation, too.

The phospholipids by means of their head groups are capable to modulate these protein-protein interactions and this way influence the enzymatic activity. Especially negatively charged phospholipids may be responsible for regulating charge dependent interactions between P–450 and reductase which has been shown by Ruckpaul et al. (1980) in the reduction reaction and by INGELMAN-SUNDBERG et al. (1980) in the p-nitro-anisole demethylation. The nonequivalence of substrate and phospholipid induced effectuation is evidenced by the finding that the substrate induced high spin content increases the reduction rate up to tenfold (Taniguchi et al., 1979) whereas phospholipids increase the reduction rate about 2-fold.

The increasing effectivity of the phospholipids with increasing acidity indicates that besides environmental effects also negative charges are important. This conclusion is supported by results on cytochrome b_5, the functionally P–450 linked endoplasmic electron-transferring system which have been recently reported by DAILEY and STRITTMATTER (1979, 1980). Previous studies by STRITTMATTER (LOVERDE and STRITTMATTER, 1968) indicated that charge interactions between lysyl residues on the reductase with complementary negatively charged carboxyl groups on the cytochrome b_5 may occur. Similar observations have been made for cytochrome b_5/cytochrome c interaction. Based on the three dimensional structure determined by x-ray crystallography SALEMME (1976) has proposed a probable complementary charge interaction between lysyl residues of cytochrome c and carboxyl residues of cytochrome b_5 including the exposed propionate of its heme group. In recent studies on interactions between cytochrome b_5 and the NADH-reductase the carboxyl groups of the heme peptide of cytochrome b_5 have been modified by using carbodiimide, methylamine, glycine ethyl ester or taurine. The kinetic data of the reaction with NADH-reductase revealed that the modification of cytochrome b_5 heme peptides lowers the affinity to the reductase. From these data it was concluded that cytochrome b_5 interacts with NADH cytochrome b_5 reductase via complementary charge pair interactions in which side chain carboxyls of glu 47, 48 and 52, the single exposed heme propionate and a fifth as yet unidentified side chain carboxyl group of cytochrome b_5 are involved.

A similar mechanism may be of importance also in the interactions between P–450 and its NADPH-dependent reductase. Selective labeling of the N-terminal methionine of P–450 LM2 with fluoresceinisothiocyanate revealed a lowered reduction extent and lowered demethylation activity via the NADPH supported P–450 reductase catalyzed reaction at retained activity in the, hydroperoxide supported reaction (BERNHARDT et al., 1980). This observation suggests a charge pair compensation to be involved in recognition and interaction processes also in the P–450 system.

What general conclusions can be drawn with respect to possible interactions within endoplasmic membranes. Phospholipids due to its highly ordered bilayer structure do not only mediate a functional linkage between proteins by its amphipathic environment but rather more important make the proteins to orient themselves in a specific manner within the bilayer. This obviously is not only valid for such proteins possessing a hydrophobic anchor but also for those which are completely embedded within the bilayer and can be performed by detergent micelles, too. Therefore the phospholipid function in mediating the electron transfer is an indirect one and rather indicates protein-protein interactions involved therein. The orientation may be related to the vectorial function of some membrane bound enzymes. Negatively charged phospholipids may be responsible for recognition and regulation processed by charge compensation. This would include the possibility to regulate enzymatic activity by metabolic processes (see for details 4.1. this chapter). These peculiarities of the bilayer may help to understand the uniqueness of action of enzymes when associated with the membrane.

3.2. *Orientation of the Components of the Monooxygenatic System within the Endoplasmic Membrane*

Besides the localization of the essential components the orientation of cytochrome P–450 and its corresponding reductase within the membrane gains increasing interest. Especially with respect to understand the function of the phospholipids as essential component in the monooxygenatic system. Recently polarized optical absorption and angular EPR spectroscopy have been successfully applied to demonstrate that the heme planes of the cytochromes in the respiratory chain are not randomly arranged within the mitochondrial membrane but are highly oriented with respect to the plane of the membrane. Basing on experimental results in which it was shown that the average orientation of the planes of the membranes in cytochrome c oxidase multilayers was parallel to the plane of its support (BLASIE et al., 1977) it was found that the hemes of cytochrome c oxidase, cytochrome c_1 and cytochrome b are oriented in a manner with the normal to their heme planes lying approximately in the plane of the mitochondrial membrane (ERECINSKA et al., 1978b, 1979).

The orientation of both hemes, cytochrome a and a_3, of the cytochrome c oxidase by optical and EPR studies on hydrated multilayers of membranous cytochrome c oxidase in various liganded states also has been determined. From dichroic ratios of the Soret and visible absorption bands of the azide, sulfide and formate complexes of the oxidized cytochrome oxidase and more clearly from the EPR studies it has been evaluated that the angle between the heme plane and the membrane plane is almost 90° meaning that the heme planes were oriented perpendicular to the membrane plane (ERECINSKA et al., 1978a). As the absorption bands of the liganded oxidase exhibit well separated maxima for cytochromes a and a_3 and both have the same optical polarization properties it was concluded that both hemes have similar orientation with respect to the plane of the membrane. The average orientations of the hemes were qualitatively similar in the oxidized and reduced form. No differences were found between oriented hydrated multilayers of membranous cytochrome c oxidase and hydrated multilayers of mitochondrial membranes (ERECINSKA et al., 1978b, 1979; BLUM et al., 1978a). The heme chromophores of the cytochrome b—c_1 complex are similarly oriented as those of cytochrome c oxidase and have the same orientation in liposomes as in hydrated multilayers of mitochondrial membranes.

Employing the techniques of angular EPR spectroscopy to cytochrome P–450 systems oriented multilayers from bovine adrenal cortex submitochondrial particles have been studied (BLUM et al., 1978b, c). It was found that the heme plane of mitochondrial cytochrome P–450 unlike the cytochromes of the respiratory chain is oriented parallel to the membrane plane. RICH et al. (1979) extended these studies on P–450 and cytochrome b_5 of rat liver microsomes. The plane of the heme group of P–450 of the low spin and the high spin species of oriented ribosome depleted rat liver microsomes was found to be significantly oriented and to lie parallel to the plane of the membrane surface similar to that found for cytochrome P–450 in adrenal cortical mitochondria. The cytochrome b_5, however, appears to be relatively unoriented indi-

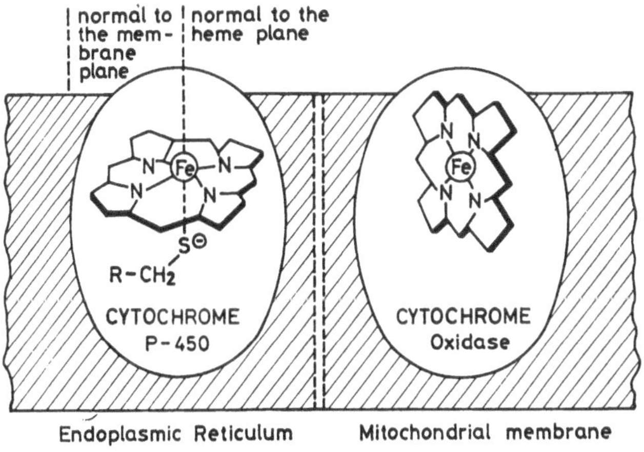

Fig. 7. Schematical drawing of the heme orientation in cytochrome P–450 and cytochrome odixase

cating a high degree of lateral mobility. Only little or no paramagnetic interaction between the heme groups of cytochromes b_5 and P–450 exists. By these results a model is favoured in which cytochrome P–450 and b_5 exists as independent entities interacting by molecular collision and do not form a tightly organized complex. From the alteration of the EPR signals by addition of water soluble paramagnetic ions (e.g. nickel EDTA) it is derived that both hemes of cytochromes b_5 and P–450 lie within 20 Å of the membrane surface.

These findings slightly differ from those obtained by RICHTER et al. (1979) who measured the rotational diffusion by flash-induced transient dichroism of the CO-complex of rat liver microsomal P–450. The experiments indicate that the heme of the P–450 CO complex does not have an orientation as has been found for the adrenal mitochondrial cytochrome P–450 (BLUM et al., 1978b, c). The heme is assumed to be inclined at an appreciable angle to the plane of the membrane (RICHTER et al., 1979).

The above mentioned results indicate that the heme orientation is not only important to understand structure-function relationships of the membrane bound monooxygenatic system but that the function of the phospholipids itself is related to these findings.

By incorporation into the bilayer membrane of the endoplasmic reticulum the heme group is oriented parallel to the surface of the membrane and as has been shown the paramagnetic iron has a distance from the surface of about 20 Å that means it is localized relatively near the membrane surface. In order to understand the functional importance of the phospholipids one has to consider that not only P–450 molecules but also the NADPH dependent reductase after incorporation into the membrane is oriented by means of its hydrophobic anchor. This orientation of both proteins with respect to the membrane surface together with the induction of conformational changes can be assumed to be a nesessary requirement for interactions between both components in order to constitute an enzymatically active system.

3.3. *Spatial Arrangement: Rigid or Non-Rigid Model*

P–450 can account for as much as 20% of the microsomal protein. The number of the molecules of this hemoprotein can be 10 to 35 times greater than that of the NADPH P–450 reductase molecules in the membrane. In pig liver microsomes a 5 times higher concentration of P–450 (0.85 nmole/mg protein) as compared with the NADPH P–450 reductase has been found which after correction of the molecular weight for reductase (from 91.000 to 79.000 daltons) increase to a ratio of about 1:9 (SCHULZE and STAUDINGER, 1975). ESTABROOK et al. (1971b) have determined a significant higher ratio in rat liver microsomes of about 1:20 which is enhanced by induction with phenobarbital to about 1:35. Similar ratios have been found in rabbits (PFEIL et al., 1980). This increased ratio is caused by a twofold increase of the reductase and a 3 times higher content of P–450 due to phenobarbital induction. From this stoichiometry it appears that one reductase molecule can interact with a certain number of P–450 molecules for efficient catalysis. The organization, the mobility and the mechanism of interaction of these enzymes within the membrane are necessary prerequisites to understand the enzymatic function. Despite great experimental efforts however, some crucial points have not been solved as yet.

Two possible mechanisms are conceivable for functional interactions among the essential protein components of the membrane bound P–450 systems. Interaction can be effected by lateral motion and subsequent collision of the proteins in the plane of the membrane including cross-linked interactions between adjacent electron transfer chains. The other possibility is that the membrane proteins constitute functional clusters in which the interaction depends on spatial organization in which electron transfer is only possible within a single functional unit without interactions between adjacent complexes. Both possibilities have been discussed with respect to the cytochrome P–450 dependent liver microsomal monooxygenatic system as being organized either in a non-rigid arrangement or as a rigid cluster.

Two lines of experimental approach have been followed in order to prove which of both models is consistent with experimental data The first line of evidence comes from the analysis of the temperature dependence of the functional behaviour of the system including activity measurements of systematically changed stoichiometry of the protein components in microsomes. The second line of experimental approach is concerned with more direct measurements basing on spin and fluorescent labeled proteins within biological or reconstituted membranes to get informations on protein mobilities from which conclusions on the organization can be drawn.

The relationship between phospholipids, their arrangement, distribution and physical state and the functional properties of the microsomal proteins have been studied by STIER and SACKMANN (1973). They found that the reduction kinetics of two spin labeled substances one of which is water soluble (phosphate ester of tetramethylpiperidine nitroxide) and the other is lipid soluble (stearic acid nitroxide) thus reflecting properties of the phospholipids within the membrane differs significantly. The activation energy of the fatty acid reduction by NADPH-dependent P–450 reductase of rabbit liver microsomes decreases abruptly at about 32 °C from 30.8 kcal/mole to 8.7 kcal/mole. No such break is observed in the Arrhenius plot of the water soluble substance yielding an activation energy of 13.8 kcal/mole in the whole temperature range investigated. In accordance with BLAZYK and STEIM (1972) who found that membrane lipids of microsomes using scanning calorimetry undergo a phase transition at 0 °C the authors demonstrated that the bulk of the microsomal lipid phase does not show any thermally induced phase transition above 5 °C. More than 80% of the microsomal phospholipids are estimated to be located in this bulk phase. From their results the authors therefore concluded that the enzyme system is enclosed in a phospholipid halo which is more rigid than the bulk phase of microsomal lipid and undergoes a crystalline — liquid crystalline phase transition at 32 °C. To a similar assumption of fluid areas and regions being in the gel state approached MORRISETT et al. (1975) studying phase transitions in *E. coli* membranes and in extracted lipids. Extending the model that lateral asymmetry leads to patches of phospholipids surrounding the membrane proteins which differ in their fluidity from the bulk phase STIER et al. (1979) basing on ^{31}P-NMR studies proposed a membrane structure in which inverted micelles of phosphatidylethanolamine surround the proteins thus forming a non-lamellar lipid phase. This assumption was derived from different lyotropic mesomorphism of individual lipids and especially from the fact that phosphatidylethanolamine prefers to form hexagonal structures as to form bilayers.

Analogous investigations basing on phase transitions of the lipids at distinct temperatures have been used by several authors to draw conclusions from functional properties on the spatial arrangement of the components within the membrane. A characteristic functional behaviour of P–450 represents its biphasic reductase and NADPH supported reduction which has been found by PETERSON et al. (1976), and supported by BLANCK et al. (1979) and OPRIAN et al. (1979) for microsomes (for details see chapter 3), but also by TANIGUCHI et al. (1979) and MIWA et al. (1979) for vesicular and by IMAI et al. (1977) for nonlamellar detergent systems, respectively.

PETERSON et al. (1976) studied the temperature function of the NADPH-dependent reduction reaction in rat microsomes. In the temperature range between 4° and 37 °C the reaction shows biphasic characteristics and is composed of two first order processes. The slow reaction phase exhibits a discontinuous temperature dependence with a break at about 20 °C which disappeared in the presence of certain substrates. The fraction of total P–450 reducible in the fast phase increases from 40% at 5 °C to a maximum of about 70% at 25 °C. These results led them to suggest that the reduction is mainly performed (50–70% rapid phase) in structurally associated clusters of P–450 and its reductase. The number of P–450 molecules intimately associated with the reductase is assumed to vary between 8 and 12 since almost 50% of the P–450 are reduced in the fast phase at low temperature. The reductase is assumed to randomly reduce the P–450 molecules immediately adjacent to it independent of lateral motion. The slow phase on the other hand, should represent diffusion dependent reduction in those P–450 molecules localized outside the clusters which have to pass into them in order to become reduced.

DUPPEL et al. (1976) observed phase transition dependent breaks in the Arrhenius plots of the 7-ethoxycoumarine and p-nitroanisole dealkylation reaction. With cumenehydroperoxide substituting the physiological oxygen supporting system, however, no breaks appeared thus indicating a mechanism dependent differentiation. YANG et al. (1977) investigating the N-demethylation of ethylmorphine, benzphetamine, aminopyrine and p-nitroanisole by rat liver microsomes observed similar breaks at around 24 °C. Activation energies of 10–12 and 19–21 kcal/mole at temperature ranges above and below the break were calculated. In the presence of 30% glycerol the breaks disappeared. Whereas DUPPEL et al. (1976) and PETERSON et al. (1976) attributed the breaks to phase transitions of the lipids YANG et al. (1977a) discuss that the phase transition may influence the lateral mobility of the enzymes.

Already FRANKLIN and ESTABROOK (1971) found the overall N-demethylation of ethylmorphine by rat liver microsomes to be inhibited in the presence of the organic mercurial mersalyl (25 nmol/mg protein) to approximately the same extent as the NADPH-cytochrome P–450 reductase (approximately 50%). This inhibition was accompanied by a small conversion of P–450 to P–420 (about 18%). 50% inhibition of the NADPH-dependent cytochrome P–450 reduction as the result of 50% inactivation of the reductase leads the authors to suggest the existence of a rigid spatial arrangement without free exchange between electron donors and acceptors. They further concluded that reducing equivalents can not be transferred between the components of different complexes. Reproducing these experiments using the same chemical modification YANG et al. (1977) gave a somewhat differing interpretation proposing that P–450 and the reductase are not rigidly organized in the membrane and that both molecules possess lateral mobility. In a series of additional papers YANG (1975; 1977a, b) provided experimental evidence for a non-rigid model of the P–450 dependent monooxygenatic enzyme system applying the same methods which have been used for the spatial arrangement of the cytochrome b_5 system which in the following are briefly summarized.

STRITTMATTER and ROGERS have accumulated experimental evidence for random distribution and translational movement of cytochrome b_5 and its NADH-dependent reductase. Detergent solubilized cytochrome b_5 was bound in a 10—15 fold molar excess over the cytochrome b_5 reductase thus increasing the normal microsomal concentration of cytochrome b_5 from 1 nmol/mg protein to nearly 20% of the total protein of the microsomal preparation. The activity of the exogeneous cytochrome b_5 was found to be indistinguishable from the endogeneous cytochrome b_5 (STRITTMATTER et al., 1972). Electron transfer between molecules of cytochrome b_5 could be excluded.

These experiments were extended to the amphipathic form of NADH-cytochrome b_5 reductase, which was bound in a 100 fold increase over the endogeneous enzyme level up to a maximum of 7.2 ± 0.5 nmoles reductase/mg microsomal protein. Also these experiments indicated (ROGERS and STRITTMATTER, 1974b) almost identical functional efficiency of extra bound and endogeneous reductase. These observations are not explicable with any model of the microsomal membrane in which cytochrome b_5 and the NADH-reductase are organized in a 'cluster-like' rigid arrangement. The experimental data rather led the authors to postulate a random distribution of both proteins on the surface of the microsomal membrane including translational diffusion prior to their interaction.

Analogous experiments have been carried out with P-450 and its respective NADPH-dependent reductase by YANG and STRICKHART (1975), YANG (1977b) and YANG et al. (1978). Detergent solubilized P-450 or purified NADPH-cytochrome P-450 reductase (30 units/mg protein, 0.3 M K-phosphate buffer, pH = 7.7) were incorporated into rat liver microsomes. The microsomes enriched with exogeneous protein were separated from non-incorporated protein by gel chromatography (Sepharose 4B). The incorporation of the exogeneous proteins occurs within 30 min at 37 °C. Similar to the cytochrome b_5 and NADH-reductase experiments also in the P-450 dependent monooxygenatic system the incorporated exogeneous proteins induced an enhancement of catalytic activity which appears as a general phenomenon with different types of monooxygenatic activities and with microsomes that contain different forms of P-450. The degree of enhancement of the hydroxylation activities depends on the respective substrate and the pretreatment of the rats and varies between 2—3 fold (MIWA et al., 1978).

From these data the authors draw the conclusion that they are consistent with a non-rigid model assuming the reductase molecule being not rigidly associated with 10—25 P-450 molecules in the membrane. As compared to the by more experimental evidence supported model of the cytochrome b_5 system there seem some points in the sequence of evidence concerned with the P-450 system which need further experimental support. First: almost the same catalytic efficiency of the purified hydrophilic fragments of the NADH-dependent reductase and cytochrome b_5 as compared to the same fragments firmly attached to the microsomes by hydrophobic peptides indicate that the cytochrome b_5 system is not drastically altered in its functional properties by binding to or incorporation into the microsomal membrane (ROGERS and STRITTMATTER, 1974a). The P-450 system on the other hand is subjected to mutual influences which alter conformation, interaction and resulting activities thus impairing conclusions concerning its structural organization. Secondly: one of the most important arguments favouring the postulation of a non-rigid model is the exclusion of an electron transfer between cytochrome b_5 molecules. For cytochrome P–450 experimentally proved evidence for the absence of an electron transfer between P–450 molecules is still lacking.

Pokrovski et al. (1977) have found that addition of a 35 fold excess of detergents solubilized NADPH-reductase (1680 nmoles cytochrome c reduced × $min^{-1} \cdot$ mg $protein^{-1}$) to rat liver microsomes and subsequent centrifugation in order to remove unbound protein effects a twofold increase of benzopyrene hydroxylation. The reduction rate of cytochrome P–450 did not change in the fast phase but increased only in the slow phase. From the findings it was concluded that endogeneous reductase with cytochrome P–450 forms associates (the exogeneous reductase does not) and that the electron in the slow phase is transferred by a mechanism of lateral mobility whereas the electron transfer in the fast phase is determined by associated molecules. These findings are somewhat different from those reported by Taniguchi et al. (1979). Partly these differences may be caused by different methodological approach. Using phosphatidylcholine vesicles the rate constant of the fast phase increases with an increasing molar ratio of reductase at a constant ratio P–450 to phosphatidylcholine. An increase of the fast phase was also found when the amount of phosphatidylcholine is decreased indicating interaction between the two proteins that is determined by their lateral mobilities on the plane of the membrane. The rate of the slow phase remains unaffected either at an altered ratio of reductase or at that of the two proteins to the phospholipid.

Summarizing all results reviewed so far there arise some difficulties to combine these apparently differing data. In order not to be confused one first has to take into account that actually the temperature dependence has been analyzed by the different authors but not of the same reaction. Stier and Sackmann (1973) investigated the reduction of spin labels, Yang et al. (1977) as well as Duppel et al. (1976) studied the hydroxylation reaction and finally in the studies of Peterson et al. (1976) the reduction reaction of P–450 in the absence and in the presence of substrates has been investigated in which only the insertion of the first electron is involved. Obviously in the first and second series of experiments dealing with the reduction of spin labelled compounds and the hydroxylation, respectively, diffusion steps are included and therefore the reactions can be affected by lipid phase transitions. The fast phase (main fraction) of the P–450 reduction reaction which determines the catalytic activity seems rather independent of phase transitions thus indicating that mainly diffusion independent interactions between reductase and P–450 govern the reaction (cluster interaction) presumably by protein protein interactions. Results of Strittmatter and Rogers (1975) on the temperature dependence of the reduction of a cytochrome b_5/NADH reductase system reconstituted in dimyristoyl lecithin liposomes support the existence of protein protein interactions. The cytochrome b_5: reductase ratio was varied at a constant reductase concentration in the range from 2:1 to 45:1. In this range the distance between the molecules varies from 100 Å to 5 Å. Assuming a random distribution the break observed at low protein concentrations can be explained by a phase transition of the phospholipids. At a ratio of 45:1 a break cannot be observed because at this concentration instead of random distribution rather a cluster like assembly is to be expected.

Despite still open questions the experimental results favour the assumption of a cluster-like arrangement of the P–450 system. Obviously besides temperature dependent diffusion processes mainly phospholipid dependent properties of the proteins as conformational changes, temperature induced spin state changes and by these changes induced alteration of interactions as well as concentration dependences may be causally connected with the described phenomena. Recent experimental data leading to the second line of evidence justify this statement.

By use of different spectroscopic methods the rotational diffusion of proteins within the membrane can be determined. From those data conclusions on the structural organization of membrane bound enzyme systems can be drawn. Measurements of the rotational correlation time by different techniques support the validity of the rigid model. RICHTER et al. (1979) first studied the rotational diffusion of P–450 in rat liver microsomes by flash induced transient dichroism of the heme CO-complex. Although the accuracy of their data did not allow a detailed analysis of the motion it was the first study demonstrating that P–450 at physiological temperatures is capable of relatively rapid diffusion. Using the eosine-type delayed fluorescence depolarization GREINERT et al. (1979) determined a rotational correlation time for liposome reconstituted P–450 LM2 of 111 μs.

By means of saturation transfer EPR techniques (ST-EPR) SCHWARZ et al. (1982a) determined a rotational correlation time of 480 μs for microsomal P–450 which was modified by a covalently bound maleimide spin label. Extending these studies the same authors (SCHWARZ et al., 1982b) evaluated a correlation time of 180 μs for P–450 LM2 incorporated into liposomes which is in fair agreement with 111 μs obtained by GREINERT et al. (1979) by use of optical methods. For a freely rotating monomer within the bilayer membrane a rotational correlation time of 20 μs was calculated which is significantly lower than the experimentally observed values. Obviously the higher value indicates that P–450 molecules in the bilayer membrane exist in aggregates. The difference between the liposomal and the microsomal state may emerge from different protein phospholipid ratios in both systems. Recent results from CHERRY and coworkers (KAWATO et al., 1982; GUT et al., 1982) by which earlier data are expanded (MC. INTOSH, 1980) evidence a dependence of the rotational diffusion on the protein concentration by monitoring the decay of absorption anisotropy after photolysis of the CO complex of P–450 by a vertically polarized laser flash. Increasing the protein to lipid ratio 30 fold (30:1 to 1:1 by weight) they found a significant increase of immobilized P–450 in vesicles which they assigned to self-aggregation of P–450. The rotational mobility was found to be influenced by NADPH P–450 reductase which significantly decreased the mean rotational relaxation time and eliminated completely the immobilized fraction (mean rotational relaxation time > 500 μs).

Considering the data available so far it is not possible to unambiguously decide whether the cytochrome P–450 dependent monooxygenatic hepatic microsomal system is functionally determined by an organization which is consistent with a non-rigid model or a rigid one, despite a great number of experimental data favouring the latter.

4. Nutritionally Induced Alterations of the Phospholipid Composition of Membranes

From the data accumulated in the last decade it has become evident that we know something about the principles of the mechanism of the P–450 dependent monooxygenases under standard conditions. But the knowledge about the influence of distinct pathological and physiological states on the drug metabolism is rather poor. We are just in the very beginning to understand the relationship between diet and drug metabolism. Especially the influence of different fats has attracted the interest of researchers to learn more about the biological regulation of the system.

4.1. *Influence of Changed Phospholipid Composition on Membrane Bound Enzymes*

The biological importance of the influence of phospholipids on P–450 dependent reactions is experimentally substantiated by several reports describing alterations of the drug metabolism in dependence on nutrition, under physiological, ontogenetic and pathological conditions (Rumack et al., 1973; Paine and McLean, 1973; Laitinen et al., 1975; Hammer and Wills, 1978; Denk, 1979; Feuer et al., 1980). From studies on the influence of certain diets on the phospholipid composition of the hepatic endoplasmic reticulum which have been correlated with the functional properties of the membrane bound enzymes metabolic regulations of drug, hydrocarbon, steroid, bile acid and fatty acid metabolism can be derived.

Due to its chemical structure there exist two different ways to modify the phospholipid (i) by a change of the fatty acid residue or (ii) by substituting the head group.

Despite a great number of contributions to the influence of diet on the P–450 content an unambiguous correlation between the saturation degree of the fatty acids fed and the P–450 content has not been found (Norred and Wade, 1972; Agradi et al., 1975; Hopkins and West, 1976).

A total fat deficiency in the diet decreases the activities of hydroxylation and glucuronidation of foreign compounds in microsomes (Norred and Wade, 1972). If, however, this diet was supplied with 3% corn oil which was found to increase the amount of linoleic (18:2) and arachidonic acid (20:4) in the phospholipids of hepatic microsomes (Wade et al., 1978) then in general the activity of aniline hydroxylase (Chaster et al., 1970) and ethylmorphine demethylase (Norred and Wade, 1972) increased. In accordance with these results is the observation that the activity of certain microsomal demethylase activities is enhanced after feeding rats unsaturated dietary triglycerides (Agradi et al., 1975; Rowe and Wills, 1976). These experiments have been extended to establish a linear relationship between the amount of polyunsaturated fatty acids and the metabolic activity of P–450 and P–448 in rats.

Fasting of rats leads to an overall increase in saturation of fatty acids in microsomal hepatic lipids and this way to a decrease in p-nitroanisole 0-dimethylation. Analogous results have been obtained investigating the activity of benzo-(α)pyrene metabolism (Becker et al., 1978). All these findings provide evidence that enhancement of unsaturated and especially of polyunsaturated fatty acids incorporated in the phospholipids of the microsomal membrane increase its ability to metabolize drugs and foreign compounds.

As has been shown by several authors (Ingelman-Sundberg, 1977; Ingelman-Sundberg et al., 1980; Kissel et al., 1979; Ruckpaul et al., 1980b) the efficiency of phospholipids to enhance activity parameters of the P–450 system (monooxygenatic activity, extent of reduction and high spin shift of spin equilibrium) increases with their acidity. Although the head groups contrary to the fatty acid residues exhibit a high degree of constancy the results indicate the possibility to modulate the activity of the system metabolically by in vivo phosphorylation of phosphatidylinositol which is contained in the endoplasmic phospholipids in an amount of approximately 5—10% . This assumption is supported (i) by the finding that phosphatidylinositol is almost equally distributed between both sides of the membrane (SUNDLER et al., 1977) and (ii) by data providing evidence for the formation of higher phosphorylated phosphatidylinositol derivatives together with kinases in the microsomal fraction of rat liver capable to convert 1-phosphatidylinositol 4-phosphate to 1-phosphatidylinositol 3,4-diphosphate (BEHAR-BANNELIER and MURRAY, 1980).

These authors have found that at invitro phosphorylation of rough and smooth membranes of the hepatic endoplasmic reticulum of mice appreciable amounts of ^{32}P from $/\gamma$-$^{32}P/ATP$ are incorporated into products migrating in the front of electropherograms. The products have been identified as being 1-phosphatidylinositol 4-phosphate. This front has already been observed in previous studies on rat liver microsomes (SHARMA et al., 1978; LAM and KASPER, 1979) and is in agreement with the observation that kinases are present in the microsomal fraction of rat liver which convert 1-phosphatidylinositol 4-phosphate to 1-phosphatidylinositol 3,4-diphosphate (HARWOOD and HAWTHORNE, 1969). It needs experimental proof to substantiate if there actually exists a metabolically controlled regulation mechanism of the activity of the hepatic monooxygenatic enzyme system in which also intermediately changed head group ratios are involved.

4.2. *Influence of Cholesterol on Structure and Function of Endoplasmic Enzymes*

Besides phospholipid induced modifications of the function of the monooxygenatic enzyme system there is an increasing amount of evidence that also the incorporation of cholesterol affects the drug metabolizing enzyme system (cf. BERNHARDT and RUCKPAUL, 1981). Changes of the cholesterol content of native membranes cause alterations in certain physico-chemical parameters of the membrane (e.g. fluidity, permeability, cell fusion properties) which lead to remarkable biochemical consequences (INOUE, 1974; BLOK et al., 1977). Different techniques have been applied to study cholesterol induced alterations in membrane fluidity: *X*-ray-analysis, calorimetry, NMR- and ESR-spectrometry, IR-, laser Raman- and uv-vis-spectroscopy (cf. OLDFIELD and CHAPMAN, 1972). Utilizing these different techniques is was found that the modifying effect of cholesterol in native and artificial membranes strongly depends both on the nature of the head group and on the fatty acid residue (DEGIER et al., 1969). When cholesterol was incorporated into a lipid phase below its transition temperature (Tc) it causes a higher mobility of the hydrocarbon chains of the fatty acids, whereas the mobility of the acyl chains is lowered when it is added above Tc thus producing a so-called 'intermediate fluidity' (LADBROOKE et al., 1968; VANDERKOOI et al., 1974; YEAGLE et al., 1977). The molecular mechanism of the interaction between cholesterol and the phospholipid bilayer, however, has not been sufficiently clarified (SEIMER and GELERINTER, 1979). Cholesterol preferably interacts with those lipid regions in membranes which are in the liquid-crystalline state (DEKRUIJFF et al., 1973). Moreover the affinity of cholesterol varies in dependence on the head group. For different phospholipids the following order of affinity has been obtained: sphingomyelin > phosphatidylserine > phosphatidic acid > phosphatidylcholine > phosphatidylethanolamine (VAN DIJCK, 1979), suggesting an asymmetrical distribution of cholesterol in biomembranes due to the phospholipid distribution. Indeed a greater affinity of the outer lipid bilayer in erythrocytes towards cholesterol has been found (JACOBSOHN and JACOBSOHN, 1976).

Although our knowledge about the influence of cholesterol on structure and function of P–450 is rather poor some data referring to different levels of integration have been elaborated. Therefore it seems to be useful to differentiate between effects concerning the whole organism and those which have been observed at a molecular level of analysis.

The normal content of cholesterol in the liver and hepatic microsomes varies considerably (cf. BERNHARDT and RUCKPAUL, 1981) in dependence on the species analyzed, the kind of nutrition or impurities of microsomal membrane preparations from 0.04 to 0.50 mol/mol phospholipid. The distribution of cholesterol exhibits a higher content in the smooth membranes of the endoplasmic reticulum (0.137 mg/mg phospholipid) as compared to the rough membranes (0.067 mg/mg phospholipid) at an average microsomal content of 0.09 mg/mg phospholipid (GLAUMANN and DALLNER, 1968).

Taking all data together there is accumulating evidence that a correlation exists between the cholesterol content in the diet and the cholesterol content in the liver (HO et al., 1974; TSAI et al., 1975; SWANN et al., 1975; GREEN et al., 1976; O'BRIEN and REISER, 1979). There is further evidence that the cholesterol content increases with age in serum and liver cells of male rats (UCHIDA et al., 1978). The increase independent of its origin is paralleled by a decrease of drug metabolizing activity (SCHMUCKER et al., 1977; MÜLLER and KLINGER, 1978).

However, in dependence on the substrate and on the length of pretreatment with a cholesterol rich diet the results obtained vary. The codein N-demethylation is inhibited in guinea pigs fed for 72 days a cholesterol rich diet like in rats fed this diet for 137 days. The codein N-demethylation increases, however, in rats which were fed for only 21 days another cholesterol rich diet (TSAI et al., 1977). The demethylation of ethoxycoumarin is described to increase also at cholesterol accumulation (HIETANEN et al., 1978). On the other hand the activity of the NADPH cytochrome P–450 reductase in rats and rabbits has been found to remain unaltered after a cholesterol rich diet (LANG et al., 1976; HIETANEN et al., 1978). This finding indicates that either P–450 itself or interactions between P–450 and the reductase are influenced by cholesterol.

The NADPH dependent lipid peroxidation which obviously is catalyzed by P–450 was inhibited after a cholesterol rich diet as compared to the standard diet whereas the P–448 dependent benzo-(α)pyrene hydroxylase which seems to be located in a more saturated patch of the microsomal membrane as compared with P–450 (BECKER et al., 1978) is not influenced by cholesterol feeding.

The P–450 dependent steroid and bile acid metabolism is also influenced by cholesterol enrichment of liver microsomes. In rats the conversion of 4-adrostene-3,17-dione and 5α-androstene-3α,17β-diol is stimulated, whereas the 5α-reduction of 4-androstene-3,17-dione has been found to be decreased. Cholesterol feeding leads further to a decrease in 12α-hydroxylation of 7α-hydroxy-4-cholesten-3-one and to an activation of the 6β-hydroxylation of lithocholic acid. The hydroxylation of cholesterol in the 7α-position which represents the rate-limiting step in the biosynthesis of bile acids, is enhanced after cholesterol insertion into liver microsomes (GUSTAFSSON et al., 1975; RAICHT et al., 1975).

Two possibilities of the cholesterol influence on the enzyme system are taken into account: (i) cholesterol forms an integral part of the lipid environment of the enzyme system and therefore its concentration may influence the catalytic activity of the respective hydroxylase; (ii) cholesterol may induce the synthesis of specific P–450 species involved in the metabolism of steroid hormones, cholesterol, 7α-hydroxy-4-cholesten-3-one and lithocholic acid.

It should be mentioned that by feeding rats a cholesterol rich diet not only cholesterol is accumulated in the liver but in addition to an enhanced production of bile acids and an increased release of neutral steroids cholesterol also causes an inhibition of cholesterol synthesis by inhibiting the 3-hydroxy-3-methylglutaryl-coenzyme A reductase (GREGORY and BOOTH, 1975). This inhibition is regulated by a feedback mechanism by which part of the synthesized bile acid reenters the liver via the enterohepatic pathway and there inhibits the endogeneous sterol synthesis. A similar control mechanism has been found also in other tissues of guinea pigs, but not in other sterol synthesizing organs as only liver of rat, monkey and dog (SWANN et al., 1975; DIETSCHY and SIPERSTEIN, 1967). In man a cholesterol-rich diet depressed biosynthesis of cholesterol (cf. DOKUSSOVA, 1977).

The analysis of the influence of phospholipids and cholesterol on the activity of membrane bound enzymes is in the beginning. It has been shown that the phospholipid composition and the incorporation of cholesterol into the bilayer membrane may induce alterations of the enzymatic activity of endoplasmic enzymes. The molecular mechanism of these processes is far from being understood. Concerning the P-450 enzyme system neither sufficiently detailed nor satisfying unambiguous results are available to understand the mechanism of this regulation. Cholesterol obviously influences the physical state of the enzyme surrounding phospholipids which results in altered lipid/P-450 and lipid mediated P-450/reductase interactions. It has still to be proved if there exists besides this indirect influence a direct interaction between P–450 and cholesterol resulting in a modified activity. There is rather poor evidence on mutual interactions between the fatty acid composition of phospholipids, free fatty acids and cholesterol on the one hand and the activity on the other. There are only few data on the P–450 activity in man dependent on age despite the fact that just old individuals are consumers of more drugs than young people. A wide field is open to systematically investigate such kind of relevant correlations in order to contribute to a scientific basis for the prescription of drugs.

5. Concluding Remarks

Comprehending all data concerning interactions between phospholipids and the membrane bound proteins which determine the functional role of the endoplasmic reticulum one may come to the conclusion that despite an abundance of experimental data there exist many contradictary aspects and interpretations. This may arise from the difficulties emerging from the multicomponent character of the monooxygenatic system, the difficulties to analyze functional/structural relationships of a membrane bound enzyme system and finally from the different sources of data. Therefore it may be useful for a certain ordering to establish what can be taken for sure, what has been studied but not unambiguously acknowledged and finally what open questions and problems remain and what conclusions can be drawn for forthcoming experimental lines.

The P–450 system has been proved to consist of three essential components with species dependent molar stoichiometries between P–450, the NADPH-dependent reductase and phospholipids far from 1 : 1 : 1. There exists a multiplicity with respect to the terminal oxidase which dependent on specific inducers can be changed with respect to the specific isozyme as well as to the total concentration. The reductase has been shown to be the same for all P–450 isozymes but is also increased at induction.

The phospholipids act as an orientational matrix for P–450 and the reductase. This orientation is a necessary prerequisite to exert catalytic activity and can in part be substituted by detergents. This linker function is phospholipid controlled. Increasing acidity of the head group has been proved to increase the reduction extent as well as the hydroxylation activity. The acyl residues of the phospholipids as well have been found to influence the hydroxylating activity.

Although not object of this article the induction as biosynthetic process is functionally linked with the endoplasmic reticulum and therefore some of the most remarkable results should be briefly mentioned. P–450 after phenobarbital treatment of rats has been found to be synthesized by microsomes bound to the endoplasmic reticulum and cotranslationally inserted (BAR-NUN et al., 1980) thus differing from cytochrome b_5 which is synthesized by free polysomes (RACHUBINSKI et al., 1980). The hydrophobic N-terminal sequence of P–450 which is comparable to the signal sequence of presecretory proteins (HAUGEN et al., 1977) may serve to insert the polypeptide chain into the membrane and contrary to secretory proteins remains in the mature protein (DU BOIS and WATERMAN, 1979; KUMAR and PADMANABAN, 1980). The data available suggest the biosynthesis of P–450 to be induced rather transcriptionally than translationally.

Basing on a stoichiometry between NADPH P–450 reductase and P–450 ranging from 1 : 20 under uninduced conditions to 1 : 35 under induced conditions a biphasic reduction reaction results the molecular mechanism of which is not understood. Different models for the structural arrangement of the system within the membrane have been proposed which are still under discussion. The participation of the cytochrome b_5 system in the hydroxylation reaction has been proved but it is still open if it acts as an electron transmitter or an effector of P–450. The molecular parameters determining the recognition and interaction process between P–450 and the NADPH reductase are not understood and also not the charge dependent role phospholipids play therein. Does exist any form of interaction between reductase and P–450 which is mediated by phospholipids besides the orientation effect via the ordered arrangement of the membrane bilayer and what is its molecular mechanism. The role of the protein concentration within the bilayer membrane determining cluster-like behaviour has not been clarified. What is maintaining transversal asymmetry of phospholipids and/or proteins and by what mechanism and force lateral asymmetry is maintained is not understood and also not by what mechanism the stoichiometry is regulated.

Almost 3 decades of still increasing investigation activity from the discovery of the system till to the present state of knowledge have accumulated such an amount of data that it becomes more and more relevant to look for possibilities to influence the system in vivo. Not only by attempts to understand the induction process in more detail on an individual level but also by looking for the decisive role of phospholipids with respect to activity. Due to its dependence on age, nutrition and environmental conditions the regulation of the activity of the system under normal and pathological conditions needs increasing interest to detect possibilities for making it being applied for human beings.

References

AGRADI, E., C. SPAGNUOLO, and C. CALLI (1975): Pharmac. Res. Commun. 7, 469–480

AXELROD, J. (1955): J. Pharmacol. **114**, 430–438

BAILEY, D. J., R. K. MURRAY, and F. S. ROLLESTON (1974): Can. J. Biochem. **52**, 1003–1012

BALLEN, I. C. and J. A. HIGGINS (1980): Biochem. J. **189**, 475–480

BAR-NUN, S., G. KREIBICH, M. ADESNIK, L. ALTERMAN, M. NEGISHI, and D. D. SABATINI (1980): Proc. Natl. Acad. Sci. U.S. **77**, 965–969

BAYER, E., H. A. O. HILL, A. RÖDER, and R. J. P. WILLIAMS (1969): Chem. Commun. 109

BEAUFAY, H., A. AMAR-COSTESEC, D. THINÈS-SEMPOUX, M. WIBO, M. ROBBI, and J. BERTHET (1974): J. Cell Biol. **61**, 213–231

BECKER, J. F., T. MEEHAN, and J. C. BARTHOLOMEW (1978): Biochim. Biophys. Acta **512**, 136–146

BEHAR-BANNELIER, M. and R. K. MURRAY (1980): Biochem. J. **187**, 147–156

BERGELSON, L. D. and L. I. BARSUKOV (1977): Science **197**, 224–230

BERNHARDT, R., N. T. NGOC DAO, J. FRIEDRICH, O. RISTAU, G.-R. JÄNIG, and K. RUCKPAUL in: Biochemistry, Biophysics and Regulation of Cytochrome P-450, eds. GUSTAFSSON, J. A., J. CARLSTEDT-DUKE, A. MODE, and J. RAFTER (1980), Elsevier/-North-Holland Biomedical Press, p. 595–598

BERNHARDT, R. and K. RUCKPAUL (1981): Pharmazie 36, 1–9

BJÖRKHEM, I., H. DANIELSSON and K. WIKVALL (1974): J. Biol. Chem. **249**, 6439–6445

BLANCK, J., K. ROHDE, J. BEHLKE, G.-R. JÄNIG, D. PFEIL, and K. RUCKPAUL (1979): Acta biol. med. germ. **38**, 399–408

BLASIE, J. K., M. ERECINSKA, S. SAMUELS, and J. S. LEIGH (1977): Biochim. Biophys. Acta **501**, 33–52

BLAZYK, J. F. and J. M. STEIM (1972): Biochim. Biophys. Acta **266**, 737–741

BLOJ, B. and D. B. ZILVERSMIT (1976): Biochemistry **15**, 1277–1283

BLOK, M. C., L. L. M. VAN DEENEN, and J. DEGIER (1977): Biochim. Biophys. Acta **464**, 509–518

BLUM, H., H. J. HARMON, J. S. LEIGH, J. C. SALERNO, and B. CHANCE (1978a): Biochim. Biophys. Acta **502**, 1–10

BLUM, H., J. S. LEIGH, J. C. SALERNO, and R. OHNISHI (1978b): Arch. Biochem. Biophys. **187**, 108–112

BLUM, H., J. S. LEIGH, J. C. SALERNO, and T. OHNISHI (1978c): Arch. Biochem. Biophys. **187**, 153–157

BLUME, A. (1979): Biochim. Biophys. Acta **557**, 32–44

BOCK, K. W. and W. LILIENBLUM (1979): Biochem. Pharmacol. **28**, 695–700

BÖSTERLING, B., A. STIER, A. G. HILDEBRANDT, J. H. DAWSON, and J. R. TRUDELL (1979): Molec. Pharmacol. **16**, 332–342

BÖSTERLING, B. and J. R. TRUDELL in: Microsomes, Drug Oxidations and Chemical Carcinogenesis, eds. COON, M. J., A. H. CONNEY, R. W. ESTABROOK, H. V. GELBOIN, J. R. GILLETTE, P. J. O'BRIEN (1980), Academic Press, New York, pp. 115–118

BRETSCHER, M. S. (1972a): Nature **238**, 11–12

BRETSCHER, M. S. (1972b): J. Molec. Biol. **71**, 523–528

BRETSCHER, M. S. (1973): Science **181**, 622 to 629

BRODIE, B., J. AXELROD, J. R. COOPER, L. GAUDETTE, B. N. LADU, C. MITOMA, and S. UDENFRIEND (1955): Science **121**, 603 to 604

BURLAKOVA, E. B. in: Lipids — Structure, Biosynthesis, Metabolism and Function, (1977) Nauka, Moscow, p. 16–27

CAPDEVILA, J., R. W. ESTABROOK, and R. A. PROUGH (1980): Arch. Biochem. Biophys. **200**, 186–195

CASTER, W. O., A. E. WADE, W. P. NORRED, and R. F. BARGMANN (1980): Pharmacol. **3**, 177–186

CHAMPION, P. M., B. R. STALLARD, G. C. WAGNER, and I. C. GUNSALUS (1982): J. Amer. Chem. Soc. **104**, 5469–5473

CHAPLIN, M. D. and G. J. MANNERING (1970): Molec.. Pharmacol. **6**, 631–640

CHIANG, Y.-L. and M. J. COON (1979): Arch. Biochem. Biophys. **195**, 178–187

CHO, A. K. and G. T. MIWA (1973): Drug Metab. Dispos. **2**, 477–483

CLAUDE, A. (1938): Proc. Soc. exp. Biol. Med. **39**, 398–403

CLAUDE, A. (1943): Biol. Symp. **10**, 111–129

COHEN, L. B., R. D. KEYNES, and B. HILLE (1968): Nature **218**, 438–441

COON, M. J., S. D. BLACK, D. R. KOOP, E. T. MORGAN, and G. E. TARR in: Microsomes, Drug Oxidations and Drug Toxicity, eds. R. SATO and R. KATO (1982), Japan Scientific Soc. Press, Tokyo and Wiley-Interscience, New York, p. 13–23

COOPER, D. Y., S. LEVIN, S. NARASIMHULU, O. ROSENTHAL, and R. W. ESTABROOK (1965): Science **147**, 400–402

CULLIS, P. R. and B. DEKRUIJFF (1978a): Biochim. Biophys. Acta **507**, 207–218

CULLIS, P. R., P. W. M. VAN DIJCK, B. DEKRUIJFF, and J. DEGIER (1978b): Biochim. Biophys. Acta **513**, 21–30

CULLIS, P. R. and B. D. DEKRUIJFF (1979): Biochim. Biophys. Acta **559**, 399–420

DAILEY, H. A. and P. STRITTMATTER (1979): J. biol. Chem. **254**, 5388–5396

DAILEY, H. A. and P. STRITTMATTER (1980): J. biol. Chem. **255**, 5184–5189

DANIELLI, J. F. and H. DAVSON (1935): J. Cell. Comp. Physiol. **5**, 495–508

DEGIER, J., J. G. MANDERSLOOT, and L. L. M. VAN DEENEN (1969): Biochim. Biophys. Acta **173**, 143–145

DEKRUIJFF, B., R. A. DEMEL, A. J. SLOTBOOM, L. L. M. VAN DEENEN, and A. F. ROSENTHAL (1973): Biochim. Biophys. Acta **307**, 1–19

DEKRUIJFF, B., A. M. H. P. VAN DEN BESSELAAR, P. R. CULLIS, H. VAN DEN BOSCH, and L. L. M. VAN DEENEN (1978) Biochim. Biophys. Acta **514**, 1–8

DEKRUIJFF, B., A. RIETVEL, and P. R. CULLIS (1980): Biochim. Biophys. Acta **600**, 343–357

DENK, H. in: Progress in Liver Disease, Vol. **6**, eds. POPPER, H. and F. SCHAFFNER (1979) Grune and Stratton, New York, p. 263–283

DEPIERRE, J. W. and G. DALLNER (1975): Biochim. Biophys. Acta **415**, 411–472

DEPIERRE, J. W. and L. ERNSTER (1975): FEBS Lett. **55**, 18–21

DEPIERRE, J. W. and L. ERNSTER (1977): Annu. Rev. Biochem. **46**, 201–262

DIEHL, H., S. CAPALNA, and V. ULLRICH (1969): FEBS Lett. **4**, 99–102

DIETSCHY, J. M. and M. D. SIPERSTEIN (1967): J. Lipid Res. **8**, 97–104

DOKUSOVA, O. K. in: Lipids — Structure, Biosynthesis, Metabolism and Function (1977) Nauka, Moscow, p. 42–53

DUBOIS, R. N. and M. R. WATERMAN (1979): Biochem. Biophys. Res. Commun. **90**, 150–157

DUPPEL, W., J. M. LEBEAULT, and M. J. COON (1973): Europ. J. Biochem. **36**, 583–592

DUPPEL, W. and V. ULLRICH (1976): Biochim. Biophys. Acta **426**, 399–407

DYATLOVITSKAYA, E. V., E. V. SINITSYNA, A. F. LEMENOVSKAYA, and L. D. BERGELSON (1978a): Biochimija **43**, 420–423

DYATLOVITSKAYA, E. V., N. G. TIMOFEEVA, and L. D. BERGELSON (1978b): Europ. J. Biochem. **82**, 463–471

EBEL, R. E., D. H. O'KEEFFE, and J. A. PETERSON (1978): J. Biol. Chem. **253**, 3888–3897

EIBL, H., and P. WOOLEY (1979): Biophys. Chem. **10**, 261–271

ERECINSKA, M., D. F. WILSON, and J. K. BLASIE (1978a): Biochim. Biophys. Acta **501**, 53–62

ERECINSKA, M., D. F. WILSON, and J. K. BLASIE (1978b): Biochim. Biophys. Acta **501**, 63–71

ERECINSKA, M., D. F. WILSON, and J. K. BLASIE (1979): Biochim. Biophys. Acta **545**, 352–364

ERNSTER, L. and S. ORRENIUS (1973): Drug Metab. Dispos. **1**, 66–73

ESTABROOK, R. W., D. Y. COOPER, and O. ROSENTHAL (1963): Biochem. Z. **338**, 741–755

ESTABROOK, R. W., A. G. HILDEBRANDT, J. BARON, K. J. NETTER, and K. LEIBMAN (1971a): Biochem. Biophys. Res. Commun. **42**, 132–139

ESTABROOK, R. W., M. R. FRANKLIN, B. COHEN, A. SHIGAMATZU, and A. G. HILDEBRANDT (1971b): Metabolism **20**, 187–199

FEUER, C. F., M. S. I. DHAMI, F. A. DELA IGLESIA, and G. FEUER in: Microsomes, Drug Oxidations and Chemical Carcinogenesis, eds. COON, M. J., A. H. CONNEY, R. W. ESTABROOK, H. V. GELBOIN, J. R. GILLETTE, and P. J. O'BRIEN (1980), Academic Press, New York, p. 559–566

FOWLER, S., J. REMACLE, A. TROUET, H. BEAUFAY, J. BERTHET, M. WIBO, and P. HAUSER (1976): J. Cell. Biol. **71**, 535–550

FRANKLIN, M. R. and R. W. ESTABROOK (1971): Arch. Biochem. Biophys. **143**, 318–329

FRENCH, J. S., F. P. GUENGERICH, and M. J. COON (1980): J. Biol. Chem. **255**, 4112–4119

FUJIIKURIYAMA, Y., Y. MIZUKAMI, K. KAWAJIRI, K. SOGAWA, and M. MURAMATSU (1982) Proceed. Natl. Acad. Sci. U.S. **79**, 2793–2797

GARFINKEL, D. (1958): Arch. Biochem. Biophys. **77**, 493–509

GIBSON, G. G., D. L. CINTI, S. G. SLIGAR, and J. B. SCHENKMAN (1980): J. biol. Chem. **255**, 1867–1873

GLAUMANN, H. and G. DALLNER (1970): J. Cell. Biol. **47**, 34–48
GRANT, C. W. M. and H. M. MCCONNELL (1973): Proc. Natl. Acad. Sci. U.S. **70**, 1238–1240
GREEN, M. H., M. GRIM, M. TRABER, and R. OSTWALD (1976): J. Nutr. **106**, 515–528
GREGORY, K. W. and R. BOOTH (1975): Biochem. J. **148**, 337–339
GREIM, H., J. B. SCHENKMAN, M. KLOTZBÜCHER, and H. REMMER (1970): Biochim. Biophys. Acta **201**, 20–25
GREINERT, R., H. STAERK, A. STIER, and A. WELLER (1979): J. Biochem. Biophys. Meth. **1**, 77–83
GRIFFIN, B. W. and J. A. PETERSON (1975): J. Biol. Chem. **250**, 6445–6451
GUNSALUS, I. C. and S. G. SLIGAR (1978): Advanc. Enzymol. **47**, 1–44
GUSTAFSSON, B. E., K. EINARSSON, and J. A. GUSTAFSSON (1975): J. Biol. Chem. **250**, 8496–8502
GUSTAFSSON, J. A., E. G. HRYCAY, and L. ERNSTER (1976): Arch. Biochem. Biophys. **174**, 440–453
GUT, J., C. RICHTER, R. J. CHERRY, K. H. WINTERHALTER, and S. KAWATO (1982) J. biol. Chem. **257**, 7030–7036
HAAPARANTA, T., J. RYDSTRÖM, and M. INGELMAN-SUNDBERG in: Microsomes, Drug Oxidations and Chemical Carcinogenesis, eds. COON, M. J., A. H. CONNEY, R. W. ESTABROOK, H. V. GELBOIN, J. R. GILLETTE, and P. J. O'BRIEN (1980), Academic Press, New York, p. 533–535
HAMMER, C. T. and E. D. WILLS (1978): Biochem. J. **174**, 585–594
HANIU, M., L. G. ARMES, M. TANAKA, K. T. YASUNOBU, B. S. SHASTRY, G. C. WAGNER, and I. C. GUNSALUS (1982): Biochem. Biophys. Res. Commun. **105**, 889–894
HANSON, L. K., S. G. SLIGAR, and I. C. GUNSALUS (1977): Croatica Chem. Acta **49**, 237–250
HARWOOD, J. L. and J. N. HAWTHORNE (1969): Biochim. Biophys. Acta **171**, 75–88
HAUGEN, D. A., L. G. ARMES, K. T. YASUNOBU, and M. J. COON (1977): Biochem. Biophys. Res. Commun. **77**, 967–973
HAUSER, H. and M. D. BARRAT (1973): Biochem. Biophys. Res. Commun. **53**, 399–405
HAUSER, H., K. HOWELL, and M. C. PHILLIPS (1977): FEBS Lett. **80**, 355–359
HAYAISHI, O., M. KATAGIRI, and S. ROTHBERG (1955): J. Am. Chem. Soc. **77**, 5450–5451
HIETANEN, E., O. HÄNNINEN, M. LAITINEN, and M. LANG (1978): Pharmacol. **17**, 163–172
HIGGINS, J. A. and R. M. C. DAWSON (1977): Biochim. Biophys. Acta **470**, 342–356
HILL, H. A. O., A. RÖDER, and R. J. P. WILLIAMS (1970): Naturwissenschaften **57**, 69–72
HO, K. J., L. C. PANG, and C. B. TAYLOR (1974): Atheroscler. **19**, 561–566
HOLTZMANN, J. L., R. R. ERICKSON, R. K. DEWAN, and V. A. BLOOMFIELD (1973): Biochem. Biophys. Res. Commun. **52**, 15–20
HOPKINS, G. J. and C. E. WEST (1976): Lipids **11**, 736–740
HRYCAY, E. G. and P. J. O'BRIEN (1972): Arch. Biochem. Biophys. **153**, 480–494
IMAI, Y., R. SATO, and T. IYANAGI (1977): J. Biochem. **82**, 1237–1246
INGELMAN-SUNDBERG, M. (1977): Biochim. Biophys. Acta **488**, 225–234
INGELMAN-SUNDBERG, M., I. JOHANSSON, and A. HANSSON (1979): Acta Biol. Med. Germ. **38**, 379–388
INGELMAN-SUNDBERG, M. and I. JOHANSSON (1980): Biochemistry **19**, 4004–4011
INGELMAN-SUNDBERG, M., H. GLAUMANN, and I. JOHANSSON in: Microsomes, Drug Oxidations and Chemical Carcinogenesis, eds. COON, M. J., A. H. CONNEY, R. W. ESTABROOK, H. V. GELBOIN, J. R. GILLETTE, and P. J. O'BRIEN (1980), Academic Press, New York, p. 513–516
INOUE, K. (1974): Biochim. Biophys. Acta **339**, 390–402
ISHIMURA, Y., V. ULLRICH, and J. A. PETERSON (1971): Biochem. Biophys. Res. Commun. **42**, 140–146
ITO, A. (1974): J. Biochem. **75**, 787–793
JACOBS, R. E., B. S. HUDSON, and H. C. ANDERSEN (1977): Biochemistry **16**, 4349–4359
JACOBSOHN, M. K. and G. M. JACOBSOHN (1976): Fed. Proc. **35**, 1431
JACOBSON, K. and D. PAPAHADJOPOULOS (1975): Biochemistry **14**, 152–161
JONES, M. N. in: Biochemical Thermodynamics ed. JONES, M. N. (1979), Elsevier Scientific Publishing Company, Amsterdam, Oxford, New York, p. 185–223
JOST, P. C., R. A. CAPALDI, G. VANDERKOOI, and O. H. GRIFFITH (1973a): J. Supramol. Structure **1**, 269–280
JOST, P. C., O. H. GRIFFITH, R. A. CAPALDI, and G. VANDERKOOI (1973b): Proc. Natl. Acad. Sci. U.S. **70**, 480–484

KANG, S. Y., H. S. GUTOWSKY, and E. OLDFIELD (1979a): Biochemistry 18, 3268–3272
KANG, S. Y., H. S. GUTOWSKY, J. C. HSUNG, R. JACOBS, T. E. KING, D. RICE, and E. OLDFIELD (1979b): Biochemistry 18, 3257–3267
KAPITULNIK, J., W. LEVIN, A. Y. H. LU, R. MORECKI, P. M. DANSETTE, D. M. JERINA, and A. H. CONNEY (1977): Clin. Pharmacol. Therap. 21, 158–165
KATAGIRI, M., B. N. GANGULI, and I. C. GUNSALUS (1968): J. Biol. Chem. 243, 3543–3546
KAWATO, S., J. GUT, R. J. CHERRY, K. H. WINTERHALTER, and C. RICHTER (1982): J. biol. Chem. 257, 7023–7029
KIMURA, T. and K. SUZUKI (1965): Biochem. Biophys. Res. Commun. 20, 373–379
KISSEL, M. A., S. A. USANOV, D. I. METELITZA, and A. A. AKHREM (1979): Bioorgan. Chem. (Moscow) 5, 1558–1563
KLAUSNER, R. D., A. M. KLEINFELD, R. L. HOOVER, and M. J. KARNOVSKY (1980): J. Biol. Chem. 255, 1286–1295
KLEEMANN, W. and H. M. MCCONNELL (1976): Biochim. Biophys. Acta 419, 206–222
KLINGENBERG, M. (1958): Arch. Biochem. Biophys. 75, 376–386
KORNBERG, R. D. and H. M. MCCONNELL (1971): Biochemistry 10, 1111–1120
KREBS, J. J. R., H. HAUSER, and E. CARAFOLI (1979): J. Biol. Chem. 254, 5308–5316
KUMAR, A. and G. PADMANABAN (1980): J. Biol. Chem. 255, 522–525
KUO, A. L. and C. G. WADE (1979): Chem. Phys. Lipids 25, 135–139
LADBROOKE, B. D., R. M. WILLIAMS, and D. CHAPMAN (1968): Biochim. Biophys. Acta 150, 333–340
LAITINEN, M., E. HIETANEN, H. VAINIO, and O. HÄNNINEN (1975): Lipids 10, 461–466
LAM, K. S. and C. B. KASPER (1979): Biochemistry 18, 307–311
LANG, M., M. LAITINEN, E. HIETANEN, and H. VAINIO (1976): Acta Pharmacol. Toxicol. 39, 273–288
LEE, A. G., N. J. M. BIRDSALL, and J. C. METCALFE in: Nuclear Magnetic Relaxation and the Biological Membrane, ed. KORN E. D. (1974), Plenum Press, New York
LEE, A. G. (1977): Biochim. Biophys. Acta 472, 285–344
LEE, T. C. and F. SNYDER (1973): Biochim. Biophys. Acta 291, 71–82
LEVIN, W., D. RYAN, S. WEST, and A. Y. H. LU (1974): J. Biol. Chem. 249, 1747–1754
LOVERDE, A. and P. STRITTMATTER (1968): J. Biol. Chem. 243, 5779–5787
LU, A. Y. H. and M. J. COON (1968): J. Biol. Chem. 243, 1331–1332
LU, A. Y. H., H. W. STROBEL, and M. J. COON (1969a): Biochem. Biophys. Res. Commun. 36, 545–551
LU, A. Y. H., K. W. JUNK, and M. J. COON (1969b): J. Biol. Chem. 244, 3714–3721
LU, A. Y. H., W. LEVIN, and R. KUNTZMAN (1974): Biochem. Biophys. Res. Commun. 60, 266–272
LU, A. Y. H. and W. LEVIN (1974): Biochim. Biophys. Acta 344, 205–240
LU, A. Y. H., D. M. JERINA, and W. LEVIN (1977): J. Biol. Chem. 252, 3715–3723
LU, A. Y. H., G. T. MIWA, and S. B. WEST in: Microsomes, Drug Oxidations and Chemical Carcinogenesis, eds. COON, M. J., A. H. CONNEY, R. W. ESTABROOK, H. V. GELBOIN, J. R. GILLETTE and P. J. O'BRIEN (1980), Academic Press, New York, p. 59–66
MABREY, S. and J. M. STURTEVANT (1976): Proc. Natl. Acad. Sci. U.S. 73, 3862–3866
MABREY, S., G. POWIS, J. B. SCHENKMAN, and T. R. TRITTON (1977): J. Biol. Chem. 252, 2929–2933
MARINETTI, G. V., A. V. SENIOR, R. LOVE, and C. I. BROADHURST (1976): Chem. Phys. Lipids 17, 353–362
MARSH, D., A. WATTS, W. MASCHKE, and P. F. KNOWLES (1978): Biochem. Biophys. Res. Commun. 81, 397–402
MASON, H. S., W. L. FOWLKS, and E. PETERSON (1955): J. Am. Chem. Soc. 77, 2914–2915
MATSUURA, S., Y. FUJII-KURIYAMA, and Y. TASHIRO (1978): J. Cell. Biol. 78, 503–519
MCINTOSH, P. R., S. KAWATO, R. B. FREEDMAN, and R. J. CHERRY (1980): FEBS Lett. 122, 54–58
MCMURCHIE, E. J. and J. K. RAISON (1979): Biochim. Biophys. Acta 554, 364–374
MCNAMEE, M. G. and H. M. MCCONNELL (1973): Biochemistry 12, 2951–2958
METCALFE, J. C., N. J. M. BIRDSALL, and A. G. LEE (1972): FEBS Lett. 21, 335–340
MIHARA, K., R. SATO, R. SAKAKIBARA, and H. WADA (1978): Biochemistry 17, 2829–2834
MIWA, G. T., S. B. WEST, and A. Y. H. LU (1978): J. Biol. Chem. 253, 1921–1929
MIWA, G. T., S. B. WEST, M. T. HUANG, and A. Y. H. LU (1979): J. Biol. Chem. 254, 5695–5700

MORIMOTO, T., S. MATSUURA, S. SASAKI, Y. TASHIRO, and T. OMURA (1976): J. Cell. Biol. **68**, 189–201

MORRISETT, J. D., H. D. POWNALL, R. T. PLUMLEE, L. C. SMITH, Z. E. ZEHNER, M. ESFAHANI, and S. J. WAKIL (1975): J. Biol. Chem. **250**, 6969–6976

MÜLLER, D. and W. KLINGER (1978): Pharmazie **33**, 397–400

NARASIMHULU, S., D. Y. COOPER, and O. ROSENTHAL (1965): Life Sci. **4**, 2101–2107

NEGISHI, M., T. SAWAMURA, T. MORIMOTO and Y. TASHIRO (1975): Biochim. Biophys. Acta **381**, 215–220

NEGISHI, M., Y. FUJII-KURIYAMA, Y. TASHIRO and Y. IMAI (1976): Biochem. Biophys. Res. Commun. **71**, 1153–1160

NILSSON, R., E. PETERSON, and G. DALLNER (1973): J. Cell. Biol. **56**, 762–776

NILSSON, O. S. and G. DALLNER (1975): FEBS Lett. **58**, 190–193

NILSSON, O. S. and G. DALLNER (1977a): Biochim. Biophys. Acta **464**, 453–458

NILSSON, O. S. and G. DALLNER (1977b): J. Cell. Biol. **72**, 568–583

NORRED, W. P. and A. E. WADE (1972): Biochem. Pharmacol. **21**, 2887–2897

O'BRIEN, B. C. and R. REISER (1979): J. Nutr. **109**, 98–104

OESCH, F. (1977): Arzneimittelforschung **27**, 1823–1835

OLDFIELD, E. and D. CHAPMAN (1972): FEBS Lett. **23**, 285–297

OMURA, T. and R. SATO (1964a): J. Biol. Chem. **239**, 2370–2378

OMURA, T. and R. SATO (1964b): J. Biol. Chem. **239**, 2379–2385

OMURA, T., P. SIEKEVITZ, and G. E. PALADE (1967): J. Biol. Chem. **242**, 2389–2396

OP DEN KAMP, J. A. F. (1979): Annu. Rev. Biochem. **48**, 47–71

OPRIAN, D. D., K. P. VATSIS, and M. J. COON (1979): J. Biol. Chem. **254**, 8895–8902

OWICKI, J. C. and H. M. MCCONNELL (1979): Proc. Natl. Acad. Sci. U.S. **76**, 4750–4754

PAINE, A. J. and A. E. M. MCLEAN (1973): Biochem. Pharmacol. **22**, 2875–2880

PALADE, G. E. and K. R. PORTER (1954): J. Exp. Med. **100**, 641–656

PALADE, G. E. and P. SIEKEVITZ (1956): J. Biophys. Biochem. Cytol. **2**, 171–199

PEDERSEN, J. I. (1978): FEBS Lett. **85**, 35–39

PETERSON, J. A., R. E. EBEL, D. H. O'KEEFFE, T. MATSUBARA, and R. W. ESTABROOK (1976): J. Biol. Chem. **251**, 4010–4016

PFEIL, D. (1980): Doctoral thesis, Berlin

PHILLIPS, M. C. and D. CHAPMAN (1968): Biochim. Biophys. Acta **163**, 301–313

PHILLIPS, M. C., R. M. WILLIAMS, and D. CHAPMAN (1969): Chem. Phys. Lipids **3**, 234–244

PHILSON, S. B. (1977): PH. D. THESIS, University of Illinois, Urbana, Illinois, USA

PINK, D. A. and D. CHAPMAN (1979): Proc. Natl. Acad. Sci. U.S. **76**, 1542–1546

POKROVSKY, A., V. MISHIN, N. RIVKIND, and V. LYAKHOVICH (1977): Biochem. Biophys. Res. Commun. **77**, 912–917

PORTER, K. R. (1953): J. Exp. Med. **97**, 727–750

PROUGH, R. A. and D. M. ZIEGLER (1977): Arch. Biochem. Biophys. **180**, 363–373

RACHUBINSKI, R. A., D. P. S. VERMA, and J. J. M. BERGERON (1980): J. Cell Biol. **84**, 705–716

RAICHT, R. F., B. I. COHEN, S. SHEFER, and E. H. MOSBACH (1975): Biochim. Biophys. Acta **388**, 374–384

REIN, H., O. RISTAU, J. FRIEDRICH, G.-R. JÄNIG, and K. RUCKPAUL (1977): FEBS Lett. **75**, 19–22

REMACLE, J., S. FOWLER, H. BEAUFAY, and J. BERTHET (1974): J. Cell Biol. **61**, 237–240

REMACLE, J., S. FOWLER, H. BEAUFAY, A. AMAR-COSTESEC, and J. BERTHET (1976): J. Cell Biol. **71**, 551–564

REMMER, H., J. SCHENKMAN, R. W. ESTABROOK, H. SASAME, J. GILLETTE, S. NARASIMHULU, D. Y. COOPER, and O. ROSENTHAL (1966): Molec. Pharmacol. **2**, 187–190

RENNEBERG, R., J. J. CAPDEVILA, N. CHACOS, R. W. ESTABROOK, and R. A. PROUGH (1981): Biochem. Pharmacol. **30**, 843–848

RICE, D. and E. OLDFIELD (1979): Biochemistry **18**, 3272–3279

RICE, D., J. C. HSUNG, T. E. KING, and E. OLDFIELD (1979a): Biochemistry **18**, 5885–5892

RICE, D., M. D. MEADOWS, A. O. SCHEINMAN, F. M. GONI, J. C. GOMEZ-FERNADEZ, M. A. MOSCARELLO, D. CHAPMAN, and E. OLDFIELD (1979b): Biochemistry **18**, 5893–5903

RICH, P. R., D. M. TIEDE, and W. D. BONNER (1979): Biochim. Biophys. Acta **546**, 307–315

RICHTER, C., K. H. WINTERHALTER, and R. J. CHERRY (1979): FEBS Lett. **102**, 151–154

RISTAU, O., H. REIN, S. GRESCHNER, G.-R. JÄNIG, and K. RUCKPAUL (1979): Acta Biol. Med. Germ. **38**, 177–185

ROBERTS, R. B. in: Microsomal Particles and Protein Synthesis, Biophys. Soc. Symp. 1 (1958) Pergamon Press, New York, Oxford

RODRIGUEZ-BOULAN, E., D. D. SABATINI, B. N. PEREYRA, and G. KREIBICH (1978): J. Cell Biol. **78**, 894–909

ROGERS, M. J. and P. STRITTMATTER (1974a): J. Biol. Chem. **249**, 895–903

ROGERS, M. J. and P. STRITTMATTER (1974b): J. Biol. Chem. **249**, 5565–5569

ROTHMAN, I. E. and E. A. DAWIDOWICZ (1975): Biochemistry **14**, 2809–2816

ROTHMAN, J. E. and J. LENARD (1977): Science **195**, 743–753

ROWE, L. and E. D. WILLS (1976): Biochem. Pharmacol. **25**, 175–179

RUCKPAUL, K., H. REIN, D. P. BALLOU, and M. J. COON (1980): Biochim. Biophys. Acta **626**, 41–56

RUCKPAUL, K., H. REIN, J. BLANCK, O. RISTAU, and M. J. COON (1982): Acta biol. med. germ. **41**, 193–203

RUMACK, B. H., G. HOLTZMANN, and N. P. CHASE (1973): J. Pharmacol. Exp. Ther. **186**, 441–446

SAGARA, Y., T. HARANO and T. OMURA (1978): J. Biochem. **83**, 807–812

SALEMME, F. R. (1976): J. Mol. Biol. **102**, 563–568

SANDERMANN, H. (1978): Biochim. Biophys. Acta **515**, 209–237

SCHACHTER, H., I. JABBAL, R. L. HUDGIN, L. PINTERIC, E. J. MCGUIRE, and S. ROSEMAN (1970): J. Biol. Chem. **245**, 1090–1100

SCHENKMAN, J. B., R. REMMER, and R. W. ESTABROOK (1967): Molec. Pharmacol. **3**, 113–123

SCHENKMAN, J. B. (1970): Biochemistry **9**, 2081–2091

SCHENKMAN, J. B., D. L. CINTI, S. ORRENIUS, P. MOLDEUS, and R. KASCHNITZ (1972): Biochemistry **11**, 4243–4251

SCHMUCKER, D. L., J. MOONEY, and A. L. JONES (1977): Science **197**, 1005–1008

SCHULZE, H. V., J. M. PÖNNIGHAUS, and H. STAUDINGER (1972): Hoppe Seyler's Z. physiol. Chem. **353**, 1195–1204

SCHULZE, H. J. and H. STAUDINGER (1975): Naturwissenschaften **62**, 331–340

SCHWARZ, D., J. PIRRWITZ, and K. RUCKPAUL (1982a): Arch. Biochem. Biophys. **216**, 322–328

SCHWARZ, D., J. PIRRWITZ, M. J. COON, and K. RUCKPAUL (1982b) Acta biol. med. germ. **41**, 425–430

SEELIG, J. and N. WAESPE-SARCEVIC (1978): Biochemistry **17**, 3310–3315

SEIDEGÅRD, J., M. S. MORON, L. C. ERIKSSON, and J. W. DEPIERRE (1978): Biochim. Biophys. Acta **543**, 29–40

SEIMER, R. and E. GELERINTER (1979): Chem. Phys. Lipids **23**, 201–211

SHARMA, R. N., M. BEHAR-BANNELIER, F. S. ROLLESTON, and R. K. MURRAY (1978): J. Biol. Chem. **253**, 2033–2043

SHEN, A. L., W. E. FAHL, and R. E. JEFCOATE (1980): Arch. Biochem. Biophys. **204**, 511–523

SHIMATAKA, T., K. MIHARA, and R. SATO (1972): J. Biochem. **72**, 1163–1174

SIEKEVITZ, P. (1972): Annu. Rev. Phys. **34**, 117–140

SINGER, S. J. and G. L. NICOLSON (1972): Science **175**, 720–731

SINGER, S. J. (1974): Annu. Rev. Biochem. **43**, 805–833

SLIGAR, S. G., K. A. KENNEDY, and D. C. PEARSON (1980): Proc. Natl. Acad. Sci. U.S. **77**, 1240–1244

SPATZ, L. and P. STRITTMATTER (1971): Proc. Natl. Acad. Sci. U.S. **68**, 1042–1046

SPATZ, L. and P. STRITTMATTER (1973): J. Biol. Chem. **248**, 793–799

STASIECKI, P., F. WAECHTER, P. BEUTLEY, and F. OESCH (1979): Biochim. Biophys. Acta **568**, 446–453

STEIM, J. M., M. E. TOURTELLOTTE, J. C. REINERT, R. N. MCELHANEY, and R. L. RADER (1969): Proc. Natl. Acad. Sci. U.S. **63**, 104–109

STIER, A. and E. SACKMANN (1973): Biochim. Biophys. Acta **311**, 400–408

STIER, A., S. A. E. FINCH, and B. BÖSTERLING (1978): FEBS Lett. **91**, 109–112

STIER, A., W. KÜHNLE, B. BÖSTERLING, and S. A. E. FINCH (1979): Symposia Medica Hoechst № 14, Schattauer Verlag, Stuttgart, New York, p. 227–241

STRITTMATTER, P., M. J. ROGERS, and L. SPATZ (1972): J. Biol. Chem. **247**, 7188–7194

STRITTMATTER, P. and M. J. ROGERS (1975): Proc. Natl. Acad. Sci. U.S. **72**, 2658–2661

STROBEL, H. W., A. Y. H. LU, J. HEIDEMA, and M. J. COON (1970): J. Biol. Chem. **245**, 4851–4854

SUNDLER, R., S. L. SARCIONE, A. W. ALBERTS, and P. R. VAGELOS (1977): Proc. Natl. Acad. Sci. U.S. **74**, 3350–3354

SUZUKI, K. and T. KIMURA (1965): Biochem. Biophys. Res. Commun. **19**, 340–345

SWANN, A., M. H. WILEY, and M. D. SIPERSTEIN (1975): J. Lipid Res. **16**, 360–366

TANIGUCHI, H., Y. IMAI, T. IYANAGI, and R. SATO (1979): Biochim. Biophys. Acta **550**, 341–356

THREADGOLD, L. T. (1969): The Ultrastructure of the Animal Cell, Pergamon Press, Oxford, London, p. 123

TOMKINS, G. M., J. F. CURRAN, and P. J. MICHAEL (1958): Biochim. Biophys. Acta **28**, 449–450

TRÄUBLE, H. and D. H. HAYNES (1971): Chem. Phys. Lipids **7**, 324–335

TRÄUBLE, H. and H. EIBL (1974): Proc. Natl. Acad. Sci. U.S. **71**, 214–219

TSAI, A. C., D. R. ROMSOS, and G. A. LEVEILLE (1975): J. Nutr. **105**, 939–945

TSAI, A. C., G. M. THIE, and C. R. S. LIN (1977): J. Nutr. **107**, 310–319

TSONG, T. Y. and C. S. YANG (1978): Proc. Natl. Acad. Sci. U.S. **75**, 5955–5959

TYSON, C. A., R. L. TSAI, J. D. LIPSCOMB, and I. C. GUNSALUS (1972): J. Biol. Chem. **247**, 5777–5784

UCHIPA, K., Y. NOMURA, M. KADOWAKI, H. TAKASE, K. TAKANO, and N. TAKEUCHI (1978): J. Lipid Res. **19**, 544–552

ULLRICH, V., H. SAKURAI, and H. H. RUF (1979): Acta Biol. Med. Germ. **38**, 287–297

VAN DER HOEVEN, T. A. and M. J. COON (1974) J. biol. Chem. **249**, 6302–6310

VANDERKOOI, J., S. FISCHKOFF, B. CHANCE, and R. A. COOPER (1974): Biochemistry **13**, 1589–1595

VANDERKOOI, G. (1974): Biochim. Biophys. Acta **344**, 307–345

VAN DIJCK, P. W. M., B. DEKRUIJFF, A. J. VERKLEIJ, L. L. VAN DEENEN, and J. DEGIER (1978): Biochim. Biophys. Acta **512**, 84–96

VAN DIJCK, P. W. M. (1979): Biochim. Biophys. Acta **555**, 89–101

VERKLEIJ, A. J., B. DEKRUIJFF, P. H. J. T. VERVERGAERT, J. F. TOCANNE, and L. L. M. VAN DEENEN (1974): Biochim. Biophys. Acta **339**, 432–437

VLASUK, G. P. and F. G. WALZ, jr. (1979): Fed. Proc. **38**, 357

WADE, A. E., W. P. NORRED and J. S. EVANS in: Nutrition and Drug Interrelations, eds. J. N. MATHCOCK and M. J. COON (1978), Academic Press, New York, p. 475–503

WELTON, A. F., T. C. PEDERSON, J. A. BUEGE, and S. D. AUST (1973): Biochem. Biophys. Res. Commun. **54**, 161–167

WHITE, R. E., S. G. SLIGAR, and M. J. COON (1980): J. Biol. Chem. **255**, 11108–11111

WILLIAMSON, D. G. and V. J. O'DONNELL (1969): Biochemistry **8**, 1289–1300

WILLS, E. D. in: Microsomes, Drug Oxidations and Chemical Carcinogenesis, eds. M. J. COON, A. H. CONNEY, R. W. ESTABROOK, H. V. GELBOIN, J. R. GILLETTE, and P. J. O'BRIEN (1980), Academic Press, New York, p. 545–548

WINQVIST, L. and G. DALLNER (1976): Biochim. Biophys. Acta **436**, 399–412

WUNDERLICH, F., W. KREUTZ, P. MAHLER, A. RONAI, and G. HEPPELER (1978): Biochemistry **17**, 2005–2010

YANG, C. S. (1975): FEBS Lett. **54**, 61–64

YANG, C. S. and F. S. STRICKHART (1975): J. Biol. Chem. **250**, 7968–7972

YANG, C. S., F. S. STRICKHART, and L. P. KICHA (1977): Biochim. Biophys. Acta **465**, 362–370

YANG, C. S. (1977a): Life Sci. **21**, 1047–1058

YANG, C. S. (1977b): J. Biol. Chem. **252**, 293–298

YANG, C. S., F. S. STRICKHART, and L. P. KICHA (1978): Biochim. Biophys. Acta **509**, 326–337

YANG, C. S. and T. Y. TSONG in: Microsomes, Drug Oxidations and Chemical Carcinogenesis, eds. M. J. COON, A. H. CONNEY, R. W. ESTABROOK, H. V. GELBOIN, J. R. GILLETTE, and P. J. O'BRIEN (1980), Academic Press, New York, p. 199–202

YEAGLE, P. L., W. C. HUTTON, C. HUANG, and R. B. MARTIN (1977): Biochemistry **16**, 4343–4349

ZILVERSMIT, D. B. and M. E. HUGHES (1977): Biochim. Biophys. Acta **469**, 99–110

ZWAAL, R. F. A., B. ROELOFSEN, and C. M. COLLEY (1973): Biochim. Biophys. Acta **300**, 159–182

Chapter 2

Structure-Function Relationships of the Essential Components of the Liver Microsomal Monooxygenase System

G.-R. JÄNIG and D. PFEIL

Contents

Abbrevations

P–450	cytochrome P–450 (EC 1.14.14.1)
P–448 (P_1–450)	P–450 hemoproteins found predominantly in livers of animals treated with certain polycyclic hydrocarbons, including 3-MC
P–420	enzymatically inactive P–450
P–450 LM	any or all forms of rabbit liver microsomal P–450
P–450 a, A, ...	homogeneous forms of rat liver microsomal P–450
P–450 CAM	soluble P–450 from *Pseudomonas putida* grown on camphor
P–450 reductase	NADPH-cytochrome P–450 (cytochrome c) reductase (EC 1.6.2.4)
PB	sodium phenobarbital
3-MC	3-methylcholanthrene
BNF	β-naphthoflavone
TCDD	2,3,7,8-tetrachlorodibenzo-p-dioxin
PCB	polychlorinated biphenyl
PCN	pregnenolone-16α-carbonitrile
SDS	sodium dodecylsulfate
PAGE	polyacrylamide gel electrophoresis
mol. wt.	molecular weight

1. Introduction

In 1968, Lu and Coon evidenced that the hepatic monooxygenase in the membranes of the endoplasmic reticulum is composed of three essential components: cytochrome P–450 (EC 1.14.14.1), NADPH-cytochrome P–450 reductase (EC 1.6.2.4), and a heat-stable lipid factor, which can be reconstituted to an enzymatically active enzyme system. It does not contain an iron-sulfur protein like the mitochondrial monooxygenase system from adrenals (Omura et al., 1966) or the hepatic mitochondrial P–450 system (Pedersen et al., 1977). Due to the membrane-bound state and the presence of many other proteins in the endoplasmic reticulum the purification of the monooxygenase system was rather difficult to achieve. Many investigations were aimed to solve the function of each individual component and the mechanism of their interaction, to establish the arrangement of this enzyme system within the membranes of the endoplasmic reticulum, and to understand the molecular mechanism of oxygen activation and the catalytic mechanism in performing monooxygenations of a great number of chemically very different compounds.

The biocatalytic capability of the monooxygenase system consists in the biotransformation of a variety of substrates both endogeneous and exogeneous. The latter function is of importance when analyzing the response to pharmacologically active agents. The system from liver microsomes exhibits the highest specific content as compared to the P–450 systems from other tissues as well as subcellular fractions. Due to the broad reaction specificity many lipophilic drugs and xenobiotics are hydroxylated to more polar products making them more readily excreted than stored.

To explain the unusually broad specificity of the P–450 system which differs in dependence on the species, strain, sex, age, and nutritional state the existence of multiple forms of P–450 has been postulated. Each cytochrome should have a different, but with other forms overlapping substrate specificity. In the last decade, growing evidence for the occurrence of many different forms has been accumulated from electrophoretic and immunochemical analyses of microsomal hemoproteins from animals and from their differential response to various inhibitors. The pattern of P–450 forms seems to be genetically controlled and will be altered both in quality and quantity by the exposition of an individuum to xenobiotics. The induction increases not only the monooxygenase activity but changes the product pattern, too. This is of special importance in the hydroxylation of polycyclic hydrocarbons in man resulting in the formation of active metabolites which are capable of binding to the genetic material of the cell leading to toxic, mutagenic or carcinogenic effects.

Extensive purification efforts are necessary to obtain as much as possible native forms of cytochrome P–450 in a pure state. This accomplishment will permit more precise studies of substrate kinetics essential to resolve the functional aspects of multiple forms of P–450. Moreover, the knowledge of the primary structure of the components requiring pure enzymes is the basis for solving the question whether the multiple forms of P–450 present are due to genetically determined different informations or the result of post-translational processing of a single protein molecule. For this purpose, the molecular identification and characterization is a necessary prerequisite, too.

In this article the progress in purifying the components of the hepatic microsomal monooxygenase system obtained mainly from rats and rabbits will be reviewed which has been gained since 1968. The separation and isolation of multiple forms of P–450 from human liver microsomes will also be included in order to emphasize the decisive role of the monooxygenase system in the degradation of drugs and xenobiotics as well as in the steroid metabolism in man.

2. Cytochrome P-450

2.1. *Purification of Liver Microsomal Cytochrome P-450*

Besides the microsomal cytochrome P–450 the occurrence and the partial purification of the enzyme system from liver mitochondria has been described (Okuda et al., 1977; Sato et al., 1977; Pedersen, 1978, Pedersen and Godager, 1978), and evidence for the existence of such a system in the nuclei and nuclear membranes is accumulating (cf. Bresnick et al., 1979; Matsuura et al., 1978). Recent investigations seem to prove that the monooxygenase activities of nuclear membranes are not due to cytoplasmic contaminations (Rogan and Cavalieri, 1978; Guengerich, 1979b). The methods used for the isolation of pure proteins originally bound to the membrane of subcellular fractions are similar in principle. In order to give a comprehensive representation of the development of the preparation methods the survey of the purification procedures will be focused on P–450 from liver microsomes of rabbits and rats.

The endoplasmic reticulum of liver cells shows a high reactivity towards several exogeneous factors. The reason is the specific induction of the components of the monooxygenase in liver microsomes. After treatment of rats, rabbits or mice with drugs, polycyclic hydrocarbons and other foreign compounds the P–450 content is changed qualitatively and quantitatively in dependence on the inducing agent. PB increases the amount of P–450 about threefold (Lu et al., 1972). The induction with 3-MC leads to the predominant synthesis of a spectrally distinct form the socalled P–448 (Alvares et al., 1967). Studies on rat microsomes suggested that multiple forms of P–450 exist — each with different substrate specificity (Nebert et al., 1973; Comai and Gaylor, 1973; Welton and Aust, 1974; Mailman et al., 1975). The isolation of certain forms with distinguishable properties after induction by various inducers as well as from untreated animals gave evidence for the occurrence of multiple forms as concluded from the analysis of microsomes (Alvares and Siekevitz, 1973; Welton and Aust, 1974; Welton et al., 1975).

The specific content of P–450 in the liver of male animals is lower than that of females. Despite of this fact, male animals are used usually due to a higher induction of P–450 in microsomes after PB-treatment than in females. The lacking response of female animals to P–450 induction by PB is not yet fully understood.

Investigations on the mechanism of induction have shown that the formation of the distinct forms P–450 LM2 and P–450 LM4 which were induced by treating the animals with PB and BNF or TCDD, respectively, is the result of de novo synthesis rather than a decreased protein degradation (Haugen et al., 1976; DuBois and Waterman, 1979; Boobis et al., 1977; Kitchin and Woods, 1978; Norman et al., 1978). The induction of at least two distinctly different forms of P–450 (P–448 and P–450) and more than twenty associated monooxygenase activities by 3-MC-like inducers in mouse, rabbit and rat is under the genetic regulation by the Ah locus. Both forms of P–450 appear to be structural gene products controlled by the Ah regulatory gene(s) (cf. Nebert and Jensen, 1979). The response of animals to different inducing agents is age-dependent and tissue-specific (Atlas et al., 1977; Klinger et al., 1979; Liem et al., 1980).

2.1.1. Solubilization of the Monooxygenase System and Partial Purification of Cytochrome P-450

The difficulties connected with the isolation of membrane-bound enzymes consist in finding most active substances with respect to solubilization which do not interfere with enzymatic activity. The solubilization of liver microsomes can be performed either by solubilizing agents or enzymatically. Using proteases, or lipases more or less altered proteins were obtained. Taking into consideration that the enzymatic activity should be almost completely retained by the solubilization procedure detergents have proved to be very suitable for a most careful solubilization.

First attempts to solubilize the microsomes by different types of detergents led to the conversion of P-450 into its inactive form (P-420) (OMURA and SATO, 1964). This disadvantage was overcome by utilizing glycerol as stabilizing agent for P-450 (ICHIKAWA and YAMANO, 1967) and by the use of sodium deoxycholate, sodium cholate or nonionic detergents for solubilization experiments in the following years. It is argued that the stabilization by glycerol is due to the strengthening effect of polyols on the pairwise hydrophobic interaction between hydrophobic groups (BACK et al., 1979). Its free energy per pair of interacting CH_2 groups is more negative in 6 M glycerol than in water (-2.63 ± 0.29 kJ/mol to -1.53 ± 0.09 kJ/mol, respectively).

In 1968, LU and COON first of all succeeded in the solubilization of the monooxygenase system from liver microsomes of untreated rabbits by using deoxycholate. The solubilized preparation was resolved into three essential components P-450, P-450 reductase measured by its activity towards cytochrome c, and a lipid component which later was identified as dioleoylphosphatidylcholine (STROBEL et al., 1970). All three fractions were required to reconstitute maximal activity in the hydroxylation of laurate. The solubilized P-450 characterized by substrate difference spectra and by EPR spectra exhibited a relatively broad substrate specificity (LU et al., 1969a, b, 1970). A variety of drugs as well as alkanes were found to serve as substrates. The spectral changes observed with the corresponding substrates were similar to those of the native microsomal form of this hemoprotein. The individual components obtained by this procedure were cross-contaminated with each other and contained other microsomal proteins. The same specific content of P-450 in the solubilized sample as in the starting microsomes of 0.6 nmol per mg protein clearly indicated this procedure to be inappropriate for getting pure components. Therefore alternative purification methods had to be developed.

In 1972, again LU and collaborators obtained P-450 fractions from male rats pretreated with PB or 3-MC with a threefold increased specific content in comparison to those of the starting microsomes (1.8 or 1.4 nmol per mg of protein, respectively) (LU et al., 1972, LEVIN et al., 1972). The solubilization by sodium cholate resulted in less P-420 than by deoxycholate. The purification was performed by ammonium sulfate fractionation followed by adsorption on calcium phosphate gel. This partially purified preparation with a specific content of 5—6 nmol P-450 per mg protein contained still 5% cytochrome b_5 and 20% NADPH-cytochrome c reductase activity with respect to the values determined in microsomes, respectively. 90% of the microsomal phospholipid remained in the supernatant of the ammonium sulfate fractionation.

A progress in the chromatographic purification was only attained when the nonionic detergent Emulgen 911 was used instead of cholate. Chromatography on DEAE-cellulose in the presence of this detergent resulted in a preparation with a specific content of 10—11 nmol per mg protein nearly free of cytochrome b_5, P–450 reductase, and phospholipids (LEVIN et al., 1974; RYAN et al., 1975b).

2.1.2. *Preparation of Highly Purified Forms of Cytochrome P–450*

2.1.2.1. Nomenclature of Cytochrome P–450

The isolation of the PB-inducible form of P–450 and of further electrophoretically homogeneous forms from untreated animals (see 2.1.3.) gave evidence for the occurrence of different proteins supported by their different properties and catalytic activities towards various substrates. At the present time a nomenclature basing on the function of the multiple forms is still insufficient due to our incomplete knowledge about the individual proteins. Therefore, the forms of P–450 have been distinguished by SDS PAGE. Those forms prepared from rabbit liver microsomes are designated LM1, LM2, etc. on the basis of their decreasing mobility and increasing molecular weights (HAUGEN et al., 1975; JOHNSON, 1980b). The P–450 form which is specifically induced by PB-treatment in rabbits was designated as P–450 LM2 (HAUGEN et al., 1976). Such a preliminary nomenclature existing for the proteins from rabbit liver microsomes is not available for P–450 forms prepared from rat liver. Therefore the original designation of the workers dealing with the purification must be used for comparison. Due to the differing procedures of several groups it is difficult to compare the preparations obtained with each other. Regarding the different designations of the P–450 isolated from PB-treated animals of the same species further analyses are necessary to identify the preparations.

2.1.2.2. Multiple Forms of Cytochrome P–450 from Rabbit Liver Microsomes

In 1974, the preparation of homogeneous P–450 fractions was simultaneously reported by two laboratories employing different procedures (VAN DER HOEVEN et al., 1974; IMAI and SATO, 1974a). Due to its greater stability with respect to the formation of P–420 as compared with those of rats microsomes from PB-induced rabbits were used. In the last 6 years, significant progress has been achieved in the isolation of P–450 from rabbits treated with PB (HAUGEN and COON, 1976; GUENGERICH, 1977b; IMAI et al., 1980; WOLF et al., 1980; HANSSON and WIKVALL, 1980).

Finally the purification from PB-treated male rabbits led to a homogeneous P–450 with a specific content of 17 nmol per mg of protein using sodium cholate for the solubilization and an affinity chromatography on ω-aminooctyl-Sepharose in the presence of cholate (IMAI and SATO, 1974a, 1974b). n-octylamine is known to be an efficient ligand for ferric P–450 with a spectral dissociation constant (K_s value) of about 0.01 mM at 25 °C (JEFCOATE et al., 1969). Today it is generally accepted that both the hydrophobic nature of the alkyl chain and the ionic group attached to the alkyl chain in Sepharose derivatives are important to achieve a good separation of the microsomal enzymes (IMAI, 1976). After partial removal of contaminating proteins P–450 was eluted from the affinity column with a buffer containing both cholate and the nonionic detergent Emulgen 913. The separation of NADH-cytochrome b_5 reductase from the P–450 was achieved by hydroxyapatite and CM-Sephadex chromatography in the presence of Emulgen 913. Subsequently the detergent used could be removed nearly completely by washing the P–450 adsorbed on CM-Sephadex. Based on the initial procedures of

Imai and Sato (1974a) a method was described by Kamataki et al., (1976a, b) which allowed the simultaneous preparation of homogeneous P–450 and a partially purified P–450 reductase from rabbit or rat liver microsomes.

Also in 1974, Coon's group published a procedure resulting in a homogeneous P–450 from PB-treated male rabbits which is based on a polyethylene glycol 6000 fractionation of cholate-solubilized microsomes (Van der Hoeven et al., 1974). After chromatography on DEAE-cellulose and two hydroxyapatite steps P–450 was obtained with a specific content of 17 nmol per mg of protein. The detergent Renex 690 was removed by treating the protein solution with the polystyrene resin Amberlite XAD-2 and calcium phosphate gel (Haugen and Coon, 1976; Coon et al., 1978).

The nonionic detergents of the polyoxyethylene nonylphenole type (Renex 690, Emulgen 911 resp. 913, or Triton N-101) can be replaced mutually as shown by the results of Kawalek et al. (1975), Guengerich (1978a), Dawson et al., (1978), Karuzina et al. (1979) and Bösterling et al. (1979).

Slight modifications of Coon's procedure (Coon et al., 1978) have been used by Dawson et al. (1978) and Bösterling et al. (1979) in preparing the P–450 from PB-induced rabbits in the presence of Triton N-101. Archakov's group has combined the procedures of Imai and Sato (1974b) and Haugen and Coon (1976) and obtained an electrophoretically homogeneous P–450 from PB-treated animals with a yield of 13% (Karuzina et al. 1979). Excess detergent was removed by precipitation of P–450 with polyethylene glycol.

By means of hydrophobic-interaction chromatography on octyl-Sepharose 4B Wolf et al., (1980) achieved a great increase in purity when isolating the homogeneous P–450 forms from liver microsomes of PB-treated rabbits or from lung microsomes of untreated animals. Despite of many chromatographic steps (DEAE-cellulose, hydroxyapatite, octyl-Sepharose, hydroxyapatite removing also the Emulgen 911 used) a yield of 17% was obtained. By this procedure it was also possible to prepare the P–450_I from lung microsomes in sufficient amounts in order to compare the structure of both forms.

The isolation of purified monooxygenases induced by PB has been described in detail. In the following sections, the procedures for preparing multiple forms from rabbits pretreated by 3-MC-like inducers or from untreated animals are reviewed. Originally, the form P–450 LM4 had been isolated from the microsomes of BNF-treated rabbits (Haugen et al., 1975; Haugen and Coon, 1976) but it has been found also in 3-MC- and PB-induced or uninduced liver microsomes with the same properties as that from BNF-treated microsomes as shown by spectral data and immunochemical methods (Dean and Coon, 1977). P–450 LM4 was also obtained from adult rabbits treated with TCDD by application of usual chromatographic purification techniques such as DEAE-cellulose and hydroxyapatite chromatography (Johnson and Muller-Eberhard, 1977a, b, c). Simultaneously, the forms P–450 LM1 and P–450 LM6 were isolated from TCDD-induced microsomes by this group using improved procedures for purification. In adult rabbits TCDD mainly causes the induction of P–450 LM4 but to a less extent also of P–450 LM6. The induction of P–450 LM4 has been proved to be age-dependent (Norman et al., 1978). In neonatal rabbits, however, the form P–450 LM6 is predominantly induced by TCDD. Analogous age-dependent changes in the induction pattern produced by BNF have been shown by means of a copper phenanthroline cross-linking reaction, too (McIntosh and Freedman, 1979; 1980). Differences in the minimal molecular weights, in the immunological properties, as well as in the fingerprints obtained after limited proteolysis have been reported for the three forms indicating differences in the primary structure of these proteins (NORMAN et al., 1978; JOHNSON et al., 1979a)

The isolation of the pure P–450 forms LM2 and LM4 was rendered by the existence of a P–450 form which had an intermediate mobility between that of P–450 LM2 and P–450 LM4 upon SDS PAGE (COON et al., 1978). This P–450 LM3 was purified to electrophoretic homogeneity with a specific content of 8 to 12 nmol P–450 per mg protein by slightly modified chromatographic procedures after polyethylene glycol fractionation (INGELMAN-SUNDBERG et al., 1979). Using similar procedures the forms P–450 LM3b, P–450 LM3, and P–450 fraction III which seem to be identical have been purified by KOOP and COON (1979), JOHNSON (1980a) and HANSSON and WIKVALL (1980), respectively. The forms were isoated from untreated, PB- and TCDD-reated rabbits with a specific content of more than 17 nmol P–450 per mg protein. The proteolytic peptide fingerprints of P–450 LM3 clearly differed from those of P–450 LM2 and P–450 LM4 indicating without doubt this form to exist as real isozyme.

A further P–450 LM3-like cytochrome which shows a high binding affinity to cytochrome b_5 has been purified from rabbit liver microsomes with a chromatography on immobilized tryptic cytochrome b_5 to a gel-electrophoretically homogeneous state (MIKI et al., 1980a, b). With respect to the overlapping substrate specificities of various P–450s it is remarkable that this P–450 was found to need cytochrome b_5 as an obligatory component for the o-demethylation of p-nitroanisole (SUGIYAMA et al., 1980). Further studies with these LM3 forms, e.g. analyses of the N- and C-terminal amino acid residues, will indicate which of the hemoproteins differently denoted are identical to each other.

The present state of the separation and purification of multiple forms of rabbit

Table 1. Isozymes of rabbit liver cytochrome P–450 induced by different agents

Isozyme	Origin	Inducing agent	Particularly active in catalyzing	Reference
LM1	constitutive	–	N,N-dimethyl-4-aminoazobenzene N-demethylation	GUENGERICH (1977b)
LM2	inducible	PB	Benzphetamine N-demethylation	HAUGEN and COON (1976)
LM3	constitutive	–	Aminopyrine N-demethylation	JOHNSON (1980a)
LM3b	constitutive	–	Testosteron 6β-hydroxylation	KOOP and COON (1979)
LM4	constitutive	–	Acetanilide hydroxylation	COON and VATSIS (1978)
LM4	inducible	PB, BNF, 3-MC	Acetanilide hydroxylation	COON and VATSIS (1978)
LM4	inducible	TCDD	2-Acetylaminofluorene N-hydroxylation	JOHNSON (1980b)
LM6	inducible	TCDD	Benzpyrene hydroxylation	NORMAN et al. (1978)
LM6	inducible	TCDD	2-Acetylaminofluorene hydroxylation	JOHNSON (1980b)
LM7	constitutive	–	N,N-dimethyl-4-aminoazobenzene N-demethylation	GUENGERICH (1977b)

liver P-450 indicates by several criteria that these forms are isozymes (see 2.3.). According to the present knowledge the different isozymes can be divided into constitutive and inducible (adaptive) isozymes as listed in Table 1. In order to characterize the functions of the isozymes which will be discussed later in detail (4.2.2.) only a few reactions have been selected which are catalyzed by the individual P-450 with high activity.

2.1.2.3. Multiple Forms of Cytochrome P-450 from Rat Liver Microsomes

The purification of electrophoretically homogeneous P-450 from PB-induced rats has been described at first by RYAN et al. (1975a). The purification procedure for rat P-450 had to be improved because the electrophoretically homogeneous forms of rat P-450 were composed of several immunological different forms (THOMAS et al., 1976a, b).

Highly purified forms of P-450 from rats treated with PB have been isolated and described in detail by GUENGERICH (1977a, 1978a). Using the affinity chromatography on aminooctyl-Sepharose the P-450 eluted has been separated by hydroxyapatite chromatography. The forms B and D eluted by increasing molarity of phosphate have been purified finally by DEAE-cellulose chromatography to a specific content of 19 nmol/mg protein. The P-450 B was homogeneous as assayed by SDS PAGE in six different gel systems. The best resolution was obtained according to the standard procedure of LAEMMLI (1970). Using different antibodies prepared against forms which have been induced by different inducers the P-450 B has been shown to be immunochemically homogeneous. The P-450 D gave few precipitation lines indicating the occurrence of distinct forms with different antigenic sites. Basing on the previous purification schemes GUENGERICH and MARTIN (1980) have recently presented a simple procedure for the simultaneous purification of P-450, epoxide hydrase and P-450 reductase from rat liver microsomes.

A rapid procedure has been developed for the purification of the PB-inducible form of P-450 from liver microsomes of PB-treated rats (WEST et al., 1979). After solubilization of microsomal proteins by sodium cholate and fractionation by polyethylene glycol the two-step chromatography on DEAE-cellulose was carried out at room temperature (WARNER and NEIMS, 1979). The resulting P-450 was nearly homogeneous on SDS PAGE with a specific content of 16 nmol/mg protein. The yield of about 14% could be further improved to homogeneity by subsequent chromatography on QAE-Sephadex. This procedure is inappropriate for obtaining P-450 LM2 from rabbits indicating species specific differences of the functionally analogous P-450 form.

In confirming both purification procedures HANSSON and WIKVALL (1980) discriminated P-450 forms with different catalytic properties towards different C_{24}- and C_{27}-steroids (see 4.2.3.). P-450 B and P-450 (WEST et al., 1979) gave very similar regioselective hydroxylation products suggesting to be identical forms (WIKVALL et al., 1980).

The ω-aminooctyl-Sepharose chromatography has been used by other workers for getting the PB inducible form of P-450 from rat liver microsomes in a homogeneous state (IMAI, 1976; KAMATAKI et al., 1976b; MASUDA-MIKAWA et al., 1979). Since in the binding of P-450 to the substituted Sepharose hydrophobic as well as biospecific interactions are involved the binding of further microsomal proteins diminished the available separation capacity decreasing the yield of this essential step. LU et al. (1980b) have been looking for further affinity ligands allowing a partial purification of different P–450 forms or the isolation of a single P–450 in a pure state. A variety of substrates and inhibitors of P–450 have been bound via a C_6—C_8 spacer to Sepharose 4B. Tryptamine, 2,4-dichloro-6-phenylphenoxyethylamine (DPEA), and p-chloramphetamine have been found to retain more than 90% of P–450 of detergent-solubilized microsomes from untreated, PB- or 3-MC-treated rats. After washing out unspecifically bound proteins P–450 could be eluted with tryptamine or higher concentrations of nonionic detergents with a specific content of 14 nmol/mg in a yield of 50—70%. Due to their poor solubility most substrates or inhibitors could not be used for the elution of P–450. A final purification to electrophoretic homogeneity can be achieved using DEAE-cellulose chromatography.

The existence of multiple forms in electrophoretically homogeneous rat P–450s was demonstrated by differences in immunochemical investigations Thomas et al., 1976a, b). This has further been evidenced by the isolation of three highly purified forms from rats treated with polychlorinated biphenyls (Aroclor 1254), PB, and 3-MC using chromatographic separation techniques (Ryan et al., 1977; 1979). The highly purified forms designated as P–450a, P–450b, and P–450c differ in their electrophoretic, spectral, immunochemical and catalytic properties. Furthermore, different peptide fragments are formed when these cytochromes are subjected to limited proteolysis in the presence of SDS. P–450b is the hemoprotein preferentially induced by PB, whereas P–450c corresponds to that P–450 predominantly induced by 3-MC. P–450a is a minor form induced by both compounds. However, these distinct P–450s are not identical with those described by Guengerich (1978a) and Hansson and Wikvall (1980) as indicated by differences in the amino acid composition and in the molecular weight (Botelho et al., 1979). By sequencing the N-terminal amino acids of P–450a, P–450b, and P–450c it became evident that different inducers cause the formation of identical forms (see 2.3.). These results give additional support for the homogeneity of the forms isolated. Therefore the somewhat confusing results obtained by different authors are rather due to preparational differences than to an increasing number of isozymes.

The three P–450s above mentioned are also induced in rat liver microsomes by isosafrole (4-propenyl-1,2-methylenedioxybenzene (Ryan et al., 1980) known as inhibitor of monooxygenase reactions. The predominant form P–450d induced by this compound was purified by ammonium sulfate fractionation, calcium phosphate gel adsorption, and ion exchange chromatography at 4 °C followed by DEAE-cellulose and immunoaffinity chromatography at room temperature. The yield of P–450d isolated as an isosafrole-metabolite complex with a specific content of 9—16 nmol of heme/mg of protein is 0.75—1.0% of total microsomal P–450.

Using similar chromatographic procedures Saito and Strobel (1981) described the isolation of a P–450 form containing up to 18.6 nmol P–450/mg of protein with a total yield of 9—15% from BNF-treated rats with the highest turnover for benzo(a)pyrene reported up to now.

Besides the two major classes of microsomal enzyme inducers — "3-MC-like" and "PB-like" generating the corresponding isozymes — there are many compounds acting as "atypical" inducers (cf. Nebert, 1979). The simple and rapid technique of West et al. (1979) was used for the isolation and purification of a P–450 form found to be a unique hemoprotein from rats treated with pregnenolone-16α-carbonitrile. This novel form purified to a specific content of 8 nmol/mg of protein was assumed to account for the high N-demethylation of ethylmorphine in microsomes prepared from

PCN-treated rats (ELSHOURBAGY and GUZELIAN, 1980).

For the isolation of P–450 isozymes from untreated rat liver microsomes an affinity chromatography analogously to that for rabbit liver microsomes has been used (GIBSON and SCHENKMAN, 1978). They applied microsomes solubilized by Emulgen 911 on a laurate-substituted 1,6-diaminohexane-Sepharose 4B. Laurate binds to P–450 with a K_s value of 0.005 mM (ORRENIUS and THOR, 1969). The latter of the two forms (P–450_{II}) eluted by a salt gradient was purified by gelfiltration to a specific content of 8.9 nmol P–450 per mg of protein with a yield of 38%.

Comparing the various procedures used for the preparation of electrophoretically homogeneous P–450 from PB-treated rats and rabbits it can be estimated that methods using the affinity chromatography on ω-aminooctyl- Sepharose are most suited for getting the PB-induced P–450 with equal high quality with respect to the content of P–420 as well as apocytochrome in more than 10% yield of the starting microsomes. These higher yields are probably related to the almost complete separation of individual forms of P–450 in the aminooctyl- Sepharose step(s) in comparison to the DEAE-cellulose chromatography used by COON and associates (1978) which is performed separately with each supernatant from the polyethylene glycol fractionation. A more rapid procedure to obtain larger quantities of the PB-induced form from PB-treated rats in a nearly homogeneous quality with a yield of 14% seems to be the two-step chromatography on DEAE-cellulose at room temperature (WEST et al., 1979). Within two days more than 1,000 nmol of highly purified P–450 could be prepared from 4 g of microsomal protein.

2.1.2.4. Preparation of Cytochrome P–450 from Human Liver Microsomes

Despite of the intensive progress in purification of the components of the monooxygenase system from laboratory animals little is known about these enzymes in man. The occurrence of P–450 and P–450 reductase in human liver microsomes has been reported by several authors as shown in Table 2. The differences in the specific contents are probably due to interindividual differences of the enzyme profile in man. On an average, the P–450 content in human microsomes is somewhat lower than that of microsomes from uninduced laboratory animals.

In 1975, the partial purification of human P–450 by a factor of 2 was reported by KASCHNITZ and COON (1975) using ammonium sulfate fractionation. Recently, 3 groups have isolated pure forms of P–450 from human liver samples using modifications and improvements of the techniques developed for the laboratory animals (BEAUNE et al., 1979; KAMATAKI et al., 1979; WANG et al., 1980). By means of the affinity chromatography on n-octylamino-Sepharose 4B well approved in the purification of P–450 from different animals and other sources with low specific content such as yeast or by hydrophobic chromatography on octyl-Sepharose CL-4B the isolation and purification of P–450 to a specific content of 3—6 nmol per mg protein could be attained. The preparations were nearly devoid of P–450 reductase, cytochrome b_5, and NADH-cytochrome b_5 reductase.

From mutual inhibition data of the 7-ethoxycoumarin dealkylation by human liver microsomes (ULLRICH and KREMERS, 1977; BEAUNE et al., 1979) a multiplicity of human P–450 has been suggested, too. This apparently holds true also for P–450 from human fetal liver (KITADA and KAMATAKI, 1979) and placenta as

Table 2. Components of the monoxygenase system in human liver samples

P–450 reductase (µmol/min · mg protein)	P–450 (nmol/mg protein)	cytochrome b_5 (nmol/mg protein)	ratio b_5/P–450	Reference
—	0.25	0.35	1.40	Alvares et al. (1969)
0.126 ± 0.011	0.29 ± 0.03	0.15 ± 0.02	0.52	Ackermann (1970)
—	0.27	0.20	0.74	Raj et al. (1971)
0.115 ± 0.030	0.59 ± 0.28	0.26 ± 0.09	0.48	Pelkonen et al. (1974)
0.103 ± 0.015	0.60 ± 0.10	0.49 ± 0.06	0.92	Ahmad and Black (1977)
0.148 ± 0.003	0.53 ± 0.07	0.52 ± 0.07	0.98	Jänig (unpublished)

reviewed recently in detail by Juchau (1980).

To solve this question unequivocally for human P–450 despite of the inter-individual variability the isolation of different forms of P–450 differing by several criteria was absolutely necessary. The use of affinity chromatography on n-octylamino-Sepharose and of further purification steps resulted in P–450 preparations with specific contents between 10 and 14 nmol P–450 per mg protein whereas the yields ranged from 2–30%. Kamataki et al. (1979) suggested from the chromatographic behaviour that there are more than one species of P–450 in human liver microsomes. This was strenghtened by the results of Wang et al. (1980) and of Bösterling and Trudell (1980). Both groups separated electrophoretically homogeneous P–450 forms differing in their subunit molecular weights. The conclusion with respect to the number of multiple forms in human liver has to be regarded tentatively due to the possible formation of artifacts from autopsy material.

Summarizing the methods for the purification of different forms of P–450 to apparent homogeneity from liver microsomes of PB, 3-MC-, BNF-, or TCDD-treated or untreated rabbits and rats gave direct evidence for the existence of multiple forms of P–450. This statement is supported by 2 experimental findings. Differences in the primary structure have been proved by partial analyses of the N- and C-terminal regions of the various forms (2.3.) and by the proteolytic peptide mapping in SDS suggesting different amino acid sequences. Together with immunochemical differences it can be stated that the different forms of P–450 are isozymes. This multiplicity of isozymes seems to be the basis for an efficient biotransformation of numerous drugs, chemical carcinogens, environmental pollutants, and other foreign compounds.

2.2. Physicochemical Properties of Highly Purified Forms of Cytochrome P–450

Some characteristics of highly purified forms of rabbit liver P–450 are listed in Table 3. The most intensively studied forms P–450 LM2 and P–450 LM4 induced by PB and BNF, respectively, exhibit molecular weights of the polypeptide chain of 48,700 and 55,300 (Haugen and COON, 1976), whereas a molecular weight of 52,000 has been determined for the constitutive form P–450 LM3b (KOOP and COON, 1979). The purified cytochromes do not contain any detectable amounts of contaminating proteins and only a very low content of phospholipids (0.3 mol per mol P–450). Although the preparations are electrophoretically and immunochemically homogeneous at the last step of the purification procedure the specific content is sometimes less than given in the table due to a loss of heme. This loss of heme during purification results in the formation of apo-P–450 and seems to be

Table 3. Biochemical properties of cytochrome P–450 fractions isolated from liver microsomes of different species

Species	Form	Inducer	Spec. content (nmol/mg pr.)	Mol. wt.	Absolute spectra λ_{max} (nm) Fe^{3+}	Fe^{2+}	$Fe^{2+}\cdot CO$	Reference
Rabbit	LM1	–	6.8	47,000	–	–	449	GUENGERICH (1977b)
	LM2	PB	17–18	48,700	418	413	451	HAUGEN and COON (1976)
	LM3b	–	17	52,000	418	412	451	KOOP and COON (1979)
	LM4	BNF, 3-MC, TCDD	17–18	55,300	394	411	448	HAUGEN and COON (1976)
	LM6	TCDD	16.1	57,000	–	–	448	NORMAN et al. (1978)
	LM7	–	4.3	60,000	–	–	449	GUENGERICH (1977b)
Rat	P–450a	PB	12–14	48,000	417	411	452	RYAN et al.
	P–450b	PB	15–17	52,000	417	410	450	(1979)
	PB-P–450B	PB	18.0	53,100	–	–	–	GUENGERICH
	MC-P–450B	3-MC	17.1	55,850	–	–	–	(1978a)
	P–450c	3-MC	15–17	56,000	417	409	447	RYAN et al. (1979)
	P–446	BNF	18.1	53,400	416	410	450	SAITO and STROBEL (1981)
	P–450d	Isosafrole	14	52,000	392	410	447	RYAN et al. (1980)
	P–450	PCN	8	51,000	420	417	450	ELSHOURBAGY and GUZELIAN (1980)
Man	Patient 6	–	9.5–13.8	53,000	416	418	450	WANG et al.
	Patient 3	–	9.4	55,500	–	–	–	(1980)

dependent on several factors, e.g. duration of the purification procedure, chromatographic materials used, and protein concentrations of the solutions worked with. The content of residual detergent amounts to about 2 mol per mol polypeptide chain, nearly independent of the procedures used for its removal (treatment with calcium phosphate gel and/or Amberlite XAD-2; adsorption to CM-Sephadex C-50 or QAE-Sephadex A-25) (HAUGEN and COON, 1976; GUENGERICH and HOLLADAY, 1979; WOLF et al., 1980; IMAI et al., 1980).

From the data summarized in Table 3, it can be seen that the individual P–450 forms isolated from rabbit liver microsomes differ in their spectral characteristics, especially in the ferric state and as CO complex. The spectra of oxidized P–450 LM2 and P–450 LM3b are typical for those of low spin hemoproteins but that of P–450 LM4 is high spin. Contrary to P–450 LM2 the spectrum of oxidized P–450 LM4 is dependent on the experimental conditions. The binding of nonionic detergents producing a reverse type I difference spectrum (K_s = 1.26 mM) leads to a conversion from the high spin to the low spin state (HAUGEN and COON, 1976; HASHIMOTO and IMAI, 1976; HASHIMOTO-YUTSUDO et al., 1980). The effect of high protein concentrations of P–450 LM4 forming a low spin spectrum, too, suggests that the preparation was not free from the detergent (HASHIMOTO-YUTSUDO et al,. 1980).

The spectral parameters of the different P–450 forms obtained from rat liver are characteristic for low spin hemoproteins but those of purified P–450d represent an isosafrole metabolite-P–450 complex (RYAN et al., 1980).

The spectrum of ferric P–450 LM2 (Fig. 1) shows in the median UV region

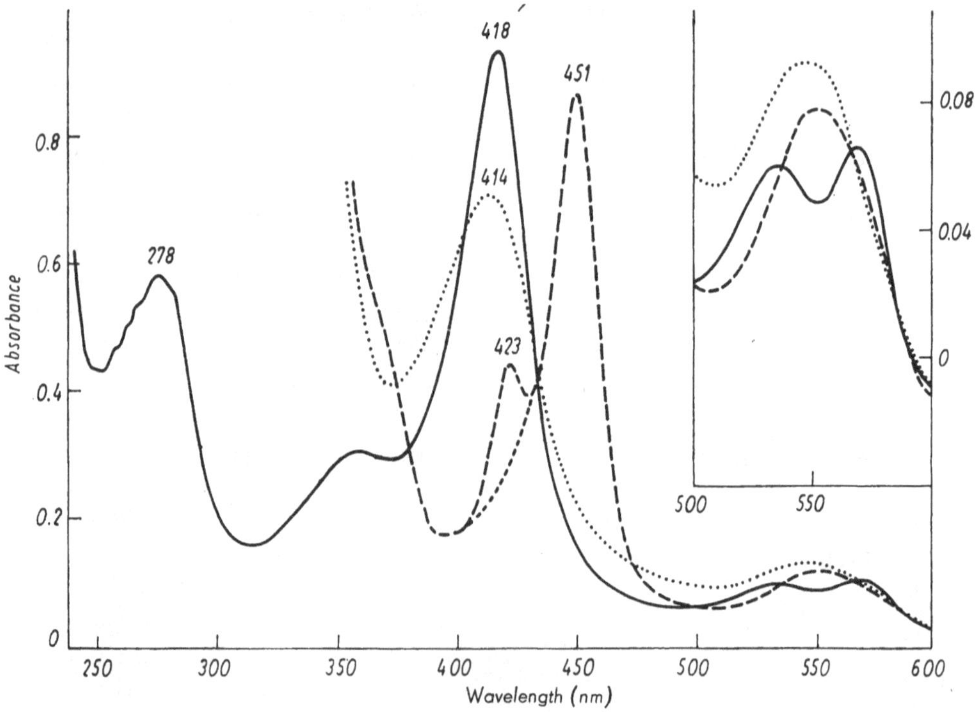

Fig. 1. Absorption spectra of cytochrome P–450 LM2 purified from rabbit liver microsomes: ferric (——), ferrous .....), and ferrous-CO complex (----). The peak at 423 nm originates from the presence of some cytochrome P–420-CO in the preparation.

a fine structure caused by the high content of 31 phenylalanine residues per chain. A higher resolution of fine structures in this spectral region could be obtained by second derivative spectroscopy clearly indicating the red shift of the chromophores of aromatic amino acids in proteins caused by the hydrophobic environment in comparison to amino acid mixtures (RUCKPAUL and REIN, 1980; RUCKPAUL et al., 1980b). The ratio of absorbances of the Soret region to the median UV region can be used to prove the degree of purity of a hemoprotein. The ratio of A_{418} to A_{278} of P-450 LM2 will be lowered in the presence of apoprotein as well as by nonionic detergents used during the preparation procedure which strongly absorb at 276 nm. A_{418}/A_{278} ratios of about 2.0 indicating a preparation with almost no apo-P-450 were reported by the Japanese groups (IMAI and SATO, 1974a; SHIMIZU et al., 1979; HASHIMOTO-YUTSUDO et al., 1980) and by WOLF et al. (1980).

The aggregation of purified membrane proteins after solubilization from membranes or in the absence of detergents is a well known phenomenon (SIMONS et al., 1978). This is true also for microsomal proteins such as cytochrome b_5, NADH-cytochrome b_5 reductase, and epoxide hydrase (SPATZ and STRITTMATTER, 1971; SPATZ and STRITTMATTER, 1973; LU and MIWA, 1980b) which might be involved in the hydroxylation of certain substrates. Likewise, the highly purified P-450 11β and P-450 SCC are aggregated after isolation from adrenal mitochondria (cf. MITANI, 1979).

Sedimentation experiments and gel filtration analyses using calibrated Sepharose 6B columns showed the formation of high molecular aggregates by purified P-450 proteins in the absence of detergents or lipids. As indicated in Table 4, the apparent molecular weight of P-450 LM2 determined by chromatography was about 300,000 (HOEVEN and COON, 1974; CHIANG and COON, 1979; IMAI et al., 1980; FRENCH et al., 1980b). Related to the minimal molecular weight the P-450 LM2 would appear to contain six polypeptide chains in aqueous solution. The results of chromatographic techniques are consistent with a value of 300,000 as calculated from sedimentation equilibrium (BEHLKE et al., 1979) but differs slightly from the value of 390,000 obtained by sedimentation experiments (GUENGERICH and HOLLADAY, 1979). All these data confirm the provisional apparent molecular weight of about 350,000 of partially purified P-450 LM determined previously by several methods (AUTOR et al., 1973). The final size of the aggregates apparently depends on the conditions of the preparation (methods used for the removal of detergent, content of aprotein, etc.). A molecular weight of 700,000 $\pm$ 30,000 published for P-450 LM2 by KARUZINA et al. (1979) supports this idea.

The P-450 LM4, however, has been reported to exist as heptamer to nonamer with molecular weights between 420,000 to 500,000 respectively (GUENGERICH and HOLLADAY, 1979; CHIANG and COON, 1979).

Analogous investigations of rat liver P-450 forms isolated from PB- or 3-MC-induced animals indicated also the aggregation of these proteins. Apparent molecular weights of about 500,000 and 408,000 have been obtained for the PB- and 3-MC-rat P-450, respectively (GUENGERICH and HOLLADAY, 1979; BASKIN and YANG, 1980). The temperature dependence of the association process suggests that hydrophobic forces predominate in the self-association of monomers to the oligomeric form in aqueous solution. It has to be assumed that the aggregation is due to the accumulation of hydrophobic amino acid residues in certain parts of the molecule (HAUGEN et al., 1977; BOTELHO et al., 1979) but nothing can

Table 4. Molecular weights of various forms of cytochrome P–450

Cytochrome	Species	Inducer	Estimated molecular weights			
			Apparent (Gelfiltration)	Apparent (Sed. equilibrium)	Minimal (SDS PAGE)	Minimal (Sed. velocity)
P–450 LM2	Rabbit	PB	270,000 a) 280,000 b) 300,000 c) 300,000 d)	300,000 e) 390,000 f)	49,000 c)	—
P–450 LM4	Rabbit	3-MC, BNF	500,000 c)	422,000 f)	55,000 c)	—
PB P–450 B	Rat	PB	520,000 h)	500,000 f)	54,000 g)	57,000 f)
MC P–450 B	Rat	3-MC	n.d.	408,000 f)	56,000 g)	55,800 f)

a) Imai et al. (1980)
b) Van der Hoeven and Coon (1974)
c) Chiang and Coon (1979)
d) French et al. (1980b)
e) Behlke et al. (1979)
f) Guengerich and Holladay (1979)
g) Guengerich (1978a)
h) Baskin and Yang (1980)

be derived from these results on the structure of P–450 within the membrane.

Besides the analysis of molecular weight, electrophoretic mobility, absorption spectra, and immunological properties also isoelectric focusing has been used for further characterization of different P–450s. In principle, the occurrence of different proteins with identical mobilities represented by a single band in SDS PAGE cannot be excluded as shown by Philpot and Arinc (1976) who isolated two P–450 forms with the same molecular weight from untreated rabbits by ion exchange chromatography. By means of electrofocusing on granulated gel P–450 LM2 has been separated into several fractions but only one of them has been proved to be active in steroid hydroxylation (Ingelman-Sundberg and Gustafsson, 1977). This finding requires careful interpretation. Investigations with homogeneous forms of P–450 and P–450 reductase demonstrated an artificial heterogeneity which is attributed to strong protein-protein interactions and perhaps to complex formation of proteins with carrier ampholytes (Dus et al., 1976; Warner et al., 1978; Guengerich, 1979a; Vlasuk and Walz, 1980).

2.3. Amino Acid Compositions and Partial Amino Acid Sequences of Cytochrome P–450 Isozymes

The basis for an understanding of the molecular mechanism of structure-function relationships is the detailed knowledge of the structure of the respective molecule. With respect to the hepatic monooxygenase the resolution of this problem is complicated due to 2 facts: (i) The enzymatic activity is exerted by a complex structure consisting of 3 interacting components. (ii) This enzyme complex is embedded in the phospholipid matrix of the endoplasmic reticulum. Therefore the tertiary structure of only one component can provide rather limited information concerning the structure-function relationships of the multicomponent complex and still less conclusions can be drawn from the primary structure with respect to the problem mentioned above. The analysis of structure-function relationships is further complicated by conformational changes of P-450 induced by the NADPH-cytochrome P-450 reductase as well as by phospholipids (see 4.1.). An understanding of these phenomena therefore requires finally the analysis of the spatial structure of a complex consisting of P-450, P-450 reductase and phospholipid. At the present stage, even the construction of a model of such a complex from the structure of the individual proteins is impossible due to the difficult crystallization of P-450 and P-450 reductase. Therefore, further physicochemical investigations are required to accumulate data from which conclusions on structure-function relationships can be derived.

The knowledge of the amino acid sequence is necessary for the decision whether the multiple forms are products of separate genes or arose by post-translational processing of one primary gene product. Further information concerning the biosynthesis of these integral membrane proteins and the functional requirements for their insertion into the membranes of the endoplasmic reticulum can be derived from the primary structure.

The amino acid composition of various forms of P-450 from microsomal and microbial origin is given in Table 5. Considering the new data of YASUNOBU et al. (1980) there exist a remarkable similarity of the amino acid composition between P-450 LM2 and the soluble P-450 CAM. P-450 LM2 exhibits a higher lysine content whereas on the other hand P-450 CAM contains more alanine and glutamic acid but less phenylalanine residues. The animal P-450 forms are of lower cysteine content than the P-450 CAM protein. They do not contain any cystine thus excluding both intrachain and interchain disulfide bonds (GUENGERICH et al., 1975).

The forms P-450 LM2 and P-450b induced by PB in rabbits and rats, respectively, as well as P-450 LM4 and P-450c induced by 3-MC in both species show similar amino acid compositions with about 35% hydrophobic residues if methionine, valine, leucine, isoleucine, tyrosine, phenylalanine and tryptophan are considered to be hydrophobic residues (NOZAKI and TANFORD, 1971). All four forms contain 18—19% acidic residues and 14—16% basic amino acids. Similar contents with respect to acidic and basic residues have also been calculated for highly purified mitochondrial P-450 forms from adrenals catalyzing specifically the 11β-hydroxylation and the side chain cleavage of cholesterol, respectively. The forms P-450 11β and P-450 SCC composed of 375 and 395 amino acids, respectively, have also a hydrophobic content of 35% (cf. MITANI, 1979).

It is evident that no conclusions regarding the different physical behaviour can be drawn from a comparison of amino acid compositions of a soluble and a membrane-bound P-450 of different origin due to its specific spatial arrangement.

Despite the similarity of the amino acid residues in the immediate vicinity of the heme group and in the immunochemical properties (DUS et al., 1974) the N-terminal sequences of P-450 CAM and P-450 LM2 show considerable differences characterizing either a soluble or a membrane-bound P-450 (HAUGEN et al., 1977; YASUNOBU et al., 1980). P-450 LM2 contains 17 hydrophobic residues in the first 20 positions, including two clusters of 5 consecutive leucines. Such unusual hydrophobic N-terminal sequences have been found previously only in precursor segments of certain secretory proteins functioning as a special signal sequence in recognizing a membrane site and being cleft after the initial transfer of the nascent protein across the bilayer membrane (cf. DAVIS and

TAI (1980)). Such N-terminal signal sequences have been identified also in P–450a, P–450b, and P–450c from rat liver microsomes (BOTELHO et al., 1979) without being cleft after complete biosynthesis of the corresponding forms. P–450 3b and 3c also contain hydrophobic N-terminal regions (COON et al. 1980). However, sequence analysis of P–450 LM4 failed in detecting a NH_2-

Table 5. Amino acid composition of different forms of cytochrome P–450 from various species

	Uninduced		Induced by Phenobarbital			Induced by 3-Methylcholanthrene		
	P–450 CAM (1)	Human patient 6 (2)	P–450b (3)	rat PB-P–450B (4)	rabbit LM2 (5)	P–450c (3)	rat MC-P–450B (4)	rabbit LM4 (5)
Lys	13	46	24	27	19	28	31	26
His	12	1	18	13	11	15	13	11
Arg	24	19	32	25	29	26	25	30
AsX	36	47	35	42	38	47	43	44
Thr	20	25	26	29	23	28	29	24
Ser	24	36	29	30	30	36	38	31
Glx	58	45	47	52	42	47	43	46
Pro	28	28	32	27	24	24	37	28
Gly	28	45	30	33	32	36	32	38
Ala	35	27	20	29	23	27	26	29
Cys	8	5	6	7	6	6	6	7
Val	24	22	22	27	26	33	28	32
Met	10	15	11	11	7	8	10	10
Ile	24	28	24	27	19	29	24	22
Leu	42	53	54	59	54	55	54	58
Tyr	10	5	12	14	9	15	14	11
Phe	18	23	35	32	31	35	28	28
Trp	4	4	1	2	1	8	6	7
Total	418	474	455	486	424	500	487	482
Mol.wt. peptide chain	46,600	—	—	—	47,600	—	—	54,200
Mol.wt. SDS PAGE	—	53,000	52,000	53,000	49,000	56,000	55,850	55,000
Acidic amino acids [%]	22.4	19.4	18.0	19.3	18.9	18.8	17.7	18.7
Basic amino acids [%]	11.7	13.9	16.3	13.4	13.9	13.8	14.2	13.9
Hydrophobic amino acids [%]	31.6	31.6	34.9	35.4	34.7	36.6	33.7	34.8

1) YASUNOBU et al. (1980)
2) WANG et al. (1980)
3) BOTELHO et al. (1979)
4) GUENGERICH (1978a)
5) HAUGEN and COON (1976)

terminal signal sequence since multiple amino acids appeared at each step, indicating a non-homogeneous enzyme preparation or a protein which has been partially degraded by proteolysis in vivo or during isolation (HAUGEN et al., 1977). The content of hydrophobic residues in the N-terminal sequences analyzed so far amounts to a maximal value of 67%. It is suggested that this part of the molecule may serve in the binding and orientation of the P–450 in the membrane of the endoplasmic reticulum.

YASUNOBU and coworkers (1980) have tested the CHOU-FASMAN procedure as a preliminary approach to the prediction of P–450 conformations. With some 20 residues known in each of the 5 P–450 proteins (P–450 CAM, P–450 LM2, P–450a, P–450b and P–450c) very small similarities are implied. Within the limited regions the calculations point mainly to the α-helix for rabbit P–450 LM2 and a higher degree of β-sheet for rat P–450a whereas the C-terminal sequences of P–450 LM2 and P–450 CAM would imply small regions of β-sheet.

From a comparison of the N-terminal sequences and the C-terminal regions of the different forms purified from rat and rabbit it can be concluded that these hemoproteins will differ in their tertiary structure. This conclusion is supported by different immunochemical reactivities towards various antibodies (THOMAS et al., 1979). In addition, from the different amino acid compositions and the partially known N- and C-terminal sequences follows conclusively that the multiple forms of P–450 are isozymes, i.e. they catalyze the same general reaction, but they are synthesized by different genes. Therefore, epigenetic factors as reason for the multiplicity of P–450 can be excluded. It should be noted that not only the overall amino acid compositions of P–450b and P–450 LM2 induced by PB in rat and rabbit, respectively, are very similar but also their N-terminal sequences. Among the 21 identified residues 14 common residues including the two clusters of consecutive leucines have been found (BOTELHO et al., 1979). This high degree of homology of the PB-induced proteins is more pronounced than the similarity of the three P–450 forms isolated from rat liver microsomes. Furthermore, the resembling catalytic activities of the hemoproteins from PB-treated animals (cf. 4.2.3.) demonstrate that independent genes for the individual isozymes arose earlier than the differentiation into single species during the evolution process.

Despite the same prosthetic group (iron protoporphyrin IX) the different function of the various hemoproteins is mostly determined by the axial heme iron ligands and the immediate surrounding of the heme formed by the spatial structure of the polypeptide chain of the respective hemoprotein. The main interest was therefore directed to comparative analyses of the amino acid side chains within the active center which contribute to determine the diverse types of catalytic reactions of P–450 LM2 and P–450 CAM. Each hemepeptide which was stabilized by 1-(4-azidophenyl) imidazole as a photoaffinity label and digested by cyanogen bromide contains at least one residue each of cysteine, histidine, and tyrosine (DUS 1980). The pH titration data of RUCKPAUL et al. (1980b) postulating the existence of a heme-linked tyrosine as one of the heme ligands are in line with this analysis. The hemepeptides (Soret maxima at 356–360 nm) are very soluble despite a higher percentage of hydrophobic residues as compared with the whole hemoprotein which has been ascertained for P–450 LM4, P–450 SCC, P–450 11β and P–450 C21. The hemepeptides are recognized by antibodies reacting with their parent proteins. From the data it has been concluded that all P–450 hemepeptides contain a tightly structured domain which includes the heme and substrate binding sites (DUS, 1980).

The amino acid composition of one electrophoretically homogeneous human P–450 preparation (WANG et al., 1980) shows some similarity to that of P–450 isozymes from PB- and 3-MC-treated rats and rabbits (Table 5). Besides smaller differences a great difference can be seen in the content of basic residues, i.e. the human P–450 contains only a small number of histidine and much more lysine than any other animal P–450. The separation and purification of electrophoretically homogeneous P–450 forms from a greater number of individuals is necessary for more reliable conclusions concerning the occurrence of isozymes in man and their similarity with the various isozymes from animal origin.

3. NADPH-Cytochrome P–450 Reductase

3.1. Purification of NADPH-Cytochrome P–450 Reductase

The isolation of a NADPH-cytochrome c reductase from acetone powder of whole pig liver has been reported at first by HORECKER (1950). WILLIAMS and KAMIN (1962) and PHILLIPS and LANGDON (1962) independently purified this enzyme. Both laboratories established the flavoprotein nature of the enyzme and its microsomal origin. The reductase could be purified to a homogeneous state after proteolytic solubilization of the microsomal fraction of various tissues and species (PHILLIPS and LANGDON, 1962; PEDERSON et al., 1973; IYANAGI and MASON, 1973), by lipases (WILLIAMS and KAMIN, 1962; FAN and MASTERS, 1974) as well as by use of ionic (LU and WEST, 1972) and nonionic detergents (DIGNAM and STROBEL, 1975, 1977; YASUKOCHI and MASTERS, 1976; VERMILION and COON, 1974, 1978a; MIWA et al., 1978; FRENCH and COON, 1979; MASTERS and OKITA, 1980).

Proteases are able to solubilize that part of microsomal proteins which is protruding the microsomal membrane surface. In this way a NADPH-dependent reductase with activity towards cytochrome c and a modified cytochrome b_5 can be obtained without solubilization of other constituents of the microsomal electron transfer chain (OMURA and TAKESUE, 1970). A complete solubilization of all membrane proteins can be achieved by nonionic detergents as well as by the combination of ionic and nonionic ones.

Electrophoretically homogeneous NADPH-cytochrome P–450 reductase preparations which could be reconstituted to an enzymatically active monooxygenase system in the presence of P–450 and phospholipid have been obtained utilizing nonionic detergents of the polyoxyethylene nonylphenol type. The purification of P–450 reductase from rat or rabbit liver microsomes to homogeneity after solubilization by Renex 690 or Triton N-101 needs its separation from other microsomal proteins by ion exchange chromatography on DEAE-cellulose or DEAE-Sephadex and further chromatographic steps in the presence of detergents (VERMILION and COON, 1974; DIGNAM and STROBEL, 1975, 1977; KNAPP et al., 1977). Due to the application of the biospecific affinity chromatography by YASUKOCHI and MASTERS (1976) it was possible to isolate the highly purified flavoprotein in high yield (30 to 50% with respect to its content in microsomes). This procedure is based on the binding of a partially purified enzyme to 2′,5′-ADP-Sepharose 4B (Sepharose 4B-bound N^6(aminohexyl)-adenosine 2′,5′-bisphophate). After removal of accompanying proteins by washing the Sepharose beads with detergent-containing buffer the P–450 reductase is eluted by 2′-AMP which is a cosubstrate analogue. The 2′-AMP can be removed in various ways such as gel-filtration, binding to hydroxyapatite, treatment with charcoal, etc. The NADPH-cytochrome P–450 reductase preparations obtained from rat or rabbit liver microsomes by this procedure exhibited highly specific activities towards cytochrome c as artificial electron acceptor (up to 65 units per mg of protein) (YASUKOCHI and MASTERS; 1976, VERMILION and COON, 1978a; MIWA et al., 1978; FRENCH and COON, 1979). One unit of activity is defined as the amount of enzyme which catalyzes the reduction of 1 μmol of cytochrome c per min in 0.3 M potassium phosphate buffer, pH 7.7, at 30°C. Lowering the ionic strength results in a decrease of the specific activity as shown by YASUKOCHI and MASTERS (1976) thus confirming the marked influence of ionic strength on the enzymatic activity observed earlier (PHILLIPS and LANGDON, 1962). The rabbit P–450 reductase is very sensitive to a proteolytic attack and therefore needs a rapid and careful operation at the early purification steps in order to attain highly specific activities.

By use of the affinity chromatography (2′,5′-ADP-Sepharose) the P–450 reductase from human liver microsomes has also been obtained in a partially purified or in a nearly homogeneous state as shown by SDS PAGE (KAMATAKI et al., 1979; WANG et al., 1980) with specific activities of about 52 units per mg protein (BÖSTERLING and TRUDELL, 1980).

A similar procedure was developed by STROBEL and collaborators (DIGNAM and STROBEL, 1977). They used a NADP-Sepharose column as specific adsorbent for the preparation of the reductase from rat liver microsomes.

The affinity chromatography on 2′,5′-ADP-Sepharose or NADP-agarose has proved successfully for the purification of the NADPH-cytochrome P–450 reductase from various organs such as lung (GUENGERICH, 1977a), *Novikoff hepatoma* (SAINE et al., 1978) and from further sources, e.g. *Musca domestica* L. (MAYER and DURRANT ,1979) and *Catharanthus roseus* seedlings (MADYASTHA and COSCIA, 1979).

3.2. Physicochemical Properties of the Flavoprotein

Highly purified NADPH-cytochrome P–450 reductase preparations in high yield enabled extensive studies of the properties and of the mechanism of the electron transfer to its native acceptor P–450. Since the report of IYANAGI and MASON (1973), it has been generally acknowledged that the hepatic P–450 reductase irrespective of the kind of solubilization from the microsomal membrane contains 1 molecule each of FMN and FAD per polypeptide chain (MASTERS et al., 1975; VERMILION and COON, 1974; DIGNAM and STROBEL, 1975, 1977; YASUKOCHI and MASTERS, 1976; IYANAGI et al., 1978; FRENCH and COON, 1979) instead of 2 molecules FAD originally reported. Both flavins are not covalently bound to the protein and therefore can be split from the enzyme under certain conditions, i.e. low pH and high ionic strength, especially FMN (MASTERS et al., 1965; TROUT, 1976). Thus it becomes plausible that in earlier preparations only the FAD was detected. It has to be assumed that the dissociation of FMN occurred during the pH precipitation or in the presence of ammonium sulfate (cf. FRENCH and COON, 1979).

The absolute spectrum of the purified P–450 reductase in the fully oxidized form exhibits absorbance maxima at 455 and 380 nm (Figure 2) which are characteristic of a flavoprotein. In the course of reduction, the oxidized enzyme proceeds through partially reduced forms finally giving a 4-electron-reduced species. The various forms have distinct spectral properties and quite different CD spectra providing evidence that the FMN and FAD are in different environments in the enzyme (VERMILION and COON, 1978b; FRENCH and COON, 1979).

The ratios of the absorbances at 277 and 455 nm for the detergent-solubilized P–450 reductase from rat and rabbit liver

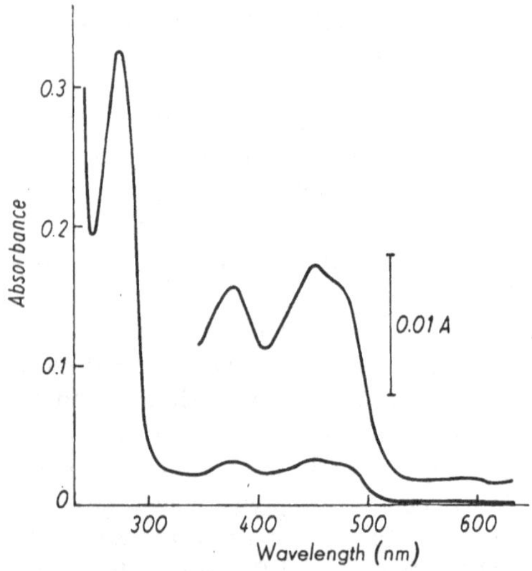

Fig. 2. Absorption spectrum of detergent-solubilized NADPH-cytochrome P-450 reductase (oxidized state) from rat liver microsomes.

are 8.7 and 7.4, respectively (VERMILION, and COON, 1978a; FRENCH and COON, 1979). The value of 6.3 observed for the trypsin-solubilized reductase from rat liver (OMURA and TAKESUE, 1970) originates from the lower content of aromatic amino acids (see Table 6).

3.3. Amino Acid Composition of NADPH-Cytochrome P-450 Reductase

Amino acid compositions of highly purified P-450 reductase preparations form rat and rabbit liver microsomes are listed in Table 6. The amino acid analysis of rat P-450 reductase shows a content of 31% hydrophobic amino acid residues, 6 cysteine residues and at least 6 tryptophan residues which is consistent with the UV and CD spectra (GUM and STROBEL, 1979; FRENCH and COON, 1979). By the latter authors an α-helical content of 34% and 29% has been calculated for the rat and rabbit P-450 reductase, respectively (FRENCH and COON, 1979).

Amino acid analyses of rabbit P-450 reductases prepared by FRENCH and COON (1979) and BLACK et al. (1979) showed compositions similar to that of the rat P-450 reductase. Both rabbit P-450 reductases contain 31% hydrophobic residues but 9 to 10 cysteine residues in comparison to 7 residues reported for rat P-450 reductase. The content of nearly twice as many glutamic and aspartic acid residues as basic ones results in isoelectric points in the pH range from 5.1 to 5.6 (GUENGERICH, 1977a; SERABJIT-SINGH et al., 1979).

In general, the discrepancies observed in amino acid analyses of P-450 reductase preparations from the same species induced by phenobarbital are difficult to understand. An explanation on the basis of different molecular weights used for the calculation of the number of amino acid residues per polypeptide chain cannot be given. Only the differences in certain amino acids such as tryptophan, threonine and serine may be attributed to methodological problems in the determination of these residues. Further studies will be required to have a more solid basis in determining subtle differences between NADPH-cytochrome P-450 reductases from various species.

Using the proteolytic peptide mapping for a comparative analysis of electrophoretically homogeneous P-450 reductase preparations from rat and rabbit liver as well as from rabbit lung GUENGERICH (1979a) has shown qualitatively that each P-450 reductase isolated from a different source gave a similar pattern of fragments. This finding is suggestive of only a little, if any differences in the primary structure existing between P-450 reductases from different origin.

The minimal molecular weight of detergent-solubilized NADPH-cytochrome P-450 reductase from rat liver has been estimated by calibrated SDS PAGE using Ferguson plots to be 79,000 (KNAPP et al., 1977). Values in the range from 76,000 to 80,000 have been obtained by various laboratories (YASUKOCHI and

Table 6. Amino acid composition of NADPH-cytochrome P-450 reductases and tryptic peptides

	Rat				Rabbit				
	Detergent		Trypsin	Δ(det.-trypsin)	Detergent		Trypsin	Δ(det.-trypsin) detm.	calc.
	(a)	(b)	(b)	(b)	(c)	(d)	(d)		(d)
Lys	36	33	32	1	30	36	30	1	6
His	21	19	20	0	15	21	18	1	3
Arg	35	37	33	4	34	39	37	2	2
Asx	62	71	58	13	64	68	64	4	4
Thr	36	38	29	9	33	37	29	3	8
Ser	45	43	36	7	41	37	35	4	2
Glx	82	87	75	12	82	84	80	9	4
Pro	31	32	32	0	29	31	29	4	2
Gly	48	44	43	1	44	48	47	4	1
Ala	50	50	46	4	56	62	58	5	4
Cys	7	6	7	0	10	9	8	0	1
Val	44	45	42	3	44	48	45	5	3
Met	16	19	15	4	14	16	15	1	1
Ile	27	24	20	4	21	24	21	3	3
Leu	62	58	52	6	57	66	59	6	7
Tyr	32	33	28	5	28	32	31	1	1
Phe	29	26	24	2	25	26	21	3	5
Trp	10	6	5	1	6	7	7	0	0
Total	673	671	597	76	633	691	634	56	57
Mol. wt. polypeptide	—	76,509	67,952	8,918	—	77,700	71,200	6,100	6,500
Mol.wt. SDS PAGE	76,000	—	—	—	74,000	—	—	—	—
Acidic amino acids (%)	21.4	23.6	22.3	32.9	23.1	22.0	22.7	23.2	—
Basic amino acids (%)	13.6	13.3	14.2	6.6	12.5	13.9	13.4	7.1	—
Hydrophobic amino acids (%)	32.7	31.4	31.2	30.3	30.8	31.7	31.4	33.9	—

a) VERMILION and COON (1978a)
b) GUM and STROBEL (1979)
c) FRENCH and COON (1979)
d) BLACK et al. (1979)

MASTERS, 1976; VERMILION and COON, 1978a; GUM and STROBEL, 1979). Sedimentation equilibrium ultracentrifugation of iodoacetamide-treated P-450 reductase in the presence of guanidine hydrochloride resulted in a minimum molecular weight of 76,500 being in good agreement with the values obtained by SDS PAGE. The minimum molecular weights for the rabbit P-450 reductase estimated by various groups (IYANAGI et al., 1978; FRENCH and COON, 1979; BLACK et al., 1979) were reported to be in the range from 74,000–78,000. These molecular weights slightly less than the values of the rat P-450 reductase might emerge from varying species specific proteolytic sensitivities (cf. MASTERS and OKITA, 1980).

For the human liver P-450 reductase minimal molecular weights of 70,000 and 67,000 have been obtained by WANG et al. (1980) and by BÖSTERLING and TRUDELL (1980), respectively, using SDS PAGE. These values are slightly less than those determined for the rat liver enzyme under identical conditions in both cases. The ability of the reductase to reduce P-450 anaerobically and to support P-450 associated hydroxylations demonstrates that the active site of human liver P–450 reductase is not significantly changed by proteolytic degradation.

As known for P–450 and other membrane proteins, also detergent-solubilized P–450 reductase from rats and rabbits forms associates in the absence of denaturating compounds or detergents with molecular weights of about 500,000 (Knapp et al., 1977; Vermilion and Coon, 1978a; French et al., 1980a). In contrast to these findings, the proteolytically solubilized reductase preparations with minimal molecular weights between 68,000 to 71,000 have lost the ability to associate (Pederson et al., 1973; Iyanagi and Mason, 1973). These results indicate NADPH-cytochrome P–450 reductase to be an amphipathic protein consisting of a hydrophilic domain protruding the membrane surface and of a hydrophobic anchor holding the molcule in the membrane.

Recent investigations on that part of the reductase remaining in the microsomal membrane after tryptic solubilization of the enzyme indicate the function of this fragment. It is necessary for the incorporation of the NADPH-cytochrome P–450 reductase into the microsomal membrane or phospholipid vesicles as evidenced by increased monooxygenase activities (Prokrovsky et al., 1977; Miwa et al., 1978; Yang et al., 1978; Kitada et al., 1979; Gum and Strobel, 1979; French et al., 1980). This part of detergent-solubilized P–450 reductase is a prerequisite for reconstituing an enzymatically active system first demonstrated by Lu et al., (1969a), Dignam and Strobel (1975), and Vermilion and Coon (1974). The tryptic-solubilized reductase, however, retains only the ability to reduce artificial electron acceptors (Ichihara et al., 1972; Pederson and Aust, 1972; Pederson et al., 1973).

From the differences in the amino acid composition of the detergent- and tryptic-solubilized reductases from rat and rabbit the amino acid compositions of the respective membrane binding domains have been calculated (Gum and Strobel, 1979 and Black et al., 1979, respectively). As shown in Table 6, the small tryptic peptides differ in their size (8,900 and 6,500 for rat and rabbit, respectively). In comparison with the native rabbit enzyme the relative hydrophobicity of the small fragment is slightly higher as shown by amino acid analysis and its unusual solubility in organic solvents (Black et al., 1979). Due to an increased content of acidic amino acid residues in this hydrophobic fragment of the reductase it was concluded that only a portion of this membrane binding domain seems to be intercalated in the lipophilic membrane of the endoplasmic reticulum. The hydrophobic domain of the P–450 reductase is localized in the N-terminal region of the enzyme as proved by the identity of C-terminal sequences of the native and tryptic-solubilized reductase (Black et al., 1979). Direct evidence for the involvement of the small peptide in the formation of a functionally active complex between P–450 and P–450 reductase in reconstituted systems has been gained from its inhibition of the NADPH oxidation as well as from the formation of a type I difference spectrum with P–450 LM2 (Black et al., 1979; French et al., 1980a).

All these data support the amphiphilic nature of the NADPH-cytochrome P–450 reductase. One domain contains the binding site for the cosubstrate transferring the electrons to the flavins, whereas the other domain is responsible to bind the P–450 reductase to the microsomal membrane in an arrangement appropriate to interact with P–450.

Contrary to P–450 the NADPH-cytochrome P–450 reductase does not exist in multiple forms (Welton and Aust, 1975). Basing on identical molecular weights and on immunochemical indifference after application of different inducers (PB, 3-MC, and TCDD) no different forms of the P–450 reductase have been detected (Prough and Burke, 1975; French and Coon, 1979). Using specific antibodies against purified P–450 reductases from various tissues and species their similarities as well as their essential role in monooxygenatic reactions have been ascertained (cf. Masters and Okita, 1980).

3.4. *Functional Importance of Selected Amino Acid Residues of NADPH-Cytochrome P–450 Reductase*

The analysis of structure-function relationships of the monooxygenase system on the basis of the spatial structure of the enzyme complex is complicated due to its peculiarities already discussed in 2.3. Therefore, investigations on the topology of functional groups in the individual proteins are very important. Such an analysis is possible by means of site-specific modifcations by group-specific modification reagents or by affinity labelling with physicochemical reporter groups such as fluorescent, spectral-active or spin-label probes. In the case of P–450 reductase the selective site-specific modifications are directed to characterize those amino acid residues which are involved in functionally linked conformational changes and to identify such residues which are involved in interaction and recognition processes.

Earlier studies of Phillips and Langdon (1962), Williams and Kamin (1962) and Masters et al., (1965) showed an inhibitory effect of p-chloromercuribenzoate on the cytochrome c reduction by proteolytically solubilized P–450 reductase. Investigations on the inhibition of this activity in microsomes by mersalyl and its protection by the addition of NADPH supported the assumption that sulfhydryl groups are essential for the binding of NADPH to P–450 reductase (Franklin and Estabrook, 1971). Investigating a proteolytically solubilized P–450 reductase from pig liver microsomes Lazar et al. (1977) showed that the modification of three accessible sulfhydryl groups by 5,5′-dithiobis(2-nitrobenzoic acid) (DTNB) caused a complete loss of the cytochrome c reduction by NADPH. However, performing the modification in the presence of $NADP^+$ or 2′-AMP only two SH-groups were changed without loss of activity. The authors obtained furthermore kinetic evidence for a competitive reaction between the inhibitor 5,5′-dithiobis(2-nitrobenzoic acid) and the substrate analogues for the same site of the enzyme suggesting that there is an essential cysteine residue at or near the NADPH binding site of the P–450 reductase. The kinetic analyses of reductase samples modified at the sulfhydryl groups by canide, tetrathionate, or N-ethylmaleimide (NEM) showed that the K_m values of NADPH and the V_{max} of the cytochrome c reduction are not drastically changed despite the modification of three groups per reductase molecule (Lumper et al., 1980). The treatment of a DTNB- modified, inactive reductase by cyanide led to the cyanylation of three SH-groups in the enzyme. This complete exchange of the modifying agent is accompanied by an almost complete restoration of the enzymatic activity with a slightly decreased affinity of the S-cyano-modified enzyme for NADPH. Obviously the binding of NADPH to its respective binding site is affected by steric hindrance or electrostatic interactions of the thiol-blocking compound. After a first modification of the enzyme by NEM in the presence of NADPH as protecting agent a second labelling is possible by NEM resulting in the modification of the essential SH-group in the NADPH binding site as shown by peptide mapping and inhibited retention on 2′,5′-ADP-Sepharose. This 'two step modification technique' results in a more than 20-fold decreased affinity of the reductase towards NADPH (LUMPER et al., 1980).

A functional role of arginine residues in NADPH binding sites of some oxidoreductases has been described by MATTHEWS et al. (1979). Due to differences in the protection of arginine residues of the reductase against modification by 2,3-butanedione or 1,2-cyclohexanedione a definite conclusion about the importance of the guanidino group for the cosubstrate binding site has not been drawn (LUMPER et al., 1980). Recent data on the modification of detergent-solubilized P–450 reductase by these reagents suggest that an arginine residue is present at the active site of the P–450 reductase which may function in the binding of NADPH via the 2'-phosphate moiety (EBEL, 1980).

The involvement of tyrosine residues in the NADPH binding site is excluded, whereas more information is needed to establish the functional role of histidine and lysine residues and of carboxyl groups in this site (LUMPER et al., 1980).

Since kinetic parameters for the binding of NADPH to the tryptic P–450 reductase are the same as for the detergent-solubilized P–450 reductase (see Table 7) the results mentioned above on the NADPH binding site of the tryptic reductase can be transferred to the native P–450 reductase.

Studies on the fluorescamine inhibition of N-demethylase reactions have revealed that the binding of this label to lysine residues of the P–450 reductase does not influence the activity indicating that lysine residues of the reductase are not involved in interactions with P–450 (WEST et al., 1980).

3.5. *Redox Properties and Mechanism of the NADPH-Cytochrome P–450 Reductase*

The monooxygenase systems from liver mitochondria (PEDERSON et al., 1977), adrenal mitochondria (cf. MITANI, 1979) and from *Pseudomonas putida* (KATAGIRI et al., 1968) contain besides cytochrome P–450 and NADPH-dependent flavoprotein a non-heme iron sulfur protein as third essential component. This ferredoxin-like protein functions in resolving the NADPH-dependent 2-electron transfer step into two sequential 1-electron transfer steps since the heme iron can accept only one electron (PETERSON, 1971; PETERSON et al., 1977, COOPER et al., 1977). On the other hand the electron transport from NADPH to P–450 in liver microsomes proceeds via the FMN-FAD-containing P–450 reductase which is able to substitute for non-heme iron protein in contrast to the other mono-FAD containing reductases of the systems mentioned above.

The NADPH-cytochrome P–450 reductase catalyzes reactions both with 1-electron acceptors such as P–450, cytochrome c, and ferricyanide, as well as with 2-electron acceptors like menadione and dichlorophenolindophenol (KAMIN et al., 1965a; STROBEL et al., 1978; VERMILION and COON, 1978a, b). The apparent turnover numbers for these artificial electron acceptors are at least one order of magnitude higher (about 4,000 min^{-1} for cytochrome *c*) than for P–450 LM2 indicating intermolecular electron transfer processes including conformational changes as rate-limiting steps. Kinetic parameters of the P–450 reductase solubilized proteolytically or by detergents are summarized in Table 7. Irrespective of the kind of solubilization, the K_m values for the electron donor NADPH and the acceptor cytochrome *c* are in the 10^{-6} M range thus excluding differences in the reduction of artificial electron acceptors between the detergent- or protease-solubilized enzyme. The K_m value for the native acceptor P–450 LM2 has been determined to be smaller than 10^{-7} M. This value is nearly two orders of magnitude lower than the values for artificial acceptors indicating the formation of a tight complex between both proteins.

Table 7. Kinetic constants for different NADPH-cytochrome P–450 reductase preparations

Enzyme source	Solubilization agent	K_m (μM)			Reference
		NADPH	cytochrome c	cytochrome P–450	
Pig liver	Steapsin	3.8	5.4	—	WILLIAMS and KAMIN (1962)
Pig liver	Trypsin	3.4	2.6	—	PHILLIPS and LANGDON (1962)
Pig liver	Bromelain	7.8 ± 0.3	n.d.	—	LUMPER et al. (1980)
Rat liver	Bromelain	6.3 ± 1.0	13.1 ± 1.1	—	CHRANKSHAW et al. (1979)
Rat liver	Renex 690	3.0	8.0	0.05	VERMILION and COON (1978a)
Rat liver	Renex 690	5.0	8.0 ± 2.0	0.09 ± 0.01	DIGNAM and STROBEL (1977)

The kinetic analyses of detergent-solubilized rat liver P–450 preparations indicated a sequential mechanism of the interaction of P–450 with NADPH and reductase. This is supported by data on the interaction with cytochrome c as artificial acceptor as well as by the competition experiments with $NADP^+$ (DIGNAM and STROBEL, 1977; STROBEL et al., 1978). These data are inconsistent with a ping-pong-mechanism for the interaction of partially purified tryptic reductase with cytochrome c and NADPH as proposed by MASTERS et al., 1965).

In addition to the kinetic studies the elucidation of the role of both different flavins is of decisive importance for an understanding of the mechanism of action of the P–450 reductase at the molecular level. The first data on the occurrence of different redox states of the flavins within the P–450 reductase have been obtained with an enzyme preparation which has been solubilized proteolytically (PHILLIPS and LANGDON, 1962; KAMIN et al., 1965b; MASTERS et al., 1965; OMURA and TAKESUE, 1970; IYANAGI and MASON, 1973; IYANAGI et al., 1974). The experiments of IYANAGI and MASON (1973) revealed the existence of partially reduced forms in the proteolytically solubilized enzyme and of an O_2-stable semiquinone which contained one reducing equivalent per two flavins more than the fully oxidized form. This result has recently been confirmed with homogeneous, detergent-solubilized P–450 reductase (VERMILION and COON, 1978a, 1978b; IYANAGI et al., 1978; MASTERS et al., 1978; YASUKOCHI et al., 1979). Contrary to the earlier interpretation according to which a cycling of the P–450 reductase flavins between fully reduced and half-reduced forms has been assumed (KAMIN et al., 1965a; MASTERS et al., 1975, YASUKOCHI and MASTERS, 1976; DIGNAM and STROBEL, 1977) it is now generally acknowledged that the reduction process involves a cycling of the P–450 reductase between the 1-electron and 3-electron reduced state (VERMILION and COON, 1978a; IYANAGI et al., 1978; GUM and STROBEL, 1979; FRENCH and COON, 1979; YASUKOCHI et al., 1979; VERMILION et al., 1981).

From the data available it can be concluded that the treatment with trypsin does not influence neither the binding site for NADPH nor the redox equilibria of both flavins forming semiquinones. Based on optical and EPR spectra and on potentiometric titrations IYANAGI et al. (1974) postulated the formation of two semiquinones which differ in their spectra and redox properties during the reaction cycle of P–450 reductase. To explain the observed three reduction steps formal potentials which fit well the following sequence of four 1-electron reduction steps have been computed to be $E'_{0,1} = -110$ mV, $E'_{0,2} = -270$ mV, $E'_{0,3} = -290$ mV, and $E'_{0,4} = -365$ mV at pH 7.0 and 25 °C (IYANAGI et al., 1974).

$$F_1 \underset{}{\overset{E'_{1,0}}{\rightleftharpoons}} F_1H\cdot \underset{}{\overset{E'_{0,2}}{\rightleftharpoons}} F_1H_2$$

$$F_2 \underset{}{\overset{E'_{0,3}}{\rightleftharpoons}} F_2H\cdot \underset{}{\overset{E'_{0,4}}{\rightleftharpoons}} F_2H_2$$

The flavins F_1 and F_2 are assigned in the order in which they are reduced by dithionite but without any regard to their identity.

According to a proposal by SIEGEL et al. (1972) for the NADPH-sulfite reductase which contains both FMN and FAD it has been suggested that the FAD cycles between the oxidized and the fully reduced forms via a semiquinone, whereas the FMN cycles between the semiquinone and the fully reduced form. Such a mechanism in which an intramolecular electron transfer between both flavins is involved has been confirmed by detailed investigations of the NADPH-dependent reduction and reoxidation of P–450 by detergent- solubilized P–450 reductase from rabbit and rat liver (IYANAGI et al., 1978; VERMILION and COON, 1978a). Each flavin may have an individual function that is, one flavin accepts two reducing equivalents from NADPH (dehydrogenase flavin) and the other one acts as an 1-electron carrier from the reductase to P–450 (flavodoxin-like flavin).

To identify the acceptor flavin the catalytic properties of FMN-depleted detergent-solubilized P–450 reductases have been studied in several laboratories (VERMILION and COON, 1978b; ALEXANDER et al., 1980; IYANAGI and MAKINO, 1980). These FMN-depleted P–450 reductases showed remarkably lowered activity towards P–450 or the artificial acceptors cytochrome c and dichlorophenolindophenol. Only slight decreases of the ferricyanide reductase activity have been observed. The initial fast reduction of the modified enzyme by NADPH remains unaltered. The lost activity of the modified reductase can be restored by the addition of FMN. These results emphasize not only the essential role of FMN but also its position in the electron transfer pathway. However, evidence for a direct interaction of FMN with the acceptor protein (compound) cannot be derived from these studies meaning that the FMN might only be necessary for an intramolecular disproportionation of a 2-electron reduced form of the P–450 reductase. Based on the results of spectrophotometric titration of the native and the FMN-depleted reductase COON and coworkers assigned the low and high midpoint potentials of the P–450 reductase to FAD and FMN, respectively (VERMILION and COON, 1978b; VERMILION et al., 1980). MASTERS and OKITA (1980) on the other hand doubt on this interpretation; they discuss the results as rather indicating the redox properties of a two-flavin system instead of individual flavin moieties.

All the data up to now presented for the reduction kinetics of P–450 by NADPH are in close agreement with the mechanism postulated by IYANAGI et al. (1974). Therefore two electrons per molecule reductase are necessary to form the O_2-stable semiquinone under aerobic conditions releasing a superoxide anion in the presence of oxygen (YASUKOCHI et al., 1979). The O_2-stable semiquinone generated in this priming reaction functions as the oxidized part in the enzymatic catalysis of the P–450 reductase. The further intermediate steps of the intramolecular electron transfer remain to be established. Furthermore, the molecular mechanism of the reduction of P–450 by the reductase is not clear yet. The elucidation of this partial reaction is further complicated due to a molar stoichiometry of about 20 : 1 between P–450 and P–450 reductase within the microsomes and due to the possible participation of cytochrome b_5 in this reaction.

4. Functional Properties of the Hepatic Monooxygenase System

4.1. *Role of Phospholipids*

In the endoplasmic reticulum the P–450 and P–450 reductase are embedded in a bilayer of membrane lipids with protein regions sticking out of the lipid. Both delipidated proteins are incapable of performing hydroxylation reactions, but they can be reconstituted to an enzymatically active monooxygenase in the presence of phospholipids (STROBEL et al., 1970; VAN DER HOEVEN and COON, 1974; JÄNIG et al., 1979) thus characterizing phospholipids as third essential component of the system.

In the last decade, several investigators contributed to the role that phospholipids might play in the structure and function of the hydroxylating system since the finding of STROBEL et al. (1970) that phospholipid (dioleoylphosphatidylcholine) is an essential constituent of the monooxygenase. Usually the autoxidable dioleoylphosphatidylcholine is substituted for reconstitution experiments by other lecithins with different fatty acids, preferentially lauric acid due to its optimal stimulation of the activity and its phase transition temperature ($T_c = -1.8\,°C$) far below the physiological temperature (MABREY and STURTEVANT, 1978). The requirement for phospholipids was not specific which was proved by substituting them by other amphipathic compounds such as nonionic detergents (LU et al., 1974a; INGELMAN-SUNDBERG, 1977; IMAI et al., 1977; BLANCK et al., 1979; TANIGUCHI et al., 1979).

In contrast to earlier experiments with partially purified P–450 (LU et al., 1970) it was repeatedly demonstrated that the binding of benzphetamine to P–450 LM2 increased about 5-fold in the presence of dilauroylphosphatidylcholine (COON et al., 1976; DAWSON et al., 1978; FRENCH et al., 1980b). Binding studies with purified P–450 proved that phospholipid is not necessary for the substrate binding (RISTAU et al., 1978) but it may change the binding affinity of P–450. The binding of substrates is paralleled by a shift of the spin increase in equilibrium to the high spin state. A shift to the high spin state can also be produced by phospholipids forming type I difference spectra (TSONG and YANG, 1978; CHIANG and COON, 1979; FRENCH et al., 1980b; RUCKPAUL et al., 1980a). At least 20 lipid molecules are bound to 1 molecule of P–450 LM2 in a non-competitive, saturable manner inducing conformational changes which are indicated by changes of the CD-spectra in the far ultraviolet region from which an increase of the α-helical content by about 10% has been evaluated (CHIANG and COON, 1979; SHIMIZU et al., 1979).

Spectrophotometric titrations as well as steady state kinetic investigations have shown that the interaction of P–450 reductase with P–450 LM2 shifts the spin equilibrium of P–450 LM2 to the high spin state in the same manner as produced by type I substrates (FRENCH et al., 1980b). In the presence of phospholipids and substrate the affinity of P–450 reductase to P–450 is further increased characterized by a change of the K_{diss} value from $1 \cdot 10^{-7}$ to $5 \cdot 10^{-8}$ M indicating mutual interactions (FRENCH et al., 1980b; LU et al., 1980a). The lack of changes in K_m observed with different substrates as well as the unchanged kinetic isotopic effect (MIWA et al., 1980) suggests that the phospholipid does not affect any kinetic step involving the substrate.

RUCKPAUL et al. (1980a) started the elucidation of this complex interaction at the molecular level by means of second-derivative spectroscopy. They could determine that tyrosine residues of P-450 LM2 are involved in the interaction of the essential components. The participation of phenylalanine residues is probable, whereas a participation of tryptophan could be excluded.

Usually, the reconstituted systems contain a 1 : 1 ratio of P-450 reductase and P-450 forming a binary complex of both proteins to obtain optimal activity in catalyzing certain hydroxylation reactions. These stoichiometric conditions used, however, differ from those within the membranes of the endoplasmic reticulum as well as the aggregational state of both proteins, and the chemical composition and the structural state of the lipid. For a simulation of microsomal conditions, therefore, reconstituted systems have been prepared in which the proteins are located in the plane of a phospholipid bilayer vesicle and may be able to undergo lateral motion and transient or permanent cluster formation. In dependence on the techniques and the chemical composition of the lipid mono- and multilamellar liposomes have been prepared for studying the interaction mechanism of the two proteins by different methods (INGELMAN-SUNDBERG and GLAUMANN, 1977; TSONG and YANG, 1978; TANIGUCHI et al., 1979; BÖSTERLING et al., 1979). It has been shown that the overall activity, the rate of the reduction of P-450 by the P-450 reductase as well as the relaxation rate of conformational changes of P-450 are greatly enhanced by lecithins with definite fatty acid residues or a mixture of phospholipids but only at a temperature above the gel to liquid crystalline phase transition (TANIGUCHI et al., 1980; TSONG and YANG, 1978; YANG and TSONG, 1980). The physical state of the lipid can even modulate the reduction rate of the P-450 reductase incorporated into liposomes with the soluble electron acceptor cytochrome c (RUF et al., 1980). The fraction of rapid phase reduction is diminished after lowering the temperature below the phase transition temperature which depends on the fatty acid residues in the phospholipid vesicles used. The breaks observed are at 23 °C for dimyristoyllecithin vesicles and between 0—5 °C for dilauroyllecithin. Besides the dependence of the protein-lipid interaction on the fatty acid chain of the phospholipid there exists a profound influence of different head groups on the specificity of this interaction (YANG and TSONG, 1980; HAAPARANTA et al., 1980; RUCKPAUL et al., 1980a). BÖSTERLING et al. (1979) therefore used a lipid mixture of phosphatidylcholine : phosphatidylethanolamine : phosphatidic acid (2 : 1 : 0.06; w/w) for the preparation of liposomes which are adapted to microsomes with respect to the bilayer structure, surface charge, and the protein lipid ratio of 1 : 5 (w/w). The negative surface charge introduced by the incorporation of phosphatidic acid seems to be responsible for preventing an aggregation of the vesicles. By this they were able to obtain liposomes with P-450 and P-450 reductase in a molar ratio of 5 : 1 reflecting properties of the enzyme system within the microsomal membrane.

Summarizing the results of the investigations on phospholipid interactions in the monooxygenase system it can be stated that (i) phospholipids are interacting in part specifically with P-450. They produce a high spin shift of the spin equilibrium and an increase in the substrate affinity indicating conformational changes within the P-450 molecule or partitioning is charged. (ii) Phospholipids are favouring the formation of the functionally active complex. (iii) Vesicular structures of phospholipids lead to more intensive interactions between the proteins resulting in higher turnover numbers for some substrates (cf. 4.2.).

4.2. Structure-Function Relationships of Different Cytochrome P–450 Forms

4.2.1. Substrate-Cytochrome P–450 Interaction

The occurrence of multiple forms of P–450 in liver microsomes led to the assumption that the individual forms have different, but overlapping substrate specificities. In order to understand the role of the different P–450 forms in metabolic processes it is necessary to clarify their inducibility towards various xenobiotics and their respective substrate specificity with regard to the binding affinity as well to its hydroxylation rate including an analysis of the dissociation and binding properties of the hydroxylation products. This problem is of special importance because oxidative metabolism of various exogeneous compounds leads to the formation of carcinogenic, mutagenic, or toxic metabolites. The molecular basis of structure-function relationships with respect to substrate specificity of different P–450 is determined by properties of the terminal oxidase especially by the stereochemical and physicochemical characteristics of its binding site, its accessibility and immediate surrounding provided by the bilayer membrane. On the other hand substrate-specific effects, i.e. electronic and steric features, have to be considered with respect to the great number of substrates with different structures capable for being metabolized.

Many attempts have been made to find common properties for the heterogeneous group of P–450 substrates. Their interaction with P-450 resulting in the formation of enzyme-substrate complexes is mostly observed by difference spectrophotometry showing the socalled type I, type II or reversed type I spectra (REMMER et al., 1966; SCHENKMAN et al., 1967; KUMAKI and NEBERT, 1978).

Systematic investigations on the binding of various homologous series of substrates and further selected compounds to P–450 of different origin revealed the hydrophobicity, the basicity and the stereochemistry of a compound are determing its spectral type (SCHENKMAN et al., 1967; IMAI and SATO, 1967; MAILMAN et al., 1974; MISSELWITZ et al., 1975). It is generally accepted that type I difference spectra emerge from hydrophobic interactions of a substrate with the active site of one or more forms of P–450. Altering the hydrophobic character by introducing polar substituents a type II binding to P–450 occurs instead of type I binding. SCHENKMAN et al. (1967) suggested type II binding spectra represent a ferrihemochrome formed between compounds containing a primary amino groups and the oxidized P–450 iron. However, as the steric access to the unbound electron pair of nitrogen is impaired by alkylation the type II binding is hindered (MAILMAN et al., 1974; ARCHAKOV et al., 1974; MISSELWITZ et al., 1975).

Early investigations indicated correlations between the lipid solubility and metabolism by the P–450 system for various groups of compounds (GAUDETTE and BRODIE, 1959; MCMAHON, 1961; MARTIN and HANSCH, 1971). It has been shown that type I and type II substrates with increasing lipid solubility (enhanced partition coefficient in the organic phase) are bound more tightly to microsomal and partially purified P–450 from rat and rabbit and to P–450 LM2 as well (PELKONEN and VAINIO, 1975; MISSELWITZ et al., 1975, 1976, 1977, 1980; SIPAL et al., 1979) indicating the lipophilicity of a molecule to be decisive for determining the affinity to P–450. This conclusion is supported by studies on the binding of various substituted aryl hydroperoxides used as artificial oxygen donors to P–450, too (BLAKE and COON, 1980).

Concerning the chemical nature of the binding sites on P–450 the temperature dependence of the K_s values for both classes of substrates have been determined (MISSELWITZ et al., 1976). The binding of type I substrates to P–450 from PB-treated rats is endothermic whereas an exothermic reaction was observed for the type II compounds. Qualitatively similar results have been obtained for rabbit P–450 independent of the integrational state (MISSELWITZ et al., 1977, 1980). The thermodynamic parameters for the substrate binding to P–450 show higher enthalpy and entropy values for rabbits than for rats. The better binding of type I substrates to rabbit P–450 compared to rat P–450 is indicative of a more hydrophobic binding in rabbit P–450 supported also by the high entropy values of the binding reaction of type I substrates. Moreover, all these data provided further evidence pointing to a difference between the binding sites of both classes of substrates.

Substantial differences in the substrate binding sites and the active centres of P–450 and P–448 from rat and rabbit have been established by proton magnetic relaxation studies on partially purified samples (MARICIC et al., 1979). The following series of increased accessibility of the heme iron emerges from these studies: rabbit P–448 $\ll$ rat P–448 $<$ rat P–450 $\ll$ rabbit P–450 LM2 (cf. chapter 5).

Likewise as in the case of non-membrane bound enzymes numerous steady-state kinetic data have been accumulated to determine the rate-limiting step(s) in the sequence of partial reactions of different hydroxylations catalyzed by microsomal P–450. Comparison of the Michaelis constants (K_m) with the spectral dissociation constants (K_s) of the corresponding substrates revealed only good correlations for some type I substrates such as aminopyrine and hexobarbital, but large differences were found which are species- and sex-dependent (SCHENKMAN et al., 1967; KATO et al., 1970; NOORDHOEK et al., 1977). Due to the complex character of K_m it depends on the rate constants of the partial reactions for the individual substrate whether K_m and K_s are in accordance with each other or deviate more or less.

$$E + S \underset{k_{-1}}{\overset{k_{+1}}{\rightleftharpoons}} ES \xrightarrow{k_{+2}} E + P$$

$$K_m = \frac{k_{-1} + k_{+2}}{k_{+1}} = \frac{k_{-1}}{k_{+1}} + \frac{k_{+2}}{k_{+1}}$$

$$= K_s + \frac{k_{+2}}{k_{+1}}$$

Similar K_s and K_m values observed for a few type I substrates point to another rate-limiting step than the formation of the enzyme-substrate complex because this reaction is a fast one (BLANCK et al., 1976; cf. chapter 3). Comparing aniline derivatives with respect to their K_s, K_m and V_{max} values ARCHAKOV et al. (1973, 1974) found the following relation $K_s \ll K_m$ and a linear proportionality between K_m and V_{max} in hydroxylation indicating a common rate constant k_{+1} in these reactions. Thus, the K_m value does not characterize the affinity of P–450 for these compounds.

Kinetic analyses of reactions on microsomal P–450 are complicated due to its heterogeneity. The presence of firmly bound endogeneous substrates and different effects of various substrates on individual reaction steps produce further difficulties in drawing general conclusions with respect to the rate-limiting step(s) in microsomal hydroxylations.

To by-pass the complexity of the microsomal system reconstituted systems containing highly purified components have proved to be appropriated in studying the relation between the structure of a substrate and the rate-limiting step in its hydroxylation. Using reconstituted systems with P–450 and the P–450 reductase as a 1 : 1 complex the different P–450 forms can be compared with respect to their substrate specificity and their metabolic profiles (MIWA et al., 1979; LU and MIWA, 1980a). The isolation of an increasing number of electrophoretically homogeneous P–450 forms during the last 6 years has induced to determine the turnover values for a great variety of substrates to understand the specificity of the isozymes. For a description of the metabolic activities of the liver microsomal monooxygenase system and for establishing structure-function relationships mainly data on the activity of constitutive forms and those induced by the two main classes of inducing agents will be presented. Most of the more detailed investigations on structure-function relationships have been performed with P–450 forms isolated from rabbits. Therefore, the following representation will be mainly based on data obtained from this species.

Further aspects which have to be considered at discussing the substrate specificity are the variations in different species as well as the influence of phospholipids and the role of cytochrome b_5 and of peroxides in monooxygenase reactions.

4.2.2. *Catalytic Activity of Cytochrome P–450 Isozymes from Rabbit Liver Microsomes*

The isozymes P–450 LM2 and P–450 LM4 are obtained after induction of male rabbits with PB or 3-MC-like inducers, respectively, whereas P–450 LM3b represents a constitutive form. The range of their catalytic activity toward various substrates are listed in Table 8. The reactions examined include N-demethylation, O-dealkylation, and aromatic ring hydroxylation of drugs. The lower part of the table contains induced enzyme activities known to be associated with the Ah locus (NEBERT and JENSEN, 1979). Large differences in their catalytic activity can be seen: P–450 LM2 $\gg$ P–450 LM4 for benzphetamine, P–450 LM2 $>$ P–450 LM4 for ethylmorphine, but P–450 LM2 is equal to P–450 LM4 in benzo(a) pyrene hydroxylation.

Differences in the apparent Michaelis constants and the maximal velocities have also been determined (Table 9). Biphenyl is hydroxylated in position 4 at comparable rates (V_{max}) by P–450 LM2 and LM4, although dissimilar K_m values were found. P–450 LM4 shows a 60 fold higher rate than P–450 LM2 in the deethylation of 7-ethoxyresorufin (JOHNSON et al., 1979b), whereas in the 7-ethoxycoumarin deethylation P–450 LM2 is 20 times more active than P–450 LM4 (SCHWAB et al., 1980). Similar V_{max} and K_m values for geraniol and nerol have been observed with P–450 LM2. The P–450 LM4 was incapable of supporting monoterpene hydroxylation (LICHT and COSCIA, 1978).

Other substrates investigated in detail were the R- and S-enantiomers of the anticoagulant warfarin. Their metabolism was studied by use of liver microsomes (FASCO et al., 1977; KAMINSKY et al., 1979) and of the purified forms of P–450 to get multiple monohydroxylated products (FASCO et al., 1978). Thus it was possible to compare the substrate specificities, regio- and stereoselectivities of P–450 LM2 and LM4 differing profoundly between both forms. P–450 LM3c showed the same activity as LM2, but P–450 LM1 and LM7 were inactive. The results indicated that P–450 LM4 independently of the inducer (PB, BNF, control) acts only on warfarin in the R-configuration when the coumarin side of the molecule was oriented to the active site of the enzyme. P–450 LM2 also acts on the coumarin side when warfarin is in the S-configuration, but also on the phenyl ring when the molecule is in the R-configuration. The metabolite

Table 8. Substrate hydroxylation in reconstituted enzyme systems containing P–450 LM2, P–450 LM3b, and P–450 LM4

Substrate	Turnover number (min^{-1})			Reference
	P–450 LM2	P–450 LM3b	P–450 LM4	
Aminopyrine	7.7	9.7	10.0	KOOP and COON (1979)
Benzphetamine	40–80	5.9	4.9–6.4	GUENGERICH (1977b), COON and VATSIS (1980), KOOP and COON (1979)
Ethylmorphine	3–15	4.9–16	1.0–4.0	KOOP and COON (1979), HANSSON and WIKVALL (1979b, 1980)
Acetanilide	0	—	6.3	COON and VATSIS (1978)
Bnzo(a)pyrene*	0.05–0.6		0.02–1.5	HAUGEN et al. (1975), RYAN et al. (1975b), GUENGERICH (1977b)
Biphenyl	5.2	—	1.4	LICHT and COSCIA (1978)
Dimethylnitrosamine	6.1	—	1.4	GUENGERICH (1977b)
N,N-Dimethylamino-azobenzene	0.5	—	4.8	GUENGERICH (1977b)
N-Methyl-4-amino-azobenzene	0.7	—	4.4	GUENGERICH (1977b)
7-Ethoxycoumarin	0.9	—	0.8	COON and VATSIS (1978)
p-Nitroanisole	5.7	—	3.4	COON and VATSIS (1978)

* Total products formed

patterns observed indicate that the regio- and stereoselectivities of P–450 forms were essentially maintained when the cytochromes were removed from the microsomal membrane and purified.

The involvement of P–450 LM2 in the formation of reactive metabolites of the insecticide parathion has been ascertained by use of a reconstituted monooxygenase system (KAMATAKI et al.,

Table 9. Kinetic parameters for various substrates in reconstituted enzyme systems containing P–450 LM2 and P–450 LM4

Substrate	app. K_m (μM)		V_{max} (min^{-1})		Reference
	LM2	LM4	LM2	LM4	
7-Ethoxycoumarin	210	19	1.01	0.05	JOHNSON et al. (1979b)
Biphenyl (4-hydroxylation)	31	9.6	4.2	3.4	SCHWAB et al. (1980)
7-Ethoxyresorufin	—	1.8	0.01	0.62	JOHNSON et al. (1979b)
Geraniol	24	—*	0.3	—*	
Nerol	19	—*	0.25	—*	LICHT and COSCIA (1978)

* no conversion

1976b). Similarly, homogeneous rat P–450 is inactivated during parathion metabolism (Halpert et al., 1980).

P–450 LM2 and P–450 LM4 are also involved in the biosynthesis and metabolism of bile acids as shown by Hansson and Wikvall (1979a, 1980). P–450 LM2 has no or very low activity in the 12α-, 25- and 26-hydroxylation of the C_{27}-steroids tested (5β-cholestance-3α, 7α-diol and 5β-cholestane-3α, 7α, 12α-triol). However, P–450 LM4 was able to perform the 12α- and 25-hydroxylation of the C_{27}-steroids with comparable rates as observed with microsomes. Both forms showed very low cholesterol 7α-hydroxylation which is the limiting step in the formation of bile acids. The lack of catalytic activity of P–450 LM2 and the activity of P–450 LM4 should be considered in view of the fact that the content of P–450 LM2 in uninduced microsomes is extremely low and that the 12α- and 25-hydroxylations are not induced by PB-treatment of rabbits (Hansson and Wikvall, 1979b). In connection with data on the testosteron hydroxylation in 6β-, 7α-, and 16α-position by P–450 LM2, LM3b and LM4 (Koop and Coon, 1979) and on androstenedione 6β-hydroxylation by the same forms (Ingelman-Sundberg et al., 1979, 1980) it is obvious that P–450 LM3 forms have a higher activity in the metabolization of endogeneous substrates than P–450 LM2 or LM4.

The highly purified P–450 LM3 can be isolated from untreated rabbits (Johnson, 1980a). Although it is present in the microsomes following the pretreatment of the animals with either PB or TCDD it is not induced by these compounds. P–450 LM3 which corresponds to the P–450 LM3b of Coon's group (Coon et al., 1980) has different activities in the benzo(a)pyrene and other hydroxylation reactions in contrast to P–450 LM2, LM4 and LM6. Thus, it differs from the three other forms inducible. A relative ranking in order of decreasing activity towards some substrates is given (Johnson, 1980a):

aminopyrine	LM3 (100%) > LM2 (30%) > LM6 (5%) ∼ LM4 (5%)
benzphetamine	LM2 (100%) > LM3 (8%) ∼LM6 (8%) >LM4 (5%)
ethoxyresorufin	LM4 (100%) > LM6 (70%) (no conversion by LM2 and LM3)
benzo(a)pyrene	LM6 (100%) > LM2 (3%) > LM3 (1%) ∼LM4 (1%)
biphenyl	LM2 (100%) > LM6 (95%) > LM4 (80%) > LM3 (70%)

Activity values of P–450 LM3b, P–450 LM3c and P–450 III with selected substrates have been reported by Koop and Coon (1979), Coon et al., (1980), and Hansson and Wikvall (1980). In hydroxylating steroids P–450 LM3b had a relatively high activity in the 6β- and 16α-position of testosterone (Koop and Coon, 1979) and the P–450 III showed only a high activity in the 12α-hydroxylation of 5β-cholestane-3α,7α-diol and in the dealkylation of ethylmorphine (Hansson and Wikvall, 1980).

The substrate specificity of P–450 LM2, LM3 and LM4 changed upon incorporation of the proteins into vesicles prepared from microsomal phospholipids and egg yolk phosphatidylcholine in comparison to the non-vesicular dilauroylphosphatidylcholine system as shown for androstenedione, aminopyrine, ethylmorphine, 7-ethoxycoumarin, benzo(a)-pyrene and acetanilide (Ingelman-Sundberg et al., 1979; Ingelman-Sundberg and Johansson, 1980). Explanations for this change could be for some substrates a shift in the NADPH-consumption for the hydroxylation and the NADPH--oxidase reaction as well as facilitated binding of the hydrophobic substrates to P–450 (see 4.1.).

By means of chromatography on immobilized cytochrome b_5 MIKI et al. (1980a, b) have isolated a P–450 form with high affinity for cytochrome b_5. The O-demethylation of p-nitroanisole by this P–450 which apparently corresponds to P–450 LM3b (KOOP and COON, 1979) is cytochrome b_5 dependent (SUGIYAMA et al., 1979; SUGIYAMA et al., 1980). Contrary to only a stimulating effect like in studies with LM2 and LM4 (BRUNSTRÖM and INGELMAN-SUNDBERG, 1980; INGELMAN-SUNDBERG et al., 1980) this P–450 obligatorily requires cytochrome b_5. The formation of p-nitrophenol could only be detected at a 1 : 1 molar stoichiometry of both proteins. Because cytochrome b_5 needs its hydrophobic anchor to be incorporated into the membrane it is plausible that the P–450 activity in the presence of trypsin solubilized cytochrome b_5 was lowered to about 10% indicating that the electron transfer to P–450 needs proteins which are incorporated into the membrane.

Considering the P–450 isozymes from rabbit liver a stimulatory effect of cytochrome b_5 has been observed in the hydroxylation reactions of following substrates: benzphetamine and dimethylaniline by P–450 LM2 (IMAI and SATO, 1977), 7-ethoxycoumarin by P–450 LM2 (IMAI, 1979), benzo(a)pyrene by P–450 LM2 and LM4 (BRUNSTRÖM and INGELMAN-SUNDBERG, 1980) and N-methylcarbazole by P–450 LM2 and LM4 (VATSIS et al., 1980b). On the other hand cytochrome b_5 is obligatory for a NADPH-dependent metabolism of acetanilide by P–450 LM2 (COON et al., 1980) and of prostaglandin E_1 by P–450 LM2 (VATSIS et al., 1980a). According to the results the role of the cytochrome b_5 in the reconstituted systems seems to be consist in the formation of a functionally complex between P–450, P–450 reductase and cytochrome b_5 and thus improving the coupling of NADPH-oxidation to hydroxylation. In the case of the substrates above mentioned it is strongly suggested that cytochrome b_5 supplies the second electron more efficiently than the P–450 reductase (IMAI, 1981). However, it remains to be established why some substrates needs cytochrome b_5 and others do not.

The occurrence of benzo(a)pyrene in the environment and its carcinogenic activity have caused a lot of studies concerning the metabolism of this polycyclic hydrocarbon and identification of its carcinogenic metabolites. Benzo(a)-pyrene may undergo different metabolic pathways (cf. GELBOIN, 1980) and so it is of interest to examine the specifity of different purified forms of P–450 converting benzo(a)pyrene to reactive intermediates. Using reconstituted systems with dilauroylphosphatidylcholine (DEUTSCH et al., 1978; BELVEDERE et al., 1980) or phospholipid vesicles containing P–450, P–450 reductase, cytochrome b_5 and epoxide hydrase in a molar ratio of 3 : 1 : 2 : 3, respectively (BRUNSTRÖM and INGELMAN-SUNDBERG, 1980), it could be shown that P–450 LM2 primarily participates in the formation of the non-mutagenic benzo(a)pyrene-4,5-diol whereas P–450 LM4 preferentially catalyzes the 9,10-epoxidation but to a less extent also the 7,8-epoxidation and thus may participate in the formation of the ultimate carcinogen of benzo(a)pyrene which is to be assumed the 7,8-diol-9,10-epoxide (cf. GELBOIN, 1980). By contrast P–450 LM3 is incapable of metabolizing benzo(a)-pyrene to epoxides.

The diols occurring in the metabolic pattern of benzo(a)pyrene are derived from arene oxides by the action of epoxide hydrase. The primary epoxidation can take place at the *K*-region or at the non-*K*-region (bay region) but only the non-*K*-region diol can be converted again by P–450 to a diol-epoxide which is a rather poor substrate for the epoxide hydrase. Thus, depending on a particular metabolic pathway epoxide hydrase can either activate or inactivate polycyclic aromatic hydrocarbon derivatives. Free arene oxides are able to bind covalently with cellular macromolecules and this reaction appears to be important in the mechanism of chemical carcinogenesis, mutagenesis, and drug toxicity (JERINA and DALY, 1974; LEVIN et al., 1977; LU and MIWA, 1980b).

It has been shown that the ratio of the formation of the primary K-region epoxide of benzo(a)pyrene (4,5-epoxide) to the non-K-region epoxide in 7,8-position is governed by the ratio of P-450 to P-448 or of P-450 LM2 to P-450 LM4 (WOOD et al., 1976; DEUTSCH et al., 1978; BELVEDERE et al., 1980). Thus, a high content of P-448 or P-450 LM4 leads primarily to the formation of the benzo(a)pyrene-7,8-epoxide which is converted to the corresponding diol by the epoxide hydrase. The ultimate carcinogen benzo(a)pyrene-7,8-diol-9,10-epoxide is formed by the action of P-450 LM4 (BELVEDERE et al., 1980).

Using benzo(a)pyrene it could be shown recently that the reaction mechanism in the presence of NADPH and P-450 reductase considerably differs from that with organic hydroperoxides (CAPDEVILA et al., 1980). Further analyses of the product pattern of different substrates converted by various oxygen-supporting agents are necessary to establish the validity of peroxide-driven reactions (cf. O'BRIEN, 1978) as representative of NADPH-supported P-450-dependent hydroxylation reactions.

Besides P-450 LM2, LM4 and the LM3 isozymes catalytic activities have been measured with P-450 LM1 and P-450 LM7 isolated from uninduced male rabbits. The enzymatic activity of both constitutive fractions was examined towards benzphetamine, benzo(a)pyrene, parathion, aflatoxin B1, N,N-dimethylaminoazobenzene, N-methyl-4-aminoazobenzene (JOHNSON and MULLER-EBERHARD, 1977a; GUENGERICH, 1978a; FASCO et al., 1978; BELVEDERE et al., 1980). The differences in the metabolic activities between P-450 LM1 and LM7 which show small turnover numbers are not significant.

P-450 LM6 which is selectively induced by TCDD during the perinatal period is an isozyme with a very high activity towards benzo(a)pyrene. In adult rabbits more forms are induced by TCDD. This differences in the response of adult and neonatal animals to TCDD suggest that an age- dependent factor controls the induction process (NORMAN et al., 1978). Variations in the composition of P-450 with development and in response to inducers will affect xenobiotic metabolism and may have important consequences on the susceptibility of an animal to e.g. chemical carcinogenesis as can be seen from a comparison of activities of P-450 LM2, LM3, LM4 and LM6 (NORMAN et al., 1978; JOHNSON, 1980a).

4.2.3. Catalytic Activity of P-450 Isozymes from Rat Liver Microsomes

The homogeneous forms of rat liver P-450 produced by the same inducer have been designated in different manner. As the distinct P-450 forms isolated by different authors have not been completely identified the original nomenclature will be used as yet. RYAN et al. (1977, 1979) described the activity of P-450 forms prepared after pretreatement of rats with Aroclor 1254 and isosafrole as inducers and compared them with forms isolated from PB- or 3-MC-induced animals. GUENGERICH obtained the forms PB-P-450 B and D as well as 3-MC-P-450 B using the commonly used inducers PB and 3-MC, respectively (1977b, 1978a). HANSSON and WIKVALL (1979a, 1980) isolated P-450 forms from PB-treated rats which are comparable to form B and D obtained by GUENGERICH. The activities from BNF-treated and uninduced animals were compared by WARNER and NEIMS (1979) and SHIVERICK and NEIMS (1979).

Turnover numbers of those P–450 isozymes well characterized (BOTELHO et al., 1979) towards selected substrates are listed in Table 10. The presented results have been determined after reconstitution of P–450 and P–450 reductase in a 1 : 1 ratio in order to have a high concentration of the catalytically active binary complex of the two proteins (MIWA et al., 1979; LU and MIWA, 1980a). Crosslinking studies provide evidence for the formation of such a complex between P–450 and P–450 reductase in a non-vesicular system (BASKIN and YANG, 1980).

The 1 : 1 stoichiometry which exceeds the microsomal one of about 20 P–450 to 1 P–450 reductase considerably results in the high turnover numbers determined (Table 10). The hydroxylation of aniline is similar in P–450a and P–450b and only slightly increased in P–450c. The dominant form for the benzphetamine N-demethylation is P–450b independent of the inducer used. P–450c is active in the hydroxylation of benzo(a)pyrene and zoxazolamine and also in the O-dealkylation of 7-ethoxycoumarin confirmed by IMAI (1979), too. However, very high turnover numbers of P–446 isolated from BNF-induced rats have been reported by SAITO and STROBEL (1981) for benzo(a)-pyrene (up to 52 nmol/min/nmol P–446).

The catalytic activities found by GUENGERICH (1977b, 1978a) and by HANSSON and WIKVALL (1890) for their P–450 forms toward benzo(a)pyrene and benzphetamine are smaller than tose of the P–450 forms presented in Table 10. This may be mainly · due to different assay conditions, that is the P–450 to P–450 reductase ratio, too.

Steroid hydroxylase activity has been detected in all forms but each P–450 shows a preferential activity to a definite position of the steroid skeleton. Thus, P–450a, P–450b and P–450c hydroxylate testosterone at the 7α-, 16α-, and 6β-position, respectively (RYAN et al., 1979). The cytochromes P–450 B and D showed small activities in the 12α-, 25- and 26-hydroxylation of C_{27}-steroids (5β-cholestane-3α-,7α-diol and 5β-cholestane-3α-,7α,12α-triol) which were markedly lower than in microsomes, whereas their N-demethylase activity was higher than in microsomes (HANSSON and WIKVALL, 1980). Both forms were inactive in the 7α-hydroxylation of cholesterol which is the limiting step in the biosynthesis of bile acids from cholesterol. A purified P–450 fraction which was active in the 7α-hydroxylation but nearly inactive in the other positions mentioned above has been obtained by the same authors from rats treated with cholestyramine which increases this hydroxylase activity in microsomes.

Table 10. Catalytic activity of P–450 isozymes from rat liver*

	Turnover number (min^{-1})		
	P–450a (PB, PCB, 3-MC)	P–450b (PCB, PB)	P–450c (PCB, 3-MC)
Aniline	5.1 —6.3	6.7—7.7	9.3—10.6
Benzphetamine	2.6 —3.2	176.6—216.6	5.0—5.1
Benzo(a)pyrene	0.04—0.3	0.2—0.5	19.2—24.5
7-Ethoxycoumarin	0.2 —1.1	12.1—13.9	48.5—67.5
Zoxazolamine	0.5 —2.0	2.2—2.4	27.9—29.7

* (Taken from RYAN et al., 1979)

The activity in the hydroxylation of 5α-androstane-3α,17β-diol and 5α-androstane-3α,17β-diol,3,17-disulfate in the 2α-, 7α-, 7β-, 12β- and 16α-position has been examined by GUSTAFSSON and INGELMAN-SUNDBERG (1976) using so-called peak fractions. Furthermore, the binding-catalytic site distances have been calculated. In dependence on the hydroxylation position differences have been found. The authors suggested that the specificity of various P–450-dependent steroid hydroxylases resides in the nature of the binding site on the enzyme (electrophilic, nucleophilic, or hydrophobic) and in the binding-catalytic site distance.

In summary it can be stated that the substrate specificity of microsomes changes dramatically after pretreatment with different inducers. The P–450b isozyme (LM2 analogue) is the main protein for the metabolism of drugs, whereas the other substrates are preferentially metabolized by P–450 forms produced by 3-MC-like inducers (LM4 analogue). Analogously to rabbit P–450 isozymes the different P–450 isozymes from rat liver are regioselective in the hydroxylation of steroids.

4.2.4. Catalytic Activity of Hepatic Human Cytochrome P–450

In the last decade also the human P–450 was included into systematic studies due to its important role in the metabolism of drugs and in the bioactivation and detoxification of an increasing number of potentially toxic xenobiotics. The main interest was directed to compare the catalytic properties of human P–450 with those of laboratory animals such as rats and rabbits at the microsomal level (cf. citation in Table 11). These studies are, however, in the very beginning. Further results have to be accumulated in order to diminish the inter-individual differences which might be caused by diseases, abuse of drugs and alcohol or smoking of an individuum.

To evaluate the relative importance of the enzymes involved in drug metabolism generally the rates of conversion of a drug and its metabolites are studies using microsomal preparations obtained from various species. In Table 11 Michaelis constants and maximal velocities for 6 substrates are presented. The apparent K_m values determined for benzo(a)pyrene and p-nitroanisole as substrates of human microsomes correspond to those of rats, whereas they are up to 10 fold higher in man for aminopyrine and codeine. The K_m value for ethylmorphine was approximately 20 fold increased in human liver microsomes as compared with rat microsomes. The V_{max} values for aminopyrine and codeine are twice as high with rats as with man, whereas the activity towards p-nitroanisole corresponds to that of rat. A smaller activity of human microsomes was determined with ethylmorphine and aniline which amounted to 20–30% of rats if male populations are compared. On the contrary, the V_{max} values determined in liver microsomes of women and female rats show no significant differences.

In microsomes of rats and rabbits the substrate specificity can be attributed to the special pattern of P–450 isozymes. This is very difficult to estimate for man due to the small amount of material which is available for the solubilization and purification. Therefore, only a few studies with purified human P–450 have been performed accumulating evidence for the occurrence of multiple forms of P–450, too.

The catalytic activity of human P–450 determined after reconstitution is no more compaıable to that of experimental animals (BEAUNE et al., 1979). No conclusions can be drawn from these results whether the P–450 isolated from human liver corresponds to a P–450 induced by PB or 3-MC. This assumption is

Table 11. Species differences in the substrate specificity of liver microsomes

Substrate	K_m (mM)		V_{max} (nmol/mg·min)		Reference
	Man	Rat	Man	Rat	
Aminopyrine	3–5	1.0	3.7	6.0	PEDERSON and AUST (1970)
	6.1	0.4	—	—	ACKERMANN (1970)
Codeine	10.0	0.5	5.7	10.9	ACKERMANN (1970)
Benzo(a)pyrene	0.015–0.025	0.006–0.02	0.003	0.006	PELKONEN et al. (1975)
	0.012–0.033	—	—	—	ALVARES et al. (1969)
Ethylmorphine (♂)	6.0	0.31	6.7	22.1	ACKERMANN (1972)
(♀)	3.5	0.27	4.7	4.7	ACKERMANN (1972)
Aniline (♂)	0.03	0.04	0.59	1.6	ACKERMANN (1972)
(♀)	0.07	0.04	0.48	0.61	ACKERMANN (1972)
p-Nitroanisole	0.02	0.02	2.2	1.3	ACKERMANN and HEINRICH (1970)

supported by WANG et al. (1980) who treated various isolated P–450 forms with antibodies against rat PB- or 3-MC-P–450. Both antibodies inhibit the benzo(a)-pyrene hydroxylation to the same extent. On the contrary, the demethylation of benzphetamine is more strongly influenced by anti-PB-rat. Using the complement fixation technique it was possible to establish immunochemical similarity of rat and human P–450. Despite of interindividual differences it could be concluded that the human P–450 is more similar to P–450 obtained from PB-induced rats than from 3-MC-treated ones. This is supported by the physicochemical properties of human P–450 (see 2.1.4.), too.

In comparison with rats it can be stated with one exception that the conversion of the selected substrates is smaller in human liver microsomes. The different K_m values obtained for human and rat microsomes indicate differences in the rate-limiting step for several substrates which may be due to different inducibility in man and rat leading to different patterns of P–450 differing between man and rat. Furthermore, the differences observed between female and male animals are not so pronounced between women and men. It is known that monooxygenase activities depend on the age and the nutrition of people resulting in a large variation of the results. The situation in determining the substrate specificity in samples from man is more complicated by these influences which can be controlled in studies on experimental animals.

In summary, the isolation of different P–450 LM forms with definite physicochemical properties after induction by various inducers (Table 1) together with the isozyme specific amino acid sequence and the immunological data provide substantial experimental evidence for the existence of genetically determined multiple forms. Their individual expression is species and tissue specific as well as age-dependent. Each of these forms shows different activities towards various substrates after reconstituting the isolated fractions with the essential components and in some cases with cytochrome b_5. The data available give increasing evidence for the assumption that the constitutive forms of P–450 are involved in the metabolism of steroids and other endogeneous substrates, but nearly inactive towards xenobiotics, whereas the inducible forms of P–450 are involved primarily in the metabolism of xenobiotics. As already shown for few cancerogenic compounds the balance between activation and detoxification will be determined by the factors which regulate the P–450 composition of a tissue as well as by the substrate specificity of each isozyme.

4.3. *Influence of Chemical Modification on Structure-Function Relationships*

The analysis of the catalytic mechanism of the P–450 system represents a complex experimental problem. Besides substrate specific parameters (4.2.1.) protein specific aspects have to be considered such as the structure of the active site, of intermediate complexes and of transient states. An understanding of the mechanism of enzymatic catalysis requires the knowledge of the rate constants of partial reactions and their interpretation in structural terms. The interaction between P-450 and P–450 reductase as 2nd essential component of the enzyme system which is located in and influenced by the membrane of the endoplasmic reticulum has further to be considered.

The substrate specificity of P–450 and the reactivity of the system can be better understood from the knowledge of structural parameters which in principle are accessible utilizing X-ray analysis of the crystallized enzyme. However, the availability of informations concerning the functional importance of certain amino acid side chains is limited by this technique as well as by amino acid sequencing. Therefore the method of chemical modification of amino acid residues by group-specific reagents has proved an appropriate means to analyze such residues which are essential for biological activity. Due to the presence of a large number of reactive side chains in a protein molecule it is necessary to use highly specific reagents. Since the site specific modification represents a kinetic process it is possible in certain cases to achieve a selective modification of one amino acid residue under definite experimental conditions. Chemical modification studies of P–450 were aimed to determine those amino acid residues of the molecule which are involved in the catalytic process within the active center as well as in interactions between P–450 and the P–450 reductase. Mainly the role of sulfhydryl, imidazole, phenolic hydroxyl and amino groups was investigated.

The unique capability of P–450 to activate molecular oxygen essentially depends on the presence of a cysteine residue evidenced as the 5th axial ligand of the heme iron by Resonance Raman spectroscopy (Champion et al., 1982). The essential role of this SH-group but also of further SH groups in P–450-mediated mixed function oxidation of xenobiotics was shown by many investigators. The inhibition of the microsomal drug metabolism by different types of sulfhydryl reagents (Conney et al., 1957; Orrenius, 1965; Kampffmeyer and Kiese, 1964; Creaven et al., 1965; Franklin and Estabrook, 1971; Illing and Netter, 1975) proceeds in a non-competitive manner. The action of mercurials usually leads to a significant conversion of P–450 into P–420 which is paralleled by the inability to produce type I binding spectra (Murakami and Mason, 1967; Franklin and Estabrook, 1971; Franklin, 1972a, b). The inhibition of the overall hydroxylation by sulfhydryl reagents has been demonstrated using microsomal suspensions. But due to the presence of other proteins in these experiments the involvement of a distinct protein is difficult to determine. Therefore it is necessary to reinvestigate the reactivity of SH-groups in highly purified P–450 preparations in order to show conclusively the state and the functional importance of these groups in P–450. The sensitivity of the P–450 reductase towards sulfhydryl reagents has already been discussed (see 3.4.).

The few data obtained with different P–450 isozymes show qualitative differences in the reactivity and functional importance of SH-groups. With exception of the SH-group acting as the 5th axial ligand no conclusion with respect to their participation in different steps of the overall reaction can be drawn. Highly purified P–450 from 3-MC-induced rabbits or rats has been shown by spectrophotometric titrations (Kawalek et al., 1977) to contain 7 sulfhydryl groups per mol of enzyme which is in agreement with the results of amino acid analyses (see Table 5). All SH-groups are accessible to different modifying agents 5,5′-dithiobis(2-nitrobenzoic acid) (DTNB), 4,4′-dipyridine disulfide (PDS), 2-nitro-5-thiocyanobenzoic acid (NTCB), p-mercuribenzoate (PMB) and N-ethylmaleimide (NEM). These results confirm preliminary findings on the absence of disulfide bridges within P–450 as concluded from investigations of P–450 LM2 (Guengerich et al., 1975). When rat or rabbit P–448 was modified with DTNB the benzo (a)pyrene hydroxylation was considerably inhibited on introduction of 2 groups per molecule which has been attributed to steric hindrance as well as to conformational changes due to the bulky, charged thionitrobenzoate substituent. Titration experiments with further reagents (PDS, NTCB and NEM) indicated that the SH-groups of rat P–448 are not essential for benzo(a)-pyrene hydroxylation since no inhibition did occur. In contrast rabbit P–448 was inhibited after modification by all four reagents suggesting that rabbit P–448 contains sulfhydryl group(s) in or near the active site. Besides a contribution to the loss of activity due to steric hindrance or conformational changes could not be excluded. The formation of P–420 as cause for the decrease in the activity could be excluded.

Recent modification studies on purified P–450 from PB-treated rats (Halpert et al., 1980) with PDS or p-bromphenacyl bromide demonstrated that three cysteine residues can be substituted without affecting the benzphetamine N-demethylase activity. A similar result has been obtained by modifying one SH-group of P–450 LM2 by DTNB (Vatsis et al., 1977). Besides the identified SH-group in the position of the 5th axial ligand further studies are necessary to clarify the functional importance of the other sulfhydryl groups in P–450.

Investigations by Gibson and Tamburini (1980) have shown that 16 histidine residues of rat P–450 can readily be modified by diethylpyrocarbonate without significant conversion to P–420. The modified P–450 was still capable of the temperature-dependent spin transition in the presence of excess substrate, but the extent of this shift was reduced. The binding affinity for benzphetamine was lowered indicating that histidine residues may be localized at or near the substrate binding site of P–450 whereas their role in the catalytic mechanism remains to be clarified.

Extending pH-titration experiments of Ruckpaul et al. (1980b) chemical modification of P–450 LM2 with N-acetylimidazole proved the accessibility of 3 tyrosine residues to O-acetylation (Jänig et al., 1982). One fast reacting tyrosine acetylated first is not obligatory for the N-demethylation of benzphetamine. However, the activity is lost by about 60% after modification of one more tyrosine residue and by further 20% after the third residue had reacted indicating different functional importance for each of them. The analysis of the kinetics of substrate binding and of K_s values revealed that modification of tyrosine residues of P–450 LM2 influences type I and type II substrates differently. The rate constant of benzphetamine binding is diminished by one order of magnitude after modification of more than one tyrosine whereas the binding affinity remained unchanged on acetylation of 3 tyrosine residues.

Contrary, a decreased rate constant for aniline is observed on acetylation of only one tyrosine residue in P–450 LM2 concomitant with an increase of the aniline binding affinity. That means the rate of the aniline dissociation reaction is significantly slowed on acetylation of the tyrosine residue reacting first. Protection from modification in the presence of the inhibitor metyrapone indicated this tyrosine to be located in the type II substrate binding site. This conclusion is in agreement with results of photoaffinity labeling of P–450 LM2 evidencing one tyrosine residue in the heme peptide of P–450 LM2 (Dus, 1980). The two more slowly reacting tyrosine residues appear to participate in catalysis at the active site and/or in interactions between P–450 LM2 and P–450 reductase.

Supporting the results on acetylation about three tyrosine residues of P–450 LM2 can be nitrated with tetranitromethane (Jänig et al., 1982). This modification results in the formation of nitrophenolate groups and leads to a broadening of the absorbance in the Soret region and a significant conversion into P–420. The decrease of the absorbance at 418 nm is paralleled by an increase of the heme ultraviolet band at 360 nm and a new band at 460 nm. Such bands are characteristic of hyperporphyrin spectra which are formed by the heme in the presence of two negatively charged ligands (Ruf et al., 1979). It is known that the nitration increases the acidity of the phenolic hydroxyl group by a shift of the pK from 9.6 to 7.0. That means that the phenolic hydroxyl group is partially deprotonated by nitration. If indeed the 6th heme iron ligand is a tyrosine its ionization would lead to weakening of the sulfur-iron bond at the 5th axial coordination position exhibiting the hyper spectra. With respect to the accumulating data that an oxygen atom may be the axial ligand trans to the thiolate (White and Coon, 1982; Yoshida et al., 1982) the results obtained by nitration suggest to assume a tyrosine residue as the 6th axial heme iron ligand.

Only few attempts have been made to elucidate the role of amino groups for the overall activity and different steps of the reaction cycle of the P–450 system. Previous studies have shown that partially purified as well as P–450 LM2 can be attached via free amino groups to Sepharose 4B with maintenance of the enzymatic activity after reconstitution with the complementary components (Baess et al., 1975; Jänig et al., 1979). The immobilized P–450 was found to be remarkably stable on storage in comparison to the free enzyme (Jänig et al., 1977). The N-demethylase activity of the immobilized P–450 amounted to about 20–30% that of the free enzyme in dependence on the degree of purification, the species, the presence of phospholipids, and the sequence of matrix binding if P–450 and the P–450 reductase are co-immobilized. The loss of activity is not caused by a considerable formation of P–420. The lowered activity of bound P–450 LM2 retained in the cumene hydroperoxide-supported reactions as well as on reconstitution indicates conformational changes of P–450 LM2 including changes of the active center which lead to disturbed interaction with the P–450 reductase. Extending these result Bernhardt et al. (1980a, b, 1983) modified amino groups of partially purified P–450 as well as P–450 LM2 using fluorescein isothiocyanate (FITC). Both P–450 preparations can be modified by increasing the molar excess of FITC to P–450 without any conversion of P–450 into P–420. The increase in the modification ratio to about 1 mol FITC bound per mol P–450 LM2 is paralleled by a decrease of the benzphetamine N-demethylase activity to about 80%. By the covalently bound FITC the N-terminal methionine of P–450 LM2 is selectively modified as demonstrated by end group analysis. This modification has no influence on the K_s values for benzphetamine and aniline and does not inhibit the cumene hydroperoxide- or H_2O_2-dependent demethylation of benzphetamine indicating a rather small if any effect on the conformation of P–450 LM2. The reduction rate of the modified P–450 by sodium dithionite or by NADPH via the P–450 reductase as electron donor however is decreased. Usually about 10% of that P–450 LM2 reducible by sodium dithionite are reduced by NADPH using a 20 : 1 stoichiometry of P–450 to P–450 reductase. The ratio of the reduction with NADPH and reductase to the sodium dithionite reduction is further decreased in the modified P–450 LM2. The diminished reducibility is ascribed to steric hindrance of groups directly involved in the interaction between P–450 LM2 and P–450 reductase or to a blockade of the charge pair interactions of amino groups of P–450 and respective negatively charged groups of P–450 reductase. It has been concluded from these experiments that an electrostatic recognition between P–450 LM2 and P–450 reductase is a necessary prerequisite for the electron transfer process.

The binding of fluorescamine to homogeneous rat P–450 and P–448 (West et al., 1980) produced a more pronounced effect on the catalytic activity of both forms than that of FITC to rabbit P–450 LM2 (Bernhardt et al., 1980b). A molar ratio of 30 : 1 (fluorescamine : P–450) results in about 75% inhibition of the N-demethylase activity with benzphetamine or the ethoxycoumarin deethylation by a reconstituted system with P–450 and P–448, respectively. The benzphetamine binding to P–450 is sligthly decreased without a loss in the P–450 content. Substrates as benzphetamine or dimethylaniline did not protect P–450 from modification by fluorescamine. In conclusion, further studies are needed to establish whether a primary amino group is involved in the substrate binding to P–450 or in the catalytic site of P–450. As generally stated the inhibition could also be due to conformational changes of the hemoprotein or steric hindrance by modified amino acid side chains. On the other hand, the modification of the P–450 reductase by fluorescamine did not result in an inhibition of benzphetamine N-demethylase activity (West et al., 1980). Therefore it can be concluded that those amino groups of the P–450 reductase accessible to a modifying fluorescamine are not involved in the interaction with P–450. The modification of lysine residues of protease-solubilized P–450 reductase by pyridoxal-5′-phosphate diminishes the cytochrome c reduction substantially by interfering with the cosubstrate binding site (Lumper et al., 1980). The results of fluorescamine binding to NADPH-reductase fit well the concept that acidic amino acid residues of the P–450 reductase membrane binding domain participate in orienting the P–450 reductase for maximal interaction with P–450 (Gum and Strobel, 1979).

Previous studies proved that the hydroxylation of drugs by rat liver microsomes is lowered in the presence of 1-anilinonaphthalene 8-sulfonate (1,8-ANS) which also effects the binding of some substrates to microsomes (DiAugustine et al., 1970; Lang, 1976). To gain more information its interaction with P–450 LM2 and P–450 LM4 was recently investigated by Chiang (1980). Four to five mol 1,8-ANS are bound per mol P–450 LM2 or P–450 LM4 with high affinity ($K_{diss} = 4\ \mu M$). In the presence of benzphetamine the binding of 1,8-ANS is further increased. From the inhibition of the benzphetamine demethylation and the NADPH-oxidase in reconstituted enzyme systems by 1,8-ANS one can suppose that hydrophobic binding sites of P–450 molecules which are occupied by 1,8-ANS may be involved in electron transfer reactions. Despite decreasing ΔA_{max} values of type I binding the K_s value for benzphetamine remains constant. Therefore, the substrate binding site is not influenced by 1,8-ANS but it is postulated that the fluorescent label may be bound proximal to the substrate binding site and perturbs the heme chromophore. Similar perturbations were observed following the binding of the fluorescent labels fluorescamine and FITC to P–450 (West et al., 1980; Bernhardt et al., 1980, respectively).

In summary, it can be suggested from the limited investigations on the functional importance of certain amino acid side chains in cytochrome P–450 hitherto performed that the side chains of histidine, tyrosine and lysine residues are involved in the binding of substrates. In addition to its essential role as 5th ligand of the heme iron a few sulfhydryl groups and at least one tyrosine residue are essential for the monooxygenatic reaction within the active center. The N-terminus and further amino groups of P–450 are involved in interactions with P–450 reductase, whereas tyrosine residues may participate also in protein-lipid interaction. Moreover, the phenolic hydroxyl group of a tyrosine residue was suggested to act as the 6th axial ligand of the heme iron of P–450.

5. Concluding Remarks

The purpose of this article was to review the progress in the development of techniques for the separation and purification of the components of the monooxygenase system from liver microsomes and in their biochemical characterization which has been performed by very different methods in the last decade. The most striking property of the hepatic monooxygenase system which consists of cytochrome P–450, NADPH-cytochrome P–450 reductase, and phospholipids is its lack of specificity which permits the metabolization of a great number of lipophilic compounds. It is now generally acknowledged that the broad substrate specificity of the system originates also from the presence of numerous P–450 isozymes each with a definite, but with other forms overlapping substrate specificity. The pattern of isozymes is also influenced by environmental factors.

By use of subtle chromatographic techniques including the elegant affinity chromatography the P–450 reductase and a great number of multiple forms of P–450 have been isolated from different species and organs despite the difficulties in purification due to the occurrence of the system in a membrane-bound state. The various purified forms of P–450 were found to be different based on the minimal molecular weights in SDS PAGE, spectral characteristics, substrate specificity, immunological reactivity with specific antibodies, amino acid analyses, and partial determination of the N- and C-terminal sequences. The lack of immunochemical identity and the differences in the amino acid sequences indicate that the primary and tertiary structures of these cytochromes are different implying that the multiple forms of P–450 are isozymes and do not arise by post-translational processing of one common precursor protein.

Future work will be directed to gain more information about the primary and spatial structure of P–450 as well as of the P–450 reductase in order to achieve a better knowledge about the essential interactions of both proteins leading to a deeper understanding of structure-function relationships. The present difficulties due to lacking progress in elucidation of the spatial structure can be substituted partly in the near future by chemical modification techniques. The modification is suitable for recognizing those amino acid side chains which are essential for the catalytic mechanism within the active center as well as for the interaction between the functionally linked proteins. By comparing the influence of chemical modification on different P–450 forms with the same modifying agent it should be possible to find out differences in the active center of various forms which might be responsible for different substrate specificities, too.

Further biochemical investigation on the monooxygenatic mechanism are expected to be directed to integrate the numerous data obtained with reconstituted systems in order to mimic microsomal conditions. For this reason the conditions for reconstitution have to be improved by incorporating the P–450 and P–450 reductase together with cytochrome b_5 (and epoxide hydrase for polycyclic hydrocarbons) in phospholipid vesicles of definite chemical composition. The incorporation into phospholipid vesicles facilitates protein-protein and protein-lipid interactions allowing the complete analysis of the metabolite profile of endogeneous substrates as well as of xenobiotics. Such an analysis is of vital importance for a lot of environmental pollutants which are activated by monooxygenases to highly reactive intermediates with toxic, mutagenic or carcinogenic properties. Since the fate of a distinct chemical is determined by the pattern of multiple forms in microsomes which is dependent on the species, sex, age, nutrition, etc. future studies will include undoubtly human samples in order to get an overview about interindividual deviations of the P–450 isozyme profile in man. The present and future research will be directed to correlate the genetically determined isozyme profile of P–450 with a biochemical marker for predicting an increased susceptibility to environmental chemicals with cancerogenic action in man.

References

Ackermann, E. (1970) Biochem. Pharmacol. **19**, 1955–1973

Ackermann, E., and J. Heinrich (1970): Biochem. Pharmacol. **19**, 327–342

Ackermann, E. (1972): Biochem. Pharmacol. **21**, 2169–2180

Ahmad, N., and M. Black (1977): J. Pharmacol. Exp. Ther. **203**, 397–408

Alexander, L. M., L. B. Hersh, and B. S. S. Masters (1980) in: Biochemistry, Biophysics and Regulation of Cytochrome P–450 (J.-Å. Gustafsson, J. Carlstedt-Duke, A. Mode, and J. Rafter, eds.) pp. 285–288, Elsevier/North-Holland, Biomedical Press, Amsterdam, New York, Oxford

Alvares, A. P., G. Schilling, W. Levin, and R. Kuntzman (1967): Biochem. Biophys. Res. Commun. **29**, 521–526

Alvares, A. P., G. Schilling, W. Levin, R. Kuntzman, L. Brand, and L. C. Mark (1969): Clin. Pharmac. Ther. **10**, 655–659

Alvares, A. P. and P. Siekevitz, (1973): Biochem. Biophys. Res. Commun. **54**, 923–929

Archakov, A. I., I. I. Karuzina, I. S. Kokareva, and G. I. Bachmanova (1973): Biochem. J. **136**, 371–379

Archakov, A. I., I. I. Karuzina, V. N. Tveritinov, and I. S. Kokareva (1974): Biochem. Pharmacol. **23**, 1053–1063

Atlas, S. A., A. R. Boobis, J. S. Felton, S. S. Thorgeirsson, and D. W. Nebert (1977): J. Biol. Chem. **252**, 4712–4721

Autor, A. P., R. M. Kaschnitz, J. K. Heidema, and M. J. Coon (1973): Mol. Pharmacol. **9**, 93–104

Back, J. F., D. Oakenfull, and M. B. Smith (1979): Biochem. **18**, 5191–5196

Baess, D., G.-R. Jänig, and K. Ruckpaul (1975): Acta Biol. Med. Germ. **34**, 1745–1754

Baskin, L. S., and C. S. Yang (1980): Biochemistry **19**, 2260–2264

Beaune, Ph., P. Dansette, J. P. Flinois, S. Columelli, D. Mansuy, and J. P. Leroux (1979): Biochem. Biophys. Res. Commun. **88**, 826–832

Behlke, J., G.-R. Jänig, and D. Pfeil (1979): Acta Biol. Med. Germ. **38**, 389–397

Belvedere, G., H. Miller, K. P. Vatsis, M. J. Coon, and H. V. Gelboin (1980): Biochem. Pharmacol. **29**, 1693–1702

Bernhardt, R., N. T. Ngoc Dao, J. Friedrich, G.-R. Jänig, M. Müller-Frohne, G. Smettan, and K. Ruckpaul (1980a): Studia Biophysica **79**, 163–164

Bernhardt, R., N. T. Ngoc Dao, J. Friedrich, O. Ristau, G.-R. Jänig, and K.

RUCKPAUL (1980b): in Biochemistry, Biophysics and Regulation of Cytochrome P-450 (J.-Å. GUSTAFSSON, J. CARLSTEDT-DUKE, A. MODE, and J. RAFTER, eds.) pp. 595–598, Elsevier/North-Holland, Biomedical Press, Amsterdam, New York, Oxford

BERNHARDT, R., N. T. NGOC DAO, H. STIEL, W. SCHWARZE, J. FRIEDRICH, G.-R. JÄNIG, and K. RUCKPAUL (1983): Biochim. Biophys. Acta **745**, 140–148

BLACK, S. D., J. S. FRENCH, C. H. WILLIAMS, and M. J. COON (1979): Biochem. Biophys. Res. Commun. **91**, 1528–1535

BLAKE, R. C., and M. J. COON (1980): J. Biol. Chem. **255**, 4100–4111

BLANCK, J., G. SMETTAN, G.-R. JÄNIG, and K. RUCKPAUL (1976): Acta Biol. Med. Germ. **35**, 1455–1463

BLANCK, J., K. ROHDE, and K. RUCKPAUL (1979): Acta Biol. Med. Germ. **38**, 23–32

BOOBIS, A. R., D. W. NEBERT, and J. S. FELTON (1977): Mol. Pharmacol. **13**, 259–268

BÖSTERLING, B., J. M. DAWSON, A. G. HILDEBRANDT, A. STIER, and J. R. TRUDELL (1979): Mol. Pharmacol. **16**, 332–342

BÖSTERLING, B., and J. R. TRUDELL (1980) in: Microsomes, Drug Oxidations and Chemical Carcinogenesis, Vol. 1 (M. J. COON, A. H. CONNEY, R. W. ESTABROOK, H. V., GELBOIN, J. R. GILLETTE, and P. J. O'BRIEN, eds.) pp. 115–118, Academic Press, New York, London, Toronto, Syndey, San Francisco

BOTELHO, L. H., D. E. RYAN, and W. LEVIN (1979): J. Biol. Chem. **254**, 5635–5640

BRESNICK, E., D. BORAKER, D. HASSUK, W. LEVIN, and P. E. THOMAS (1979): Mol. Pharmacol. **16**, 324–331

BRUNSTRÖM, A., and M. INGELMAN-SUNDBERG (1980): Biochem. Biophys. Res. Commun. **95**, 431–440

CAPDEVILA, J., R. W. ESTABROOK, and R. A. PROUGH (1980): Arch. Biochem. Biophys. **200**, 186–195

CHAMPION, P. M., STALLARD, B. R., WAGNER, G. C. and GUNSALUS, I. C. (1982): J. Amer. Chem. Soc. **104**, 5469–5473

CHIANG, Y.-L., and M. J. COON (1979): Arch. Biochem. Biophys. **195**, 178–187

CHIANG, Y.-L. (1980) in: Microsomes, Drug Oxidations and Chemical Carcinogenesis, Vol. 1 (M. J. COON, A. H. CONNEY, R. W. ESTABROOK, H. V. GELBOIN, J. R. GILLETTE, and P. J. O'BRIEN, eds.) pp. 123–126, Academic Press, New York, London, Toronto, Sydney, San Francisco

CHRANKSHAW, D. L., K. HETNARSKI, and C. F. WILKINSON (1979): Biochem. J. **181**, 593–605

COMAI, K., and J. L. GAYLOR (1973): J. Biol. Chem. **248**, 4947–4955

CONNEY, A. H., E. C. MILLER, and J. A. MILLER (1957): J. Biol. Chem. **228**, 753–766

COON, M. J., D. A. HAUGEN, F. P. GUENGERICH, J. L. VERMILION, and W. L. DEAN (1976) in: The Structural Basis of Membrane Function (Y. HATEFI, and L. DJAVADI-OHANIANCE, eds.) pp. 409–427, Academic Press, New York

COON, M. J., T. A. VAN DER HOEVEN, S. B. DAHL, and D. A. HAUGEN (1978): Methods Enzymol. **52**, 107–117

COON, M. J., and K. P. VATSIS (1978): Polycyclic Hydrocarbons and Cancer **1**, 335–360

COON, M. J., D. R. KOOP, A. V. PERSSON, and E. T. MORGAN (1980) in: Biochemistry, Biophysics and Regulation of Cytochrome P-450 (J.-Å. GUSTAFSSON, J. CARLSTEDT-DUKE, A. MODE, and J. RAFTER, eds.) pp. 7–15, Elsevier/North-Holland Biomedical Press, Amsterdam, New York, Oxford

COOPER, D. Y., M. D. CANNON, H. SCHLEYER, and O. ROSENTHAL (1977): J. Biol. Chem. **252**, 4755–4757

CREAVEN, P. J., D. V. PARKE, and R. T. WILLIAMS (1965): Biochem. J. **96**, 390–398

DAVIS, B. D., and P.-C. TAI (1980): Nature **283**, 433–437

DAWSON, J. H., J. R. TRUDELL, R. E. LINDER, G. BARTH, E. BUNNENBERG, and C. DJERASSI (1978): Biochemistry **17**, 33–42

DEAN, W. L., and M. J. COON (1977): J. Biol. Chem. **252**, 3255–3261

DEUTSCH, J., J. C. LEUTZ, S. K. YANG, H. V. GELBOIN, Y. L. CHIANG, K. P. VATSIS, and M. J. COON (1978): Proc. Natl. Acad. Sci. U.S. **75**, 3123–3127

DIAUGUSTINE, R. P., T. E. ELING, and J. R. FONTS (1970): Chem. Biol. Interact. **2**, 17–27

DIGNAM, J. D., and H. W. STROBEL (1975): Biochem. Biophys. Res. Commun. **63**, 845–852

DIGNAM, J. D., and H. W. STROBEL (1977): Biochemistry **16**, 1116–1123

DUBOIS, R. N., and M. R. WATERMAN (1979): Biochem. Biophys. Res. Commun. **90**, 150–157

Dus, K., W. L. Litchfield, A. G. Miguel, T. A. van der Hoeven, D. A. Haugen, W. L. Dean, and M. J. Coon (1974): Biochem. Biophys. Res. Commun. **60**, 15–21

Dus, K., R. Goewert, C. C. Weaver, D. Carey, and C. A. Appleby (1976): Biochem. Biophys. Res. Commun. **69**, 437–445

Dus, K. (1980) in: Biochemistry, Biophysics and Regulation of Cytochrome P-450 (J.-Å. Gustafsson, J. Carlstedt-Duke, A. Mode, and J. Rafter, eds.) pp. 129–132, Elsevier/North-Holland, Biomedical Press, Amsterdam, New York, Oxford

Ebel, R. E. (1980) in: Microsomes, Drug Oxidations and Chemical Carcinogenesis, Vol. 1, (M. J. Coon, A. H. Conney, R. W. Estabrook, H. V. Gelboin, J. R. Gillette, and P. J. O'Brien, eds.) pp. 315–318, Academic Press, New York, London, Toronto, Sydney, San Francisco

Elshourbagy, N. A., and P. S. Guzelian (1980): J. Biol. Chem. **255**, 1279–1285

Fan, L. L., and B. S. S. Masters (1974): Arch. Biochem. Biophys. **165**, 665–671

Fasco, M. J., K. P. Vatsis, L. S. Kaminsky, and M. J. Coon (1978): J. Biol. Chem. **253**, 7813–7820

Fasco, M. J., L. J. Piper, and L. S. Kaminsky (1977): Arch. Biochem. Biophys. **182**, 379–389

Franklin, M. R., and R. W. Estabrook (1971): Arch. Biochem. Biophys. **143**, 318–329

Franklin, M. R. (1972a): Mol. Pharmacol. **8**, 711–721

Franklin, M. R. (1972b): Mol. Pharmacol. **8**, 722–730

French, J. S. and M. J. Coon (1979): Arch. Biochem. Biophys. **195**, 565–577

French, J. S., S. D. Black, C. H. Williams, and M. J. Coon (1980a) in: Microsomes, Drug Oxidations and Chemical Carcinogenesis, Vol. 1 (M. J. Coon, A. H. Conney, R. W. Estabrook, H. V. Gelboin, J. R. Gillette, and P. J. O'Brien, eds.) pp. 387–390, Academic Press, New York, London, Toronto, Sydney, San Francisco

French, J. S., F. P. Guengerich, and M. J. Coon (1980b): J. Biol. Chem. **255**, 4112–4119

Gaudette, L. E., and B. B. Brodie (1959): Biochem. Pharmacol. **2**, 89–96

Gelboin, H. V. (1980): Physiol. Rev. **60**, 1107–1166

Gibson, G. G., and J. B. Schenkman (1978): J. Biol. Chem. **253**, 5957–5963

Gibson, G. G., and P. P. Tamburini (1980) in: Biochemistry, Biophysics and Regulation of Cytochrome P-450 (J.-Å. Gustafsson, J. Carlstedt-Duke, A. Mode, and J. Rafter, eds.) pp. 133–136 Elsevier/North-Holland, Biomedical Press, Amsterdam, New York, Oxford

Guengerich, F. P., D. P. Ballou, and M. J. Coon (1975): J. Biol. Chem. **250**, 7405–7414

Guengerich, F. P. (1977a): Mol. Pharmacol. **13**, 911–923

Guengerich, F. P. (1977b): J. Biol. Chem. **252**, 3970–3979

Guengerich, F. P. (1978a): J. Biol. Chem. **253**, 7931–7939

Guengerich, F. P. (1978b): Biochem. Biophys. Res. Commun. **82**, 820–827

Guengerich, F. P. (1979a): Biochim. Biophys. Acta **577**, 132–141

Guengerich, F. P. (1979b): Biochem. Pharmacol. **28**, 2883–2890

Guengerich, F. P., and L. A. Holladay (1979): Biochemistry **18**, 5442–5449

Guengerich, F. P., and P. S. Mason (1979): Mol. Pharmacol. **15**, 154–164

Guengerich, F. P., and M. V. Martin (1980): Arch. Biochem. Biophys. **205**, 365–379

Gum, J. R., and H. W. Strobel (1979): J. Biol. Chem. **254**, 4177–4185

Gustafsson, J.-Å., and M. Ingelman-Sundberg (1976): Eur. J. Biochem. **64**, 35–43

Haaparanta, T., J. Rydström, and M. Ingelman-Sundberg (1980) in: Microsomes, Drug Oxidations and Chemical Carcinogenesis, Vol. 1 (M. J. Coon, A. H. Conney, R. W. Estabrook, H. V. Gelboin, J. R. Gillette, and P. J. O'Brien, eds.) pp. 533–536, Academic Press, New York, London, Toronto, Syndey, San Francisco

Halpert, J., D. Hammond, and R. A. Neal (1980): J. Biol. Chem. **255**, 1080–1089

Hansson, R., and K. Wikvall (1979a): Acta Chem. Scand. B **33**, 611–612

Hansson, R., and K. Wikvall (1979b): Eur. J. Biochem. **93**, 419–426

Hansson, R., and K. Wikvall (1980): J. Biol. Chem. **255**, 1643–1649

Hashimoto, C., and Y. Imai (1976): Biochem. Biophys. Res. Commun. **68**, 821–827

Hashimoto-Yutsudo, C., Y. Imai, and R. Sato (1980): J. Biochem. Tokyo **88**, 505–516

HAUGEN, D. A., T. A. VAN DER HOEVEN, and M. J. COON (1975): J. Biol. Chem. **250**, 3567–3570

HAUGEN, D. A., and M. J. COON (1976): J. Biol. Chem. **251**, 7929–7939

HAUGEN, D. A., M. J. COON, and D. W. NEBERT (1976): J. Biol. Chem. **251**, 1817–1827

HAUGEN, D. A., L. G. ARMES, K. T. YASUNOBU, and M. J. COON (1977): Biochem. Biophys. Res. Commun. **77**, 967–973

VAN DER HOEVEN, T. A., and M. J. COON (1974): J. Biol. Chem. **249**, 6302–6310

VAN DER HOEVEN, T. A., D. A. HAUGEN, and M. J. COON (1974): Biochem. Biophys. Res. Commun. **60**, 569–576

HORECKER, B. L. (1950): J. Biol. Chem. **183**, 593–605

ICHIHARA, K., E. KUSUNOSE, and M. KUSUNOSE (1972): FEBS Lett. **20**, 105–107

ICHIKAWA, Y., and T. YAMANO (1967): Biochim. Biophys. Acta **131**, 490–492

ILLING, H. P. A., and K. J. NETTER (1975): Xenobiotica **5**, 1–15

IMAI, Y. (1976): J. Biochem. Tokyo **80**, 267–276

IMAI, Y. (1979): J. Biochem. Tokyo **86**, 1697–1707

IMAI, Y. (1981): J. Biochem. Tokyo **89**, 351–362

IMAI, Y., and R. SATO (1967): Eur. J. Biochem. **1**, 419–426

IMAI, Y., and R. SATO (1974a): Biochem. Biophys. Res. Commun. **60**, 8–14

IMAI, Y., and R. SATO (1974b): J. Biochem. Tokyo **75**, 689–697

IMAI, Y., and R. SATO (1977): Biochem. Biophys. Res. Commun. **75**, 420–426

IMAI, Y., R. SATO, and T. IYANAGI (1977): J. Biochem. Tokyo **82**, 1237–1246

IMAI, Y., C. HASHIMOTO-YUTSUDO, H. SATAKE, A. GIRARDIN, and R. SATO (1980): J. Biochem. Tokyo **88**, 489–503

INGELMAN-SUNDBERG, M. (1977): Biochim. Biophys. Acta **488**, 225–234

INGELMAN-SUNDBERG, M., and H. GLAUMANN (1977): FEBS Lett. **78**, 72–76

INGELMANN-SUNDBERG, M., and J.-Å. GUSTAFSSON (1977): FEBS Lett. **74**, 103–106

INGELMAN-SUNDBERG, M., I. JOHANSSON, and A. HANSSON (1979): Acta Biol. Med. Germ. **38**, 379–388

INGELMAN-SUNDBERG, M., I. JOHANSSON, A. BRUNSTRÖM, G. EKSTRÖM, T. HAAPARANTA, and J. RYDSTRÖM (1980) in: Biochemistry, Biophysics and Regulation of Cytochrome P-450 (J.-Å. GUSTAFSSON, J. CARLSTEDT-DUKE, A. MODE, and J. RAFTER, eds.) pp. 299–306, Elsevier/North-Holland, Biomedical Press, Amsterdam, New York, Oxford

INGELMAN-SUNDBERG, M., and I. JOHANSSON (1980): Biochemistry **19**, 4004–4011

INGELMAN-SUNDBERG, M., and H. GLAUMANN (1980): Biochim. Biophys. Acta **599**, 417–436

IYANAGI, T., and H. S. MASON (1973): Biochemistry **12**, 2297–2308

IYANAGI, T., N. MAKINO, and H. S. MASON (1974): Biochemistry **13**, 1701–1710

IYANAGI, T., F. K. ANAN, Y. IMAI, and H. S. MASON (1978): Biochemistry **17**, 2224–2230

IYANAGI, T., and R. MAKINO (1980) in: Microsomes, Drug Oxidations and Chemical Carcinogenesis, Vol. 1 (M. J. COON, A. H. CONNEY, R. W. ESTABROOK, H. V. GELBOIN, J. R. GILLETTE, and P. J. O'BRIEN, eds.) pp. 407–410, Academic Press, New York, London, Toronto, Sydney, San Francisco

JÄNIG, G.-R., D. PFEIL, H. HONECK, and K. RUCKPAUL (1977): Acta Biol. Med. Germ. **36**, 35–44

JÄNIG, G.-R., D. PFEIL, and K. RUCKPAUL (1979): Acta Biol. Med. Germ. **38**, 409–422

JÄNIG, G.-R., SMETTAN, G., FRIEDRICH, J., BERNHARDT, R., RISTAU, O. and RUCKPAUL, K. (1982) in: Cytochrome P-450 – Biochemistry, Biophysics and Environmental Implications (E. HIETANEN, M. LAITINEN, and O. HÄNNINEN, eds) pp. 513–516, Elsevier/North-Holland Biomedical Press, Amsterdam

JEFCOATE, C. R. E., J. L. GAYLOR, and R. L. CALABRESE (1969): Biochemistry **8**, 3455–3463

JERINA, D. M., and J. W. DALY (1974): Science **185**, 573–582

JOHNSON, E. F., and U. MULLER-EBERHARD (1977a): Biochem. Biophys. Res. Commun. **76**, 644–651

JOHNSON, E. F., and U. MULLER-EBERHADR (1977b): Biochem. Biophys. Res. Commun. **76**, 652–659

JOHNSON, E. F., and U. MULLER-EBERHADR (1977c): J. Biol. Chem. **252**, 2839–2845

JOHNSON, E. F., M. C. ZOUNES, and U. MULLER-EBERHARD (1979a): Arch. Biochem. Biophys. **192**, 282–289

JOHNSON, E. F., G. E. SCHWAB, and U. MULLER-EBERHARD (1979b): Mol. Pharmacol. **15**, 708–718

JOHNSON, E. F. (1980a): J. Biol. Chem. **255**, 304–309

JOHNSON, E. F. (1980b) in: Biochemistry, Biophysics and Regulation of Cytochrome P-450(J.-Å. GUSTAFSSON, J. CARLSTEDT-DUKE, A. MODE, and J. RAFTER, eds.) pp. 65–72, Elsevier/North–Holland, Biomedical Press, Amsterdam, New York, Oxford

JUCHAU, M. R. (1980): Pharmacol. Therap. A 8, 501–524

KAMATAKI, T., D. H. BELCHER, and R. A. NEAL (1976a): Mol. Pharmacol. **12**, 921–932

KAMATAKI, T., M. C. M. LEELIN, D. H. BELCHER, and R. A. NEAL (1976b): Drug Metabol. Disp. **4**, 180–189

KAMATAKI, T., M. SUGIURA, Y. YAMAZOE, and R. C. KATO (1979): Biochem. Pharmacol. **28**, 1993–2000

KAMIN, H., B. S. S. MASTERS, Q. J. GIBSON, and C. H. WILLIAMS (1965a): Fedn. Proc. **24**, 1164–1171

KAMIN, H., B. S. S. MASTERS, Q. J. GIBSON, and C. H. WILLIAMS (1965b): J. Biol. Chem. **240**, 921–931

KAMINSKY, L. S., M. J. FASCO, and F. P. GUENGERICH (1979): J. Biol. Chem. **254**, 9657–9662

KAMPFFMEYER, H., and M. KIESE (1964): Arch. Exp. Path. Pharmak. **246**, 397–412

KARUZINA, I. I., G. I. BACHMANOVA, D. E. MENGAZETDINOV, K. N. MYASOEDOVA, V. O. ZHIKHAREVA, G. P. KUZNETSOVA, and A. I. ARCHAKOV (1979): Biochimija (russ.) **44**, 1049–1057

KASCHNITZ, R. M. and M. J. COON (1975): Biochem. Pharmacol. **24**, 295–297

KATAGIRI, M., B. N. GANGULI, and I. C. GUNSALUS (1968): J. Biol. Chem. **243**, 3543–3546

KATO, R., A. TAKANAKA, and M. TAKAYANAGHI (1970): J. Biochem. Tokyo **68**, 395–413

KAWALEK, J. C., W. LEVIN, D. RYAN, P. E. THOMAS, and A. Y. H. LU (1975): Mol. Pharmacol. **11**, 874–878

KAWALEK, J. C., and A. Y. H. LU (1975): Mol. Pharmacol. **11**, 201–210

KAWALEK, J. C., W. LEVIN, D. RYAN, and A. Y. H. LU (1977): Arch. Biochem. Biophys. **183**, 732–741

KITADA, M. and T. KAMATAKI (1979): Biochem. Pharmacol. **28**, 793–797

KITADA, M., H. KITAWAGA, and T. KAMATAKI (1979): Biochem. Pharmacol. **28**, 2670–2673

KITCHIN, K. T., and J. S. WOODS (1978): Mol. Pharmacol. **14**, 890–899

KLINGER, W., D. MÜLLER, and U. KLEEBERG (1979) in: The Induction of Drug Metabolism (R. W. ESTABROOK and E. LINDENLAUB, eds.) pp. 517–554, Schattauer, Stuttgart, New York

KNAPP, J. A., J. D. DIGNAM, and H. W. STROBEL (1977): J. Biol. Chem. **252**, 437–443

KOOP, D. R. and M. J. COON (1979): Biochem. Biophys. Res. Commun. **91**, 1075–1081

KUMAKI, K., and D. W. NEBERT (1978): Pharmacology **17**, 262–279

LAEMMLI, U. K. (1970): Nature **227**, 680–685

LANG, M. (1976): Biochem. Biophys. Acta **455**, 947–960

LAZAR, T., H. EHRIG, and C. LUMPER (1977): Eur. J. Biochem. **76**, 365–371

LEVIN, W., A. Y. H. LU, D. RYAN, W. WEST, R. KUNTZMAN, and A. H. CONNEY (1972): Arch. Biochem. Biophys. **153**, 543–553

LEVIN, W., D. RYAN, S. WEST, and A. Y. H. LU (1974): J. Biol. Chem. **249**, 1747–1754

LEVIN, W., A. W. WOOD, A. Y. H. LU, D. RYAN, S. WEST, A. H. CONNEY, D. R. THAKKER, H. YAGI, and D. M. JERINA (1977) in: Drug Metabolism Concepts (ed. D. M. JERINA) pp. 99–122, American Chem. Soc., Washington, D. C.

LICHT, H J., and C. J. COSCIA (1978): Biochemistry **17**, 5638–5646

LIEM, H. H., U. MULLER-EBERHARD, and E. F. JOHNSON (1980): Mol. Pharmacol. **18**, 565–570

LU, A. Y. H., and M. J. COON (1968): J. Biol. Chem. **243**, 1331–1332

LU, A. Y. H., K. W. JUNK, and M. J. COON (1969a): J. Biol. Chem. **244**, 3714–3721

LU, A. Y. H., H. W. STROBEL, and M. J. COON (1969b): Biochem. Biophys. Res. Commun. **36**, 545–551

LU, A. Y. H., H. W. STROBEL, and M. J. COON (1970): Mol. Pharmacol. **6**, 213–220

LU, A. Y. H., and W. LEVIN (1972): Biochem. Biophys. Res. Commun. **46**, 1334–1339

LU, A. Y. H., and S. B. WEST (1972): Mol. Pharmacol. **8**, 490–500

LU, A. Y. H., W. LEVIN, and R. KUNTZMAN (1974): Biochem. Biophys. Res. Commun. **60** 266–272

LU, A. Y. H., and G. T. MIWA (1980a) in: Biochemistry, Biophysics and Regulation of Cytochrome P-450 (J.-Å. GUSTAFSSON, J. CARLSTEDT-DUKE, A. MODE, and J. RAFTER eds.) pp. 125–128, Elsevier/North-Holland, Biomedical Press, Amsterdam, New York, Oxford

LU, A. Y. H., and G. T. MIWA (1980b): Ann. Rev. Pharmacol. Toxicol. **20**, 513–521

LU, A. Y. H., G. T. MIWA, and S. B. WEST (1980a) in: Microsomes, Drug Oxidations and Chemical Carcinogenesis, Vol. 1 (M. J. COON, A. H. CONNEY, R. W. ESTABROOK, H. V. GELBOIN, J. R. GILLETTE, and P. J. O'BRIEN, eds.) pp. 59–67, Academic Press, New York, London, Toronto, Sydney, San Francisco

LU, A. Y. H., J. W. MORIN, and M. WARNER (1980b) in: Biochemistry, Biophysics and Regulation of Cytochrome P-450 (J.-Å. GUSTAFSSON J., CARLSTEDT-DUKE, A. MODE and J. RAFTER, eds.) pp. 17–24 Elsevier/North-Holland, Biomedical Press, Amsterdam, New York, Oxford

LUMPER, L., F. BUSCH, S. DZELIC, J. HENNING, and T. LAZAR (1980): Int. J. Pept. Prot. Res. **16**, 83–96

MABREY, S., and J. M. STURTEVANT (1978): Methods Membr. Biol. **9**, 237–274

MADYASTHA, K. M., und C. J. COSCIA (1979): J. Biol. Chem. **254**, 2419–2427

MAILMAN, R. B., A. P. KULKARNI, R. C. BAKER and E. HODGSON (1974): Drug. Metab. Disp. **2**, 301–308

MAILMAN, R. B., L. G. TATE, K. E. MUSEM, L. B. COONS, and E. HODGSON (1975): Chem. Biol. Interact. **10**, 215–228

MARICIC, S., B. BENKO, G. BUTSCHAK, G. SCHEUNIG, G.-R. JÄNIG, H. REIN, and K. RUCKPAUL (1979): Acta Biol. Med. Germ. **38**, 217–233

MARTIN, Y. C., and C. HANSCH (1971): J. Med. Chem. **14**, 777–779

MASTERS, B. S. S., H. KAMIN, Q. H. GIBSON, and C. H. WILLIAMS (1965): J. Biol. Chem. **240**, 921–931

MASTERS, B. S. S., R. A. PROUGH, and H. KAMIN (1975): Biochemistry **14**, 607–613

MASTERS, B. S. S. and R. T. OKITA (1980): Pharmacol. Therap. A **9**, 227–244

MASUDA-MIKAWA, R., Y. FUJII-KURIYAMA, M. NEGISHI, and Y. TASHIRO (1979): J. Biochem. Tokyo **86**, 1383–1394

MATSUURA, S., Y. FUJII-KURIYAMA, and Y. TASHIRO (1978): J. Cell Biol. **78**, 503–519

MATTHEWS, D. A., R. A. ALDEN, S. T. FREER, N. XUONG, and J. KRAUT (1979) J. Biol. Chem. **254**, 4144–4151

MAYER, R. T., and J. L. DURRANT (1979): J. Biol. Chem. **254**, 756–761

MCINTOSH, P. R., and R. B. FREEDMAN (1979): FEBS Lett. **105**, 217–223

MCINTOSH, P. R., and R. B. FREEDMAN (1980): Biochem. J. **187**, 227–237

MCMAHON, R. E. (1961): J. Med. Pharmacol. Chem. **4**, 67–78

MIKI, N., T. SUGIYAMA, and T. YAMANO (1980a) in: Microsomes, Drug Oxidations and Chemical Carcinogenesis, Vol. 1 (M. J. COON, A. H. CONNEY, R. W. ESTABROOK, H. V. GELBOIN, J. R. GILETTE, and P. J. O'BRIEN, eds.) pp. 27–36, Academic Press, New York, London Toronto, Sydney, San Francisco

MIKI, N., T. SUGIYAMA, and T. YAMANO (1980b): J. Biochem. Tokyo **88**, 307–316

MISSELWITZ, R., G.-R. JÄNIG, H. REIN, E. BUDER, D. ZIRWER, and K. RUCKPAUL (1975): Acta Biol. Med. Germ. **34**, 1755–1766

MISSELWITZ, R., G.-R. JÄNIG, H. REIN, E. BUDER, D. ZIRWER, and K. RUCKPAUL (1976) Acta Biol. Med. Germ. **35**, K19–K25

MISSELWITZ, R., G.-R. JÄNIG, H. REIN, E. BUDER, D. ZIRWER, and K. RUCKPAUL (1977): Acta Biol. Med. Germ. **36**, K35–K41

MISSELWITZ, R., H. REIN, O. RISTAU, G.-R. JÄNIG, D. ZIRWER, and K. RUCKPAUL (1980): Stud. Biophysica **79**, 165–166

MITANI, F. (1979) Mol. Cell Biochem. **24**, 21–43

MIWA, G. T., S. B. WEST, and A. Y. H. LU (1978): J. Biol. Chem. **253**, 1921–1929

MIWA, G. T., S. B. WEST, M.-T. HUANG, and A. Y. H. LU (1979): J. Biol. Chem. **254** 5695–5700

MIWA, G. T., J. S. ZWIEG, J. S. A. WALSH, and A. Y. H. LU (1980) in: Microsomes, Drug Oxidations and Chemical Carcinogenesis, Vol. 1 (M. J. COON, A. H. CONNEY, R. W. ESTABROOK, H. V. GELBOIN, J. R. GILLETTE, and P. J. O'BRIEN, eds.) pp. 363–366, Academic Press, New York, London, Toronto, Sydney, San Francisco

MURAKAMI, K., and H. S. MASON (1967): J. Biol. Chem. **242**, 1102–1110

NEBERT, D. W., J. R. ROHINSON, and H. KON (1973): J. Biol. Chem. **248**, 7637–7647

NEBERT, D. W. (1979) Mol. Cell Biochem. **27**, 27–46

NEBERT, D. W., and N. M. JENSEN (1979): CRC Crit. Rev. Biochem. 6, 401–437
NORMAN, R. L., E. F. JOHNSON, and U. MULLER-EBERHARD (1978): J. Biol. Chem. 253, 8640–8647
NOORDHOEK, J., A. P. VAN DEN BERG, E. M. SAVENIJE-CHAPEL, and E. KOOPMAN-KOOL (1977) in: Microsomes and Drug Oxidations (V. ULLRICH, I. ROOTS, A. HILDEBRANDT, R. W. ESTABROOK, and A. H. CONNEY, eds.) pp. 534–542, Pergamon Press, Oxford
NOZAKI, Y., and C. TANFORD (1971): J. Biol. Chem. 246, 2211–2217
O'BRIEN, P. J. (1978): Pharmacol. Therap. A 2, 517–536
OKUDA, K., P. WEBER, and V. ULLRICH (1977): Biochem. Biophys. Res. Commun. 74, 1071–1076
OMURA, T. and R. SATO (1964): J. Biol. Chem. 239, 2370–2378
OMURA, T., E. SANDERS, R. W. ESTABROOK, D. Y. COOPER, and O. ROSENTHAL (1966): Arch. Biochem. Biophys. 117, 660–673
OMURA, T., and S. TAKESUE (1970): J. Biochem. Tokyo 67, 249–256
ORRENIUS, S. (1965): J. Cell. Biol. 26, 713–724
ORRENIUS, S., and H. THOR (1969): Eur. J. Biochem. 9, 415–418
PEDERSON, J. I., M. OFTEBRO, and T. VANNGÅRD (1977): Biochem. Biophys. Res. Commun. 76, 666–673
PEDERSEN, J. I. (1978): FEBS Lett. 85, 35–39
PEDERSEN, J. I., and H. K. GODAGER (1978): Biochim. Biophys. Acta 525, 28–36
PEDERSON, T. C., and S. D. AUST (1970): Biochem. Pharmacol. 19, 2221–2230
PEDERSON, T. C., and S. D. AUST (1972): Biochem. Biophys. Res. Commun. 48, 789–795
PEDERSON, T. C., J. A. BUEGE, and S. D. AUST (1973): J. Biol. Chem. 248, 7134–7141
PELKONEN, O., E. H. KALTIALA, T. K. I. LARMI, and N. T. KÄRKI (1974): Chem. Biol. Interact. 9, 205–216
PELKONEN, O., E. H. KALTIALA, N. T. KÄRKI, K. JALONEN, and K. PYÖRÄLÄ (1975): Xenobiotica 5, 501–509
PELKONEN, O., and H. VAINIO (1975): FEBS Lett. 51, 11–14
PETERSON, J. A. (1971): Arch. Biochem. Biophys. 144, 678–693
PETERSON, J. A., R. E. WHITE, Y. YASUKOCHI, M. L. COOMES, D. H. O'KEEFFE, R. E. EBEL, B. S. S. MASTERS, D. P. BALLOU, and M. J. COON (1977): J. Biol. Chem. 252, 4431–4434
PHILLIPS, T. C., and R. G. LANGDON (1962): J. Biol. Chem. 237, 2652–2660
PHILPOT, R. M., and E. ARINC (1976): Mol. Pharmacol. 12, 483–493
PROKROVSKY, A., V. MISHIN, N. RIFKIND, and V. LYAKHOVICH (1977): Biochem. Biophys. Res. Commun. 77, 912–917
PROUGH, R. A., and M. D. BURKE (1975): Arch. Biochem. Biophys. 170, 160–168
RAJ, P., E. B. NELSON, and R. W. ESTABROOK (1971): Chem. Biol. Interact. 3, 303–304
REMMER, H., J. SCHENKMAN, R. W. ESTABROOK, H. SASAME, J. GILLETTE, S. NARASIMHULU, D. Y. COOPER, and O. ROSENTHAL (1966): Mol. Pharmacol. 2, 187–190
RISTAU, O., H. REIN, G.-R. JÄNIG, and K. RUCKPAUL (1978): Biochim. Biophys. Acta 536, 226–234
ROGAN, E., and E. CAVALIERI (1978): Mol. Pharmacol. 14, 215–219
RUCKPAUL, K., and H. REIN (1980): Acta Biol. Med. Germ. 39, 55–69
RUCKPAUL, K., H. REIN, J. BLANCK, O. RISTAU, and M. J. COON (1980a) in: Biochemistry, Biophysics and Regulation of Cytochrome P-450 (J.-Å. GUSTAFSSON, J. CARLSTEDT-DUKE, A. MODE, and J. RAFTER, eds.) pp. 539–549, Elsevier/North-Holland, Biomedical Press, Amsterdam, New York, Oxford
RUCKPAUL, K., H. REIN, D. P. BALLOU, and M. J. COON (1980b): Biochim. Biophys. Acta 626, 41–56
RUF, H. H., H. REINHARDT, H. GRAF, and J. POENSGEN (1980) in: Microsomes, Drug Oxidations and Chemical Carcinogenesis, Vol. 1 (M. J. COON, A. H. CONNEY, R. W. ESTABROOK, H. V. GELBOIN, J. R. GILLETTE, and P. J. O'BRIEN, eds.) pp. 529–532, Academic Press, New York, London, Toronto, Sydney, San Francisco
RUF, H. H., P. WENDE, and V. ULLRICH, (1979) J. Inorg. Biochem. 11, 189–204
RYAN, D., A. Y. H. LU, J. KAWALEK, S. B. WEST, and W. LEVIN (1975a) Biochem. Biophys. Res. Commun. 64, 1134–1141
RYAN, D., A. Y. H. LU, S. B. WEST, and W. LEVIN (1975b): J. Biol. Chem. 250, 2157–2163
RYAN, D., P. E. THOMAS, and W. LEVIN (1977): Mol. Pharmacol. 13, 521–532

RYAN, D., P. E. THOMAS, D. KORZENIOWSKI, and W. LEVIN (1979): J. Biol. Chem. **254**, 1365—1374

RYAN, D., P. E. THOMAS, and W. LEVIN (1980): J. Biol. Chem. **255**, 7941—7955

SAINE, S. E., W.-F. FANG, and H. W. STROBEL (1978): Biochim. Biophys. Acta **526**, 345—358

SAITO, T., and H. W. STROBEL (1981): J. Biol. Chem. **256**, 984—988

SATO, R., Y. ATSUTA, Y. IMAI, S. TANIGUCHI, and K. OKUDA (1977): Proc. Natl. Acad. Sci. U.S. **74**, 5477—5481

SCHENKMAN, J. B., H. REMMER, and R. W. ESTABROOK (1967): Mol. Pharmacol. **3**, 113—123

SCHWAB, G. E., R. L. NORMAN, U. MULLER-EBERHARD, and E. F. JOHNSON (1980): Mol. Pharmacol. **17**, 218—224

SERABJIT-SINGH, C. J., C. R. WOLF, and R. M. PHILPOT (1979): J. Biol. Chem. **254**, 9901—9907

SHIMIZU, T.; T. NOZAWA, M. HATANO, H. SATAKE, Y. IMAI, C. HASHIMOTO, and R. SATO (1979): Biochim. Biophys. Acta **579**, 122—133

SHIVERICK, K. T., and A. H. NEIMS (1979): Drug Metab. Disp. **7**, 290—295

SIEGEL, L. M., E. J. FAEDER, and H. KAMIN (1972): Z. Naturforschung B **276**, 1087—1089

SIMONS, K., A. HELENIUS, K. LEONARD, M. SARVAS, and M. J. GETHING (1978): Proc. Natl. Acad. Sci. U.S. **75**, 5306—5310

SIPAL, Z., P. ANZENBACHER, P. PUTZ, J. CHLUMSKY, and O. KRIVANOVA (1979): Acta Biol. Med. Germ. **38**, 483—494

SPATZ, L., and P. STRITTMATTER (1971): Proc. Natl. Acad. Sci. U.S. **68**, 1042—1046

SPATZ, L., and P. STRITTMATTER (1973): J. Biol. Chem. **248**, 793—799

STROBEL, H. W., A. Y. H. LU, J. HEIDEMA, and M. J. COON (1970): J. Biol. Chem. **245**, 4851—4854

STROBEL, H. W., J. D. DIGNAM, S. E. SAINE, W.-F. FANG, and P. M. FENNELL (1978): Mol. Cell Biochem. **22**, 79—91

SUGIYMA, T., N. MIKI, and T. YAMANO (1979): Biochem. Biophys. Res. Commun. **90**, 715—720

SUGIYAMA, T., N. MIKI, and T. YAMANO (1980): J. Biochem. Tokyo **87**, 1457—1467

TANIGUCHI, H., Y. IMAI, T. IYANAGI, and R. SATO (1979): Biochim. Biophys. Acta **550**, 341—356

TANIGUCHI, H., Y. IMAI, and R. SATO (1980): in: Microsomes, Drug Oxidations and Chemical Carcinogenesis, Vol. 1 (M. J. COON, A. H. CONNEY, R. W. ESTABROOK, H. V. GELBOIN, J. R. GILETTE, and P. J. O'BRIEN, eds.) pp. 537—540, Academic Press, New York, London, Toronto, Sydney, San Francisco

THOMAS, P. E., A. Y. H. LU, D. RYAN, S. B. WEST, J. KAWALEK, and W. LEVIN (1976a): Mol. Pharmacol. **12**, 746—758

THOMAS, P. E., A. Y. H. LU, D. RYAN, S. B. WEST, J. KAWALEK, and W. LEVIN (1976b): J. Biol. Chem. **251**, 1385—1391

THOMAS, P. E., D. KORZENIOWSKI, D. RYAN, and W. LEVIN (1979): Arch. Biochem. Biophys. **192**, 524—532

TROUT, G. E. (1977): Eur. J. Biochem. **71**, 533—537

TSONG, T. Y., and C. S. YANG (1978): Proc. Natl. Acad. Sci. U.S. **75**, 5955—5959

ULLRICH, V., and P. KREMERS (1977): Arch. Toxicol. **39**, 41—50

VATSIS, K. P., A. D. THEOHARIDES, D. KUPFER and M. J. COON (1980a) in: Biochemistry, Biophysics and Regulation of Cytochrome P-450 (J.-Å. GUSTAFSSON, J. CARLSTEDT-DUKE, A. MODE, and J. RAFTER, eds.) pp. 73—76, Elsevier/North-Holland Biomedical Press, Amsterdam, New York, Oxford

VATSIS, K. P., D. P. GURKA, and P. F. HOLLENBERG (1980b) in: Biochemistry, Biophysics and Regulation of Cytochrome P-450 (J.-Å. GUSTAFSSON, J. CARLSTEDT-DUKE, A. MODE, and J. RAFTER, eds.) pp. 347—350 Elsevier/North-Holland Biomedical Press, Amsterdam, New York, Oxford

VATSIS, K. P., Y. L. CHIANG, and M. J. COON (1977): Fedn. Proc. **36**, 832

VERMILION, J. L., and M. J. COON (1974): Biochem. Biophys. Res. Commun. **60**, 1315—1322

VERMILION, J. L., and M. J. COON (1978a): J. Biol. Chem. **253**, 2694—2704

VERMILION, J. L., and M. J. COON (1978b): J. Biol. Chem. **253**, 8812—8819

VERMILION, J. L., V. MASSEY, D. P. BALLOU, and M. J. COON (1980): in: Microsomes, Drug Oxidations and Chemical Carcinogenesis, Vol. 1 (M. J. COON, A. H. CONNEY. R. W. ESTABROOK, H. V. GELBOIN, J. R.

GILLETTE, and P. J. O'BRIEN, eds.) pp. 379–382, Academic Press, New York, London, Toronto, Sydney, San Francisco
VERMILION, J. L., D. P. BALLOU, V. MASSEY, and M. J. COON (1981: J. Biol. Chem. **256**, 266–277
VLASUK, G. P., and F. G. WALZ (1980): Anal. Biochem. **105**, 112–120
WANG, P., P. S. MASON, and F. P. GUENGERICH, (1980): Arch. Biochem. Biophys. **199**, 206–219
WARNER, M., M. LAMARCA, and A. H. NEIMS (1978): Drug Metab. Dispos. **6**, 353–362
WARNER, M., and A. H. NEIMS (1979): Drug Metab. Dispos. **7**, 188–193
WELTON, A. F., and S. D. AUST (1974): Biochem. Biophys. Res. Commun. **56**, 898–906
WELTON, A. F., and S. D. AUST (1975): Biochem. Pharmacol. **24**, 1641–1644
WELTON, A. F., F. O. O'NEAL, L. C. CHANEY, and S. D. AUST (1975): J. Biol. Chem. **250**, 5631–5639
WEST, S. B., M.-T. HUANG, G. T. MIWA and A. Y. H. LU (1979): Arch. Biochem. Biophys. **193**, 42–50
WEST, S. B., G. T. MIWA, and A. Y. H. LU (1980) in: Microsomes, Drug Oxidations and Chemical Carcinogenesis, Vol. **1** (M. J. COON, A. H. CONNEY, R. W. ESTABROOK, H. V. GELBOIN, J. R. GILLETTE, and P. J. O'BRIEN, eds.) pp. 191–194, Academic Press, New York, London, Toronto, Sydney, San Francisco
WHITE, R. E. and M. J. COON, (1982): J. Biol. Chem. **257**, 3073–3083
WIKVALL, K., H. DANIELSSON, and R. HANSSON (1980) in: Biochemistry, Biophysics and Regulation of Cytochrome P-450 (J.-Å. GUSTAFSSON, J. CARLSTEDT-DUKE, A. MODE, and J. RAFTER, eds.) pp. 499–504 Elsevier/North-Holland, Biomedical Press, Amsterdam, New York, Oxford
WILLIAMS, C. H., and H. KAMIN (1962): J. Biol. Chem. **237**, 587–595
WOLF, C. R., S. R. SLAUGHTER, J. P. MARCINISZYN, and R. M. PHILPOT (1980): Biochim. Biophys. Acta **624**, 409–419
WOOD, A. W., W. LEVIN, A. Y. H. LU, M. YAGI, O. HERNANDEZ, D. M. JERINA, and A. H. CONNEY (1976): J. Biol. Chem. **251**, 4882–4890
YANG, C. S., F. S. STRICKHART, and L. P. KICHA (1978): Biochem. Biophys. Acta **509**, 326–337
YANG, C. S., and T. Y. TSONG (1980): Microsomes, Drug Oxidations, and Chemical Carcinogenesis, Vol. 1 (M. J. COON, A. H. CONNEY, R. W. ESTABROOK, H. V. GELBOIN, J. R. GILETTE, and P. J. O'BRIEN, eds.) pp. 199–202, Academic Press, New York, London, Toronto, Sydney, San Francisco
YASUKOCHI, Y., and B. S. S. MASTERS (1976): J. Biol. Chem. **251**, 5337–5344
YASUKOCHI, Y., J. A. PETERSON, and B. S. S. MASTERS (1979): J. Biol. Chem. **254**, 7097–7104
YASUNOBU, K. T., L. G. ARMES, J. W. CRABB, M. HANIU, I. C. GUNSALUS, and M. J. COON (1980): Biochemistry, Biophysics and Regulation of Cytochrome P-450 (J.-Å. GUSTAFSSON, J. CARLSTEDT-DUKE, A. MODE, and J. RAFTER, eds.) pp. 49–56, Elsevier/North-Holland Biomedical Press, Amsterdam, New York, Oxford
YOSHIDA, Y., Y. IMAI, and C. HASHIMOTO-YUTSUDO (1982): J. Biochem. Tokyo **91**, 1651–1659

Chapter 3

The Cytochrome P-450 Reaction Mechanism — Kinetic Aspects

J. BLANCK, G. SMETTAN and S. GRESCHNER

Contents

Introduction

The kinetics of the reaction mechanism of cytochrome P-450 (P-450) has been extensively investigated. The studies were applied to the native enzyme of different species as well as to reconstituted systems. Either the overall function — as studied in the substrate conversion process (K_M, V_{max}) — and partial reactions in the catalytic cycle have been analysed. The literature has accumulated, the size of which is determined by the structural and functional complexity of the manyfold systems. General aspects of the reaction mechanism have been recently reviewed (BJÖRKHEM, 1977; DEPIERRE and ERNSTER, 1977; BLANCK and SMETTAN, 1978; CHANG and DOLPHIN, 1978; GUNSALUS and SLIGAR, 1978; ISHIMURA 1978; JUNG and RISTAU, 1978; LU and WEST, 1978; O'BRIEN, 1978; REIN and RISTAU, 1978; ESTABROOK et al., 1979b; GROVES, 1979; GUENGERICH, 1979;

Abbreviations

P-450	Cytochrome P-450 (EC 1.14.14.1)
P-450 LM	Liver Microsomal Cytochrome P-450

P–450 LM2, LM4 ...	Subfractions of Rabbit Liver Microsomal Cytochrome P–450
P–450 a, A ...	Subfractions of Rat Liver Microsomal Cytochrome P–450
P–450 A_1, A_2 ... C_1, C_2 ..	Subfractions of Mouse Liver Microsomal Cytochrome P–450
P–450 CAM	Bacterial Camphor Hydroxylating Cytochrome P–450 (*Pseudomonas putida*)
P–450 AC	Adrenocortical Mitochondrial Cytochrome P–450
P–450 SCC	Adrenocortical Cytochrome P–450, Side Chain Cleavage Subfraction
P–450 11β	Adrenocortical Cytochrome P–450, 11β Hydroxylating Subfraction
P–448 (P_1–450)	Cytochrome P–448
P–420	Cytochrome P–420, Catalytically Inactive
P–450_0	Cytochrome P–450, oxidized (Fe^{3+})
P–450_r	Cytochrome P–450, reduced (Fe^{2+})
P–450 Reductase	NADPH-dependent Cytochrome P–450 (Cytochrome c) Reductase (EC 1.6.2.4)
(F_1, F_2)	Cytochrome P–450 Reductase (FAD, FMN — Constituting Flavoproteins)
b_5	Cytochrome b_5
b_5 Reductase	NADH-dependent Cytochrome b_5 Reductase (EC 1.6.2.2)
F_p	Cytochrome b_5 Reductase (F_p-Flavoprotein)
hs, ls	Cytochrome P–450 High-Spin State, Low-Spin State Configuration
e_1, e_2	Electron Transfer Reactions 1, 2
K_1, K_2 ...	Equilibrium Constants
k_1, k_2 .. k_{11}, k_{12} ..	Rate Constants of Partial Reactions
RH, ROH	Substrate, Product
ROOH, XOOH	Oxygen Supporting Agent
BP	Benzphetamine
CHP	Cumene Hydroperoxide
3-MC, MC	3-Methylcholanthrene
NEM	N-Ethylmaleimide
PB	Sodium Phenobarbital

Coon and White, 1980; Gander and Mannering, 1980; White and Coon, 1980).

Main emphasis of the present review is directed to the kinetics of the respective processes, which control oxygen activation and electron transfer, as well. It will refer predominantly to the mammalian liver system, supporting data from other species (systems) are exceptionally treated too. In general the results of specified investigations must be checked by parallel studies with supplementary methods. This holds especially true with respect to kinetics. Despite of such restrictions there is no chance to understand the function of enzymes without kinetic analysis of elementary steps and partial reactions. These points of view should be applied to the discussion and evaluation of the presented results.

P–450 systems catalyze the biotransformation of drugs, xenobiotics, steroids, fatty acids, bile acids, and hydrocarbons, i.e. the conversion of endogeneous as well as of exogeneous substrates, by means of hydroxylation. The reaction proceeds via the activation of molecular oxygen which binds to the Fe(II) atom of the terminal oxidase, the P–450 molecule; this is attained by provision of 2 electrons from a flavoprotein, the NADPH-dependent P–450 reductase. The distinct couple of oxygen activation and electron transfer represents the specific feature of any P–450 system.

The kinetic analysis of mechanisms of such systems should be directed predominantly to the following topics:

a) overall mechanism
b) partial reactions
c) rate-determining steps
d) specificity of component interactions.

The proposed questions will be treated in the context; subdivisions are rather determined by the respective reaction step. The overall reaction with respect to the substrate hydroxylation has been established by MASON (1957)

$$RH + O_2 + NADPH + H^+ \rightarrow ROH + H_2O + NADP^+ \quad (1)$$

The enzyme system generates this way the hydroxylated product but is furthermore capable of "uncoupling" and of formation of hydrogen peroxide

$$O_2 + NADPH + H^+ \rightarrow H_2O_2 + NADP^+ \quad (2)$$

The degree of uncoupling depends specifically on the respective substrate (ESTABROOK and WERRINGLOER, 1977; NORDBLOM and COON, 1977).

Alkylhydroperoxides (included H_2O_2), peracids, sodium chlorite, iodosobenzene can be used to by-pass the physiologic NADPH dependent O_2 activation; these compounds can directly provide the required oxygen (HRYCAY et al., 1975a; DANIELSSON and WIKVALL, 1976; GUSTAFSSON and BERGMAN, 1976)

$$RH + XOOH \rightarrow ROH + XOH \quad (3)$$

The general reaction cycle and its constituting partial reactions is represented in Fig. 1. The sequence contains the binding processes for either substrate ① and O_2 ③ in the formation of the

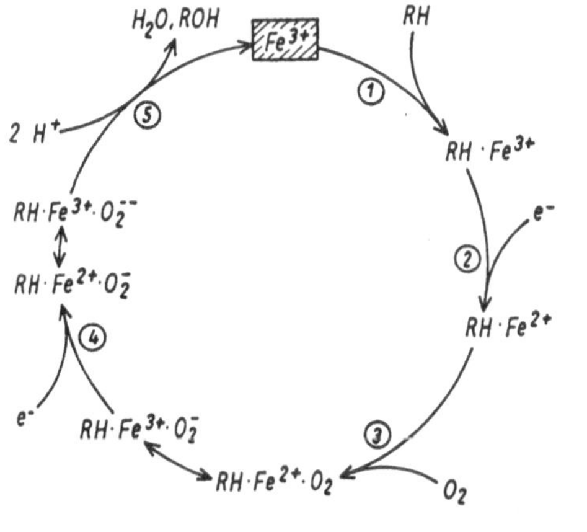

Fig. 1. Reaction Cycle of the P–450 Catalyzed Substrate Hydroxylation

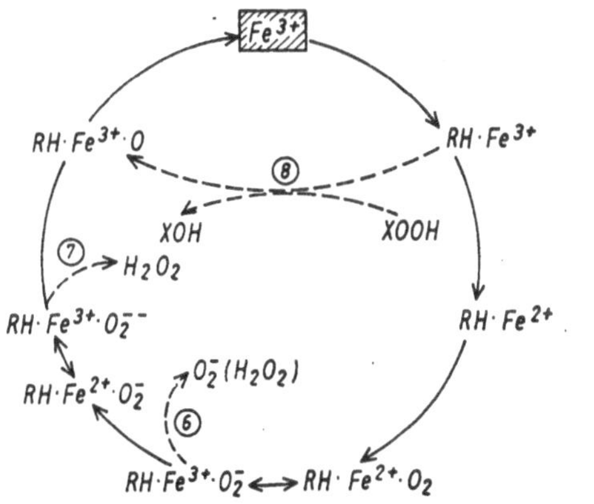

Fig. 2. Side Reactions of the P–450 Cycle (Uncoupling, Shunt)

respective P–450 binary- and ternary complexes, the electron transfer steps ② and ④ (introduction of the first and second electron), and the composed substrate conversion process ⑤. The extended Fig. 2 further considers the above drawn reactions (2) (uncoupling) and (3) (shunt) as partial steps ⑥, ⑦, ⑧.

The reaction cycle generally refers to P–450 systems of different species. Specifications occur with respect to structure and function of the two main proteins (P–450, P–450 reductase), to the requirement for further components (ferredoxins), to the demand for phospholipid, and to stoichiometry and organization of the system components (OMURA, 1978).

The liver enzyme system (LM) as obtained from the endoplasmic reticulum of mammalian liver cells is composed of 3 main components, the cytochrome P–450 (EC 1.14.14.1), the NADPH-dependent cytochrome P–450 reductase (EC 1.6.2.4), and of phospholipid (~55% phosphatidylcholine, ~25% phosphatidylethanolamine) (LU and COON, 1968; COON and WHITE, 1980). The composition stoichiometry has been determined with about 1:20/40 for the reductase:P–450 ratio, and to 1:150/180 for the P–450: phospholipid ratio (ESTABROOK et al., 1971b; DEPIERRE and DALLNER, 1975).

The enzyme system is membrane-bound, its function is specifically phospholipid controlled (STROBEL et al., 1971; COON et al., 1976); that holds further true for the XOOH shunt pathway ⑥ (NORDBLOM et al., 1976). The liver cells further contain a second electron transfer system, the NADH-dependent cytochrome b_5 reductase and cytochrome b_5 as electron acceptor. The dependence of P–450 hydroxylation reactions on the NADH donation function evidences some interference with the b_5 system; possible partial reactions of this synergism are discussed in section 6.

The overall hydroxylation process and partial steps in the P–450 LM system, respectively, have been predominantly investigated at microsomes, i.e. at the "biochemical equivalent" of the endoplasmic reticulum. By means of alterations in the components stoichiometry (incorporation of distinct components) further insight in rate-limiting constituents is obtained (MIWA et al., 1978). Alterations in the integrational state of the system (solubilization, detergent treatment) represent another access for the correlation of functional and structural properties.

The components of the LM system can be isolated with high purity (P–450: IMAI and SATO, 1974; HASHIMOTO and IMAI, 1975; RYAN et al., 1975; HAUGEN and COON, 1976; P–450 reductase: VERMILION and COON, 1974; DIGNAM and STROBEL, 1975, 1977; YASUKOCHI and MASTERS, 1976). This procedure enables the reconstitution of systems of different stoichiometric constitution — especially with P–450 subfractions, as well as with different phospholipids. This way is much important with respect to the kinetic analysis, because of the attainable broad range of concentration variations — which is required for mechanistic studies.

The P–450 LM system exhibits rather less specificity towards substrates. In contrast to the specific P–450 CAM system from *Pseudomonas putida* or to the adrenal cortex P–450 11β and P–450 SCC systems, the liver system is capable of the conversion of numerous compounds. The respective agents modify somewhat the basic reactions, i.e. the LM-system shows at least substrate specific properties. The analysis of these modifications is a further possibility to elucidate components interactions.

The physiologic relevance of kinetic studies should be rather high with respect to the investigation of microsomal systems. The analysis of artificial systems, on the other hand, may be impaired by isolation and purification procedures as well as by reconstitution techniques. Therefore results from reconstituted systems must be carefully checked with regard to its generalization to and relevance for the physiologic system, respectively.

1. Substrate and Ligand Binding

The substrate binding process represents the first step ① in the P–450 reaction sequence. This partial reaction is drawn in Fig. 3 in the CLELAND notation. That notation is further used with respect to the following steps and intermediate mechanisms, respectively.

Substrate binding to the cytochrome P–450 leads to the formation of characteristic difference spectra, the differentiation of which being due to specificities of the binding geometry (first investigations: IMAI and SATO, 1966; REMMER et al., 1966; SCHENKMAN et al., 1967; reviews: ORRENIUS et al., 1972; YOSHIDA and KUMAOKA, 1975; JEFCOATE, 1978). Three types of substrate complexes can be distinguished:

— Predominantly apolaric substances (type I) are bound by interactions of apolaric (hydrophobic) groups with respective structures of the cytochrome receptor (WHITE et al., 1980a). The difference spectra exhibit extrema at about 390 nm (max) and 420 nm (min),

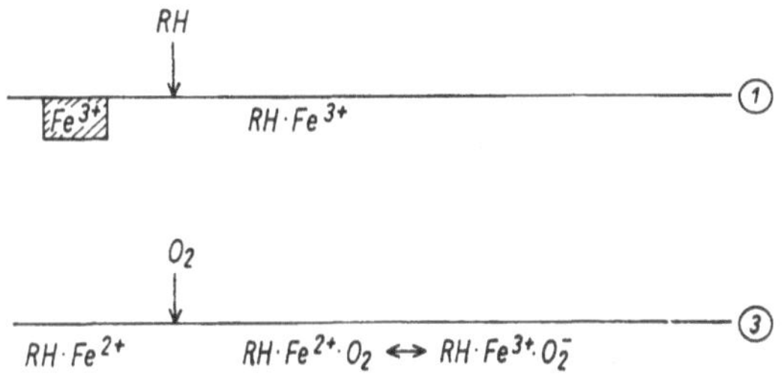

Fig. 3. Partial Reactions ① (Substrate Binding) and ③ (Dioxygen Binding) in the P–450 Cycle (CLELAND Notation)

respectively. Such spectral changes are paralleled by the conversion towards the high-spin state of a spin equilibrium. The spin state changes have been proved by ESR methods (TSAI et al., 1970; ESTABROOK, 1973; WATERMAN et al., 1973; GRIFFIN et al., 1975). The observed spin shift is probably due to the displacement of the sixth ligand in the Fe(III) low-spin configuration and thus to the establishment of a five-coordinated high-spin structure (COON and WHITE, 1980). Most of the hydroxylatable substrates belong to this group.

— The type II group contains especially primary amines, e.g. aniline. The spectral behaviour of these complexes is characterized by 390 nm minima and 425 nm maxima in the difference spectra, i.e. the spectral alterations indicate a spin shift towards the low-spin configuration. Since typical iron ligands like imidazole and pyridine, respectively, exhibit these properties, the binding mode is evidenced to be iron coordination in position 6 of the complex structure. That position, on the other hand, is the site for the O_2 molecule in the ternary complex of the hydroxylation sequence. Therefore type II compounds must be additionally capable of further binding to the porphyrin or protein moiety of the cytochrome in order to become converted. That should be realized again by means of apolaric interactions, as indicated by the increase of the respective binding constant for alkylisocyanides with increasing chain length (ICHIKAWA and YAMANO, 1968). A dual binding mode could be further evidenced for aniline by means of CN^- competition investigations (MATSUBARA and TOCHINO, 1976).

— The third group of substrates exhibits "reversed type I" or "modified type II" behaviour, the extrema of the difference spectra correspond reversely to the extrema localization of type I substrates. The binding mode of these substrates is supposed to originate from competition with an endogeneous substrate and participation of an OH-group (YOSHIDA and KUMAOKA, 1975). Like type II compounds these substrates give rise to respective low-spin shifts. Based on the above specifications the kinetics of substrate binding to the cytochrome may be analyzed under the following aspects:

- general characterization of the binding process (reaction homogeneity, order, rate constant(s))
- substrate classification (type characteristics)
- Fe(III)/Fe(II) differentiation (reaction mechanism)
- spin conversion kinetics
- Fe(II) ligandation (CO, O_2).

1.1. General Characterization of the Binding Process

The binding of substrates to the oxidized form of cytochrome P–450 proceeds quite rapidly. First investigations of the camphor complex formation of the *Pseudomonas putida* enzyme system revealed a second order reaction with rate constants $k(4°C) = 4.1 \cdot 10^6\ M^{-1}\,s^{-1}$ and $k(25°C) = 15.8 \cdot 10^6\ M^{-1}s^{-1}$ (GRIFFIN and PETERSON, 1972). A similar binding rate was observed

Table 1. Rate Constants of Cytochrome P–450 Substrate Binding Reactions[1])

Iron State	Substrate	Type	Rate Constant $[M^{-1} s^{-1}]$
Fe(II)	CO	—	$6.6 \cdot 10^5$
Fe(III)	Benzphetamine	I	$9.3 \cdot 10^4$
	Benzylephedrine		$4.2 \cdot 10^4$
	Hexobarbital		$1.8 \cdot 10^4$
	Clobenzorex		$1.1 \cdot 10^4$
	Aminopyrine	II	$6.4 \cdot 10^3$
	Aniline		$2.3 \cdot 10^3$
	Pyridine	II	$1.1 \cdot 10^4$
	Imidazole		$4.3 \cdot 10^2$
Fe(II)	Benzphetamine	I	$\sim 1 \cdot 10^3$
	Aniline	II	$\sim 1 \cdot 10^0$
	Pyridine	II	$1.2 \cdot 10^4$
	Imidazole		$2.1 \cdot 10^1$

[1]) Blanck et al. (1976)

for the competitive inhibitor metyrapone, $k(4°C) = 9.5 \cdot 10^5\ M^{-1}s^{-1}$ (Griffin and Peterson, 1972). The binding of hexobarbital (type I) and aniline (type II) to the liver P–450 system was characterized with half times of about 20 ms (at 5 mM substrate concentration), that corresponds to second order rate constants of about $10^4\ M^{-1}s^{-1}$ (Estabrook et al., 1972). These data point already to some species differentiation in that the bacterial system exhibits the higher values. In line with this statement is the benzphetamine (type I) complex formation in a reconstituted LM system, the rate constant of which amounts to about $10^5\ M^{-1}s^{-1}$ (Imai et al., 1977). More detailed investigations have been performed with a partially purified solubilized LM system (Blanck et al., 1976, 1977). Tab. 1 shows the binding reaction data for characteristic type I and type II compounds and the Fe(III)- and Fe(II)-P–450, respectively. The second order rate constants for the overall process amount to $10^3 - 10^5\ M^{-1}s^{-1}$ with respect to the oxidized enzyme.

Concerning the question of rate-limitation, the substrate binding process is evidently more rapid in general. The turnover data for characteristic substrates (Tab. 5, 9, 10, 11) clearly show, that the maximum rates amount to about $60\ min^{-1}$ in reconstituted systems. The endoplasmic reticulum (or the microsomes), however, functions at lower reductase/P–450 stoichiometry, i.e. the substrate conversion proceeds even less rapidly. Some restriction of the general line may refer to especially type II substrates in the mM concentration range. Under such conditions the binding rate may reach the time scale of the overall reaction. Dependencies on substrate preincubation have been observed (Burstein et al., 1972; Holtzman and Rumack, 1973).

The above statement is confirmed by investigations of the temperature function. The Arrhenius plots for the overall reaction (product formation) exhibit distinct breaks (Annayev et al., 1972; Duppel and Ullrich, 1976; Yang et al., 1977), which are refered to structural changes of the phospholipid matrix of the proteins (phase alterations). Rate-limiting partial reactions, therefore, should likewise show these characteristics. That has never been observed, neither in K_S-titrations nor in binding kinetics (Misselwitz et al., 1976; Jänig et al., 1977; Smettan et al., 1978). Despite of phospholipid K_S control (French, 1979; French et al., 1980), evidently no control is exerted with respect to some supramolecular actions.

Substrate binding to the solubilized cytochrome proceeds inhomogeneously. That holds true for both Fe-oxidation states, both groups of substrates (I, II), as well as for the iron ligand CO. The differences in reactivity are obiously not related to heterogenous microdistribution (spatial arrangement) in microsomes or the solubilized system. Disintegration by means of detergents preserves inhomogeneity. This behaviour may indicate reactivity specificities of P–450 subfractions. The problem of unnatural fractionation by means of preparational damage, however, cannot be eliminated. Heterogeneities have been likewise observed for the CO binding and dissociation process (Debey et al., 1973a; Debey and Douzou, 1974; Balny and Debey, 1976). A detailed analysis of the phase distribution points further to P–450 subspecies (Gray, 1978). Different CN-affinity e.g. of three subfractions in rat liver microsomes could be proved (Comai and Gaylor, 1973).

1.2. *Substrate Classification*

The binding of type I substrates — as far as data are accessible — proceeds about one order of magnitude more rapidly as compared to type II compounds. Evidently specificities of the receptor site/substrate interactions, i.e. of the binding mode, control this differentiation. Respective specifications have been observed for binding constants (K_S) (Misselwitz et al., 1975) as well as for the conversion parameters (cf. Tab. 5, 9, 10, 11). The thermodynamic parameters of the substrate binding likewise differentiate the two groups: type I substrates are more entropically driven associated, thus pointing to a rather hydrophobic binding mode; type II compounds, on the other hand, are predominantly enthalpy controlled (Jänig et al., 1977). The activation parameters for the benzphetamine (I) and aniline (II) binding process are in line with these results (Smettan et al., 1978). Respective substrate classification is observed in the reduction reaction (e_1), too (cf. Tab. 2,5). The significance of the observed substrate classification with respect to the overall reaction is rather less, due to the more rapid binding process. However, some rate interference with type II compounds seems resonable.

1.3. *Fe(III)/Fe(II) Differentiation*

The binding of substrates to the Fe(II) cytochrome has been studied for some selected substrates (Tab. 1). Further investigations are lacking. It could be evidenced, that the iron redox state change causes a significant decrease of the respective rate constants. This refers to either group of compounds. With the liver system deviations from linear Lineweaver-Burk plots have been observed; such peculiarities led to the proposal of a random overall reaction (Gillette et al., 1973). That pathway is included in Fig. 4 in parallel to the sequential

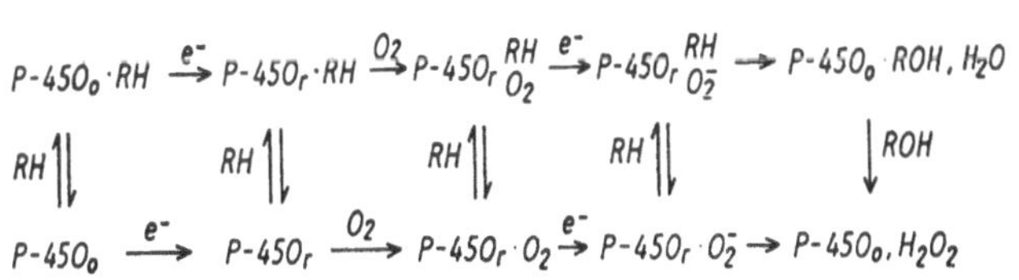

Fig. 4. P–450 Reaction Sequence (Gillette et al. (1973))

process. The scheme considers the substrate binding in all distinct reaction steps. The random mechanism has been disproved, at least with respect to the steps prior to the formation of the ternary complex (RH, O_2). The substrate binding rate constants for the reduced P-450 are about two orders of magnitude decreased, furthermore type I substrates control the first reduction process ② (rate-enhancement). Therefore the sequential mechanism is rather likely.

1.4. *Spin State Conversion*

The binding of substrates to the oxidized form of P-450 is coupled with a change of the spin state of the respective species dependent spin equilibrium (KATAGIRI et al., 1968; PETERSON, 1971; WATERMAN et al., 1973; SLIGAR, 1976; LANGE et al., 1977; REIN et al., 1977; WERRINGLOER et. al., 1979). The spin equilibrium is a sensitive indicator of structural implications especially with respect to the electron transfer in the P-450 cycle: high-spin compounds are preferably reduced (Tab. 2, MATSUBARA et al., 1976).

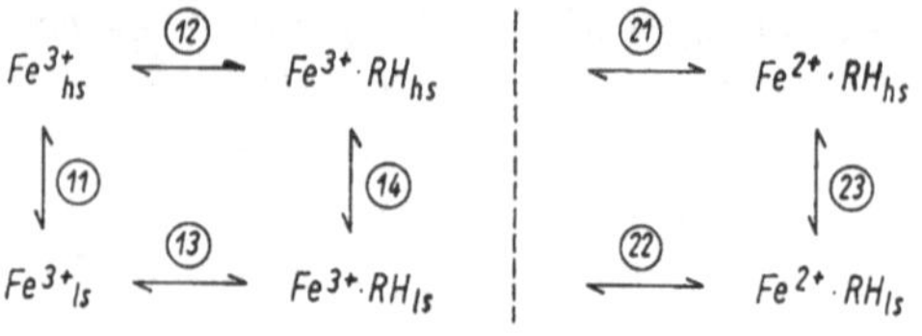

Fig. 5. Spin State Conversion Processes in Partial Reactions ① (Substrate Binding) and ② (First Electron Transfer e_1)

From the kinetic point of view the question arises in what way do spin states and transitions, respectively, control the substrate binding process (this question will be treated further with respect to the first reduction step (e_1)). The reaction scheme of the coupled process is drawn in Fig. 5. In either the camphor- and the liver-enzyme system, the cytochrome is predominantly in the low-spin configuration before substrate binding, and is then converted to the mainly high-spin complex in the camphor system, and to a significantly increased amount of high-spin configuration in the liver system, respectively. Therefore the substrate binding reaction (⑫, ⑬) could be controlled by

Table 2. Effect of Various Substrates and Pretreatment on the Cytochrome P-450 Reduction Reaction (e_1)[1]

Treatment	Substrate	Rate Constant [s^{-1}]		Fast phase (%)
		Fast Phase	Slow Phase	
None	None	0.51	0.10	45
	Hexobarbital	0.78	0.15	48
	Ethylmorphine	1.08	0.21	47
Phenobarbital	None	0.64	0.22	53
	Hexobarbital	1.06	0.39	58
	Ethylmorphine	1.74	0.17	46
	Benzphetamine	1.17	0.31	80
	Aniline	0.53	0.10	48
3-Methylcholanthrene	None	0.20	0.07	80
	Hexobarbital	0.34	0.09	56
	Ethylmorphine	0.74	0.19	22

[1]) MATSUBARA et al. (1976); reproduced with permission from Academic Press. Inc.

pre- or reequilibration (⑪, ⑭) if preferable substrate interaction with the high-spin state P–450 would occur (⑫) or if the spin shift would rather follow the initial binding process to the low-spin configuration (⑬), respectively.

The established second order substrate binding process proves the pre- and reequilibration steps ⑪, ⑭ to be not rate determining. That result is supported by direct investigation of the respective partial reactions by means of temperature-jump methods.

In the CAM system a relaxation rate constant $k_{obs} > 10^5\,s^{-1}$ was determined for process ⑭ (SLIGAR, 1976). This value is in the order of respective reactions at cytochrome c, metmyoglobin (OH, N_3) (BEATTIE and WEST, 1974). High resolution laser techniques even proved for metmyoglobin-hydroxide values in the order of $10^7\,s^{-1}$ (DOSE et al., 1977). Such rate constants correspond to intramolecular changes in spin multiplicity of protein-free model compounds (DOSE et al., 1978). Moreover, pure electronic transitions may proceed in the time range of electron motion frequencies ($10^{16}\,s^{-1}$) (MARCUS, 1964). Somewhat decreased rates for the P–450 CAM relaxation in presence of substrate resulted from [K^+]- and pa_H-jump investigations: at 4°C rate constants of $\sim 50\,s^{-1}$ have been determined, which indicate some intramolecular implications (conformational changes) (LANGE et al., 1979).

For the camphor-free P–450 a relatively slow transition was observed with respect to ⑪. The rate constant $k_{obs} = 58\,s^{-1}$ was interpretated as being due to the dissociation of the sixth ligand (i.e. protein) (SLIGAR, 1976). A model for that reaction is taken from bisimidazole-mesotetraphenyl-ironporphyrin, this compound exhibited a relaxation rate constant of about $60\,s^{-1}$ (LAMAR and WALKER, 1972).

A recent reinvestigation of the P–450 CAM relaxation process confirmed the data of the substrate-free P–450 and revealed a somewhat decreased rate constant $k_{obs} = 365\,s^{-1}$ in presence of the substrate (COLE and SLIGAR, 1981).

The liver system P–450 was observed to relax quite rapidly too, irrespective of substrate. Rate constants exceding $10^4\,s^{-1}$ could be evidenced (RISTAU et al., 1979). Recently by use of laser temperature jump techniques relaxation processes in the nanosecond range could be resolved, the rate constants of which amounted to about $10^6\,s^{-1}$ with and without substrate, respectively (ZIEGLER et al., 1982).

As to ⑫ and ⑬, equilibrium data indicate that step ⑫ is thermodynamically favoured as compared to step ⑬: camphor associates 200 times more favourably to the high-spin form (SLIGAR, 1976); the LM2 fraction of the liver system, on the other hand, binds e.g. benzphetamine but only 5 fold favoured to the high-spin configuration (RISTAU et al., 1979). Thus if no alternative predominates even both partial reactions may contribute to the process of substrate binding. Further investigations may clarify the distinct pathway; in either case, however, the partial steps ⑪, ⑫, ⑬, ⑭ are rapid as compared to the product formation (overall reaction), i.e. neither sequence in the substrate binding process will be rate-limiting.

1.5. Ligand Binding to the Fe(II) State P–450

Step ③ in the hydroxylation sequence of the enzyme system (Fig. 1, 3) is likewise a ligand binding process on the cytochrome. The O_2 complex formation represents an iron coordination in position 6 (ISHIMURA et al., 1971; PETERSON et al., 1972; TYSON et al., 1972). The reaction is quite insensitive with respect to spectral changes, nevertheless it has been resolved for the camphor system (PETERSON et al., 1972) and likewise for the liver P–450 (RÖSEN and STIER, 1973; GUENGERICH et al., 1976). The second order rate constants have been determined with about $10^6-10^7\ M^{-1}\ s^{-1}$; in parallel to the substrate binding process, the liver system binds O_2 less rapidly. The rate constants evidence, that this partial reaction is not rate-limiting in the cycle.

Extended investigations have been performed with CO because of the high sensitivity of the concomitant absorption change of the reaction (OMURA et al., 1965 (AC system); PETERSON et al., 1972; PETERSON and GRIFFIN, 1972; GUNSALUS et al., 1974 (CAM system); DEBEY et al., 1973b; RÖSEN and STIER, 1973; BALNY and DEBEY, 1976; BLANCK et al., 1976 (LM system)). This ligand has often been used as O_2 model compound in hemoprotein reactions. CO is likewise bound in position 6 of the iron complex and therefore acts as competitive inhibitor of the enzyme system. CO inhibition of a substrate conversion process indicates the participation of P–450. The inhibition can be reversed photolytically (ESTABROOK et al., 1963), that property provided further evidence for P–450 being the terminal oxidase of the hydroxylation reaction. The rate constant of the CO binding process is in the order of those for other hemoproteins ($k \sim 10^5-10^6\ M^{-1}s^{-1}$), that holds further true with respect to the activation energies (SMETTAN et al., 1978). Thus the reaction may be predominantly controlled by the Fe/CO interaction, whereas the specific influence of the protein should be less important. That is further demonstrated by the fact that even inactivation of the oxidase by means of its conversion into the P–420 form does not change the overall reactivity towards CO (BLANCK et al., 1976).

The CO binding reaction is rather heterogeneous. Detailed analyses of the phase distributions have been performed, and P–450 subfractionation (isozymes and conformers) has been evidenced (GRAY, 1978; DAVYDOV et al., 1980). (For detailed treatment cf. section 3).

2. NADPH-dependent Reduction Reaction

The second step in the P–450 reaction cycle represents the reduction of the enzyme-substrate complex $Fe^{3+} \cdot RH$ (②). This partial reaction has been extensively studied. The accumulated results, however, require critical examination because of size and heterogeneity of the data material. The NADPH-dependent reaction is controlled by interactions including all components of the system, it exhibits structural specificities, is tightly coupled to substrate control and spin shift phenomena, and is further strongly related to the rate-limiting step in the cycle and even represents this limitation in specified cases.

The reaction scheme (Fig. 1, 4, 6) contains two reduction processes in the cycle (②, ④). In these steps 1 electron each is transfered to P–450 (PETERSON, 1971; COOPER et al., 1977; PETERSON et al., 1977). Primary electron donor in P–450 systems are either NADPH and NADH, respectively. Both pyridine nucleotides are used, the bacterial camphor system is NADH dependent (KATAGIRI et al., 1968; TSAI et al., 1971; GUNSALUS et al., 1974), the liver system receives the electrons from NADPH and is further controlled by NADH synergism (GARFINKEL, 1958; KLINGENBERG, 1958; OMURA and SATO, 1962). The donor electrons are transfered to the cytochrome via the NADPH-dependent cytochrome P–450 reductase which is responsible for the transformation of one 2-electron acceptor reaction into two 1-electron donor reactions towards the cytochrome. The participation of the NADH-dependent cytochrome b_5 system, especially with respect to the donation of the second electron (④) is evidenced. The NADH support exhibits substrate specificities. Since reconstituted enzyme systems without b_5 system components are functionally active, the NADPH pathway is evidently the more important e⁻-donor chain (cf. section 6).

The NADPH-dependent P–450 reductase transfers the electrons directly to the oxidase (Lu and Coon, 1968; VanderHoeven and Coon, 1974; Vermilion and Coon, 1974; Yasukochi and Masters, 1976; Imai et al., 1977; Iyanagi et al., 1978). An interrelated ferredoxin compound — as proved in the camphor system (Cushman et al., 1967), and in the mitochondrial adrenocortical or hepatic transfer chain (Nakamura et al., 1966; Imura et al., 1966; Kimura and Suzuki, 1967; Simpson and Boyd, 1967a, b; Sato et al., 1977; Pedersen, 1978; Hodgson, 1979) — is deficient. The liver enzyme reductase contains in contrast to the forementioned systems two flavin molecules (FMN, FAD) per polypeptide chain; the FMN is obviously functionally related to the ferredoxins in its electron transfer properties (Iyanagi and Mason, 1973; VanderHoeven and Coon, 1974; Vermilion and Coon, 1974; Dignam and Strobel, 1975; Masters et al., 1975). The reductase, if FMN depleted, is incapable to reduce the cytochrome (Vermilion and Coon, 1978a, b; Alexander et al., 1980; Vermilion et al., 1980, 1981). The electron probably is transfered via the chain NADPH/FAD/FMN/P–450, at least to the cytochrome bound oxygen to become activated. The kinetic peculiarities of this sequence will be treated subsequently at the level of NADPH/flavoprotein- and of flavoprotein/P–450 interaction, respectively.

2.1. NADPH/Flavoprotein Interaction

The NADPH-dependent cytochrome P–450 reductase from liver microsomes has been isolated as early as in the 60ies (Phillips and Langdon, 1962; Williams and Kamin, 1962). Originally the enzyme has been discovered as electron transfering protein to cytochrome c (Horecker, 1950). The functional properties of the enzyme, therefore, mainly have been studied on this protein. First investigations of Horecker proved a respective reduction turnover of 1140 min^{-1}.

The mechanism of electron transfer across the reductase is rather complex (Iyanagi et al., 1978; Vermilion and Coon, 1978a, b; Yasukochi et al., 1979; Masters and Okita, 1980). The protein molecule is capable to accept 4 electrons (2 electrons per flavin) at most, therefore up to 9 redox states might exist in dependence on the distinct electron uptake.

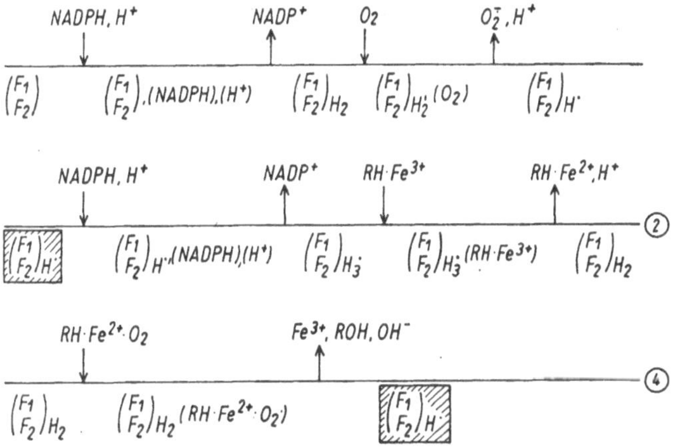

Fig. 6. Electron Transfer Reactions ② (First Electron Transfer e_1) and ④ (Second Electron Transfer e_2) in the P–450 Cycle

Aerobically two states are preferably achieved. The first one is a one-electron reduced state of the enzyme molecule, i.e. one flavin radical is formed. This state is air-stable, it has been observed aerobically as well as anaerobically. The formation proceeds via a two-step mechanism, the first of which is the uptake of 2 electrons from NADPH to produce one fully reduced flavin in the protein molecule. In a second step one electron is transfered to oxygen to form an O_2^- radical. Anaerobically intermolecular electron transfer is supposed to achieve that state too (MASTERS and OKITA, 1980). The 1-equivalent reduced state of the reductase is taken to be the basic species ("oxidized") in the electron transfer couple with P–450, since it does not transfer electrons to cytochrome c (MASTERS et al., 1965). The uptake of further 2 electrons leads to the 3-equivalent reduced state, i.e. to the "reduced" species in the catalytic cycle. This 1—3 shuttle mechanism which is coupled to intramolecular F_1F_2 electron transfer is considered to represent the 2 electron donation process towards P–450 (IYANAGI and MASON, 1973; IYANAGI et al., 1978; YASUKOCHI et al., 1979; MASTERS and OKITA, 1980) (Fig. 6).

Recent studies of OPRIAN et al. (1979), IYANAGI and MAKINO (1980), VERMILION et al. (1980, 1981), IYANAGI et al. (1981), OPRIAN and COON (1982) suggest the following electron transfer sequences with respect to the distinct flavins:

- FMN depleted reductase is reduced by NADPH to give the $FADH_2$ species, which is in successive one-electron steps reoxidized (e.g. by O_2)

$$FAD \xrightarrow{NADPH} FADH_2 \xrightarrow{O_2} FADH^{\cdot} \xrightarrow{O_2} FAD$$

- The O_2-stable semiquinone of the holoenzyme as obtained by NADPH reduction in the presence of O_2 is a one-electron reduced $FAD \cdot FMNH^{\cdot}$ species. Further reduction by NADPH results in the semiquinone formation of the other flavin ($FADH^{\cdot} \cdot FMNH_2$) via the $FADH_2 \cdot FMNH^{\cdot}$ species

$$FAD \cdot FMNH^{\cdot} \xrightarrow{NADPH} FADH_2 \cdot FMNH^{\cdot} \rightarrow FADH^{\cdot} \cdot FMNH_2$$

- The reaction sequence for the fully oxidized reductase is

$$FAD \cdot FMN \xrightarrow{NADPH} FADH_2 \cdot FMN \rightarrow FADH^{\cdot} \cdot FMNH^{\cdot} \rightarrow FAD \cdot FMNH_2 \xrightarrow{NADPH} FADH_2 \cdot FMNH_2$$

The transfer of electrons to P–450 in the catalytic cycle thus proceeds via the reaction

$$FMNH_2 \rightarrow FMNH^{\cdot}$$

The 1—3 shuttle enables this way the support of both electrons. Likewise the 4-electron reduced form could donate 2 electrons to P–450. The 2-electron reduced molecule, on the other hand, after donation of 1-electron to the oxidized cytochrome would achieve the air-stable semiquinone state. It remains open if this 1-electron reduced molecule is capable of donating electrons to oxycytochrome.

Kinetic investigations of reductase reactions refer mainly to the enzyme turnover, less information is available for distinct electron transfer steps. The initial turnover data in the cytochrome c reduction reaction could be confirmed. Anaerobically determined values amount to 1200 to 1400 min^{-1}. The 1. and 2. NADPH reduction process proved to proceed with an 1300 min^{-1} turnover. The respective value for the total diminution of the oxidized enzyme was 2700 min^{-1} (KAMIN et al., 1965, MASTERS et al., 1965). Respective aerobic data (FMN turnover) have been determined for cytochrome c (4010 min^{-1}), ferricyanide (5310 min^{-1}), DCIP (1960 min^{-1}) (VERMILION and COON, 1978a, b). Recent aerobic stopped-flow studies evidenced, that at low NADPH concentration (equal to or less than 2 electrons per 2 flavins) only one of the, two flavins is fully reduced with $k \doteq 180 s^{-1}$, and that the reaction of that species with oxygen proceeds rather slowly with $k = 0.026 s^{-1}$ (YASUKOCHI et al., 1979).

The reduction of cytochrome c follows a ping-pong mechanism with respect to the reductase, the formation of a NADPH/reductase/cytochrome c ternary complex is excluded (KAMIN et al., 1965, MASTERS et al., 1965). Cytochrome P–450 is supposed to react likewise (GILLETTE et al., 1973).

The kinetic data clearly show, that the NADPH-dependent P–450 reductase is capable of relatively high catalytic turnover as compared to the overall P–450 reduction, i.e. no rate-limitation in that reaction should be due to the general reductase function. The resolved distinct reduction processes at the reductase are in line with these statements. It remains, however, to be clarified if electron donation steps towards P–450 may be controlled by the flavoprotein. The P–450 reduction (②), which proceeds much less rapidly as compared to cytochrome c (about one order of magnitude) should be probably due to interaction specificities which originate from the cytochrome. Since rather the second electron transfer is catalytically rate-limiting it must be questioned if that restriction is possibly due to the donor function of the reductase.

2.2. NADPH-Dependent Cytochrome P–450 Reduction (e_1)

The reduction of the oxidase (②) represents the second partial reaction of the electron transfer from NADPH to cytochrome P–450. This step requires the interaction of two proteins — of the reductase and the cytochrome, respectively — i.e. its intermolecular electron transfer should be controlled not only by the physicochemical properties of the respective prosthetic groups, but further by some direct or indirect protein participation on the way. Therefore macromolecular specificities — e.g. conformational changes, association equilibria — may exert control or even rate-limitation to this step. This is evidenced by strong dependencies of the reaction on the third main component of the enzyme system, the phospholipid (STROBEL et al., 1971; AUTOR et al., 1973a, b; LU and LEVIN, 1974; COON et al., 1976; INGELMAN-SUNDBERG et al., 1980a, b; INGELMAN-SUNDBERG, 1976, 1977; INGELMAN-SUNDBERG and GLAUMANN, 1977; INGELMAN-SUNDBERG and JOHANSSON, 1980) or detergents as its "substituent" (LEVIN et al., 1974; GUENGERICH and COON, 1975; PETERSON et al., 1976; BLANCK et al., 1979a, b; TANIGUCHI et al., 1979). The specific substrate control of this partial reaction may likewise depend on protein mediation.

The analysis of the kinetics of P–450 reduction by NADPH has been frequently studied with microsomes (GIGON et al., 1969; SASAME and GILLETTE, 1969; DIEHL et al., 1970; SCHENKMAN, 1972; GILLETTE et al., 1973; HOLTZMAN and RUMACK, 1973; MATSUBARA et al., 1976; PETERSON et al., 1976; MIWA et al., 1978; BLANCK et al., 1979a, b; RUF, 1980). Solubilized systems followed (COON et al., 1971, 1973; BLANCK et al., 1979a, b), and after the evidence of catalytically active reconstitution of the enzyme system from the isolated components (LU and COON, 1968; VAN DER HOEVEN and COON, 1974; VERMILION and COON, 1974), especially detailed investigations with modified component compositions and stoichiometries, have been performed (IMAI et al., 1977; MIWA et al., 1978, 1979; BLANCK et al., 1979a, b; OPRIAN et al., 1979; TANIGUCHI et al., 1979). Just these studies are presently favoured, but criticism with respect

Table 3. Kinetic Parameters of the Microsomal State P–450 LM Reduction Reaction (NADPH)[1])

NADPH/ $Na_2S_2O_4$ Reduction (%)	Phase Distribution (%)	$k_{app.}$ [s^{-1}]		Conditions (Reference)
		Fast Phase	Slow Phase	
> 95	66/34	$\sim 1 \cdot 10^{0}$	$2.3 \cdot 10^{-2}$	Rat, anaerobic GILLETTE et al. (1973)
> 95	73/27	$9.9 \cdot 10^{-1}$	$3.3 \cdot 10^{-2}$	Rat, anaerobic 30 °C
	70/30	$1.6 \cdot 10^{-1}$	$3.2 \cdot 10^{-3}$	Rat, anaerobic 8 °C, PETERSON et al. (1976)
~ 95	Different Models	$5.9 \cdot 10^{-1}$	Different Models	Rat, anaerobic 30 °C, DIEHL et al. (1970)
52.5	75/25	$3.7 \cdot 10^{-1}$	$\sim 6.2 \cdot 10^{-2}$	Rat, aerobic 22 °C
31.0	70/30	$2.5 \cdot 10^{-1}$	$\sim 5.5 \cdot 10^{-2}$	Rabbit, aerobic 22 °C

[1]) BLANCK et al. (1979a)

to the physiologic relevance is opportune throughout.

The NADPH reduction of P–450 is predominantly investigated under anaerobic conditions in the presence of CO. This way O_2 influences, e.g. reoxidation etc., are excluded. The subsequently formed binary P–450 · CO complex serves as indicator for the reduction process, since the CO binding step is rapid and does not interfere with the electron transfer (cf. section 1.5).

The anaerobic microsomal reduction reaction exhibits biphasic kinetics with two first order processes. The phase ratio (φ_1/φ_2) amounts to about 70/30%. The reaction comprehends about 90—95% of the total protein. The rate constant of the rapid partial reaction is in the order 1 s^{-1} (30°C), it depends strongly on temperature and is controlled by substrates. The respective second phase constant is about 10—20 times diminished (Tab. 3). The observed differentiation obviously originates from structural arrangements of the reactants in the membrane (heterogeneity of distribution of components, stoichiometric relations, cf. section 2.4.). Intermicrosomal interactions between the respective components do not occur, the reaction process is governed by intramicrosomal steps (PETERSON et al., 1976).

The microsomal reduction reaction in the presence of oxygen, i.e. under physiologic conditions, exhibits similar characteristics. The respective rate constants are in the main unchanged, the extent of the reduction (with respect to total P–450) is decreased by reoxidation (Tab. 3).

The aerobic reduction has been analysed in detail using a partially purified solubilized enzyme system (with nearly unchanged microsomal P–450/reductase ratio but decreased phospholipid) (BLANCK et al., 1979a, b). This system again exhibits biphasic reduction kinetics with about 70/30% phase distribution, but only 10—15% reaction extent. The rate constants correspond to the microsomal data. Addition of detergents (Triton N-101, sodium cholate) causes disintegration of the solubilized system; by means of sedimentation analysis the change of the mean molecular weight from about $1 \cdot 10^6$ down to 50000—70000 has been ascertained. The gradual disintegration on detergent titration is correlated with a decrease in the amount of the rapid phase reduction in favour of the slow process. A completely disintegrated system ex-

Table 4. Kinetic Parameters of the Solubilized State P–450 LM Reduction Reaction (NADPH)[1])

Detergent (Triton N-101) (%)	NADPH/ $Na_2S_2O_4$ Reduction (%)	Phase Distribution (%)	$k_{app.}$ [s^{-1}] Fast Phase	Slow Phase
–	9.7	70/30	$2.3 \cdot 10^{-1}$	$1.2 \cdot 10^{-2}$
0.0025	15.3	70/30	$2.3 \cdot 10^{-1}$	$1.2 \cdot 10^{-2}$
0.0050	16.0	40/60	$2.0 \cdot 10^{-1}$	$1.2 \cdot 10^{-2}$
0.010	15.3	10/90	$2.8 \cdot 10^{-1}$	$1.4 \cdot 10^{-2}$
0.020	11.1	–/100	–	$1.1 \cdot 10^{-2}$

[1]) BLANCK et al. (1979a)

hibits only a slow monophasic reaction (Tab. 4). The disintegration procedure is reversible (by dilution of detergent concentration).

In order to get access to the basic interaction phenomena in the physiologic membrane bound (microsomal) enzyme system, the distinct steps of the reductase/P–450 electron transfer have been analyzed in the completely disintegrated system. This way definite concentration dependencies on either protein could be easily achieved and studied, respectively. The investigations proved an O_2-dependent MICHAELIS-MENTEN type mechanism. The complex formation between reductase and P–450 to form the interaction species is rather rapid and not rate-limiting in the reduction process; similar results have been obtained with the soluble camphor system and the interaction of the respective electron transfer components putidaredoxin and P–450 (anaerobic studies (PETERSON and MOCK, 1979)). Rate-limiting is obviously a first order process, which follows the formation of the MICHAELIS-MENTEN complex. That could be the electron transfer itself or some concomitant conformational change. An at least two-step reduction process could be resolved by means of low-temperature investigations (DAVYDOV et al., 1978, 1979; GRESCHNER et al., 1979). The investigations further exclude some diffusion limitation. A model treatment of the reaction fits well to the experimental data and revealed characteristic parameters of the system (K_M, individual k-values, original reductase/P–450 stoichiometry).

The method disintegration and reassociation by means of detergent treatment was used to reconstitute integrated systems with distinct stoichiometries of the components. An increase of the relative reductase content led to an increase of the rapid phase reduction and thus to an increased aerobic reduction extent. The slow phase, however, did not completely dissappear under the experimental conditions (Fig. 7).

Highly purified components were likewise used in mechanistic studies (MIWA et al., 1978, 1979; OPRIAN et al., 1979; TANIGUCHI et al., 1979; FRENCH et al.,

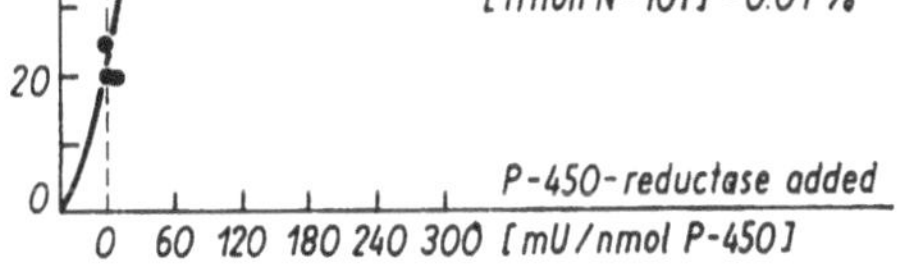

Fig. 7. Aerobic Reduction (NADPH) of Reconstituted P–450 LM at Different P–450/Reductase Ratios (BLANCK et al., 1979a)

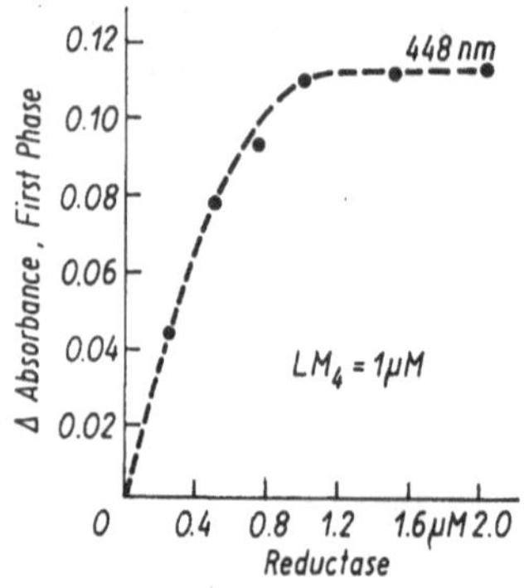

Fig. 8. Anaerobic Reduction (NADPH) of Reconstituted P–450 LM4 as a Function of Reductase Concentration (OPRIAN et al., 1979); reproduced with permission from The American Socicty of Biological Chemists, Inc.

1980). Reconstitutions with different stoichiometries have been performed with the LM2 and the LM4 protein as well. The LM4 fraction has been prefered because this P–450 species is deficient of substrate control and spin transitions in the reduction process.

The anaerobic reaction (LM4 system, OPRIAN et al., 1979) exhibits 3 first order phases with rate constants $k_1 = 1.0\ s^{-1}$, $k_2 = 0.2\ s^{-1}$, $k_3 \sim 10^{-2}\ s^{-1}$. The phase distribution (φ_1, φ_2, φ_3) depends strongly on the reductase/P–450 ratio. Increase of the relative reductase content favours the φ_1 and φ_2 contribution at the expence of φ_3 (Fig. 8). At 1 : 1 stoichiometry and excess NADPH φ_2 amounts to about 22–24% of the φ_1 contribution. At low NADPH concentration φ_1/φ_2 changes in favour of φ_2. Only φ_3 kinetics is observed if preincubation of the components is excluded before starting the reactions. The results are discussed as follows:

- P–450 and reductase establish a catalytically active complex of 1 : 1 stoichiometry which is responsible for the φ_1 and φ_2 contribution, the φ_1/φ_2 differentiation is supposed to be due to multiple oxidation-reduction states of the reductase rather than to P–450 specificities.
- φ_3 represents the interaction of free P–450 with the reductase.

Since this reaction is of first order, the formation of the P–450/reductase interaction complex (transient) must be rate-limiting preceded or followed.

Similar results were obtained with an LM2 system (FRENCH, 1979; FRENCH et al., 1980); a P–450/reductase complex formation of 1 : 1 stoichiometry (6 molecules of each component) could be evidenced by sedimentation analysis. The formation of a P–450/reductase complex could be likewise proved by MIWA et al. (1979).

The following common properties of the different systems could be evidenced by the distinct investigations (Fig. 9):

1. Systems which are capable of P–450/reductase complex formation exhibit the more rapid P–450 reduction with some fractional differentiation (φ_1, φ_2).
2. The reductase/P–450 ratio determines the amount of the respective complex and by this the extent of the rapid reduction phases (φ_1, φ_2).
3. The uncomplexed P–450 exhibits the less rapid interaction process (φ_3). The rate is not limited by the second order interaction complex formation (transient) nor by diffusion.
4. Oxygen perturbs the anaerobic kinetics pattern. The equilibrium state of the reduction process is decreased, obviously because of P–450 reoxidation. The reaction exhibits nearly unchanged apparent

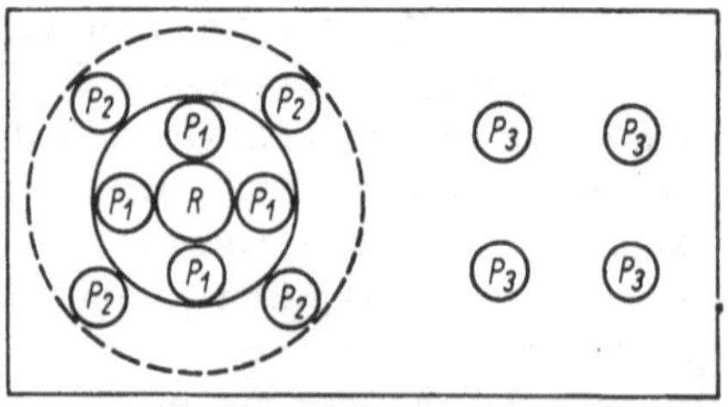

Fig. 9. Modes of P–450/Reductase Interaction in the Electron Transfer Reaction ② (cf. Text): P_1, P_2 — Reductase Associated (Bound) P–450, P_2 Structurally Disfavoured (φ_1, φ_2); P_3 — Nonassociated (Disintegrated) P–450 (φ_3)

rate constants and phase distribution, as well, thus indicating a rather complex mode of action of the oxygen. The distinct aerobic reduction mechanism of the microsomal P–450 system is unresolved till now.

The reduction characteristics of the reconstituted and disintegrated systems, respectively, join to the microsomal data. The most serious discrepancy that remains is obviously the large rapid reduction contribution in the microsomal reaction (50–70%) despite of the low reductase/P–450 ratio of only 1 : 20–40. If only a preformed 1 : 1 interaction complex is capable of rapid electron transfer, then but 2.5–5% of the P–450 could exhibit that peculiarity. The microsomal data may be understood, on the other hand, if cluster formation — i.e. formation of reductase/P–450 complexes of less than 1 : 1 ratio — is suggested. Even reconstituted systems with nearly microsomal reductase/P–450 stoichiometry exhibited up to 15% rapid reduction kinetics (BLANCK et al., 1979a, b).

2.3. *Function and Structure in the Electron Transfer*

The analysis of the distinct reaction steps in the electron transfer between the NADPH-dependent P–450 reductase and P–450 in the first reduction reaction ② clearly shows, that the NADPH reduction of P–450 obviously strongly depends on structural implications with respect to the participants and its respective complexes. That implies, on the other hand, that reaction specifities of this process may be indicative of specifically interacting structures. Two interaction modes are alternatively discussed, rigid association of reductase and P–450 to form functional clusters (FRANKLIN and ESTABROOK, 1971; STIER and SACKMANN, 1973; PETERSON et al., 1976; STIER, 1976; FRENCH et al., 1980) and, on the contrary, a nonrigid distribution of the reactants in the membrane (YANG, 1975, 1977; DUPPEL and ULLRICH, 1976; YANG et al., 1977, 1978). With respect to the cytochrome b_5 system the nonrigid distribution of the cytochrome and its reductase is the more probable version (ROGERS and STRITTMATTER, 1974a, b), concerning the P–450 system this question is still open.

The assumption of a nonrigid distribution is favoured by mainly two arguments:

1. Partial inhibition of reductase leads to a decrease in reaction rate, but all of the P–450 molecules become reduced by the remaining still active reductase molecules, i.e. the reductase is evidently accessible to all P–450 molecules (YANG et al., 1978).
2. Denaturation of P–450 by H_2O_2 does not affect the reductase. If afterwards the denatured cytochrome is restituted by external P–450, then the catalytic function is restored. That indicates that the reductase is furthermore accessible for external P–450 and is, on the contrary, not sterically inactivated by associated denatured P–450 (YANG, 1977).

A nonrigid but heterogeneous distribution is alternatively discussed, the proteins could be enclosed in phospholipid halos but remain in high lateral mobility (YANG, 1977).

The rigid-association version is supported by phospholipid dependent temperature-inhomogeneities (breaks) in the reduction process (PETERSON et al., 1976). Rapid phase effects are referred to cluster association of reductase and P–450, from which rapid reduction originates, whereas phospholipid dependent diffusion of randomly distributed P–450 should restrict its access to the reductase and therefore decrease the reduction rate.

The investigations of the reconstituted LM4-, LM2- and partially purified systems support the assumption that the physiologic relevant rapid reduction process is due to cluster structures. These interaction structures are strongly dependent on phospholipid. Diminution of that compound leads to the gradual dissappearence of the rapid reduction (Autor et al., 1973a, b; Lu and Levin, 1974). Phospholipid can be substituted by detergents (Levin et al., 1974; Guengerich and Coon, 1975). At high concentration, on the contrary, detergents disintegrate the system (Blanck et al., 1979a, b).

Investigations of phospholipid specificities are at the very beginning, homogeneous as well as heterogeneous fractions have been successfully used in reconstitution. It seems, that either the phospholipid species and a general phospholipid-protein linker function, respectively, are acting in the functional clusters (Ingelman-Sundberg et al., 1980b; Ruckpaul et al., 1980, 1982).

The reconstitution capability of the enzyme system implicates that a predetermined membrane structure is not the precondition of catalytic activity, the respective interaction structure and function, as well, should be rather thermodynamically controlled. That is supported by recent investigations on the mitochondrial P–450 AC system where the reductase/adrenodoxin/P–450 electron transfer is controlled by a distinct couple of partial association equilibria which depend on the respective redox state (Hanukoglu and Jefcoate, 1980). Association equilibria of the proteins in the liver system have been proved to be lipid controlled (French, 1979; French et al., 1980). The restitution of denatured P–450 by external cytochrome — as argued for the nonrigid model — could be understood in following this line. The slow partial reaction, on the other hand, has been evidenced to originate in randomly distributed protein. The disintegrated state investigations prove that the reaction is not diffusion limited. The observed breaks in the temperature function therefore should be rather due to phospholipid control of the conformational state of the proteins, than to diffusion control. The results further evidence the accessibility of the cluster reductase for randomly distributed P–450 molecules.

In summary, physiologic relevant rapid reduction should originate in phospholipid controlled association structures of the respective proteins, which must not be of 1 : 1 stoichiometry. At excess of P–450 either intermolecular electron transfer between cytochrome molecules (rigid model) or high mobility — perhaps inside a phospholipid halo (nonrigid model) — is prerequisite for the experimental data. Any rigid or halo model, however, must include the accessibility of randomly distributed P–450 to the associated reductase. Recently in microsomes and liposomal LM2 systems, as well, clusterlike structures could be evidenced by EPR saturation transfer investigations (Schwarz et al., 1982a, b) (cf. chapter 1).

2.4. *Substrate Control*

The NADPH reduction of P–450 is substrate controlled in rate. Type I compounds enhance the process, especially the rapid phase reaction; type II compounds diminish the rate, on the other hand (Gigon et al., 1969; Sasame and Gillette, 1969; Diehl et al., 1970; Schenkman, 1972; Gillette et al., 1973; Holtzman and Rumack, 1973; Matsubara et al., 1976; Blanck et al., 1977), (Tab. 2). Such alterations have been observed likewise under anaerobic and aerobic conditions, as well. The early findings of this function led to the assumption that corresponding substrate conversion data could be originated by this phenomenon, i.e. that the catalytic process is rate limited by the substrate controlled reduction of P–450. This pro-

Table 5. Substrate Control of P–450 (P–448) Reduction and NADPH Oxidation[1])

Substrate	$[S]_{mM}$	P–450 Reduction $k_{app.}$ $[s^{-1}]$	%	NADPH Oxidation [nmol/min · mol]	%
Benzphetamine[2])	1	11.5	100	64.4	100
N,N-Dimethyl-aniline	9	9.2	80	50.6	79
Anisole	10	4.9	43	29.4	46
Ethylmor-phine	10	2.8	24	19.7	31
Aminopyrine	10	1.1	10	9.5	15
None	—	0.9	8	6.0	9
Benzphetamine[3])	1	0.31		11.9	
None	—	0.35		11.1	
Benzphetamine[4])	1	0.39		14.8	
None	—	0.30		13.9	

[1]) Imai et al. (1977) [2]) P–450, Rabbit (PB) [3]) P–448, Rabbit (MC)
[4]) P–448, Rabbit (PB); reproduced with permission from The Japanese Biochemical Society

posal was supported by apparent correlations in the rates of the two processes (Imai et al., 1977). Later investigations proved, however, relatively more rapid reduction, as compared to the substrate conversion (Matsubara et al., 1976; Imai et al., 1977; Coon and White, 1980). This behaviour will be discussed in 2.5. with respect to the second reduction process ④ (Tab. 5, 6).

The experimental proof of the incompetence of the first reduction process ② in rate-limitation has been verified moreover by the detection of the $P\text{–}450 \cdot O_2$ complex as an accumulated intermediate compound in the steady-state. This compound has been evidenced for the bacterial, the adrenocortical, and the liver microsomal system (cf. section 3). Rate divergencies between reduction and conversion can be sensitively monitored this way. The steady-state amount of such ternary complexes enables distinct studies of the oxygen activation and of substrate conversion, and furthermore of substrate control in the respective reaction steps (Werringloer et al., 1979).

Substrate control has been observed in microsomal, solubilized, disintegrated and in reconstituted systems, as well. The effect is strongly dependent on the reductase/P–450 stoichiometry. Thus the microsomal and solubilized state data amount to maximal 3 fold rate increase;

Tabelle 6. Rate Constants of Partial Reactions of the Benzphetamine N-Demethylation Reaction Sequence[1])

Benzphetamine Binding	$k_{app.} \sim 50\ s^{-1}$	[Benzphetamine] = 0.5 mM
P–450 Reduction	$k_{app.} \sim 4-12\ s^{-1}$	
Oxygen Binding	$k_{app.} \geqq 1000\ s^{-1}$	(Guengerich et al. 1976)
	$= 1300\ s^{-1}$	(Rösen and Stier, 1973)
NADPH Oxidation	$v = 0.8-1.2$	[nmol/s · nmol P–450]
HCHO Formation	$v = 0.5-0.7$	[nmol/s · nmol P–450]

[1]) Imai et al. (1977); reproduced with permission from The Japanese Biochemical Society

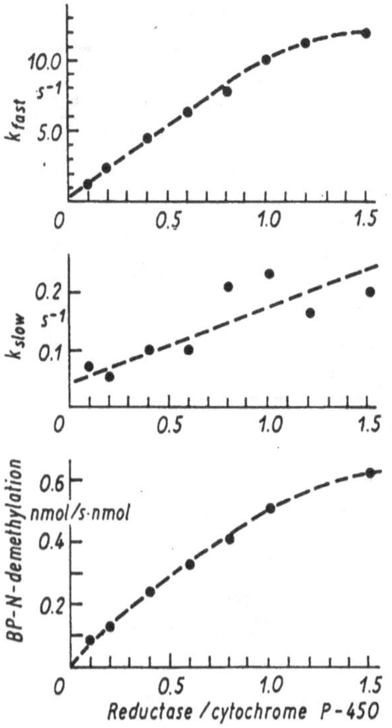

Fig. 10. Dependence of the Kinetic Parameters of the First Electron Transfer Reaction ② (φ_1, φ_2) and of the Benzphetamin-N-Demethylation in Reconstituted Systems on the Molar Ratio of Reductase/P-450 (TANIGUCHI et al., 1979); reproduced with permission from Elsevier/North Holland

in reconstituted systems, on the other hand, the rate enhancement increases with the relative reductase content and exceeds even 10 fold (Fig. 10) (IMAI et al., 1977; TANIGUCHI et al., 1979). The investigations with benzphetamine as substrate prove the effect to be strongly correlated with the substrate binding equilibrium (K_s) towards the cytochrome. They further disprove earlier studies which refer the substrate control to some reductase activation (HOLTZMAN and RUMACK, 1973).

Substrate activation of the reduction reaction is evidently composed of two parts. Binding of the substrate to P-450 may induce an activating conformational change in the cytochrome, which facilitates the electron transfer between the two proteins (intrinsic effect). The observed rate-enhancement, furthermore, depends on the reductase/P-450 ratio in the respective reconstituted systems (TANIGUCHI et al., 1979) (stoichiometric effect). The background of that dependency is rather unclear, since without any subtrate the reduction rate constant(s) seems to be less effected by the reductase/P-450 stoichiometry.

The observed shift in the high-spin/low-spin equilibrium of P-450, which is coupled to the substrate binding, correlates with respective alterations of the reduction rate (REIN et al., 1979). Therefore such correlations have been taken to suggest that the spin equilibrium may regulate the reduction process by rate-determining preequilibration. A rate-limiting spin conversion reaction, on the other hand, is discussed as originating the heterogeneity of the reduction process (BACKES et al., 1980, 1982).

The reduction of the oxidized P-450 (Fig. 5) could proceed via steps ⑭, ㉑ where ⑭ represents the preceding spin conversion process, or, on the other hand, by reduction of the low-spin species via step ㉒ and followed by the spin conversion reaction ㉓. The latter pathway is rather hypothetically because of no experimental proof of the existence of a reduced P-450 low-spin species. Any direct (simultaneous) conversion of the low-spin oxidized species to the high-spin reduced compound must be unfavourable because of being spin-forbidden.

The problem thus concentrates to reaction ⑭. This process has been evidenced to be quite rapid, the relaxation rate constant exceeds $10^4\ s^{-1}$ (RISTAU et al., 1979). Distinct rate constants for the individual on- and off-reactions in the order of $10^6\ s^{-1}$ could be resolved recently by means of laser temperature-jump experiments (ZIEGLER et al., 1982). The results thus disprove any rate limitation of the spin conversion process towards the reduction reaction, but are in line of a preequilibrium regulation. Furthermore, by use of a series of benzphetamine derivatives a clear cut correlation of high-spin content and reduction rate constant in rat microsomes could be evidenced (Blanck et al., 1982). The experiments further indicate that the reduction of the oxidized low-spin species is rather unfavourable, since the observed correlation function proves the rate constant to approach zero for that spin state.

The outlined spin state regulation of the P–450 reduction must not exclude isozyme specificities. The intrinsic rate constants of the respective high-spin species could depend on protein specificities. That must be further proved by investigation of reconstituted isozyme systems. In general, however, may be stated that a substrate induced high-spin shift favours the reduction rate of the respective P–450 complex via an increase in amount of the reactive compound in the spin preequilibrium (cf. chapter 4).

2.5. *Second Reduction Reaction* (e_2)

The second electron transfer ④ to the cytochrome could be discriminated at the camphor system (Debey et al., 1977). The reaction proceeds less rapid as compared to the first one ②, modifications by conformational change processes have been observed. Analogous investigations with respect to the liver system are lacking as yet. The pathway between the formation of the ternary complex ③ and substrate conversion ⑤ has not been resolved till now — except for intermediate steps in the oxygen compound transformation (Guengerich et al., 1976). The interference of the b_5 system as donor of the second electron further complicates the problem (cf. section 6).

As to the P–450 reductase, the uptake of electrons from NADPH enables the twofold electron supply towards P–450. In cytochrome b_5 deficient reconstituted systems the reductase really donates the required first and second electron, as well (②, ④). The correlation of NADPH consumption, first electron transfer rate, and substrate turnover, as observed in dependence on substrate and on reductase/P–450 ratio (Miwa et al., 1978, 1979) evidences a rate control of the electron transfer ② or ④, respectively. The rather rapid process ② as compared to the catalytic function in LM2 and LM4 systems (White and Coon, 1980) leaves process ④ to control the turnover and thus delivers a rate estimation of that step. Since in microsomes respective correlations (Miwa et al., 1978, 1979; Blanck et al., 1982) and rate discriminations (Matsubara et al., 1976) have been observed same circumstances may be valid. A close resemblance in rate of steps ② and ④ could be proved by recent investigations of Werringloer and Kawano (1980b) and by Ruf (1982).

3. Ternary Complex (P-450/Substrate/O_2)

P–450 in the reduced state is capable of binding O_2. This peculiarity is in general exhibited by reduced high-spin state hemoproteins. The oxygen is bound to the sixth coordination position of the heme iron; this binding mode refers likewise to the ligands CO and NO, respectively (Antonini and Brunori, 1971).

The ligand binding characteristics of P–450 is modified by preceding substrate binding to the cytochrome. Thus in the hydroxylation sequence oxygen donor and acceptor, respectively, exhibit distinct interactions at the ternary complex level. This statement refers not only to the mutual influence of the binding parameters of both partners (equilibrium, kinetics) but likewise to its P–450 controlled alignment for the oxygen transfer process.

The ternary complex in summary provides the structural basis for the distinct couple of oxygen activation and electron transfer processes in representing the locus of

- binding interaction of the oxygen donor and the acceptor substrate
- proper structural fixation of the oxygen transfer components
- complex oxygen activation processes
- distinct oxygen transfer steps
- distinct substrate conversion steps

3.1. *Functional Peculiarities*

3.1.1. *Steady-State Occurrence*

Oxygen complexes of P-450 have been evidenced in the main hydroxylation systems. ISHIMURA et al. (1971) and GUNSALUS et al. (1971) first observed the ternary complex in the camphor system, the Soret band of which being located at 418 nm; without camphor no oxygen complex could be observed. ESTABROOK et al. (1971a, c) detected the respective compound in liver microsomes after NADPH reduction in the presence of hexobarbital. The complex could not be proved, on the other hand, in the presence of type II compounds. The steady-state difference spectrum is characterized by absorption maxima at 440 and 590 nm. A similar complex could be isolated from the adrenal cortex P-450 system in the presence of desoxycorticosterone by SCHLEYER et al., (1972a, b; 1977).

The oxycytochrome is assumed to exist in an electronic state equilibrium like oxyhemoglobin, i.e. either as the ferrous oxygen compound or as the ferric superoxide complex, as well. The latter compound may dissociate to release the O_2^- anion, which by protonation will at least be reacted to hydrogen peroxide (⑥, Fig. 2, 11). This pathway could be evidenced (ESTABROOK et al., 1979a). Likewise could be proved, that the H_2O_2 generation via peroxycytochrome — the two electron reduced state of the ternary complex — is of minor importance during the cyclic function of P-450 (⑦, Fig. 2, 11).

3.1.2. *O_2/CO Competition*

The capability of carbon monoxide to compete with the physiologic oxygen donor for the binding site in P-450 has first been evidenced by ESTABROOK et al. (1963). They could show that steroid C21-hydroxylation as catalyzed by the microsomal adrenocortical system could be inhibited by CO. The CO action could be reversed by photochemical reactivation. The respective action spectrum was centered at 450 nm, i.e. at the absorption spectrum maximum of the CO-complex (OMURA et al., 1965). CO inhibition of P-450 therefore has been proved to be indicative for its participation as terminal oxidase in the hydroxylation process.

The strongly related binding properties of both ligands are suited to substitute CO for the oxygen in studies on structural implications concerning the function of hemoproteins. This way much information

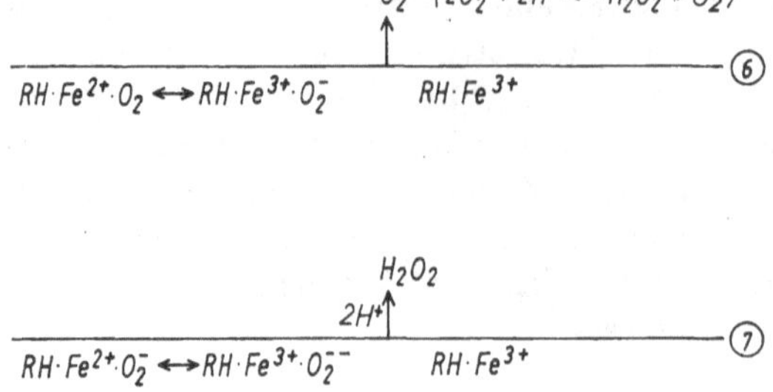

Fig. 11. Hydrogen Peroxide Generating Side Reactions ⑥ and ⑦ in the P-450 Cycle

has been obtained, e.g. with respect to heme-heme-interaction mechanisms in hemoglobin (GIBSON, 1959; ANTONINI and BRUNORI, 1971). Essential progress analogously has originated from corresponding studies on the P–450 · CO complex. Already in the investigations of GARFINKEL (1958) and KLINGENBERG (1958) P–450 was recognized as a CO binding pigment, and by this led to establish the hemoprotein nature of the cytochrome (OMURA and SATO, 1964a, b).

3.2. *Complex Formation*

3.2.1. *O_2 Binding Kinetics*

The analysis of the O_2 binding kinetics is quite restricted by the only small spectral sensitivity of the reaction and by the rapid complex formation rate, as well. PETERSON et al. (1972) performed first investigations on the camphor system. A rate constant of $7.7 \cdot 10^5$ M^{-1} s^{-1} (4°C) was determined for the substrate-bound P–450, the order of which being in range with those of other hemoproteins. Thus that process seems to be rather less specific. That is somewhat conflicting, on the other hand, with the observed substrate requirement of the reaction. These stopped-flow studies were continued by flash-photolysis experiments by RÖSEN and STIER (1973) on the liver system. They evidenced an 445 nm oxygen complex in the presence of hexobarbital for phenobarbital induced rabbit liver microsomes. The monophasic O_2-binding kinetics proceeded with a rate constant of 4.4×10^6 M^{-1} s^{-1} at 25°C.

More detailed investigations even proved a multistep O_2-interaction with reconstituted liver microsomal systems (GUENGERICH et al., 1976). Steady-state

Table 7. Rate Constants of the Ternary Complex Reactions[1])

Substrate	Rate Constants [min^{-1}] k_{31}	k_{32}	k_{33}
None	> 60000	210	12
Benzphetamine	> 60000	270	3
Cyclohexane	> 60000	215	7
p-Nitroanisole	> 60000	205	27
b_5 added			90
P–450 Reductase added			23
Reconstituted System Benzphetamine Turnover			24

[1]) GUENGERICH et al. (1976); P–450 LM2, pH 7.5, 12 °C; reproduced with permission from Academic Press, Inc.

experiments with an enzymatically reduced system evidenced an O_2-ternary complex with difference spectrum maxima at 442, 560 and 588 nm, thus resembling the microsomal complex (ESTABROOK et al., 1971a, c). A rather complex interaction kinetics, however, has been observed for oxygen binding to photoreduced P–450 systems. By means of stopped-flow techniques a 3-step reaction could be resolved (Fig. 12, Tab. 7). At 12°C at least 2 oxygen compounds have been observed, complex I formation (maxima at 430 nm and 450 nm in the difference spectrum) precedes the formation of complex II (maximum at 440 nm), the latter compound corresponding to the steady-state complex. Complex II decomposes to the oxidized cytochrome and products, as well. The resolved steps ㉜ and ㉝ clearly evidence the reaction to be controlled by either the substrate or by added reductase and cytochrome b_5, respectively.

By means of sub-zero-techniques similar reaction intermediates could be obser-

$$P\text{-}450r \xrightarrow{k_{31}(O_2)} \text{Complex I} \xrightarrow{k_{32}} \text{Complex II} \xrightarrow{k_{33}} P\text{-}450_0 + \text{Products}$$

Fig. 12. Reaction Scheme of the Interaction of Dioxygen with P–450 in the Ternary Complex (GUENGERICH et al., 1976) ved and even isolated (microsomes, purified P–450) (Balny et al., 1976; Debey et al., 1976; Begard et al., 1977; Bonfils et al., 1979). In contrast to the 12 °C, water reaction (Guengerich et al., 1976) the decay of the oxyferro intermediate of LM2 at subzero-temperature and in water/glycerol proved to be a biphasic process (Bonfils et al., 1979; Maurel et al., 1980). A more detailed analysis of the autoxidation process (2 °C, water) evidenced even three reaction phases ($k \sim 20\ s^{-1}$, $2.5\ s^{-1}$, $0.25\ s^{-1}$) with respect to purified LM2 (Bonfils et al., 1980). Incorporation into liposomes resulted in only monophasic decay ($k \sim 15\ s^{-1}$), i.e. the phospholipid vesicle obviously accelerates the autoxidation process and by this may increase the substrate conversion activity of the enzyme. Addition of reduced cytochrome b_5 further increases the rate of the complex decay (about twofold). Since the b_5 reoxidation does not correlate with the decay kinetics, b_5 is supported to be an effector in the process but no redox partner (cf. Guengerich et al,, 1976). The oxygen interaction with LM4 was likewise investigated (Oprian and Coon, 1980). An oxygenated intermediate, very similar in spectral properties to the LM2 and the camphor system species, was observed. The 10 °C, water autoxidation of the ferrous dioxygen compound was again found to be a biphasic process ($k \sim 0.99\ s^{-1}$, $0.24\ s^{-1}$).

As to the functional significance of the oxygen complexes no distinct proof is available. The formation kinetics of both compounds is rapid with respect to the rate-limitation of the substrate conversion. Thus they may represent intermediates in the formation of the "active oxygen compound", or they themselves could likewise be such species.

3.2.2. CO Model Investigations

CO binding studies on hemoproteins in general are much advantageous with regard to the moderate rates and to sufficient spectral sensitivity, as well. The binding affinity for P–450 ranges between that for hemoglobin and cytochrome oxidase; the CO/O_2 concentration ratio for 50% inhibition amounts to about 0.02 : 1 : 10 for the respective hemoproteins (Estabrook, 1965; Archakov, 1975). Flash-photolysis investigations even evidenced a tenfold increased CO affinity, i.e. the respective P–450 CO/O_2 affinity ratio amounts to 0.1 (Balny and Debey, 1976).

The binding process, as compared to porphyrin model compounds, is modified by the protein via trans- and cis-interaction at the iron site and by additional energy barriers on the way to the heme site (Cannon et al., 1976; Austin, 1977; Sono et al., 1977; Case and Karplus, 1978, 1979). Thus protein specificities of the process may reflect conformational implications.

First investigations by Peterson and Griffin (1972) proved monophasic CO binding kinetics for the camphor system. The rate constant was observed to be strongly dependent on the presence of the substrate: camphor decreased the value of $5.1 \cdot 10^6\ M^{-1}\ s^{-1}$ (substrate-free) to only $3.8 \cdot 10^4\ M^{-1}\ s^{-1}$. The same behaviour was observed with the adrenal cortex system (Akhrem et al., 1979); after flash-photolysis the overall CO-recombination reaction rate decreases in the presence of cholesterol. The liver system, on the other hand, did not exhibit any hexobarbital control (Rösen and Stier, 1973).

Further studies disproved the monophasic character of the reaction (Tab. 8). By means of improved investigation techniques up to 4 reaction phases have been discriminated, the rate constants of which ranging between $10^4 - 10^8\ M^{-1}\ s^{-1}$. The heterogeneity furthermore was dependent on the respective investigation techniques (stopped-flow, flash-photolysis).

As to the heterogeneity, a partially purified solubilized system was investi-

Table 8. Rate Constants of the CO Binding Reaction

P–450 System	Method	Temperature	Substrate	k [M^{-1} s^{-1}]	Reference
Pseudomonas putida	SF[1])	4 °C	—	$5.1 \cdot 10^6$	Peterson et al. (1972)
			Camphor	$3.8 \cdot 10^4$	
LM, Rabbit[2])	F[1])	25 °C	—	$2.4 \cdot 10^5$	Rösen and Stier (1973)
LM, Rat	SF	4 °C	—	$4.5 \cdot 10^5$	Debey et al. (1973 b)
		20 °C	—	$\sim 10^6$	
	F	4 °C	—	$2.3 \pm 0,7 \cdot 10^6$	Debey et al. (1973 a)
				$4 \pm 1.5 \cdot 10^5$	
LM, Rat	SF	25 °C	—	$6.8 \cdot 10^5$ ($\sim$ 70%)	Blanck et al. (1976)
				$\sim 3.4 \cdot 10^4$ ($\sim$ 30%)	
LM, Rat (solub.)				$6.6 \cdot 10^5$ ($\sim$ 70%)	
				$\sim 3.3 \cdot 10^4$ ($\sim$ 30%)	
LM, Rat	SF	20 °C	—	$8.7 \pm 1.7 \cdot 10^6$	Gray (1978)
				$1.59 \pm 0.18 \cdot 10^6$ (1: 1.25: 1.39)	
				$2.23 \pm 0.35 \cdot 10^5$	
	F			$12 \pm 3.4 \cdot 10^6$ (35%)	
				$2.3 \pm 0.38 \cdot 10^6$ (41%)	
				$2.3 \pm 0.05 \cdot 10^5$ (24%)	
LM, Rabbit (solub.)	F	20 °C	—	$1.0 \pm 0.2 \cdot 10^7$ (30%)	Davydov et al. (1980)
				$2.2 \pm 0.4 \cdot 10^6$ (40%)	
				$2.0 \pm 0.2 \cdot 10^5$ (30%)	Greschner et al. (1980 a, b)
			Benzphetamine	$2.3 \pm 0.2 \cdot 10^6$ (73%)	
				$2.6 \pm 0.2 \cdot 10^5$ (27%)	
			Spasman^R	$3.1 \pm 0.3 \cdot 10^6$ (68%)	
				$3.2 \pm 0.3 \cdot 10^5$ (32%)	
			Aniline	$3.7 \pm 0.3 \cdot 10^6$ (68%)	
				$3.8 \pm 0.3 \cdot 10^5$ (32%)	
LM2	F	20 °C	—	$4.3 \pm 2.0 \cdot 10^6$ (40%)	Greschner et al. (1982 b)
				$1.1 \pm 0.5 \cdot 10^6$ (45%)	
				$1.7 \pm 0.5 \cdot 10^5$ (15%)	
			Spasman^R	$4.0 \pm 2.0 \cdot 10^6$ (85%)	
				$5.0 \pm 1.0 \cdot 10^5$ (15%)	
			Aniline	$4.0 \pm 2.0 \cdot 10^6$ (85%)	
				$6.0 \pm 2.0 \cdot 10^5$ (15%)	

$$\text{P–450} + \text{CO} \underset{k_{-1}}{\overset{k_1}{\rightleftharpoons}} \text{P–450}\cdots\text{CO} \underset{k_{-2}}{\overset{k_2}{\rightleftharpoons}} \text{P–450} \cdot \text{CO}$$

				k_1 [M^{-1} s^{-1}]	k_{-1} [s^{-1}]	k_2 [s^{-1}]	k_{-2} [s^{-1}]	
LM2	SF	20°C	—	$3.2 \cdot 10^6$	18.2	53	0.11	Gray (1982)
LM4				$1.6 \cdot 10^5$	3.6	5.6	0.055	
P–450b				$1.9 \cdot 10^6$	20	48	0.050	

[1]) SF-Stopped-Flow; F-Flash-Photolysis; [2]) LM-Liver Microsomes

gated with respect to dependence on species (rat, rabbit), wavelength (subspecies), integrational state (component interaction) (Blanck et al., 1976, Smettan et al., 1978). The studies imply the functional differences to be due to differences in intrinsic reactivity of subspecies (isozymes). Gray (1978) oberved likewise in stopped-flow and flash-photolysis studies a multiphasicity in rat microsomes. 3 second order processes could be resolved, the heterogeneity was related to different P–450 fractions which have been proved to exhibit specific CN binding properties. P–450 subfractions with respective CN affinities have been isolated by Comai and Gaylor (1973). Davydov et al. (1980) and Greschner et al. (1980a, b) supported the data of Gray (1978) on a partially purified solubilized P–450 preparation from rabbit microsomes The flash-photolysis investigations likewise proved 3 second order reactions, which in the presence of substrates (benzphetamine, spasman[R], aniline) diminished to only two processes. The data were further refered to isozyme specificities in reactivity, which is controlled by substrates.

As to the reaction inhomogeneity, the origin remains to be questioned. Only isozyme specificity seems to be insufficient for an unambiguous interpretation, since the isolated fractions (LM2, LM4, P–450b) exhibit likewise an inhomogeneous CO binding reaction. Gray (1982) evidenced a predominantly 2-phase CO binding process which was analysed to represent a two-step consecutive reaction where the ligand binding step is followed by a conformational change of the CO complex (LM2, LM4, P–450b). A similar result has been obtained with the camphor system by Bazin et al. (1982). Greschner et al. (1982b), on the other hand, by means of flash-photolysis investigations discriminated 3 second order partial reactions (LM2). Substrates were shown to shift the phase distribution as already observed in the solubilized P–450 preparation. It was suggested that conformational isomers (equilibrium conformers) were responsible for the observed functional specificities, and that non-equilibrium states of the reduced P–450 after flash-photolysis could further modify the CO recombination pattern. The observed conformational changes in the investigation of Gray (1982) and Bazin et al. (1982) could support this view. The existence of non-equilibrium conformational states in P–450 has been first proved in the reduction reaction (e_1) at low temperature (Davydov et al., 1978, 1979; Greschner et al., 1979; Greschner, 1982a). By means of trapped electrons the cytochrome could be reduced at the heme site, reequilibration at room temperature (protein relaxation) established the respective reduced hemoprotein state.

3.3. Complex Decay

The P–450 binary complex (O_2) and the ternary complex (O_2, substrate), respectively, decay quite rapidly by reoxidation of the cytochrome and by the formation of either hydroxylated product and of H_2O_2 (Guengerich et al., 1976; Begard et al., 1977; Bonfils et al., 1979). The autoxidation reaction again depends considerably on substrate. The substrate-free oxycytochrome of the camphor system decays nearly 2 orders of magnitude more rapid as compared to the camphor ternary complex (Gunsalus et al., 1974). Tab. 7 evidences the substrate specificity of the process in the liver system. Benzphetamine and cyclohexane exert some stabilization in comparison to the substrate-free P–450. On the other hand, addition of reductase and cytochrome b_5 even increase in the photoreduced system the decay rates of the complex to proper values with respect to the physiologic process; cytochrome b_5 stimulation is not paralleled by its oxidation, i.e. this cytochrome does rather act as an effector but not in a redox reaction.

The ternary complex decay thus turns out to reflect the distinct couple of manyfold interactions of the components (O-donor, substrate inclusively). Unfortunately the intracomplex partial reactions — including the second electron transfer — are unresolved till now. Just this topic needs further investigations in the future.

4. Substrate Conversion

The final reaction in the hydroxylation sequence is the substrate conversion step. Just this process is again composed of several individual reactions, which may proceed either in a sequence or even in a concerted manner. Unfortunately the discrimination and characterization is rather derived from overall kinetic data, especially with respect to regio- and stereoselectivity, but less from the kinetic analysis of partial steps. That restriction makes the interpretation of mechanisms more speculative and emphasizes, on the other hand, the urgent requirement for detailed investigations at the resolved level.

4.1. *Overall Characteristics*

The overall conversion data (V_{max}) reflect twofold characteristics. They can be indicative of peculiarities of the conversion process itself, but further of preceding steps, as well. Thus the interpretation of V_{max} data is intimately related to the discussion of rate-limiting steps and by this is "kinetically" in nature.

Tab. 9 depicts selected V_{max} values which cover important peculiarities of the conversion reaction, they may be summarized as follows:

1. Physiologic substrate conversion by means of NADPH and O_2 proceeds with maximum rates of about 60 min^{-1} (V_{max}). That value assigns the rate-limiting step into the order of seconds and beyond, respectively.
2. The liver enzyme system exhibits — in contrast to the highly specific camphor system, and to the 11β- or SCC-system of the adrenal cortex mitochondria, as well — rather less substrate specificity. Despite of distinct substrate differentiation, a broad variety of compounds is converted.
3. Stereo- and regioselective product formation enables a conceptional elaboration of partial mechanisms.
4. Subfraction specificities are exerted by the respective electrophoretically isolated isozymes.
5. Rather less species dependences are exhibited (liver system); the order of substrates with respect to conversion capability is preserved.
6. By-passing of the physiologic process by means of different O-donors further modifies the reaction and leads in part to increased reaction specificity and rate, as well.

As to the preceding events which may govern the overall hydroxylation, substrate or O_2 ligandation has been ruled out as being substantial.

The features enumerated prove that the first electron transfer (e_1) may be rate-limiting only in the most rapid hydroxylation reactions — e.g. with benzphetamine or ethylmorphine. These substrates, however, could be excluded to be limited by that step (Matsubara et al., 1976; Imai et al., 1977).

Therefore rather the second electron — being delivered either by the P-450 reductase or by cytochrome b_5 — may refer to the rate-restriction, but no experimental proof has been obtained till now. The relatively decreased aniline turnover is compatible with a decreased reduction rate. Investigations of Akhrem et al.

Table 9. Substrate Specificity of Various Forms of Liver Microsomal Cytochrome P-450

Substrate	Hydroxyl Group Position in Product	Rabbit[1]			Rat[2]		Mouse[4]			
		LM2	LM4	LM1.7	P-450	P-448	A_1	A_2	C_1	C_2
Benzphetamine		66	3.0	7.5	40	3.2	6	5	45	8
Ethylmorphine		6.1	3.0	3.0	5–7[3]		1	5	3	12
Ethoxycoumarin					6.6	34				
p-Nitroanisole		6.2	0.6	0.6						
Aniline		1.0	0.4	0.6						
Benzo/a/pyrene		0.04	Trace	0.5	0.19	3.9	0.27	0.26	0.1	0.08
Biphenyl	2	0.7	0.3	0.4						
Biphenyl	4	5.4	0.4	0.6						
Testosterone	$7\alpha - OH$	0.02	0.04	0.02	0.97	0.69	0.06	0.73	0.08	0.15
Testosterone	$16\alpha - OH$	0.43	0.02	0.07	1.5	0.17	< 0.005	0.03	1.64	0.06
Testosterone	$6\beta - OH$	0.02	0.02	0.32	0.12	0.25	0.09	0.01	0.93	0.88

[1]) HAUGEN et al. (1975); [2]) THOMAS et al., (1976); [3]) LEVIN (1977); [4]) HUANG et al. (1976); Activity [nmol/min · nmol P-450]

(1977), however, doubt to the importance of the electron transfer in that case. Cyclohexane, cyclohexene, dimethylaniline, aminopyrine, quinoline and monomethylaniline could be evidenced to exhibit the same kinetic and thermodynamic (activation) parameters in the NADPH/O_2 — and in by-passing reactions by means of organic peroxides, as well.

It remains questionable, however, if oxygen cleavage exerts the rate-limitation. Respective donor specificities have been observed by the GUSTAFSSON group in steroid investigations (HRYCAY et al., 1975a, b, c, 1976; DANIELSSON and WIKVALL, 1976; GUSTAFSSON and BERGMAN, 1976). Most effective proved $NaIO_4$, followed by cumene hydroperoxide, NADPH/O_2, linoleic acid hydroperoxide, $NaClO_2$, tert.butyl hydroperoxide, and H_2O_2. The different affinities of the hydroxylating agents towards the P-450 species (subfractions) are supposed as being responsible for the differentiation. The hydroxylation of fatty acids further supports these statements. As compared to the physiologic NADPH/O_2 process the ω2-conversion of laurate, palmitate, and stearate by $NaIO_4$ and $NaClO_2$ is about 100 fold accelerated, the laurate reaction via iodosobenzene diacetate even 3 orders of magnitude (Tab.10). Iodosobenzene has been further proved to be active in 7-ethoxycoumarin-0-dealkylation; after phenobarbital and benzpyrene pretreatment the substrate is being converted with an about 2- and 8-fold rate in comparison to the NADPH/O_2 reaction (LICHTENBERGER et al., 1976).

The somewhat generalized results indicate the rate-restriction of the physiologic hydroxylation sequence by electron transfer and O_2-cleavage, respectively. The significance of the second step — which is part of the conversion reaction — is further treated below.

4.2. Reaction Steps

Prerequisite for the substrate conversion reaction is the formation of the substrate/O_2 ternary complex of the cytochrome. The introduction of the second electron then initiates "oxygen activation", i.e. the formation of the active species towards the substrate. The conversion reaction is composed of at least 3 steps

Table 10. $\omega 1$- and $\omega 2$-Hydroxylation of Laurate, Palmitate and Stearate[1])

Hydroxylating Agent	Reaction Rates (nmol/mg · min)				Fatty Acid Substrate	
	Laurate		Palmitate		Stearate	
	$\omega 1$	$\omega 2$	$\omega 1$	$\omega 2$	$\omega 1$	$\omega 2$
NADPH	0.72	0.47	0.34	0.09	0.14	0.03
NaCl O_2	0[2])	46.0	0	11.9	0	4.50
$NaIO_4$	0	151	0	34.0	0	3.75
φIO	0	188	—[3])	—	—	—
$\varphi I(OAc)_2$	0	930	—	—	—	—
φIO_2	0	0	—	—	—	—

[1]) GUSTAFSSON and BERGMAN (1976)
[2]) < 0.01
[3]) Not analyzed; reproduced with permission from Elsevier/North Holland

(Fig. 1, 13):

- the O_2 cleavage reaction, concomitant with the formation of an iron-oxo-species (RH · Fe(III) · O), and the possible release of H_2O,
- the RH cleavage reaction, followed by O insertion,
- the product dissociation process.

— As to the first step, the preceding electron transfer towards the oxygen molecule in the ternary complex at the expense of formal iron oxidation is evidenced with respect to the first added electron: the system is capable of generating O_2^- which in turn disproportionates to H_2O_2 (ESTABROOK et al., 1979a). The second electron is supposed to be likewise translocated towards the molecular oxygen, thus additionally weakening the O—O bond (OCHIAI, 1975). Cleavage of the dioxygen molecule and water release should then lead to the proposed iron-oxo-complex (oxene state) which resembles compound I in the peroxidase mechanism (LIPSCOMB et al., 1976). This oxygen species may further on initiate the substrate hydroxylation via oxygen insertion, i.e. P–450 acts as an "oxene transferase" (LICHTENBERGER et al., 1976; ULLRICH, 1977).

This scheme is modified and likewise extended to nonphysiologic oxygen supporting agents (WHITE and COON, 1980; WHITE et al., 1980b). The postulated mechanism is based on two proposals: the oxygen atom to be inserted into the substrate originates from either O_2 or ROOH, respectively; an iron-coordinated peroxide is an intermediate in the reaction. Thus the oxygen cleavage reaction is that of a peroxide bond. The reaction may proceed by two different pathways, either by heterolytic cleavage ("oxenoid" mechanism) or by homolytic rupture ("quasi-FENTON" mechanism), in which the thiolate ligand serves as a one-electron donor to the peroxide, and the established thiyl ferric hydroxide is one-electron oxidized relative to the native enzyme (ferric iron) thus being equivalent to the peroxidase compound II.

In the physiologic reaction (NADPH/O_2) both alternatives require an exogeneously supported electrophilic center for the conversion of the oxygenated P–450. The formation of an acylperoxide-iron intermediate together with the acid radical could be elucidated for the P–450 CAM interaction with lipoic acid (SLIGAR et al., 1980a, b): using ^{18}O, the isotope incorporation into the substrate concomitant with a single oxygen labelling

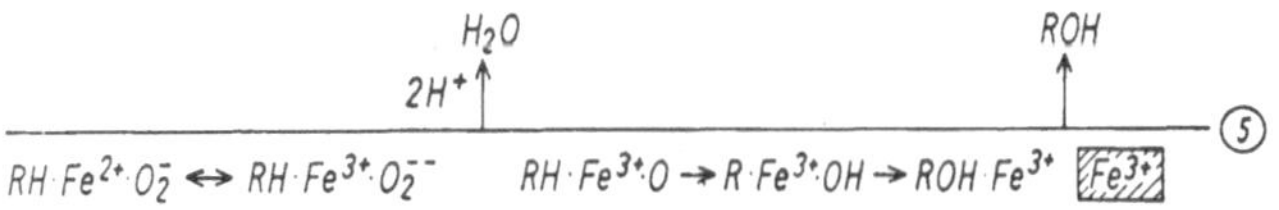

Fig. 13. Substrate Conversion Reaction ⑤ in the P–450 Cycle

of the COOH group of the acid could be evidenced. This result supported the proposal of putidaredoxin to be an acylating species in the camphor system (SLIGAR et al., 1977). Alternatively the participation of an amide via formation of a peroxiimidic acid is discussed with respect to the "quasi-FENTON" pathway (WHITE and COON, 1980).

Just recently the homolytic (radical) cleavage mechanism was supported. WHITE et al. (1980b) studied selected P-450 (LM2, LM4, CAM) supported hydroxylation reactions by means of peroxy-phenylacetic acid as O-donor. Using the ^{18}O labelled compound they could evidence that the O—O bond cleavage proceeded via phenylacetoxy radical formation ($C_6H_5—CH_2—CO_2^{\cdot}$) and following decarboxylation; the resulting benzyl radicals could be proved as benzyl alcohol in the product mixture.

As to the formation rate of the active oxygen species by splitting the O—O bond no discrete information is at hand. The dependence of the overall hydroxylation on the respective O-donor may indicate some rate-determination of this partial reaction, but conclusive proof is absent till now.

— The second step in the substrate conversion process is the RH cleavage and oxygen insertion which is supposed to proceed by a two-step mechanism: preceding RH cleavage by hydrogen abstraction and radical formation is followed by a radical recombination step. In the "oxenoid" pathway the iron-oxo-complex accepts the substrate hydrogen to form the respective hydroxyl radical; the hydrogen acceptor in the "quasi-FENTON" process is the supporting acid radical in the $NADPH/O_2$ reaction and the specific O-donor radical in the peroxide track, respectively.

The preceding hydrogen abstraction step is evidenced by the product patterns of methylcyclohexane hydroxylation (WHITE et al., 1979). High regioselectivity has been observed which is only compatible with a radical pair cleavage (homolytically) of the RH bond. Direct oxygen insertion, on the other hand, should fail in selective 1°, 2°, 3° hydrogen abstraction. Large isotope effects in the product patterns of norbornanes further elucidated favourable hydrogen abstraction as compared to deuterium (GROVES et al., 1978).

Despite of intramolecular isotope effects the overall hydroxylation rates failed to exhibit any isotope effect till now (ULLRICH, 1969; THOMPSON and HOLTZMAN, 1974; GROVES et al., 1978; MIWA et al., 1979). Therefore hydrogen abstraction is ruled out as being rate-limiting in these investigations. The recent studies of BLAKE and COON (1980a), on the other hand, elucidate rate determination of that process in cumene hydroperoxide dependent toluene conversion. The respective rate constants significantly exceed the turnover values of the physiologic $NADPH/O_2$ hydroxylation reaction. Therefore peroxide supported reactions may depend on the hydrogen abstraction process whereas the physiologic reaction does not.

The above investigations furthermore favour the "quasi-FENTON" pathway. Two intermediate CHP-complexes could be resolved, only the first of which (complex C) is directly involved in the hydroxylation (cf. section 5). Its conversion into products proved to be rate-limiting. p-substitution of either substrate and peroxy compound, respectively, influenced the product formation rate. The dependence of rate-determining hydrogen abstraction on the respective oxidant implies its participation as radical species and thus supports the homolytic cleavage mechanism.

The product formation is terminated by the recombination of the carbon radical with the hydroxyl radical species ("oxygen rebound", GROVES and McCLUSKY, 1976). Such radical reactions are known to proceed quite rapidly without any need for high activation energy. Therefore this step should be excluded as being rate-limiting.

— The last step to be discussed is the product dissociation. This process is evidently inevitable for recycling, because of the required substrate binding site and the O-donor site, as well, which may be both involved in product binding. Substrate binding proceeds with second order rate constants of about 10^3-10^5 $M^{-1}s^{-1}$; respective binding constants of 10^3-10^5 M^{-1} suggest dissociation rates in the order of seconds. Of course, product formation alters the original binding parameters of the respective substrate, but its order of magnitude indicates that product dissociation should not be excluded for sure as rate-interfering or even rate-limiting step. The release of oxygen donor products and the respective rate constants are unknown, on the other hand.

5. Shunt Reaction (Nonphysiologic Oxygen Supporting Agents)

5.1. *Characteristics of the Reaction*

Hrycay and O'Brien (1972) observed the conversion of organic and of steroid hydroperoxides, respectively, during the oxidation of N, N, N', N'-tetramethyl-p-phenylendiamine by rat liver and bovine adrenal cortex P–450, as well. The peroxidatic activity of the liver microsomes could be increased by means of phenobarbital- and 3-methylcholanthrene-induction, respectively. Progesterone and pregnenolone could be likewise metabolized into respective hydroxy compounds via 17α-hydroperoxide derivatives. The investigations were extended by several groups and resulted in the following characteristics of the "peroxide-dependent P–450 substrate conversion reaction":

— The active hemoprotein is P–450; after conversion into P–420 and removal of the apoprotein, respectively, no product formation was observed (Hrycay and O'Brien, 1972; Nordblom et al., 1976). Other hemoproteins like catalase, horseradish peroxidase, myoglobin, and cytochrome b_5 failed to exhibit any measurable product formation in the cumene hydroperoxide dependent oxidation of coumarin and aniline, respectively (Rahimtula and O'Brien, 1974). C-oxidation of dimethylaniline, aminopyrine, and benzphetamine, on the other hand, proved catalase to be an effective catalyst (Kadlubar et al., 1973). The hydrogen peroxide dependent N-demethylation of aminopyrine proceeded likewise with respective turnover (protoheme 7.8, metmyoglobin 44, horseradish peroxidase $2.5 \cdot 10^4$, [mol formaldehyde/mol heme $\times$ min]) (Griffin, 1977; Griffin and Ting, 1978). Comparable data were obtained for the N,N-dimethylaniline and the N,N-dimethyl-p-toluidine conversion (Griffin, 1978).

The results so far available indicate that the reaction specificity depends on the catalyst, the substrate, and the O-donor, as well, thus reflecting significant structural specificities of the reaction complex (ternary complex) as established by the respective participants.

— The NADPH-dependent P–450 reductase is not required in the reaction. Whereas proteolytic digestion inhibits the physiologic $NADPH/O_2$ substrate conversion, the peroxide dependent reaction, on the other hand, is only less effected (Kadlubar et al., 1973).

— Any CO inhibition, as observed in the $NADPH/O_2$ reaction, does not interfere the shunt process (Rahimtula and O'Brien, 1975; Nordblom et al., 1976).

— Oxygen donor of the reaction is the respective peroxide, any ^{18}O incorporation from the solvent could be excluded (Nordblom et al., 1976; Rahimtula et al.,

1978). O-donor consumption and substrate conversion is linked by 1:1 stoichiometry (RAHIMTULA and O'BRIEN, 1974, 1975; NORDBLOM et al., 1976).

— The peroxide dependent reaction resembles the $NADPH/O_2$ process in the product patterns. That has not only evidenced for exogeneous substrates but further for specific hydroxylations of steroids (HRYCAY et al., 1976); on the contrary, significant differences in both pathways could be evidenced recently in the benzpyrene conversion (RENNEBERG et al., 1981). Comparable NIH shifts have been observed (RAHIMTULA et al., 1978). Similar rate constants as well as activation parameters in the metabolism of cyclohexane, cyclohexene, dimethylaniline, aminopyrine, chinoline, and monomethylaniline could be proved (AKHREM et al., 1977).

The O-donor function in the P-450 shunt reaction is not only restricted to peroxides. Likewise $NaIO_4$ and $NaClO_2$ have been evidenced to be active with respect to fatty acid steroid hydroxylation, as well (HRYCAY et al., 1975a, b, 1976; DANIELSSON and WIKVALL, 1976; GUSTAFSSON and BERGMAN, 1976; GUSTAFSSON et al., 1976). Of special importance proved the iodosobenzene mediated dealkylation of 7-ethoxycoumarin (LICHTENBERGER et al., 1976). This oxygen supporting agent and respective derivatives were catalytically active with respect to steroid hydroxylation, as well. Just these investigations support the assumption that the active oxygen species in the conversion of the substrate contains atomic oxygen, i.e. it exists in a ferric oxene state in resonance with a ferryl iron compound, thus being comparable with compound I in the peroxidase reaction.

The investigations with the different oxygen supporting agents evidence, that the physiologic $NADPH/O_2$ reaction cycle can be by-passed in that the support of electrons becomes unnessessary. Obviously they only contribute to the splitting of the molecular oxygen. The by-pass (shunt reaction) is drawn in Fig. 2 (⑧). For further details cf. chapter 4.

5.2. *Kinetic Aspects*

The analysis of the shunt reaction is predominantly concerned with P-450-oxygen and -substrate intermediates, with the product patterns, and with accessible reaction parameters of the overall reaction and of partial steps, as well. The first mentioned problems are treated in chapters 4 (oxygen species) and 9 (application). Questions of the reactivity of reaction participants and of intermediates shall be discussed in this chapter. The subsequent topics have been selected:

1. Are respective reaction parameters (V_{max}) of both pathways (i.e. $NADPH/O_2$, shunt) of comparable magnitude or do they exhibit pathway specificity.
2. What conclusions can be drawn from V_{max} with respect to rate-limiting steps in either pathway.
3. What specificities with respect to the different O-donors and to P-450 subfractions are observed.
4. Can accessible partial reaction steps be compared in both processes, or do they indicate pathway specificity.

5.2.1. *Substrate Conversion Reaction*

Turnover data for a variety of substrates and O-donors, especially with respect to the microsomal P-450, have been elaborated. The shunt reaction of reconstituted systems and isolated components, as well, has been investigated too.

In the microsomal shunt reaction the respective substrate conversion data (V_{max}) for the C-oxidation of amines by cumene hydroperoxide correspond to those of the NADPH/O_2 reaction; the dimethylaniline turnover exceptionally proceeds about tenfold enhanced in the shunt reaction (Tab. 11, KADLUBAR et al., 1973).

Similar activation thermodynamics in both pathways could be evidenced for aniline derivatives, cyclohexane and further substrates (AKHREM et al., 1977). The authors therefore suggested the rate-limiting step to be the O-insertion process, modified by the respective substrate RH bond.

Systematic investigations of steroid and fatty acid oxidation, on the other hand, evidenced O-donor- as well as substrate-specificities (HRYCAY et al., 1975a, b, 1976; DANIELSSON and WIKVALL, 1976; GUSTAFSSON and BERGMAN, 1976; GUSTAFSSON et al., 1976):

Donor specificities have been proved in the conversion of androstenedione, testosterone, progesterone, and 17β-estradiole, which exhibit high sensitivity in stereospecific reactions. $NaJO_4$ revealed maximum activity in O-donation; the respective androstene hydroxylation in position 7α proceeded about 50 fold more rapidly in comparison to the NADPH/O_2 reaction. The donor activity decreases in the sequence $NaIO_4$, cumene hydroperoxide, NADPH/O_2, $NaClO_2$, pregnenolone-17α-hydroperoxide, tert.butyl hydroperoxide, and linoleic acid hydroperoxide. H_2O_2 was of only low activity. The product pattern varied likewise with the O-donor. Product distributions could be further modified by e.g. phenobarbital induction, thus suggesting different reactivities of the respective donors towards multiple forms (subfractions) of P–450.

Table 11. Enzymatic C-Oxidation of Amines [nmol Formaldehyde/mg Protein · min][1])

Substrate	NADPH/ O_2 LM[2])	Cumene LM[2])	Hydroperoxide Catalase[3])
Dimethylaniline	3.4	35.1	1.51
Aminopyrine	3.5	5.8	312
Benzphetamine	3.3	4.2	10
Propoxyphene	4.7	3.3	—
Ethylmorphine	6.3	2.2	—
Metamphetamine	2.0	0.4	—

[1]) KADLUBAR et al. (1973)
[2]) Pig Liver Microsomes
[3]) Bovine Catalase; reproduced with permission from Academic Press, Inc.

Substrate specificity could be evidenced in the $NaIO_4$ supported reaction. Steroids were converted with high activity; the donor failed, on the other hand, in the O-dealkylation of 7-ethoxycoumarin. The ω-hydroxylation of fatty acids proved further significant specificities. Lauric, palmitic, stearic acid did not exhibit ω1-hydroxylation. Lauric acid could be ω2 converted. This process with $NaIO_4$ and $NaClO_2$ is about 100 and 300 fold enhanced as compared to the NADPH/O_2 reaction. Iodosobenzene and especially the diacetate derivative increase the reaction rate even up to 1000 fold, whereas neither of the other acids are converted at all (Tab. 10).

The microsomal data are supported by investigations with reconstituted systems (NORDBLOM et al., 1976). It could be shown that P–450 LM2 likewise catalyses peroxide-, peracid-, and $NaClO_2$ dependent reactions. Cumene hydroperoxide again proved to be highly active towards e.g. N,N-dimethylaniline, N-methylaniline, and cyclohexane (~200 nmol/nmol P–450.min). Whereas NADPH, P–450 reductase, O_2, and other electron transfer proteins (cytochrome b_5) were not constitutive in the shunt reaction, phospholipid, on the other hand, was of influence. Removal of this component inhibited the benzphetamine and the dimethylaniline oxidation up to 50% and even 75% respectively (Tab. 12).

Also the bacterial P–450 CAM enzyme system is capable to react via the shunt mechanism (SLIGAR et al., 1977). Organic

Table 12. Substrate Hydroxylation by Hydroperoxides[1])

System	Benzphetamine Demethylation (H_2O_2)	N-Methylaniline Demethylation (Cumene Hydroperoxide)
P-450, Donor, Substrate	27[2])	84
P-450 Heat inactivated	2	
No P-450	0	0
P-420	0	
Hemin Chloride	0	
$FeCl_3$	0	
No Peroxide	0	0
No Dilauroyl-GPC	14	24
No O (N_2)	27	
No O_2 (CO)	28	

[1]) NORDBLOM et al. (1976)
[2]) Formaldehyde Formation (nmol/min · nmol P-450); reproduced with permission from Academic Press, Inc.

peroxides, H_2O_2, chlorite, and periodate likewise caused a P-450 catalyzed substrate conversion in the absence of P-450 reductase and under anaerobic conditions.

In summary it can be stated, that the overall kinetics of the shunt reaction exhibits significant specificities with regard to O-donor and substrate, as well. As compared to the NADPH/O_2 reaction in part highly increased turnover is observed; that indicates the by-passing of rate-limiting steps — especially of electron transfer reactions — in the physiologic process. It can not be excluded, that electron transfer is the only reason; instead of, the differences in both pathways may depend on O-donor and RH-specificities, respectively, as is generally shown for the different oxygen supporting agents.

5.2.2. Partial Reactions

The shunt reaction ⑧ is drawn in Fig. 2 and 14 with respect to distinct steps. It contains additionally reaction ① (substrate binding) and the second part of reaction ⑤ (substrate conversion) of the general sequence; binding of the O-donor as well as possible intermediate reactions on the way towards the common state RH·Fe(III)·O (in the oxenoid pathway, cf. section 4) represent thus the shunt-specific part of the reaction, i.e. from these steps emerge the specificities of the shunt process.

Studies on the spectral and kinetic properties of P-450/O-donor complexes started only recently. By means of stopped-flow techniques the formation of a P-450 LM2 complex with m-chloroperbenzoic acid could be evidenced (COON et al., 1979). The respective maxima in absorption were obtained at 375, 425, and 540 nm — at 370, 436, 540 nm in the difference spectrum. Other donors established respective complexes with similar but not identical spectral properties. Same characteristics have been observed with microsomes and different oxygen supporting agents too in earlier investigations (RAHIMTULA et al., 1974; HRYCAY et al., 1976; LICHTENBERGER et al., 1977).

The kinetics of the complex formation between P-450 and the respective O-donor could be resolved for a series of peroxy-compounds, e.g. perbenzoic acids, cumene hydroperoxides, benzyl hydroperoxides (Tab. 13, BLAKE and COON,

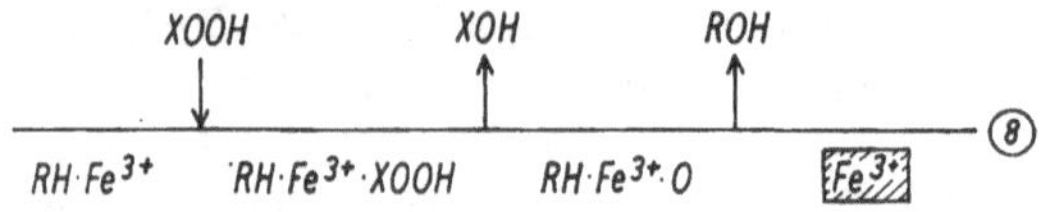

Fig. 14. Shunt Reaction ⑧, Supported by Nonphysiologic Oxygen Donors

1980b). A two-step mechanism fits to the experimental data

$$\text{P–450 LM2} + \text{O-donor} \underset{k_2}{\overset{k_1}{\rightleftharpoons}} \text{C} \underset{k_4}{\overset{k_3}{\rightleftharpoons}} \text{D} \quad (4)$$

The reaction proceeds via a dissociable enzyme-oxidant complex C which is rapidly converted to the observed product D. The first partial reaction is driven by hydrophobic binding of the O-donor to the cytochrome; the rate of the second step increases with the respective Hammett constant σ, i.e. with the electron withdrawing capability of the substituent.

The investigations may be summarized as to show that

1. the formation of the intermediates is reversible, thus contrasting the peroxidase mechanism,
2. the spectra of the intermediates depend on structural peculiarities of the O-donor. Some resemblence to the peroxidase compounds in absorption behaviour is obvious, but significant dependencies on substituent structure are exhibited,
3. O-donor specificities are likewise evident in the kinetic constants. The k_1 range — 10^1—$10^6\ M^{-1}s^{-1}$ — and the k_3 variance — 10^{-2}—$10^3\ s^{-1}$ — indicate the possibility of rate-interference with or-even rate-limitation of the overall substrate conversion process.

The investigations were continued by product formation studies (overall reaction) with p-substituted toluenes and cumene hydroperoxides, respectively. The experiments implicated the participation of only complex C in the reaction sequence, whereas complex D formation proved to be a side reaction. The rate-limiting transition of complex C into product was suggested to be originated in the abstraction of a hydrogen atom from the respective toluene by the active oxygen species. The apparent rate constant ($\log k'$) decreased with increasing Hammett constant σ of the substituent. Since the reaction was furthermore dependent on the structure of the oxidant, the homolytic pathway for the peroxide cleavage was supported (Blake and Coon, 1980a) (cf. section 4).

5.3. *Heme Degradation*

The investigation of partial reactions of the P–450 system are restricted by heme degradation reactions, especially if nonphysiologic O-donors are used, i.e. in the shunt reaction. Heme degradation already occurs in the NADPH/O_2 reaction if the "uncoupled" pathway (H_2O_2 generation) proceeds (Guengerich, 1978). Hydroxyl radicals as well as singulet oxygen are excluded from originating this effect. The H_2O_2 attack is mainly directed to the Fe(II) state of the cytochrome. Investigations of the oxidative degradation of hemoglobin support this statement (Tomoda et al., 1978). The heme destruction in the oxidase is diminished by substrates, by means of a decrease of the uncoupled reaction in favour of the substrate conversion process.

Nevertheless the Fe(III) state is subjected to a degradation by oxidative heme destruction. Cumene hydroperoxide, H_2O_2, $NaIO_4$ and $NaClO_2$ produce significant changes of the heme chromophore in the mM concentration range (Hrycay et al., 1976). Microsomal P–450 exhibits high cumene hydroperoxide sensitivity; for the adrenal cortex system e.g. 80% degradation was observed after cumene hydroperoxide incubation (1.3 mM CHP, 25°C, 5 min). Steroids — androsterone, progesterone, 17α-hydroxyprogesterone — diminished the damage to about 35% (Gustafsson et al., 1976). The heme oxidation by means of unphysiologic O-donors (shunt reaction) leads to a linear substrate conversion (min range) for only a short time, and thus to a restricted

Table 13. Kinetic Parameters of P-450 Complexes with Peroxy Compounds (cf. Equation (4))[1])

Substituent	K_d µM	k_2/k_1 mM	$k_1 \cdot 10^{-3}$ $M^{-1}s^{-1}$	k_2 s^{-1}	k_3 s^{-1}	k_4 s^{-1}
Perbenzoic Acids						
H	16	3.9	79	310	70	0.28
$p-CH_3$	17	1.4	120	170	41	0.50
$p-CH_3CH_2CH_2$	4.0	0.14	440	62	52	1.5
m-F	15	2.0	220	440	150	1.1
Cumene Hydroperoxides						
H	73	10	5.0	50	1.5	0.011
$p-CH_3$	48	3.0	15	45	0.49	0.0075
m-F	68	3.9	28	110	15	0.26
Benzyl Hydroperoxides						
H	340	19	0.041	0.78	0.033	0.0006
$p-CH_3$	180	5.5	0.006	0.033	0.015	0.0005
m-F	270	7.8	0.11	0.83	0.49	0.017

[1]) BLAKE II and COON (1980b), Selected Data; reproduced with permission from The American Society of Biological Chemists, Inc.

analytical access to the shunt process and its kinetic characteristics. In vitro studies therefore are preferably performed at subzero temperatures (~−20°C) in order to reduce perturbation by heme degradation (LICHTENBERGER et al., 1976; LICHTENBERGER and ULLRICH, 1977; SLIGAR et al., 1977).

6. NADH-dependent Cytochrome b_5 System

6.1. *Structural and Distribution Implications in the Membrane*

A second electron transfer system in the endoplasmic reticulum of mammalian liver cells is established by the NADH-dependent cytochrome b_5 reductase (EC 1.6.2.2) and cytochrome b_5. Both proteins exhibit amphipathic structural properties (SPATZ and STRITTMATTER, 1971, 1973; MIHARA and SATO, 1975). The molecular weight of the b_5 reductase amounts to 33000 (MIHARA and SATO, 1975), the molecule contains 1 FAD as functional flavoprotein group (STRITTMATTER and VELICK, 1956). The cytochrome b_5 weight depends on the solubilization procedure, it was determined with 11100 (proteolytic digestion) and 16000 (detergent extraction), respectively; i.e. the hydrophobic part of the molecule amounts to about 25% of the total protein content.

The activity of the b_5 system is coupled to the membrane, as evidenced by different investigations (STRITTMATTER and ROGERS, 1975; DUPPEL et al., 1976). Density gradient centrifugation evidences an asymmetric distribution of the components in the membrane (DALLNER et al., 1968; DALLMANN et al., 1969). That result suggested the existence of cluster-like arrangement of the structural elements. With respect to this question the NADH-dependent cytochrome b_5 reduction reaction has been extensively studied (ROGERS and STRITTMATTER, 1974a). It could be shown, that the reduction characteristics of endogeneous cytochrome b_5 is likewise exhibited by external b_5 after being additionally bound to the microsomes. Since an intermolecular b_5 electron transfer has been excluded the result disproves the existence of fixed cluster structures. Furthermore, the inactivation of the b_5 reductase by NEM causes a decreased rate of b_5 reduction, but all of the cytochrome was reduced. That investigation, too, proved the inconsistence of a specific distribution of the microsomal proteins; instead of, a random distribution as well as functional control by translational diffusion in the two-dimensional membrane phase is indicated.

6.2. *Electron-Transfer Pathways*

The function of the cytochrome b_5 system in microsomal electron transfer is rather complex. Fig. 15 shows the respective pathways:

1. The main function of the b_5 system is the electron donation in fatty acid Δ^9 desaturation reactions (SCHENKMAN et al., 1976; OSHINO, 1978). Acceptor protein is a specific desaturase (HOLLAWAY and KATZ, 1972; SHIMAKATA et al., 1972; STRITTMATTER et al., 1974). The electron transfer evidently proceeds via the NADH, b_5 reductase, b_5 couple, but additionally the NADPH, P–450 reductase track may serve as electron donor towards b_5 (OSHINO et al., 1971; OSHINO and OMURA, 1973; JANSSON and SCHENKMAN, 1977).
2. The reducibility of P–450 via the b_5 system, on the other hand, has been established already by KLINGENBERG and GARFINKEL (1958, 1958) (Fig. 15, pathway 2′). The NADH supported reduction of P–450 could be additionally evidenced by later investigations (HRYCAY and PROUGH, 1974; SASAME et al., 1974; JANSSON and SCHENKMAN, 1977).

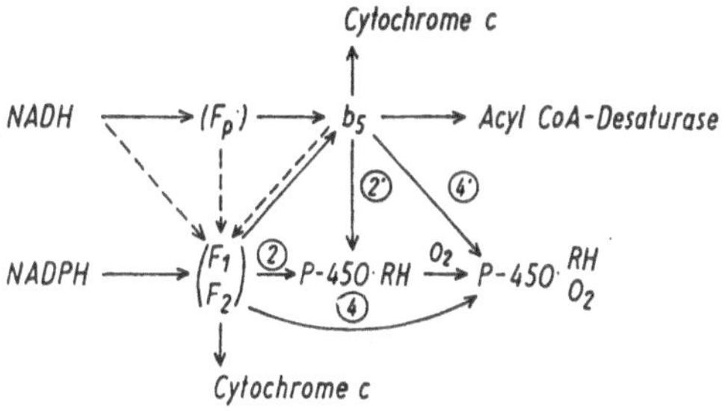

Fig.15. Microsomal Electron Transfer Pathways

3. The catalytic properties of the NADPH-dependent P–450 system, furthermore, can be altered by an NADH-dependent b_5 system synergism (pathway 4′). Extended studies on this topic have been performed and will be discussed below.

6.3. *Participation in P–450 Substrate Conversion*

6.3.1. *Microsomal Couple*

The relative content of the electron transfer proteins in the microsomal membrane (Tab. 14) seems to justify the question for some interaction of both systems. The evidenced NADH reducibility of P–450 was a first proof of a couple of both cytochromes. Additionally CONNEY et al. (1957) got evidence for a synergistic NADH action on the hydroxylation of substrates by the P–450 system. The

Table 14. Microsomal Contents of Electron Transfer Proteins[1])

Protein	nmol/mg Protein	% of Total Protein
P–450	1.0	5.0
P–450 Reductase	0.05	0.5
b_5	0.5	1.0
b_5 Reductase	0.05	0.2

[1]) OMURA, T. (1978); reproduced with permission from Kodansha, Ltd.

investigation of the synergistic couple were extended by studies on the oxidative demethylation of aminopyrine and other substrates, as well, and on concomitant cytochrome b_5 oxidation (COHEN and ESTABROOK, 1971a, b, c; ESTABROOK et al., 1971c; HILDEBRANDT and ESTABROOK, 1971). The results of the ESTABROOK group prove that in liver microsomes the NADPH/O_2 substrate conversion is accompanied by an increase in the rate of NADH oxidation which correlates with the N-demethylase activity. This process is paralleled by a decrease of the reduced cytochrome b_5 level in the steady-state. The results imply a supporting electron transfer of the b_5 system in the substrate conversion process.

The functional linkage of both systems could be confirmed by other groups (e.g. CORREIA and MANNERING, 1973a, b). For the type I substrates p-nitroanisole and hexobarbital, respectively, likewise an NADH stimulation of the turnover up to 50% was observed; the only NADH addition to the microsomes was without any effect; hexobarbital and ethylmorphine stimulated the NADH oxidation, type II compounds (aniline, acetanilide) had no influence (NETTER and ILLING, 1974). A stimulating effect of NADH has been further evidenced in the P–450 catalyzed hydroxylation of ethylmorphine and dimethylaniline, respectively (ARCHAKOV and DEVICHENSKY, 1975; ARCHAKOV et al., 1975a, b). The authors suggest some but fractional interaction of the b_5 system.

Immunochemical investigations of the microsomal N-demethylation of ethylmorphine of the oxidation of fatty acids show, that immunological inhibition of the P–450 reductase is followed by an about 65% inhibition of the NADPH pathway and by only slight alterations of the NADH track; at antibody b_5 inhibition, on the other hand, either pathway was about 30—50% blocked. This result provides evidence for the participation of cytochrome b_5 in the NADPH hydroxylation reaction, too (SASAME et al., 1973, 1974).

6.3.2. Coupling in Reconstituted Systems

The synergistic function of the b_5 system in the NADPH-dependent P–450 substrate conversion has been investigated in detail on reconstituted systems, as well (LU et al., 1974a, b; IMAI and SATO, 1977):

- Benzphetamine and 3,4-benzopyrene, respectively, are hydroxylated by a system which only consists of P–450, P–450 reductase, and lipid with the same turnover values as compared to the complete microsomal system (including the b_5 system). Obviously cytochrome b_5 is rather ineffective under these conditions (LU et al., 1974b).
- The substrate conversion in the benzphetamine system decreases after addition of cytochrome b_5; after further addition of b_5 reductase and NADH a respective compensation is obtained. This result is refered to some competition for the P–450 reductase between P–450 and b_5 (LU et al., 1974b). Supporting results have been further obtained (STAUDT et al., 1974).
- 3,4-benzopyrene conversion by a P–448 system has not been observed. The process could be activated, however, by addition of the b_5 system (b_5, b_5 reductase, NADH (LU et al., 1974b).
- Maximum turnover of chlorobenzene is only obtained in the presence of b_5. The respective P–448 reaction, on the other hand, does not require b_5 (LU et al., 1974a).
- By means of cytochrome b_5 as an affinity adsorbent an electrophoretically homogeneous P–450 subspecies could be isolated, which differs from the PB- and 3-MC-inducible form, respectively (SUGIYAMA et al., 1979, 1980). In the O-demethylation of p-

nitroanisole by this subspecies cytochrome b_5 is an obligatory component. The benzphetamine-N-demethylation, the aminopyrine-N-demethylation, and the ethoxyresorufin-O-deethylation, on the other hand, did not show this b_5 dependence (Miki et al., 1980).

- Cytochrome b_5 synergism is likewise exhibited in the N-methyl-carbazole hydroxylation in a reconstituted system which contains P–450, P–450 reductase, and lipid. LM2 and LM3b proved to be one order of magnitude more active as compared to the LM4 fraction (Vatsis et al., 1980a). The NADPH-dependent hydroxylation of prostaglandin E_1 by means of LM2 is at least 20 times increased after b_5 addition to the reconstituted system. Cytochrome b_5 is further obligatory in the LM2 catalyzed conversion of acetanilide (Coon et al., 1980; Vatsis et al., 1980b).
- Significantly increased 7-ethoxycoumarin and benzopyrene hydroxylation, respectively, could be evidenced in reconstituted LM2 and LM4 systems which contained cytochrome b_5; LM3 failed to exhibit this effect (Ingelman-Sundberg et al., 1980a). Optimal synergism in the O-demethylation of p-nitroanisole in an LM2 system was obtained at 1:1:1 molar ratio between LM2, P–450 reductase, and b_5. The authors suggest some functional complex of the 3 proteins in the membrane. Cytochrome b_5 incorporation did not increase NADPH oxidation despite of increased product formation. The P–450 reduction is likewise not enhanced. Since b_5 addition results in decreased O_2^- and H_2O_2 release, b_5 is supposed to be a more efficient donor of the second electron in the cycle as compared to the P–450 reductase (Ingelman-Sundberg, et al., 1980b).
- The stoichiometry of product formation and NADPH oxidation has been likewise investigated on a reconstituted rabbit liver system (Imai and Sato, 1974, 1977). The respective ratio formaldehyde formation/NADPH consumption amounts to 0.6 and 0.91, respectively, without and after addition of cytochrome b_5 to the benzphetamine metabolizing system. The data for N,N-dimethylaniline were 0.24 and 0.84. Cytochrome b_5 is taken to act as an effector in the couple of NADPH oxidation and product formation.
- Some effector function of cytochrome b_5 has been already proved in the decay of the ternary complex (P–450, substrate, O_2) where b_5 did not participate as redox couple (Guengerich et al., 1976).

  The results so far obtained can be summarized as to indicate that interferences of the b_5 system depend on the respective reconstituted system, i.e. on the P–450 species, the stoichiometry of the constituting components, and on the substrate, respectively.

6.3.3. Kinetics of Partial Reactions

Hildebrandt and Estabrook (1971) in earlier investigations have been shown that the NADPH reduction of microsomal P–450 exhibits the same characteristics with respect to rate and extent, irrespective of NADH addition. That result restricts the probability of the NADH supported pathway 2′ to compete with the NADPH pathway ② (Fig. 15) in the insertion of the first electron. Direct proof was obtained by the investigation of the respective partial steps (Hara and Minakami, 1971; Archakov et al., 1975a; Miyake et al., 1975; Ohnishi et al., 1975; Jansson and Schenkman, 1978). The electron transfer between NADH and the NADH-dependent b_5 reductase

Table 15. Microsomal Reduction Reactions

Species	Reaction	Rate Constant	Reference
Rat	NADH/b_5	10.6 s^{-1} (37°C)	JANSSON and SCHENKMAN (1978)
Hen	NADH/b_5	206 s^{-1} (25°C)	OHNISHI et al. (1975)
Rabbit	NADH/b_5	11.2 s^{-1} (20°C)	HARA and MINAKAMI (1971)
Rat	NADH/b_5	0.6 s^{-1} (12°C)	ARCHAKOV et al. (1975a)
Rat	NADPH/b_5	~0.6 s^{-1} (37°C) (35 min^{-1})	JANSSON and SCHENKMAN (1978)
Rat	NADPH/b_5	~0.1 s^{-1} (12°C) (7 min^{-1})	ARCHAKOV et al. (1975a)
Rat	NADH/P–450	0.2 min^{-1} (12°C)	ARCHAKOV et al. (1975a)
Hen	NADH/P–450	0.126 min^{-1}	MIYAKE et al. (1975)

and further towards cytochrome b_5 proceeds relatively rapid (Tab. 15). The data for different species prove this partial sequence to be capable of P–450 reduction in the physiologic time range. The consecutive reaction 2′ between b_5 and P–450, however, is rate-limiting and makes the composed NADH reduction of P–450 improbable as compared to the NADPH pathway ②.

The discussion of a coupled function of both electron transfer systems is therefore focussed on the origin of the second electron, i.e. does the transfer proceed via pathway ④ (P–450 reductase/P–450) or is the b_5/P–450 transfer 4′ - either by NADPH or NADH donated electrons — favoured.

SCHENKMAN and coworkers contributed interesting details to this topic (JANSSON and SCHENKMAN, 1973, 1978; SCHENKMAN and JANSSON, 1975; SCHENKMAN et al., 1976). With respect to the decreased steady-state levels of reduced b_5 in the microsomal substrate conversion, they could evidence:

— The ratio of the rate constants of the b_5 reduction (k_{red}) and reoxidation (k_{ox}) via NADPH and NADH, respectively, amounts to 13.3 and 259 (Tab. 16), i.e. in both cases the reduced b_5 state is favoured. After addition of substrates (aminopyrine, aniline) the reoxidation process is somewhat enhanced (~25% NADPH, ~60% NADH), but no indication of kinetically originated decrease of reduced b_5 is obtained. Thus the authors suggest that some interaction with other electron acceptors may be responsible for the observed diminution.

— The kinetics of the b_5 reduction and oxidation, respectively, were used to calculate the participation of b_5 in different substrate conversion processes. If NADPH was used as electron donor the b_5 passage in presence of different substrates was less than 10%; in the case of NADH about 60% (control) or even 100% (aminopyrine) of the electrons arrived at b_5. The results imply that the donation of the

Table 16. Cytochrome b_5 Redox-Kinetics in Microsomes (Untreated Rats)[1])

Reductant	Rate Constant	Control	Aniline	Aminopyrine
NADPH	$k_{red}(s^{-1})$	0.61	0.59	0.76
	k_{ox} (s^{-1})	0.046	0.057	0.061
	k_{red}/k_{ox}	13.3	10.4	12.5
NADH	$k_{red}(s^{-1})$	10.60	10.64	10.36
	k_{ox} (s^{-1})	0.041	0.064	0.068
	k_{red}/k_{ox}	259	166	152

[1]) JANSSON and SCHENKMAN (1978), Selected Data; reproduced with permission from Academic Press, Inc.

second electron from NADPH via b_5 is highly improbable, it should rather use pathway ④. NADH electrons, on the other hand, are rapidly transfered to cytochrome b_5; therefore if 4′ is in favour of ④ the b_5/P–450 track may be used, thus originating the observed NADH synergism in hydroxylating systems.

Some further differentiation was observed by recent studies of NOSHIRO and OMURA (1978). By means of immunochemical methods it could be shown that the NADH dependent conversion of 7-ethoxycoumarin, benzopyrene, benzphetamine, and aniline could be strongly inhibited by the blockade of the NADPH-dependent P–450 reductase. The product formation is furthermore restricted at cytochrome b_5 inhibition, but aniline being the exception. Finally, the inhibition of the NADH-dependent b_5 reductase diminished the conversion of 7-ethoxycoumarin and benzopyrene, but failed to perturb the benzphetamine and aniline reaction, as well.

The participation of the P–450 reductase in the NADH-dependent electron transfer is further evidenced by the increase of the 7-ethoxycoumarin O-dealkylation by pork liver microsomes after incubation with P–450 reductase. Antibody action towards the reductase on the other hand, decreased the enzymatic activity for about 30% at high NADH levels. This was taken as to indicate a 30% participation of the P–450 reductase passage track as compared to the major part reaction via b_5 reductase and b_5 (ULLRICH et al., 1980).

NOSHIRO et al. (1980) could further evidence, that already the NADH-dependent reduction of P–450 is strongly inhibited by P–450 reductase antibodies, whereas anti b_5 could only slightly perturb the reaction.

Recent investigations proved an ionic strength and pH control as well of the b_5 contribution in electron transfer. The donation towards the oxygen complex of P–450 is further established but a rather strong dependence on the experimental conditions, including the P–450 species and substrates, was evidenced (NOSHIRO et al., 1981).

These results focus on the following electron transfer pathways:

NADH supported electrons are transfered — at least in part — via the P–450 reductase. The immediate donor may be NADH, b_5 reductase, and b_5, as well. The first electron transfer — i.e. the P–450 reduction — is evidenced to follow this pathway; the second electron, on the other hand, could likewise pass the P–450 reductase to enter pathway ④, but furthermore the 4′ track (b_5/P–450) could be favoured (cf. above). Only the direct investigation of that step will answer the open question. The kinetic differentiation of the respective tracks should be rather low because of the substrate and species dependence of the antibody action.

WERRINGLOER and KAWANO (1980a) studied the interaction of P–450 and b_5 in their native environment of the microsomal membrane by means of their reduction by NADPH and NADH, respectively. At limiting amounts of the respective reducing agents NADH rapidly reduces b_5 via the b_5 reductase, the reaction is then followed by P–450 reduction at the expense of reduced b_5; NADPH, on the other hand, is capable of reducing both proteins rapidly, but again followed by reequilibration. The data suggest a reversible oxidation-reduction equilibrium between P–450 and b_5, which is further modified by substrates. The rapid b_5, reduction is taken to imply its function as reductant for oxy-P-450. This is supported by the determination of the b_5 turnover in the microsomal NADPH/NADH supported N-demethylation of ethylmorphine, which amounts to about 40–50 min^{-1}.

In recent studies of BONFILS et al. (1981) a direct evidence for the electron transfer from ferrous cytochrome b_5 to the oxyferrous P–450 LM2 intermediate could be evidenced. The reoxidation rate constant for either P–450 or b_5 in micelles at pH 7.5 and 2°C increased from 2.5 s^{-1} to about 4–7 s^{-1} in presence of benzphetamine. The data are compatible with a b_5 participation in the catalytic process.

7. Summary and Conclusions

The reviewed data on the P–450 reaction sequence and its partial reactions, respectively, evidence that the overall substrate conversion by means of oxygen activation and coupled electron transfer represents a complex interaction scheme of the interrelated compounds. Each reaction step is especially controlled by the respective participants, thus each step may be rate-interfering or even rate-limiting in the cycle. Fig. 16 represents the five reviewed partial reactions as points of attack of the interacting components, and the mode of the reaction control, as well. P–450 subfraction specificities have been omitted.

To begin with the substrate binding reaction. The substrate itself exerts specificities with respect to the binding mode (type I/type II differentiation). The oxygen donor may compete with the substrate for a common binding site or for allosteric reasons; such interference has been observed e.g. in the camphor/CO binding kinetics of the bacterial system (PETERSON et al., 1972). Besides the physiologic donor O_2, other oxygen supporting agents (shunt donors) may likewise being interrelated to or even may compete with the substrate. Phospholipids have been observed to produce type I difference spectra, too. This kind of interaction should be rather based on a supramolecular mechanism, which may lead to conformational changes of the cytochrome, and by this may shift substrate equilibria.

The first electron transfer to the cytochrome strongly depends on the electron donor. In the physiologic system the NADPH-cytochrome P–450 reductase is evidenced to donate that electron. $Na_2S_2O_4$ is likewise active, the respective reaction exhibits similar kinetics as compared to NADPH (rate, inhomogeneity). The NADH-cytochrome b_5 reductase/cytochrome b_5 couple donates the electron less rapidly (ARCHAKOV and DEVICHENSKY, 1975; MIYAKE et al., 1975). Relatively increased rates have been observed at the camphor system, the rates of methyl viologen radical mediated reduction amount to about $2 \cdot 10^3$ and $2 \cdot 10^2$ s^{-1} in the absence and in the presence of camphor, respectively, and at a radical concentration of 100 µM (DEBEY et al. 1979). The cytochrome, on the other hand, specifically accepts the electron from the

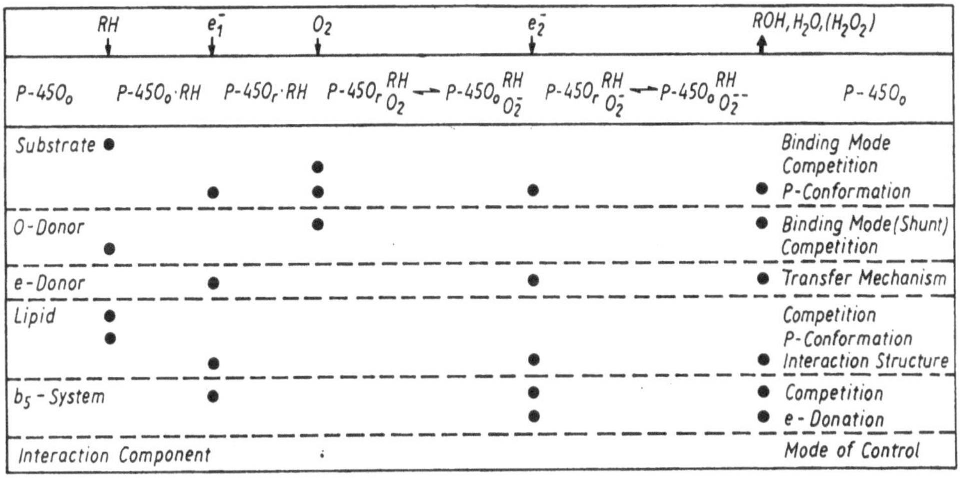

Fig. 16. Control Interrelations in the P–450 System

reductase as compared to other acceptor systems (Coon and White, 1980). Thus the transfer mechanism may contain specifications of the electron transfer function of either protein (P–450, P–450 reductase), or may originate from reductase/P–450 interaction peculiarities. The phospholipid control has been discussed as being due to the formation of special reductase/P–450 interaction structures (clusters). Competition with the b_5 system for the electron has been observed, additionally.

The oxygen donor in the physiologic sequence is the molecular species. It may possibly interact with the substrate and this way even may influence the P–450 conformation; there is, however, no experimental proof of such behaviour till now. In the case of other oxygen donors (H_2O_2, organic hydroperoxides, peracids), the additional electron support to the cytochrome can be by-passed in the substrate conversion. The product distribution then depends on the respective oxygen source, thus indicating some binding mode (molecular) specificities.

With respect to the second electron transfer reaction the above statements can be likewise adapted. In this step the b_5 system may donate the required electron, thus acting as a further control. Both electron transfer reactions (e_1, e_2) may even contribute in a "counterpoise"-regulation to the limitation of the overall reaction rate. The pH-function of the P–450 catalyzed hydroxylation indicates that rate-limitation shifts from the reduction of ferric heme protein-substrate complex to the reduction of oxy-P–450 with increasing pH (Werringloer and Kawano, 1980b).

The substrate conversion, lately, will reflect the control interaction as discussed, in respective rates and product pattern, as well. The process represents a further composed step, the specifications of which will depend on oxygen-bond cleavage, intracomplex (ternary complex) electron transfer, RH-bond cleavage etc. Unfortunately no detail has been kinetically resolved till now. But just the ternary complex kinetics should represent a key to the understanding of specifications in substrate conversion and product pattern — inclusively in the differentiation of oxygenatic, oxidatic and peroxidatic pathways, as well. Therefore future investigations will focus on this particular reaction step.

References

AKHREM, A. A., B. M. DSHAGAROV, V. K. RUMAS, YU. V. TIMINSKIJ, V. M. SHKUMATOV, and V. L. TSHASHTSHIN (1979): Dokl. Akad. Nauk SSSR **244**, 1256–1260

AKHREM, A. A., D. I. METILITZA, S. M. BIELSKI, P. A. KISELEV, M. E. SKURKO, and S. A. USANOV (1977): Croat. Chem. Acta **49**, 223 to 235

ALEXANDER, L. M., L. B. HERSH, and B. S. S. MASTERS (1980): in Microsomes, Drug Oxidations, and Chemical Carcinogenesis (M. J. COON, A. H. CONNEY, R. W. ESTABROOK, H. V. GELBOIN, J. R. GILLETTE, and P. J. O'BRIEN, eds.), Vol. I, pp. 285–288, Academic Press, New York

ANNAYEV, B., V. K. KOLTOVER, O. N. MAMEDNIYATOV, L. M. RAIKHMAN, and E. G. ROZANTSEV (1972): Biofizika **17**, 224–230

ANTONINI, E., and M. BRUNORI (1971): in Hémoglobin and Myoglobin in their Reactions with Ligands (Frontiers of Biology, Vol. 21), North-Holland, Amsterdam, London

ARCHAKOV, A. I. (1975): Mikrosomalnoje Okislenije, Isd. Nauka, Moskau

ARCHAKOV, A. I., and V. M. DEVICHENSKY (1975): Arch. Biochem. Biophys. **166**, 313 to 317

ARCHAKOV, A. I., V. M. DEVICHENSKY, and A. V. KARJAKIN (1975a): Arch. Biochem. Biophys. **166**, 295–307

ARCHAKOV, A. I., V. M. DEVICHENSKY, I. I. KARUZINA, and A. V. KARJAKIN (1975b): Arch. Biochem. Biophys. **166**, 308–312

AUSTIN, R. H. (1977): Croat. Chem. Acta **49**, 287–294

AUTOR, A. P., D. P. BALLOU, and M. J. COON (1973a): Federat. Proc. **32**, 762

AUTOR, A. P., R. M. KASCHNITZ, J. K. HEIDEMA, T. A. VAN DER HOEVEN, W. DUPPEL, and M. J. COON (1973b): Drug Metab. Dispos. **1**, 156–161

BACKES, W. L., S. G. SLIGAR, and J. B. SCHENKMAN, (1980): Biochem. Biophys. Res. Commun. **97**, 860–867

BACKES, W. L., S. G. SLIGAR, and J. B. SCHENKMAN, (1982): Biochemistry **21**, 1324–1330

BALNY, C., and P. DEBEY (1976): (FEBS Lett. **62**, 198–201

BALNY, C., P. DEBEY, and P. DOUZOU (1976): FEBS Lett. **69**, 236–239

BAZIN, M., J. PIERRE, P. DEBEY, and R. SANTUS (1982) Europ. J. Biochem. **124**, 539–544

BEATTIE, J. K., and R. J. WEST (1974): J. Amer. Chem. Soc. **96**, 1933–1935

BEGARD, E., P. DEBEY, and P. DOUZOU (1977): FEBS Lett. **75**, 52–54

BJÖRKHEM, I. (1977): Pharmacol. Ther. A **1**, 327–348

BLAKE, II, R. C., and M. J. COON (1980a): in Biochemistry, Biophysics, and Regulation of Cytochrome P-450 (J. A. GUSTAFSSON, J. CARLSTEDT-DUKE, A. MODE, and J. RAFTER, eds.), pp. 315–322, Elsevier/North-Holland, Amsterdam, New York, Oxford

BLAKE II, R. C., and M. J. COON (1980b): J. Biol. Chem. **255**, 4100–4111

BLAKE II, R. C., and COON, M. J. (1981a): J. Biol. Chem. **256**, 5755–5763

BLAKE II, R. C., and COON, M. J. (1981b): J. Biol. Chem. **256**, 12127–12133

BLANCK, J., and G. SMETTAN (1978): Pharmazie **33**, 321–324

BLANCK, J., J. BEHLKE, G.-R. JÄNIG, D. PFEIL, and K. RUCKPAUL (1979a): Acta Biol. Med. Germ. **38**, 11–21

BLANCK, J., H. REIN, M. SOMMER, O. RISTAU, G. SMETTAN, and K. RUCKPAUL (1983): Biochem. Pharmacol. **32**, 1683–1688

BLANCK, J., K. ROHDE, and K. RUCKPAUL (1979b): Acta Biol. Med. Germ. **38**, 23–32

BLANCK, J., G. SMETTAN, G.-R. JÄNIG, and K. RUCKPAUL (1976): Acta Biol. Med. Germ. **35**, 1455–1463

BLANCK, J., G. SMETTAN, G.-R. JÄNIG, and K. RUCKPAUL (1977): Croat. Chem. Acta **49**, 271–277

BONFILS, C., C. BALNY, P. DOUZOU, and P. MAUREL (1980): in Biochemistry, Biophysics, and Regulation of Cytochrome P-450 (J. A. GUSTAFSSON, J. CARLSTEDT-DUKE, A. MODE, and J. RAFTER, eds.), pp. 559–564, Elsevier/North-Holland, Amsterdam, New York, Oxford

BONFILS, C., C. BALNY, and P. MAUREL (1981): J. Biol. Chem. **256**, 9457–9465

BONFILS, C., P. DEBEY, and P. MAUREL (1979): Biochem. Biophys. Res. Commun. 88, 1301–1307

BURSTEIN, S., J. DINH, N. CO, M. GUT, H. SCHLEYER, D. Y. COOPER, and O. ROSENTHAL (1972): Biochemistry **11**, 2883–2891

CANNON, J., J. GEIBEL, M. WHIPPLE, and T. G. TRAYLOR (1976): J. Amer. Chem. Soc. **98**, 3395–3396

CASE, D. A., and M. KARPLUS (1978): J. Mol. Biol. 123, 697–701
CASE, D. A., and M. KARPLUS (1979): J. Mol. Biol. 132, 343–368
CHANG, C. K., and D. DOLPHIN (1978): in Bioorganic Chemistry (E. E. VAN TAMELEN, ed.), Vol. 4, pp. 37–80, Academic Press, New York
COHEN, B. S., and R. W. ESTABROOK (1971a): Arch. Biochem. Biophys. 143, 37–45
COHEN, B. S., and R. W. ESTABROOK (1971b): Arch. Biochem. Biophys. 143, 46–53
COHEN, B. S., and R. W. ESTABROOK (1971c): Arch. Biochem. Biophys. 143, 54–65
COLE, P. E., and S. G. SLIGAR, (1981): FEBS Lett. 133, 252–254
COMAI, K., and J. L. GAYLOR (1973): J. Biol. Chem. 248, 4947–4955
CONNEY, A. H., R. R. BROWN, J. A. MILLER, and E. C. MILLER (1957): CANCER Res. 17, 628–633
COON, M. J., and R. E. WHITE (1980): in Metal Ion Activation of Dioxygen (T. G. SPIRO, ed.), pp. 73–123, John Wiley and Sons, New York
COON, M. J., A. P. AUTOR, R. F. BOYER, R. T. LODE, and H. W. STROBEL (1973): in Oxidases and Related Redox Systems (T. E. KING, H. S. MASON, and M. MORRISON, eds.), Vol. 2, pp. 527–553, Univ. Park Press, Baltimore
COON, M. J., A. P. AUTOR, and H. W. STROBEL (1971): Chem. Biol. Interact. 3, 248–250
COON, M. J., D. P. BALLOU, D. A. HAUGEN, S. O. KREZOSKI, G. D. NORDBLOM, and R. E. WHITE (1977): in Microsomes and Drug Oxidations (V. ULLRICH, I. ROOTS, A. HILDEBRANDT, R. W. ESTABROOK, and A. W. CONNEY, eds.), pp. 82–94, Pergamon Press, Oxford
COON, M. J., R. C. BLAKE II, D. D. OPRIAN, and D. P. BALLOU (1979): Acta Biol. Med. Germ. 38, 449–458
COON, M. J., D. A. HAUGEN, F. P. GUENGERICH, J. L. VERMILION, and W. L. DEAN (1976): in The Structural Basis of Membrane Function [Y. HATEFI, and DJAVADI-OHANIANCE, eds.), pp. 409–427, Academic Press, New York
COON, M. J., D. R. KOOP, A. V. PERSSON, and E. T. MORGAN (1980): in Biochemistry, Biophysics, and Regulation of Cytochrome P-450 (J. A. GUSTAFSSON, J. CARLSTEDT-DUKE, A. MODE, and J. RAFTER, eds.), pp. 7–15, Elsevier/North-Holland, Amsterdam, New York, Oxford
COOPER, D. Y., M. D. CANNON, H. SCHLEYER, and O. ROSENTHAL (1977): J. Biol. Chem. 252, 4755–4757
CORREIA, M. A., and G. J. MANNERING (1973a): Mol. Pharmacol. 9, 455–469
CORREIA, M. A., and G. J. MANNERING (1973b): Mol. Pharmacol. 9, 470–485
CUSHMAN, D. W., R. L. TSAI, and I. C. GUNSALUS (1967): Biochem. Biophys. Res. Commun. 26, 577–583
DALLMANN, P. R., G. DALLNER, A. BERGSTRAND, and L. ERNSTER (1969): J. Cell. Biol. 41, 357–377
DALLNER, G., A. BERGSTRAND, and R. NILSSON (1968): J. Cell. Biol. 38, 257–276
DANIELSSON, H., and K. WIKVALL (1976): FEBS Lett. 66, 299–302
DAVYDOV, R. M., S. GRESCHNER, S. N. MAGONOV, K. RUCKPAUL, and L. A. BLUMENFELD (1978): Dokl. Acad. Nauk SSSR 241, 707–709
DAVYDOV, R. M., S. GRESCHNER, and K. RUCKPAUL (1979): Molekularnaya Biologia 13, 1397–1406
DAVYDOV, R. M., S. GRESCHNER, S. V. STEPANOV, and K. RUCKPAUL (1980): Molekularnaya Biologia 14, 685–693
DEBEY, P., and P. DOUZOU (1974): FEBS Lett. 39, 271–274
DEBEY, P., C. BALNY, and P. DOUZOU (1973a): FEBS Lett. 35, 86–90
DEBEY, P., C. BALNY, and P. DOUZOU (1976): FEBS Lett. 69, 231–235
DEBEY, P., G. HUI BON HOA, and P. DOUZOU (1973b): FEBS Lett. 32, 227–230
DEBEY, P., G. HUI BON HOA, and I. C. GUNSALUS (1977): Croat. Chem. Acta 49, 309–321
DEBEY, P., E. J. LAND, R. SANTUS, and A. J. SWALLOW (1979): Biochem. Biophys. Res. Commun. 86, 953–960
DEPIERRE, J. W., and G. DALLNER (1975): Biochim. Biophys. Acta 415, 411–472
DEPIERRE, J. W., and L. ERNSTER (1977): Annu. Rev. Biochem. 46, 201–262
DIGNAM, J. D., and H. W. STROBEL (1975): Biochem. Biophys. Res. Commun. 63 845–852
DIGNAM, J. D., and H. W. STROBEL (1977): Biochemistry 16, 1116–1123
DIEHL, H., J. SCHÄDELIN, and V. ULLRICH (1970): Hoppe-Seyler's Z. Physiol. Chem. 351, 1359–1371

Dose, E. V., M. A. Hoselton, N. Sutin, M. F. Tweedle, and L. J. Wilson (1978): J. Amer. Chem. Soc. **100**, 1141—1147

Dose, E. V., M. F. Tweedle, L. J. Wilson, and N. Sutin (1977): J. Amer. Chem. Soc. **99**, 3886—3888

Duppel, W., and V. Ullrich (1976): Biochim. Biophys. Acta **426**, 399—407

Duppel, W., J. Poensgen, and V. Ullrich (1976): Hoppe-Seyler's Z. Physiol. Chem. **357**, 1025

Estabrook, R. W. (1965): in Oxidases and Related Redox Systems (T. E. King, H. S. Mason, and M. Morrison, eds.), Vol. 2, pp. 832—833, Univ. Park Press, Baltimore

Estabrook, R. W. (1973): in Oxidases and Related Redox Systems (T. E. King, H. S. Mason, and M. Morrison, eds.), Vol. 2 p. 511, Univ. Park Press, Baltimore

Estabrook, R. W., and J. Werringloer (1977): in Microsomes and Drug Oxidations (V. Ullrich, I. Roots, A. G. Hildebrandt, R. W. Estabrook, and A. U. Conney, eds.), pp. 748—757, Pergamon Press, Oxford

Estabrook, R. W., J. Baron, and A. G. Hildebrandt (1971a): Chem. Biol. Interact. **3**, 260—261

Estabrook, R. W., J. Baron, J. A. Peterson, and Y. Ishimura (1972): in Biological Hydroxylation Mechanisms (G. S. Boyd, and R. M. S. Smellie, eds.), pp. 159—185, Academic Press, London, New York

Estabrook, R. W., D. Y. Cooper, and O. Rosenthal (1963): Biochem. Z. **338**, 741—755

Estabrook, R. W., M. R. Franklin, B. Cohen, A. Shigamatzu, and A. G. Hildebrandt (1971b): Metabolism **20**, 187—199

Estabrook, R. W., A. G. Hildebrandt, J. Baron, K. J. Netter, and K. Leibaman (1971c): Biochem. Biophys. Res. Commun. **42**, 132—139

Estabrook, R. W., S. Kawano, J. Werringloer, H. Kuthan, H. Tsuji, H. Graf, and V. Ullrich (1979a): Acta Biol. Med. Germ. **38**, 423—434

Estabrook, R. W., J. Werringloer, and J. A. Peterson (1979b): in Xenobiotic Metabolism: In vitro Methods (G. D. Paulson, D. S. Frear, and E. P. Marks, eds.), Amer. Chem. Soc. Symposium Series No. **97**, pp. 149—179, Washington

Franklin, M. R., and R. W. Estabrook (1971): Arch. Biochem. Biophys. **143**, 318—329

French, J. S. (1979): Doctoral Thesis, Univ. Michigan, USA

French, J. S., F. P. Guengerich, and M. J. Coon (1980): J. Biol. Chem. **255**, 4112—4119

Gander, J. E., and G. J. Mannering (1980): Pharmacol. Ther. **10**, 191—221

Garfinkel, D. (1958): Arch. Biochem. Biophys. **77**, 493—509

Gibson, Q. H. (1959): Progr. Biophys. Biophysical Chem. **9**, 1—53

Gigon, P. L., T. E. Gram, and J. R. Gillette (1969): Mol. Pharmacol. **5**, 109—122

Gillette J. R., H. Sasame, and B. Stripp (1973): Drug Metab. Dispos. **1**, 164—175

Gray, R. D. (1978): J. Biol. Chem. **253**, 4364—4369

Gray, R. D. (1982) J. Biol. Chem. **257**, 1086—1094

Greschner, S., R. M. Davydov, G.-R. Jänig, K. Ruckpaul, and L. A. Blumenfeld (1979): Acta Biol. Med. Germ. **38**, 443—448

Greschner, S., R. M. Davydov, S. V. Stepanov, and K. Ruckpaul (1980a): in Biochemistry, Biophysics, and Regulation of Cytochrome P-450 (J. A. Gustafsson, J., Carlstedt-Duke, A. Mode, and J. Rafter, eds.), pp. 587—590, Elsevier/North-Holland Amsterdam, New York, Oxford

Greschner, S., R. M. Davydov, S. V. Stepanov, and K. Ruckpaul (1980b): Studia Biophys. **81**, 133—134

Greschner, S. (1982a) Biophys. Struct. Mech. **9**, 29—34

Greschner, S., R. M. Davydov, G.-R. Jänig, K. Ruckpaul, and L. A. Blumenfeld (1982b): J. Phys. Chem. (USSR) **56**, 2884—2887

Griffin, B. W. (1977): FEBS Lett. **74**, 139—143

Griffin, B. W. (1978): Arch. Biochem. Biophys. **190**, 850—853

Griffin, B. W., and J. A. Peterson (1972): Biochemistry **11**, 4740—4746

Griffin, B. W., and P. L. Ting (1978): Biochemistry **17**, 2206—2211

Griffin, B. W., J. A. Peterson, J. Werringloer, and R. W. Estabrook (1975): Ann. N.Y. Acad. Sci. **244**, 107—131

Groves, J. T. (1979): in Advances in Inorganic Biochemistry (G. L. Eichhorn, and L. G. Marzilli, eds.), Vol. 1, pp. 119—142, Elsevier/North-Holland, New York

GROVES, J. T., and G. A. MCGLUSKY (1976): J. Amer. Chem. Soc. 98, 859–861
GROVES, J. T., G. A. MCCLUSKY, R. E. WHITE, and M. J. COON (1978): Biochem. Biophys. Res. Commun. 81, 154–160
GUENGERICH, F. P. (1978): Biochemistry 17, 3633–3639
GUENGERICH, F. P. (1979): Pharmacol. Ther. A 6, 99–121
GUENGERICH, F. P., and M. J. COON (1975): Federat. Proc. 34, 622
GUENGERICH, F. P., D. P. BALLOU, and M. J. COON (1976): Biochem. Biophys. Res. Commun. 70, 951–956
GUNSALUS, I. C., and S. G. SLIGAR (1978): Advances Enzymol. 47, 1–44
GUNSALUS, I. C., J. R. MEEKS, J. D. LIPSCOMB, P. DEBRUNNER, and E. MÜNCK (1974): in Molecular Mechanisms of Oxygen Activation (O. HAYAISHI, ed.), pp. 559–613, Academic Press, New York
GUNSALUS, I. C., C. A. TYSON, R. TSAI, and S. D. LIPSCOMB (1971): Chem. Biol. Interact. 4, 75–78
GUSTAFSSON, J. A., and J. BERGMAN (1976): FEBS Lett. 70, 276–280
GUSTAFSSON, J. A., E. G. HRYCAY, and L. ERNSTER (1976): Arch. Biochem. Biophys. 174, 440–453
HANUCOGLU, I., and C. R. JEFCOATE (1980): J. Biol. Chem. 255, 3057–3061
HARA, T., and S. MINAKAMI (1971): J. Biochem. 69, 317–324
HASHIMOTO, C., and Y. IMAI (1975): Biochem. Biophys. Res. Commun. 68, 821–827
HAUGEN, D. A., and M. J. COON (1976): J. Biol. Chem. 251, 7929–7939
HAUGEN, D. A., T. A. VAN DER HOEVEN, and M. J. COON (1975): J. Biol. Chem. 250, 3567–3570
HILDEBRANDT, A. G., and R. W. ESTABROOK (1971): Arch. Biochem. Biophys. 143, 66–79
HODGSON, E. (1979): Drug. Metab. Rev. 10, 15–33
HOLLAWAY, P. W., and J. T. KATZ (1972): Biochemistry 11, 3689–3696
HOLTZMAN, J. L., and B. H. RUMACK (1973): Biochemistry 12, 2309–2313
HORECKER, B. L. (1950): J. Biol. Chem. 183, 593–605
HRYCAY, E. G., and P. J. O'BRIEN (1972): Arch. Biochem. Biophys. 153, 480–494
HRYCAY, E. G., and R. A. PROUGH (1974): Arch. Biochem. Biophys. 165, 331–339
HRYCAY, E. G., J. A. GUSTAFSSON, M. INGELMAN-SUNDBERG, and L. ERNSTER (1975a): Biochem. Biophys. Res. Commun. 66, 209–216
HRYCAY, E. G., J. A. GUSTAFSSON, M. INGELMAN-SUNDBERG, and L. ERNSTER (1975b): FEBS Lett. 56, 161–165
HRYCAY, E. G., J. A. GUSTAFSSON, M. INGELMAN-SUNDBERG, and L. ERNSTER (1976): Europ. J. Biochem. 61, 43–52
HRYCAY, E. G., H. G. JONEN, A. Y. H. LU, and W. LEVIN (1975c): Arch. Biochem. Biophys. 166, 145–157
HUANG, M.-T., S. B. WEST, A. Y. H. LU (1976): J. Biol. Chem. 251, 4659–4665
ICHIKAWA, Y., and T. YAMANO (1968): Biochim. Biophys. Acta 153, 753–765
IMAI, Y., and R. SATO (1966): Biochem. Biophys. Res. Commun. 22, 620–626
IMAI, Y., and R. SATO (1974): Biochem. Biophys. Res. Commun. 60, 8–14
IMAI, Y., and R. SATO (1977): Biochem. Biophys. Res. Commun. 75, 420–426
IMAI, Y., R. SATO, and T. IYANAGI (1977): J. Biochem. 82, 1237–1246
INGELMAN-SUNDBERG, M. (1976): Hoppe-Seyler's Z. Physiol. Chem. 357, 1031
INGELMAN-SUNDBERG, M. (1977): Biochim. Biophys. Acta 488, 225–234
INGELMAN-SUNDBERG, M. and H. GLAUMANN (1977): FEBS Lett. 78, 72–76
INGELMAN-SUNDBERG, M., and I. JOHANSSON (1980): Biochem. Biophys. Res. Commun. 97, 582–589
INGELMAN-SUNDBERG, M., H. GLAUMANN, and I. JOHANSSON (1980a): in Microsomes, Drug Oxidations, and Chemical Carcinogenesis (M. J. COON, A. H. CONNEY, R. W. ESTABROOK, H. V. GELBOIN, J. R. GILLETTE, and P. J. O'BRIEN, eds.), Vol. I, pp. 513–516, Academic Press, New York
INGELMANN-SUNDBERG, M., I. JOHANSSON, A. BRUNSTRÖM, G. EKSTRÖM, T. HAAPARANTA, and J. RYDSTRÖM (1980b): in Biochemistry, Biophysics, and Regulation of Cytochrome P-450 (J. A. GUSTAFSSON, J. CARLSTEDT-DUKE, A. MODE, and J. RAFTER, eds.), pp. 299–306, Elsevier/North-Holland, Amsterdam, New York, Oxford
ISHIMURA, Y. (1978): in Cytochrome P-450 (R. SATO, and T. OMURA, eds.), pp. 209–227, Kodansha Ltd., Tokyo, and Academic Press, New York
ISHIMURA, Y., V. ULLRICH, and J. A. PETER-

SON (1971): Biochem. Biophys. Res. Commun. 42, 140–146
IYANAGI, T., and R. MAKINO (1980): in Microsomes, Drug Oxidations, and Chemical Carcinogenesis (M. J. COON, A. H. CONNEY, R. W. ESTABROOK, H. V. GELBOIN, J. R. GILLETTE, and O'BRIEN, eds.), Vol. I, pp. 407–410, Academic Press, New York
IYANAGI, T., and H. S. MASON (1973): Biochemistry 12, 2297–2308
IYANAGI, T., F. K. ANAN, Y. IMAI, and H. S. MASON (1978): Biochemistry 17, 2224–2230
IYANAGI, T., R. MAKINO, and F. K. ANAN (1981): Biochemistry 20, 1722–1730
JÄNIG, G.-R., R. MISSELWITZ, D. ZIRWER, E. BUDER, H. REIN, and K. RUCKPAUL (1977): Croat. Chem. Acta 49, 263–270
JANSSON, I., and J. B. SCHENKMAN (1973): Mol. Pharmacol. 9, 840–845
JANSSON, I., and J. B. SCHENKMAN (1977): Arch. Biochem. Biophys. 178, 89–107
JANSSON, I., and J. B. SCHENKMAN (1978): Arch. Biochem. Biophys. 185, 251–261
JEFCOATE, C. R. (1978): Methods Enzymol. 52, 258–279
JUNG, C., and O. RISTAU (1978): Pharmazie 33, 329–331
KADLUBAR, F. F., K. C. MORTON, and D. M. ZIEGLER (1973): Biochem. Biophys. Res. Commun. 54, 1255–1261
KAMIN, H., B. S. S. MASTERS, Q. H. GIBSON, and C. H. WILLIAMS jr. (1965): Federat. Proc. 24, 1164–1171
KATAGIRI, M., B. M. GANGULI, and I. C. GUNSALUS (1968): J. Biol. Chem. 243, 3543–3546
KIMURA, T., and K. SUZUKI (1967): J. Biol. Chem. 242, 485–491
KLINGENBERG, M. (1958): Arch. Biochem. Biophys. 75, 376–386
LAMAR, G., and F. WALKNER (1972): J. Amer. Chem. Soc. 94, 8607–8608
LANGE, R., P. DEBEY, and P. DOUZOU (1977): Croat. Chem. Acta 49, 279–285
LANGE, R., G. HUI BON HOA, P. DEBEY, and I. C. GUNSALUS (1979): Europ. J. Biochem. 94, 491–496
LEVIN, W. (1977): in Microsomes and Drug Oxidations (V. ULLRICH, O. ROOTS, A. G. HILDEBRANDT, R. W. ESTABROOK, and A. H. CONNEY, eds.), pp. 735–747, Pergamon Press, Oxford
LEVIN, W., D. RYAN, S. B. WEST, and A. Y. H. LU (1974): J. Biol. Chem. 249, 1747–1754
LICHTENBERGER, F., and V. ULLRICH (1977): in Microsomes and Drug Oxidations (V. ULLRICH, I. ROOTS, A. G. HILDEBRANDT, R. W. ESTABROOK, and A. H. CONNEY, eds.), pp. 218–223, Pergamon Press, Oxford
LICHTENBERGER, F., W. NASTAINCZYK, and V. ULLRICH (1976): Biochem. Biophys. Res. Commun. 70, 939–946
LIPSCOMB, J., S. G. SLIGAR, M. NAMTVEDT, and I. C. GUNSALUS (1976): J. Biol. Chem. 251, 1116–1120
LU, A. Y. H., and M. J. COON (1968): J. Biol. Chem. 243, 1331–1332
LU, A. Y. H., and W. LEVIN (1974): Biochim. Biophys. Acta 344, 205–240
LU, A. Y. H., and S. B. WEST (1978): Pharmacol. Ther. A 2, 337–358
LU, A. Y. H., W. LEVIN, H. SELANDER, and D. M. JERINA (1974a): Biochem. Biophys. Res. Commun. 61, 1348–1355
LU, A. Y. H., S. B. WEST, M. VORE, D. RYAN, and W. LEVIN (1974b): J. Biol. Chem. 249, 6701–6709
MARCUS, R. A. (1964): Annu. Rev. Phys. Chem. 15, 155–196
MASON, H. S. (1957): Advances Enzymol. 19, 79–233
MASTERS, B. S. S., and R. T. OKITA (1980): Pharmacol. Ther. 9, 227–244
MASTERS, B. S. S., H. KAMIN, Q. H. GIBSON, and C. H. WILLIAMS jr. (1965): J. Biol. Chem. 240, 921–931
MASTERS, B. S. S., R. A. PROUGH, and H. KAMIN (1975): Biochemistry 14, 607–613
MATSUBARA, T., and Y. TOCHINO (1976): Arch. Biochem. Biophys. 172, 696–705
MATSUBARA, T., J. BARON, L. L. PETERSON, and J. A. PETERSON (1976): Arch. Biochem. Biophys. 172, 463–469
MAUREL, P., C. BONFILS, and P. DEBEY (1980): in Microsomes, Drug Oxidations, and Chemical Carcinogenesis (M. J. COON, A. H. CONNEY, R. W. ESTABROOK, H. V. GELBOIN, J. R. GILLETTE, and P. J. O'BRIEN, eds.), Vol. I, pp. 359–362, Academic Press, New York
MIHARA, K., and R. SATO (1975): J. Biochem. 78, 1057–1073
MIKI, N., T. SUGIYAMA, and T. YAMANO (1980): in Microsomes. Drug Oxidations, and Chemical Carcinogenesis (M. J. COON, A. H. CONNEY, R. W. ESTABROOK, H. V. GELBOIN, J. R. GILLETTE, and P. J. O'BRIEN,

eds.), Vol. I, pp. 27–36, Academic Press, New York

Misselwitz, R., G.-R. Jänig, H. Rein, E. Buder, D. Zirwer, and K. Ruckpaul (1975): Acta Biol. Med. Germ. **34**, 1755–1766

Misselwitz, R., G.-R. Jänig, H. Rein, E. Buder, D. Zirwer, and K. Ruckpaul (1976): Acta Biol. Med. Germ. **35**, K 19

Miwa, G. T., S. B. West, and A. Y. H. Lu (1978): J. Biol. Chem. **253**, 1921–1929

Miwa, G. T., S. B. West, M. Tuan Huang, and A. Y. H. Lu (1979): J. Biol. Chem. **254**, 5695–5700

Miyake, Y., Y. Nakamura, and K. Horike (1975): J. Biochem. **78**, 773–783

Nakamura, Y., H. Otsuka, and B. Tamaoki (1966): Biochim. Biophys. Acta **122**, 34–42

Netter, K. J., and H. P. A. Illing (1974): Xenobiotica **4**, 549–561

Nordblom, G. D., and M. J. Coon (1977): Arch. Biochem. Biophys. **180**, 343–347

Nordblom, G. D., R. E. White, and M. J. Coon (1976): Arch. Biochem. Biophys. **175**, 524–533

Noshiro, M., and T. Omura (1978): J. Biochem. **83**, 61–77

Noshiro, M., H. H. Ruf, and V. Ullrich (1980): in Biochemistry, Biophysics, and Regulation of Cytochrome P-450 (J. A. Gustafsson, J. Carlstedt-Duke, A. Mode, and J. Rafter, eds.), pp. 351–354, Elsevier/North-Holland, Amsterdam, New York, Oxford

Noshiro, M., V. Ullrich, and T. Omura (1981): Europ. J. Biochem. **116**, 521–526

O'Brien, P. J. (1978): Pharmacol. Ther. A **2**, 517–536

Ochiai, E.-I. (1975): J. Inorg. Nucl. Chem. **37**, 1503–1509

Ohnishi, T., Y. Imai, and J. Yamazaki (1975): Arch. Biochem. Biophys. **167**, 488–495

Omura, T. (1978): in Cytochrome P-450 (R. Sato, and T. Omura, eds.), pp. 138–163, Kodansha Ltd., Tokyo, and Academic Press, New York

Omura, T., and R. Sato (1962): J. Biol. Chem. **237**, PC 1375–1376

Omura, T., and R. Sato (1964a): J. Biol. Chem. **239**, 2370–2378

Omura, T., and R. Sato (1964b): J. Biol. Chem. **239**, 2379–2385

Omura, T., E. Sanders, R. W. Estabrooks, D. Y. Cooper, and O. Rosenthal (1966): Arch. Biochem. Biophys. **117**, 660–673

Omura, T., R. Sato, D. Y. Cooper, O. Rosenthal, and R. W. Estabrook (1965): Federat. Proc. **24**, 1181–1189

Oprian, D. D., and M. J. Coon (1980): in Biochemistry, Biophysics, and Regulation of Cytochrome P-450 (J. A. Gustafsson, J. Carlstedt-Duke, A. Mode, and J. Rafter, eds.), pp. 323–330, Elsevier/North-Holland, Amsterdam, New York, Oxford

Oprian, D. D., K. P. Vatsis, and M. J. Coon (1979): J. Biol. Chem. **254**, 8895–8902

Oprian, D. D., M. J. and Coon (1982): J. Biol. Chem. **257**, 8935–8944

Orrenius, S., B. J. Wilson, C. v. Bahr, and J. B. Schenkman (1972): in Biological Hydroxylation Mechanisms (G. S. Boyd, and R. M. S. Smellie, eds.), pp. 55–77, Academic Press, New York

Oshino, N. (1978): Pharmacol. Ther. **2**, 497–515

Oshino, N., and T. Omura (1973): Arch. Biochem. Biophys. **157**, 395–404

Oshino, N., Y. Imai, and S. Sato (1971): J. Biochem. **69**, 155–167

Pedersen, J. I. (1978): FEBS Lett. **85**, 35–39

Peterson, J. A. (1971): Arch. Biochem. Biophys. **144**, 678–693

Peterson, J. A., and B. W. Griffin (1972): Arch. Biochem. Biophys. **151**, 427–433

Peterson, J. A., and D. M. Mock (1979): Acta Biol. Med. Germ. **38**, 153–162

Peterson, J. A., R. E. Ebel, D. H. O'Keeffe, T. Matsubara, and R. W. Estabrook (1976): J. Biol. Chem. **251**, 4010–4016

Peterson, J. A., Y. Ishimura, and B. W. Griffin (1972): Arch. Biochem. Biophys. **149**, 197–208

Peterson, J. A., R. E. White, Y. Yasukochi, M. L. Coomes, D. H. O'Keeffe, R. E. Ebel, B. S. S. Masters D. P. Ballou, and M. J. Coon (1977): J. Biol. Chem. **252**, 4431–4434

Phillips, A. H., and R. G. Langdon (1962): J. Biol. Chem. **237**, 2652–2659

Rahimtula, A. D., and P. J. O'Brien (1974): Biochem. Biophys. Res. Commun. **60**, 440–447

Rahimtula, A. D., and P. J. O'Brien (1975): Biochem. Biophys. Res. Commun. **62**, 268–275

Rahimtula, A. D., P. J. O'Brien, E. G. Hrycay, J. A. Peterson, and R. W. Estabrook (1974): Biochem. Biophys. Res. Commun. **60**, 695–702

RAHIMTULA, A. D., P. J. O'BRIEN, H. E. SEIFRIED, and D. M. JERINA (1978): Europ. J. Biochem. **89**, 133–141

REIN, H., and O. RISTAU (1978): Pharmazie **33**, 325–328

REIN, H., O. RISTAU, J. FRIEDRICH, G.-R. JÄNIG, and K. RUCKPAUL (1977): Croat. Chem. Acta **49**, 251–261

REIN, H., O. RISTAU, R. MISSELWITZ, E. BUDER, and K. RUCKPAUL (1979): Acta Biol. Med. Germ. **38**, 187–200

REMMER, H., J. B. SCHENKMAN, F. W. ESTABROOK, H. SASAME, J. R. GILLETTE, D. Y. COOPER, S. NARASIMHULU, and O. ROSENTHAL (1966): Mol. Pharmacol. **2**, 187–190

RENNEBERG, R., J. CAPDEVILA, N. CHACOS, R. W. ESTABROOK, and P. A. PROUGH (1981): Biochem. Pharmacol. **30**, 843–848

RISTAU, O., H. REIN, S. GRESCHNER, G.-R. JÄNIG, and K. RUCKPAUL (1979): Acta Biol. Med. Germ. **38**, 177–185

ROGERS, M. J., and P. STRITTMATTER (1974a): J. Biol. Chem. **249**, 895–900

ROGERS, M. J., and P. STRITTMATTER (1974b): J. Biol. Chem. **249**, 5565–5569

RÖSEN, P., and A. STIER (1973): Biochem. Biophys. Res. Commun. **51**, 603–611

RUCKPAUL, K., H. REIN, J. BLANCK, O. RISTAU, and M. J. COON (1980): in Biochemistry, Biophysics, and Regulation of Cytochrome P-450 (J. A. GUSTAFSSON, J. CARLSTEDT-DUKE, A. MODE, and J. RAFTER, eds.), pp. 539–549, Elsevier/North-Holland, Amsterdam, New York, Oxford

RUCKPAUL, K., H. REIN, J. BLANCK, O. RISTAU, and M. J. COON, (1982) Acta Biol. Med. Germ. **41**, 193–203

RUF, H. H. (1980): in Biochemistry, Biophysics, and Regulation of Cytochrome P-450 (J. A. GUSTAFSSON, J. CARLSTEDT-DUKE, A. MODE, and J. RAFTER, eds.), pp. 355–358, Elsevier/North-Holland, Amsterdam, New York, Oxford

RUF, H. H. (1982): in Cytochrome P-450, Biochemistry, Biophysics and Environmental Implications (E. HIETANEN, M. LAITINEN, and O. HÄNNINEN, eds.), pp. 597–600, Elsevier/North Holland, Amsterdam, New York, Oxford

RYAN, D., A. Y. H. LU, L. KAWALEK, S. B. WEST, and W. LEVIN (1975): Biochem. Biophys. Res. Commun. **64**, 1134–1141

SASAME, H. A., and J. R. GILLETTE (1969): Mol. Pharmacol. **5**, 123–130

SASAME, H. A., J. R. MITCHELL, S. S. THORGEIRSSON, and J. R. GILLETTE (1973): Drug Metab. Dispos. **1**, 150–160

SASAME, H. A., S. S. THORGEIRSSON, J. R. MITCHELL, and J. R. GILLETTE (1974): Life Sci. **14**, 35–46

SATO, R., Y. ATSUTA, Y. IMAI, S. TANIGUCHI, and K. OKUDA (1977): Proc. Nat. Acad. Sci. U.S.A. **74**, 5477–5481

SCHENKMAN, J. B. (1972): Mol. Pharmacol. **8**, 178–187

SCHENKMAN, J. B. and I. JANSSON (1975): Advances Exp. Med. Biol. **58**, 387–404

SCHENKMAN, J. B., I. JANSSON, and K. M. ROBIE-SUH (1976): Life Sci. **19**, 611–624

SCHENKMAN, J. B., H. REMMER, and R. W. ESTABROOK (1967): Mol. Pharmacol. **3**, 113–123

SCHLEYER, H., D. Y. COOPER, S. S. LEVIN, and O. ROSENTHAL (1972a): in Biological Hydroxylation Mechanisms (G. S. BOYD, and R. M. S. SMELLIE, eds.), pp. 187–206, Academic Press, New York

SCHLEYER, H., D. Y. COOPER, and O. ROSENTHAL (1972b): Federat. Proc. **31**, 484

SCHLEYER, H., D. Y. COOPER, O. ROSENTHAL, and P. CHEUNG (1977): Croat. Chem. Acta **49**, 179–206

SCHWARZ, D., J. PIRRWITZ, and K. RUCKPAUL (1982a): Arch. Biochem. Biophys. **216**, 322–328

SCHWARZ, D., J. PIRRWITZ, M. J. COON, and K. RUCKPAUL (1982b): Acta Biol. Med. Germ. **41**, 425–430

SHIMAKATA, T., K. MIHARA, and R. SATO (1972): J. Biochem. **72**, 1163–1174

SIMPSON, E. R., and G. S. BOYD (1967a): Biochem. Biophys. Res. Commun. **28**, 945–950

SIMPSON, E. R., and G. S. BOYD (1967b): Europ. J. Biochem. **2**, 275–285

SLIGAR, S. G. (1976): Biochemistry **15**, 5399–5406

SLIGAR, S. G., M. BESMAN, M. GELB, P. GOULD, D. HEIMBROCK, and D. PEARSON (1980a): in Biochemistry, Biophysics, and Regulation of Cytochrome P-450 (J. A. GUSTAFSSON, J. CARLSTEDT-DUKE, A. MODE, and J. RAFTER, eds.), pp. 379–382, Elsevier/North-Holland, Amsterdam, New York, Oxford

SLIGAR, S. G., K. A. KENNEDY, and D. C. PEARSON (1980b): Proc. Nat. Acad. Sci. U.S.A. **77**, 1240–1244

SLIGAR, S. G., B. S. SHASTRY, and I. C. GUNSALUS (1977): in Microsomes and Drug

Oxidations (V. ULLRICH, I. ROOTS, A. G. HILDEBRANDT, R. W. ESTABROOK, and A. H. CONNEY, eds.), pp. 202–209, Pergamon Press, Oxford

SMETTAN, G., J. BLANCK, G.-R. JÄNIG, and K. RUCKPAUL (1978): Acta Biol. Med. Germ. **37**, 537–544

SONO, M., J. A. MCCRAY, and T. ASAKURA (1977): J. Biol. Chem. **252**, 7475–7482

SPATZ, L., and P. STRITTMATTER (1971): Proc. Nat. Acad. Sci. U.S.A. **68**, 1042–1046

SPATZ, L., and P. STRITTMATTER (1973): J. Biol. Chem. **248**, 793–799

STAUDT, H., F. LICHTENBERGER, and V. ULLRICH (1974): Europ. J. Biochem. **46**, 99–106

STIER, A. (1976): Biochem. Pharmacol. **25**, 109–113

STIER, A., and E. SACKMANN (1973): Biochim. Biophys. Acta **311**, 400–408

STRITTMATTER, P., and M. J. ROGERS (1975): Proc. Nat. Acad. Sci. U.S.A. **72**, 2658–2661

STRITTMATTER, P., and S. F. VELICK (1956): J. Biol. Chem. **221**, 277–286

STRITTMATTER, P., L. SPATZ, D. CORCORAN, M. J. ROGERS, B. SETLOW, and R. REDLINE (1974): Proc. Nat. Acad. Sci. U.S.A. **71**, 4565–4569

STROBEL, H W., and M. J. COON (1971): J. Biol. Chem. **246**, 7826–7829

STROBEL, H. W., A. Y. H. LU, J. HEIDEMA, and M. J. COON (1971): J. Biol. Chem. **245**, 4851–4854

SUGIYAMA, T., N. MIKI, and T. YAMANO (1979): Biochem. Biophys. Res. Commun. **90**, 715 to 720

SUGIYAMA, T., N. MIKI, and T. YAMANO (1980): J. Biochem. **87**, 1457–1467

TANIGUCHI, H., Y. IMAI, T. IYANAGI, and R. SATO (1979): Biochim. Biophys. Acta **550**, 341–356

THOMAS, P. E., A. Y. H. LU, D. RYAN, S. B. WEST, J. KAWALEK, and W. LEVIN, (1976): Mol. Pharmacol. **12**, 746–758

THOMPSON, J. A., and J. L. HOLTZMAN (1974): Drug Metab. Dispos. **2**, 577–582

TOMODA, A., K. SUGIMOTO, M. SUHARA, M., TAKESHITA, and Y. YONEYAMA (1978): Biochem. J. **171**, 329–335

TSAI, R. L., I. C. GUNSALUS, and K. DUS (1971): Biochem. Biophys. Res. Commun. **45**, 1300–1306

TSAI, R. L., C.-G. YU, I. C. GUNSALUS, J. PEISACH, W. BLUMBERG, W. H. ORME-JOHNSON, and H. BEINERT (1970): Proc. Nat. Acad. Sci. U.S.A. **66**, 1157–1163

TYSON, C. A., J. D. LIPSCOMB, and I. C. GUNSALUS (1972): J. Biol. Chem. **247**, 5777–5784

ULLRICH, V. (1969): Hoppe-Seyler's Z. Physiol. Chem. **350**, 357–365

ULLRICH, V. (1977): in Microsomes and Drug Oxidations (V. ULLRICH, I. ROOTS, A. G. HILDEBRANDT, R. W. ESTABROOK, and A. H. CONNEY, eds.), pp. 192–201, Pergamon Press, Oxford

ULLRICH, V., P. WEBER, and H. GRAF (1980): in Microsomes, Drug Oxidations, and Chemical Carcinogenesis (M. J. COON, A. H. CONNEY, R. W. ESTABROOK, H. V. GELBOIN, J. R. GILLETTE, and P. J. O'BRIEN, eds.), Vol. I, pp. 95–102, Academic Press, New York

VAN DER HOEVEN, T. A., and M. J. COON (1974): J. Biol. Chem. **249**, 6302–6310

VATSIS, K. P., D. P. GURKA, and P. F. HOLLENBERG (1980a): in Biochemistry, Biophysics, and Regulation of Cytochrome P-450 (J. A. GUSTAFSSON, J. CARLSTEDT-DUKE, A. MODE, and J. RAFTER, eds.), pp. 347–350, Elsevier/North-Holland, Amsterdam, New York, Oxford

VATSIS, K. P., A. D. THEOHARIDES, D. KUPFER, and M. J. COON (1980b): in Biochemistry, Biophysics, and Regulation of Cytochrome P-450 (J. A. GUSTAFSSON, J. CARLSTEDT-DUKE, A. MODE, and J. RAFTER, eds.), pp. 73–76, Elsevier/North-Holland, Amsterdam, New York, Oxford

VERMILION, J. L., and M. J. COON (1974): Biochem. Biophys. Res. Commun. **60**, 1315–1322

VERMILION, J. L., and M. J. COON (1978a): J. Biol. Chem. **253**, 2694–2704

VERMILION, J. L., and M. J. COON (1978b): J. Biol. Chem. **253**, 8812–8819

VERMILION, J. L., D. P. BALLOU, V. MASSEY, and M. J. COON (1981): J. Biol. Chem. **256**, 266–277

VERMILION, J. L., V. MASSEY, D. P. BALLOU, and M. J. COON (1980): in Microsomes, Drug Oxidations, and Chemical Carcinogenesis (M. J. COON, A. H. CONNEY, R. W. ESTABROOK, H. V. GELBOIN, J. R. GILLETTE, and P. J. O'BRIEN, eds.), Vol. I, pp. 379–382, Academic Press, New York

WATERMAN, M. R., V. ULLRICH, and R. W. ESTABROOK (1973): Arch. Biochem. Biophys. **155**, 355–360

WERRINGLOER, J., and S. KAWANO (1980a): in Microsomes, Drug Oxidations, and Chemical Carcinogenesis (M. J. COON, A. H. CONNEY, R. W. ESTABROOK, H. V. GELBOIN, J. R. GILLETTE, and P. J. O'BRIEN, eds.), Vol. I, pp. 469–478, Academic Press, New York

WERRINGLOER, J., and S. KAWANO (1980b): in Biochemistry, Biophysics, and Regulation of Cytochrome P-450 (J. A. GUSTAFSSON, J. CARLSTEDT-DUKE, A. MODE, and J. RAFTER, eds.), pp. 359–362, Elsevier/North-Holland, Amsterdam, New York, Oxford

WERRINGLOER, J., S. KAWANO, and R. W. ESTABROOK (1979): Acta Biol. Med. Germ. **38**, 163–175

WHITE, R. E., and M. J. COON (1980): Annu. Rev. Biochem. **49**, 315–356

WHITE, R. E., J. T. GROVES, and G. A. MCCLUSKY (1979): Acta Biol. Med. Germ. **38**, 475–482

WHITE, R. E., D. D. OPRIAN, and M. J. COON (1980a): in Microsomes, Drug Oxidations, and Chemical Carcinogenesis (M. J. COON, A. H. CONNEY, R. W. ESTABROOK, H. V. GELBOIN, J. R. GILLETTE, and P. J. O'BRIEN, eds.), Vol. I, pp. 243–251, Academic Press, New York

WHITE, R. E., S. G. SLIGAR, and M. J. COON (1980b): in Biochemistry, Biophysics, and Regulation of Cytochrome P-450 (J. A. GUSTAFSSON, J. CARLSTEDT-DUKE, A. MODE, and J. RAFTER, eds.), pp. 307–314, Elsevier/North-Holland, Amsterdam, New York, Oxford

WILLIAMS, C. H., and H. KAMIN (1962): J. Biol. Chem. **237**, 587–595

YANG, C. S. (1975): FEBS Lett. **54**, 61–64

YANG, C. S. (1977): J. Biol. Chem. **252**, 293–298

YANG, C. S., F. S. STRICKHART, and L. P. KICHA (1977): Biochim. Biophys. Acta **465**, 362–370

YANG, C. S., F. S. STRICKHART, and L. P. KICHA (1978): Biochim. Biophys. Acta **509**, 326–337

YASUKOCHI, Y., and B. S. S. MASTERS (1976): J. Biol. Chem. **251**, 5337–5344

YASUKOCHI, Y., J. A. PETERSON, and B. S. S. MASTERS (1979): J. Biol. Chem. **254**, 7097 to 7104

YOSHIDA, Y., and H. KUMAOKA (1975): J. Biochem. **78**, 455–468

ZIEGLER, M., J. BLANCK, and K. RUCKPAUL (1982): FEBS Lett. **150**, 219–222

Chapter 4

Biophysical Properties of Cytochrome P-450, Analysis of the Reaction Mechanism — Thermodynamic Aspects

H. Rein, Ch. Jung, O. Ristau, J. Friedrich

Contents

1. Introduction

Most spectral and the magnetic properties of hemoproteins are caused by the prosthetic group, iron protoporphyrin IX, being the same in most hemoproteins and also present in P–450 (Omura and Sato, 1964; Maines and Anders, 1973). However, comparing the properties of P–450 with those of other hemoproteins the monooxygenase exhibits spectral and magnetic peculiarities: The red-shifted Soret band of the CO complex of the monooxygenase to 450 nm and the electron spin resonance spectrum of P–450 exhibiting striking low spin g-values which indicates a strong distortion of symmetry of the heme iron.

These peculiarities are caused by specific interactions of the prosthetic group with the apoprotein. Different sequences of amino acids result in a characteristic structure of the protein and therefore also in specific interactions of amino acid residues with the prosthetic group in the so-called heme pocket.

Several types of interaction can be differentiated: (i) a great number of van der WAALS contacts could be derived from crystal structure analysis of several hemoproteins, (ii) the coordinative binding of certain amino acid residues to the heme iron (His, Met, Cys), (iii) the covalent binding of side chains of the porphyrin to amino acid residues in cytochrome c.

By these specific interactions the electronic and structural properties of the iron porphyrin complex are modulated thus determining the specific function of the respective hemoprotein. From synthetic heme complexes with mercaptides as one ligand and thiol complexes of metmyoglobin it was suggested that a heme iron coordinated sulfur of the amino acid residue cysteine could be the reason for the peak at 450 nm as well as for the strongly distorted symmetry of the ligand field of the heme iron (RÖDER and BAYER, 1969; STERN and PEISACH, 1974; COLLMAN and SORRELL, 1975; CHANG and DOLPHIN, 1976a; TANG et al., 1976; ULLRICH et al., 1977). However, further peculiarities of the heme protein contact in P–450 cannot be ruled out. In this chapter the essential electronic properties of the heme group of P–450 in the different states of the reaction cycle are discussed on the basis of the electronic properties of the heme group itself.

Due to lacking *X*-ray structure of P–450 it is impossible to elucidate structure-function relationships at the molecular level. Therefore data from other hemoproteins well known in structure, function and related spectroscopic properties and especially data from P–450 from Pseudomonas putida are included in the discussion. The known quantitative and mechanistic parameters from the soluble P–450 CAM facilitate the interpretation of the data of the membrane bound P–450 LM. Moreover, the 5-exo-methylene hydroxylation of D(+)-camphor catalyzed by P–450 CAM can be regarded as a model for the hydroxylation reaction catalyzed by mammalian P–450 isozymes. With some exceptions the properties of microsomal liver P–450 are described including studies on the purified enzyme. In some cases the methods used for analysis of the electronic properties as well as theoretical remarks necessary for the interpretation of the spectral data are briefly introduced. Moreover, the reaction cycle of P–450 is analyzed considering thermodynamic data of the whole cycle as well as of known single steps. Finally the mechanism of the hydroxylation, the oxygen activation, is discussed.

2. Spectroscopic Behaviour of Cytochrome P-450

2.1. *Electronic and Vibrational Spectroscopic Properties*

2.1.1. *Electronic Absorption Spectra*

As mentioned before similar to other hemoproteins the absorption spectra of P–450 are determined by the heme chromophore

Abbreviations

P–450, cytochrome P–450
P–448, cytochrome P–448
3-MC, 3-methylcholanthrene
PB, phenobarbital
SKF 525-A, 2,2-diphenylvalerianic acid-2-diethylaminoethylester · HCl
EPR, electron paramagnetic resonance
CD, circular dichroism
MCD, magnetic circular dichroism
EXAFS, extended *X*-ray absorption fine structure
ENDOR, electron nuclear double resonance
P–450 LM, cytochrome P–450 from liver microsomes
P–450 CAM, cytochrome P–450 from *Pseudomonas putida*

and by the protein moiety. In the visible and near ultraviolet region the absorption bands are due to the electronic transitions of the heme group — the protoporphyrin IX iron complex (Figure 1). The electronic transitions of the protein in the middle ultraviolet region around 280 nm are caused by the aromatic amino acids whereas the transitions in the far ultraviolet region between 190 nm and 220 nm originate from the peptide amide chromophore.

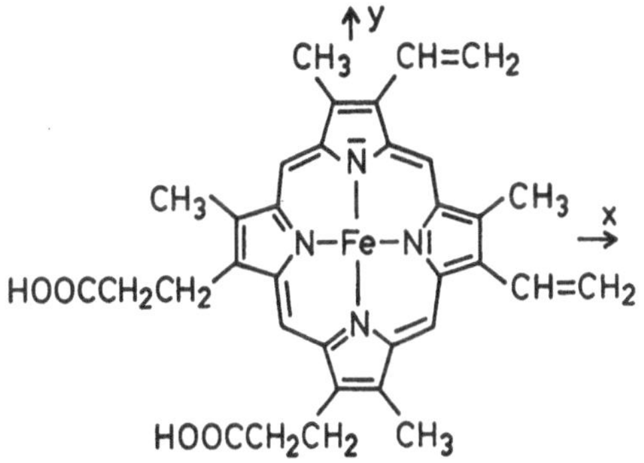

Fig. 1. Structure of iron protoporphyrin IX

Extensive experimental and theoretical studies have shown that the absorption bands of the heme group are predominantly caused by $\pi \rightarrow \pi^*$ transitions of the porphyrin. Unsubstituted neutral prophyrin (porphin) reveals four bands in the visible region — the Q bands or the bands I, II, III, IV (Figure 2). These four bands are reduced to two bands (α and β bands) for the dication, the dianion and the metal complexes. In the near ultraviolet region around 400 nm the absorption spectrum of the porphyrin exhibits a characteristic intense band — the Soret or B band — which is very broad and asymmetric in the case of the neutral porphyrin. At lower wavelengths further weak bands (N and L) exist. From fluorescence measurements (GURINOVITCH et al., 1963) it was concluded that the α and the β bands represent a vibronic structure of one electronic transition (Q). The α band is the $Q(0-0)$ and the β band the $Q(0-1)$ transition. Resonance RAMAN studies (SPIRO, 1975; MENDELSOHN et al., 1975; PLUS and LUTZ, 1974) have shown that the $Q(0-1)$ band corresponds to a vibration of the symmetry B_{1g} or B_{2g}.

The fundamental spectral properties of the porphyrin — the line shape and the number of separated absorption bands — are determined by the symmetry of the molecule. The neutral porphyrin has the symmetry D_{2h}. The porphyrin dication and dianion, however, possess the symmetry D_{4h} (Figure 3). Polarization measurements (KNOP and KNOP, 1970a, b) indicate that the transitions Q (α, β), B (Soret), N and L of the porphyrin with the symmetry D_{4h} are x,y-polarized that means energetically twofold degenerated. Therefore, these transitions transform according to the irreducible representation E_u of the symmetry group D_{4h}. They are the only allowed electronic transitions of the porphyrin. The other $\pi \rightarrow \pi^*$ transitions transforming on irreducible representations A_{1g}, A_{2g}, B_{1g} and B_{2g} are z-polarized and forbidden.

In contrast to the neutral porphyrin and the dication, the dianion and the porphyrin metal complexes show a well resolved absorption band in the UV region between 340 and 370 nm designated as UV band. In the case of the dianion this band orignates from the N-transition (JUNG, 1979). However, for metal complexes possessing a strong π-bond between the metal and the porphyrin (for example iron(II) complexes) this band consists of charge-transfer transitions from the axial π-system to the porphyrin-π-system (JUNG, 1980). Depending on the nature of the metal and the axial ligands the energies and intensities of the porphyrin transitions Q, B, N, and L are changed and new transitions (charge-transfer-transitions) occur (WILLIAMS, 1956; BUCHLER, 1975).

BUCHLER (1975) has classified the spectra of porphyrin metal complexes phenomenologically into three groups:

1. "normal spectrum: This spectrum is similar to the spectrum of the dication and the dianion spectrum and is shown by d^0 and d^{10} metal complexes.
2. "hypso" spectrum: In comparison to the "normal" spectrum the "hypso" spectrum shows a hypsochromic shift of about 25 nm and is found for d^6 to d^9 metals.
3. "hyper" spectrum: This spectrum has shifted α, β and Soret (B) bands and reveals new bands. d^1 to d^5 metals cause this spectrum.

Neglecting the small influence of the peripheral substituents of the porphyrin

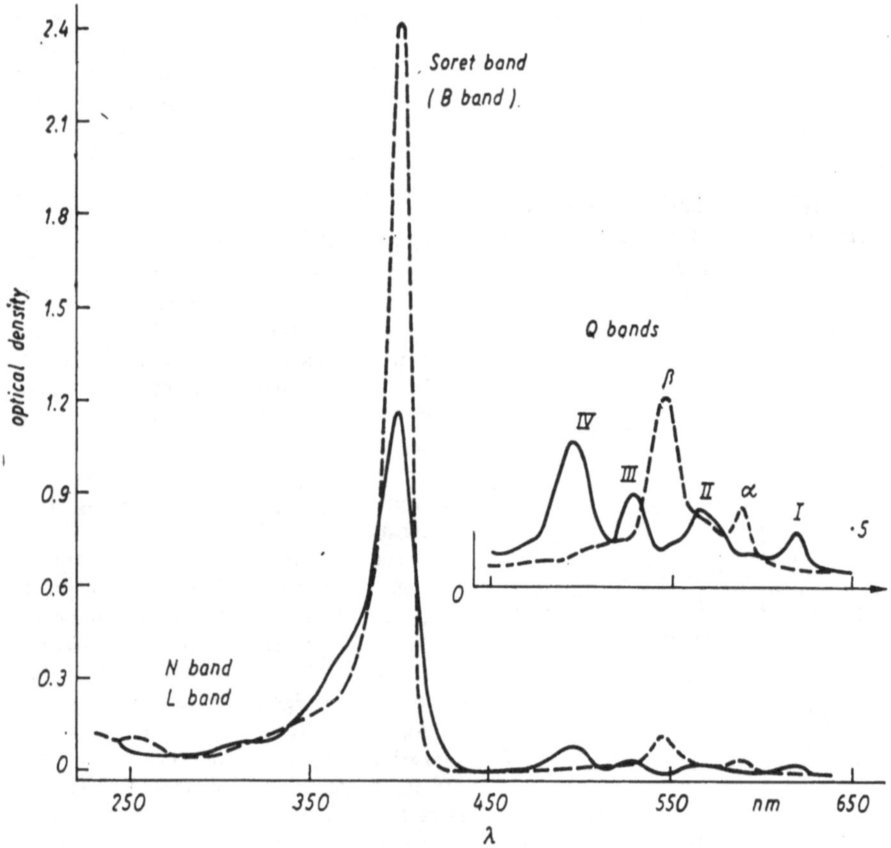

Fig. 2. Absorption spectrum of deuteroporphyrin IX dimethylester; $c = 0.75 \cdot 10^{-5}$ M, $d = 1.0$ cm

(———) neutral form solved in ethanol (symmetry D_{2h})

(- - - -) dication solved in 0.1 M aqueous hydrochloric acid (symmetry D_{4h})

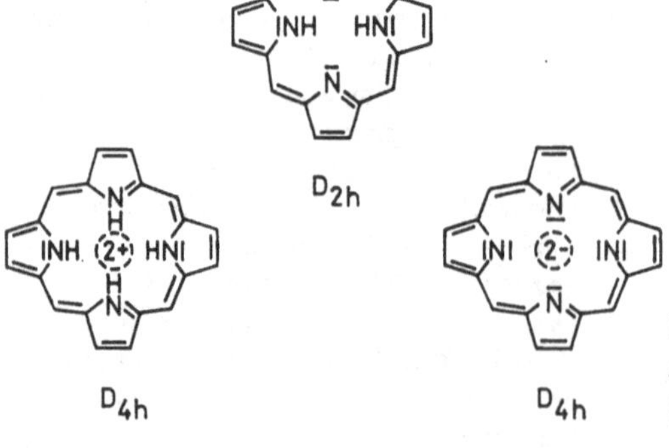

Fig. 3. Structure of neutral porphin (symmetry D_{2h}) and of porphin dication and dianion (symmetry D_{4h})

the fundamental spectral properties of the metal complexes as well as the hemoproteins are determined by the porphyrin-π-system. The axial ligands, however, give rise to the complex-specific spectral peculiarities. Therefore a comparison of the absorption spectra of the P-450 complexes with those of other well-defined hemoproteins or protein-free porphyrin iron complexes (model complexes) provides insight into the nature of the axial ligands (Table 1). The absorption spectrum of the ferric P-450 in the low-spin state (HAUGEN and COON, 1976) (Figure 4) matches that of a model complex with a mercaptide and an oxygen containing group as axial ligands ($S^- - Fe^{3+} - O \cdot R$) (TANG et al., 1976). P-450 CAM + $C_6H_5CH_2S^-$ reveals a spectrum similar to the spectrum of a dimercaptide model complex exhibiting a large red-shift of the Soret band with simultaneous enhancement and red-shift of the UV band, denoted as "hyper" spectrum (RUF et al., 1979). These comparisons indicate a mercaptide as fifth axial heme iron ligand in P-450. The absorption spectrum of the ferric P-450 in the high spin state is presented by the P-450 LM4 (HAUGEN and COON, 1976) (Figure 4) and also by the camphor bound P-450 CAM (GUNSALUS and SLIGAR, 1978). The spectra are superposition of bands of low spin and high spin spectra. It was found that 6% and 25% of the total heme iron of P-450 CAM and P-450 LM4, respectively, are in the low spin state (SLIGAR, 1976; HAUGEN and COON, 1976). The Soret maximum of the superimposed spectra is red shifted up to 394 nm, whereas the estimated Soret band maximum for the pure high spin spectrum is 387 nm (RISTAU et al., 1979). In the visible region the absorption spectra of P-450 LM4 and the camphor bound P-450 CAM show the typically "hyper" spectral behaviour according to Buchler's classification meaning the appearance of the charge-transfer transitions at 646 nm and 510 nm, characteristic for ferric high spin heme complexes (WILLIAMS, 1956). The high spin complex of the ferric P-450 can be very well simulated by the five-coordinated mercaptide model complex iron-protoporphyrin IX-dimethylester-$C_6H_5NO_2S^-$ (TANG et al., 1976) supporting a mercaptide as fifth heme iron ligand in P-450. Polarization measurements of the electronic transitions of camphor bound P-450 CAM crystals by HANSON et al. (1977) have shown the x, y-polarization of the Soret band and detectable z-polarized transitions at about 580 nm and 330 nm. They are likely to originate either from charge-transfer transitions between the axial ligand system and the porphyrin or from transitions among the d-orbitals of the iron mixed with the $3p$ orbitals of the supposed axial mercaptide sulfur ligand (HANSON et al., 1977). The ferrous P-450 is in the high spin state (CHAMPION et al., 1975a) and exhibits the Soret band at 408 nm, 411 nm and 413 nm for P-450 CAM, P-450 LM4 and P-450 LM2, respectively (Table 1 and Figure 4). As compared with P-450 CAM ferrous P-450 LM2 reveals a small red-shift of both the Soret and of the Q bands. This small red-shift of the whole spectrum is observed also for the CO and oxygen complexes of P-450 LM. The red-shift of the whole spectrum may indicate a small enlargement of the porphyrin-π-system likely induced by a decrease of the twisting angle of the vinyl groups of the porphyrin in P-450 LM2 compared to P-450 CAM. Polarization measurements of ferrous P-450 CAM crystals yielded some z-polarized absorption at about 420 nm explained with $d-d$ transitions of the heme iron influenced by sulfur p-orbitals (HANSON et al., 1977).

The binding of CO, O_2, imidazole and other "low-spin" ligands to ferrous P-450 is accompanied with the transition of the high-spin to the low spin state. The CO complex shows an anomalous absorp-

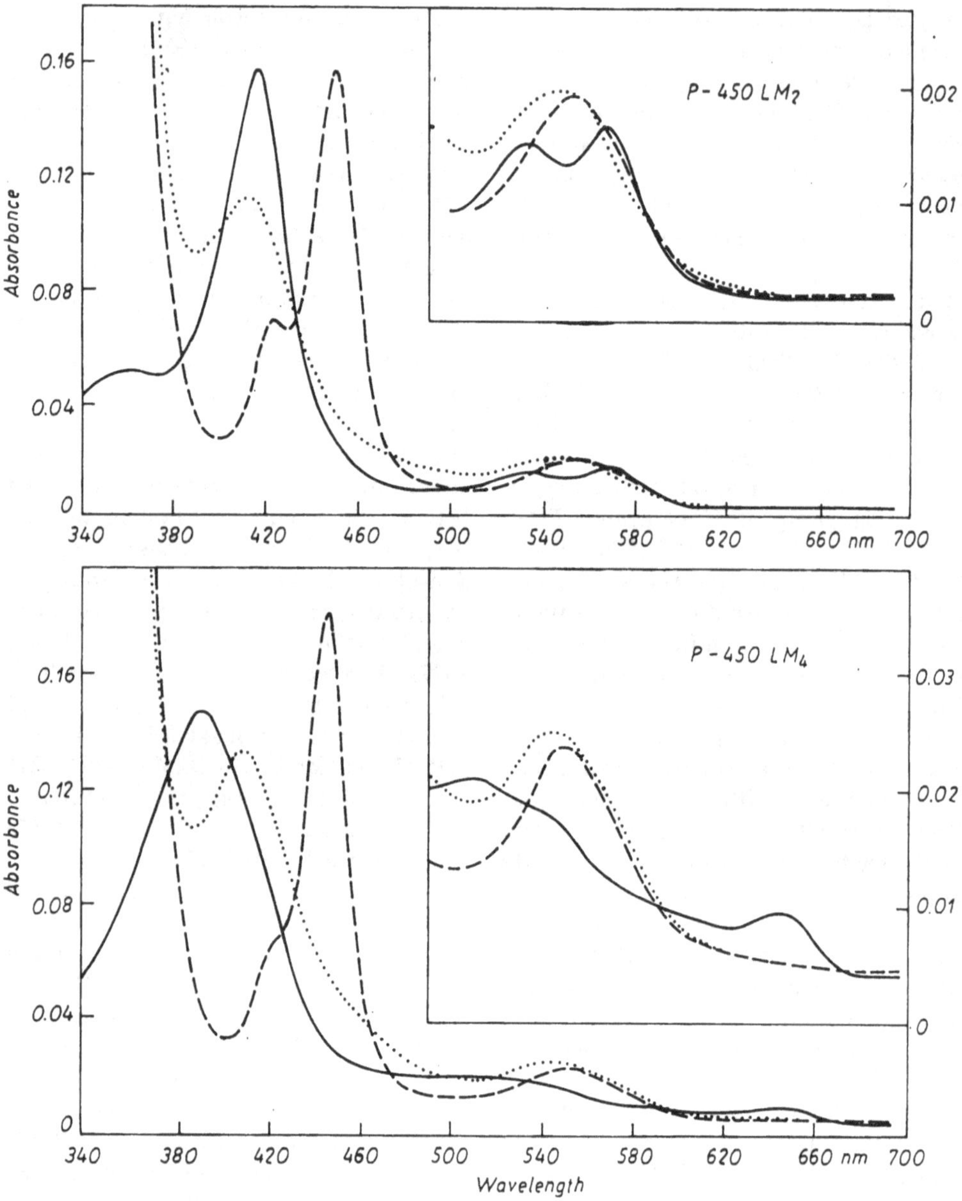

Fig. 4. Absorption spectra of purified P–450 LM2 (phenobarbital induced) and P–450 LM4 (β-naphthoflavone induced) from rabbit liver, taken from HAUGEN and COON (1976). The concentration of the cytochromes, based on heme analysis, was 1.4 nmol of P–450 LM2 and 1.53 nmol of P–450 LM4 per ml of 0.1 M potassium phosphate buffer, pH 7.4, containing 20% glycerol. (——) oxidized, (....) reduced and (----) reduced CO complex; reproduced with permission from the Amer. Soc. Biol. Chem., Inc.

Table 1: Electronic absorption spectra of P–450 and model complexes

Compound	Q bands $\lambda(\varepsilon_{mM})$	Soret band $\lambda(\varepsilon_{mM})$	UV band $\lambda(\varepsilon_{mM})$	References
Ferric low spin				
P–450$_{CAM}$	569 (11.9) 535 (11.6)	417 (115)	360 (36.7)	Gunsalus and Sligar, 1978
P–450$_{LM2}$	568 (11.2) 535 (10.4)	418 (110)	360	Haugen and Coon, 1976
Fe(PPIXDME)–($SC_6H_4NO_2$)–(2-MeTHF) 77 K	566	418	—	Tang et al., 1976
methemoglobin	575 (9.2) 540 (11.0)	410 (120)	—	Antonini and Brunori, 1971
chloroperoxidase 77 K	—	424	—	Champion et al., 1973
P–450$_{CAM}$ + $C_6H_5CH_2$–S^-	557	465	377	Ruf et al., 1979
Fe(PPIXDME)–($C_6H_5CH_2S^-$)$_2$ } 195 K	561	470	376	Ruf et al., 1979
Fe(PPIXDME)-(n-bu-S^-)$_2$	565	475	377	Ruf et al., 1979
Ferric high spin				
P–450$_{CAM}$ + camphor	646 (5.4) 540 (11.2) 510 (13.0)	391 (102)	—	Gunsalus and Sligar, 1978
P–450$_{LM4}$	644 (5.9) 547 (13.3) 512 (11.1)	394 (91)	—	Haugen and Coon, 1976
Fe(PPIXDME)–($SC_6H_4NO_2$)	646 (4.8) 549 (10.4) 517 (11.2)	391 (86.8)	—	Tang et al., 1976
methemoglobin, pH = 6.4	631 (4.4)	405 (179)	—	Antonini and Brunori, 1971
chloroperoxidase	—	396	—	Champion et al., 1973
Ferrous high spin				
P–450$_{CAM}$	540 (15.1)	408 (76.7)	—	Gunsalus and Sligar, 1978
P–450$_{CAM}$ + camphor	542 (16.0)	408 (86.5)	—	Gunsalus and Sligar, 1978
P–450$_{LM2}$	544 (15.1)	413 (78)	—	Haugen and Coon, 1976
P–450$_{LM4}$	542 (16.6)	411 (84)	—	Haugen and Coon, 1976
chloroperoxidase	—	406	—	Champion et al., 1973
Ferrous low spin				
P–450$_{CAM}$–CO	550 (14.3)	447 (121)	363	Gunsalus and Sligar, 1978; Hanson et al., 1976
P–450$_{LM2}$–CO	552 (13.4)	451 (110)	370	Haugen and Coon, 1976
P–450$_{LM4}$–CO (BNF-induced)	550 (15.3)	448 (118)	—	Haugen and Coon, 1976
P–450$_{LM4}$–CO (PB-induced)	550 (14.4)	448 (119)	—	Haugen and Coon, 1976

Table 1 (continued)

Compound	Q bands $\lambda(\varepsilon_{mM})$	Soret band $\lambda(\varepsilon_{mM})$	UV band $\lambda(\varepsilon_{mM})$	References
$P\text{-}450_{LM_4}-CO$ (not induced)	550 (18.3)	448 (120)	—	HAUGEN and COON, 1976
Fe(PPIXDEE) + [Na]SCH_3 + CO	—	449	—	COLLMAN et al., 1976
Fe(PPIXDME) + bu-S^- + CO	550	458	376	CHANG and DOLPHIN, 1976b
Fe(Meso-DME) + bu-S^- + CO	540	442	370	CHANG and DOLPHIN, 1976b
chloroperoxidase — CO	—	443	—	CHAMPION et al., 1973
Hb — CO (man)	569 (13.4) 540 (13.4)	419 (191)	344 (28)	ANTONINI and BRUNORI, 1971
$P\text{-}450_{CAM}-O_2$	552 (11.0)	418 (70)	—	EISENSTEIN et al., 1977
		418 (62)	355 (37)	DAWSON and CRAMER 1978
$P\text{-}450_{LM}-O_2$	558 (14.7)	420 (81.4)	—	BONFILS et al., 1979
$Hb-O_2$ (man)	577 (14.6) 541 (13.8)	415 (125)	344 (27)	ANTONINI and BRUNORI, 1971

Abbreviations: PPIXDME = protoporphyrin IX dimethylester; PPIXDEE = protoporphyrin IX diethylester; Meso-DME = mesoporphyrin IX dimethylester; MeTHF = 2-methyl-tetrahydrofuran; bu = n-butyl; [Na]SCH_3 = Na-methylmercaptide, solubilized by dibenzo-18-crown-6 or 18-crown-6 ether; BNF = β-naphthoflavone; PB = phenobarbital

λ values are given in nm; ε_{mM} is given in mM^{-1} cm^{-1}

tion spectrum exhibiting the red-shifted Soret band at 450 nm and the appearance of the UV band having a strongly increased intensity at about 370 nm (Fig. 5a). From polarization studies of CO-P-450 CAM crystals HANSON et al. (1976) could show that the *Q*, Soret and UV bands are *x*, *y*-polarized. The CO-P-450 is easily converted to CO-P-420 by different external influences such as pH, temperature (ICHIKAWA et al., 1967) and certain compounds (FRANKLIN, 1972; MURAKAMI et al., 1967; YU and GUNSALUS, 1974). CO-P-420 shows a "normal" absorption spectrum like CO-hemoglobin (Fig. 5c). This fact is interpreted by substitution of the mercaptide ligand by imidazole of a histidine as fifth heme iron ligand (STERN et al., 1973; BÖHM et al., 1977) or by an increase of the iron-sulfur distance originating from conformational changes of the protein (JUNG et al., 1979). The anomalous absorption spectrum of CO-P-450 can be simulated by the model complex iron-protoporphyrin IX having CO and mercaptide as axial ligands (CHANG and DOLPHIN, 1976b).

In contrast to CO-P-450 oxy-P-450 shows no red-shifted Soret band. The position of the Soret band is almost the same as that in oxyhemoglobin (Fig. 5b). Only the UV band is red-shifted and broader than in oxyhemoglobin. In order to explain the different spectral behaviour of CO and O_2 complexes of P-450 compared with those of hemoglobin several quantum chemical investigations were carried out. HANSON et al. (1976) used the Iterative Extended HÜCKEL Theory (IEHT) method which yields energies and composition of the molecular orbitals as linear combination of

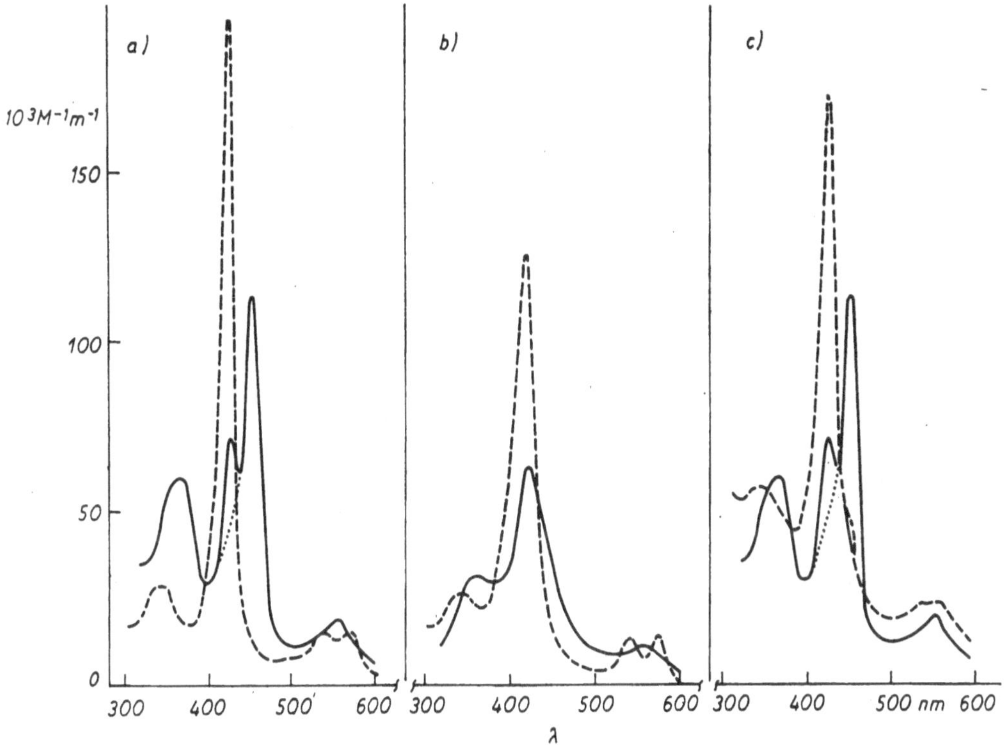

Fig. 5. Absorption spectra of (a) CO-hemoglobin (----) and partially purified CO-P-450 from phenobarbital induced male rabbit liver (———), (The peak at 420 nm in the P-450 spectrum originates from the presence of 7.8% CO-P-420 and 3.3% reduced cytochrome b_5); (b) Oxyhemoglobin (----) and oxy-P-450 CAM (———), (The oxy-P-450 CAM spectrum was digitalized and replotted from BALNY et al. (1976)) and (c) CO-P-420 (----) and partially purified CO-P-450 from phenobarbital induced male rabbits (———), (The shoulder at 450 nm in the CO-P-420 spectrum originates from the presence of some CO-P-450. The conversion of P-450 to P-420 was carried out using sodium deoxycholate), taken from JUNG et al. (1979)

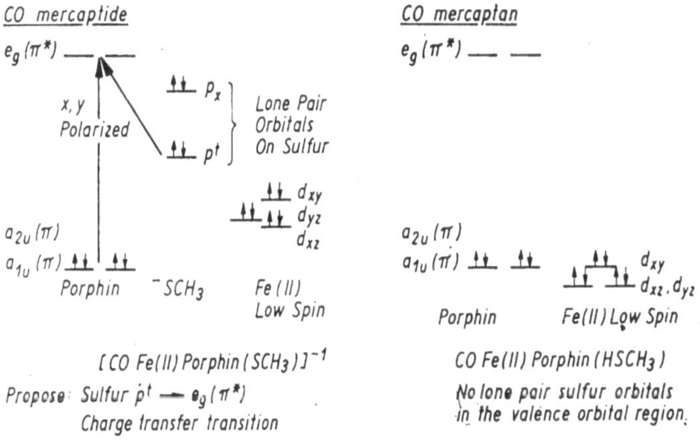

Fig. 6. Results of the IEHT calculations for CO-P-450 (HANSON et al., 1977). The calculations were carried out with both the negatively charged mercaptide (left molecular orbital scheme) and with the neutral mercaptane (right molecular orbital scheme) as fifth heme iron ligand

atomic orbitals. Assuming mercaptide as fifth heme iron ligand IEHT calculations of the iron porphyrin complex show energetically high-lying molecular orbitals originating predominantly from sulfur 3p orbitals (Figure 6). One of them has the correct symmetry to allow an x,y-polarized transition to the virtual porphin molecular orbitals e_g^*. HANSON et al. (1976) postulate that such transition $S_p^+ \rightarrow e_g^*$ leads to the intense UV band as well as to the red-shifted Soret band. The high participation of the sulfur $3p_z$ orbital in the molecular orbital S_p^+ was the reason to classify the CO-P–450 spectrum as "hyper" spectrum of the p-type.

By means of the IEHT method a quantum chemical explanation of BUCHLER's classification (BUCHLER, 1975) of the absorption spectra of porphyrin metal complexes was given. The Q and Soret (B) bands of the "normal" spectrum are predominantly produced by the molecular orbital transitions a_{1u}, $a_{2u} \rightarrow e_g^*$ (GOUTERMAN et a., 1975). In the "hypso" spectrum these four transitions are influenced by the completely occupied $d\pi$ orbitals of the metal (HANSON et al., 1976). The "hyper" spectrum is divided into the d-type and the p-type. Metals with empty or half occupied d-orbitals (for example, Fe(III)) effects the d-type which allows the new transitions a_{1u}, $a_{2u} \rightarrow d\pi$. Main group metals in low oxidation states (for example Sn(II), Pb(II), As(III), Sb(III) and Bi(III) (SCHAFFER and GOUTERMAN, 1970; GOUTERMAN et al., 1973, HANSON et al., 1976)) produce the p-type and allow transitions $a_{2u}(np_z \rightarrow e_g^*)$.

For the O_2-P–450 complex HANSON et al. (1977) assume a mercaptan as fifth heme iron ligand because only in this state the IEHT calculations show no energetically high-lying sulfur orbitals (Fig. 7). The IEHT calculations by HANSON et al. (1976, 1977) are the first step for a theoretical explanation of spectral properties of P–450. From the quantum chemical point of view a configuration interaction (CI) calculation is a necessary condition for the interpretation of electronic absorption spectra, especially for porphyrin systems (WEISS et al., 1965) and have been performed for CO-P–450 and other hemoprotein complexes by JUNG et al. (1977, 1979), using an π-Intermediate Neglected Differential Overlap-Configuration Interaction (π-INDO-CI-) method extended for metal complexes. A comparative analysis of

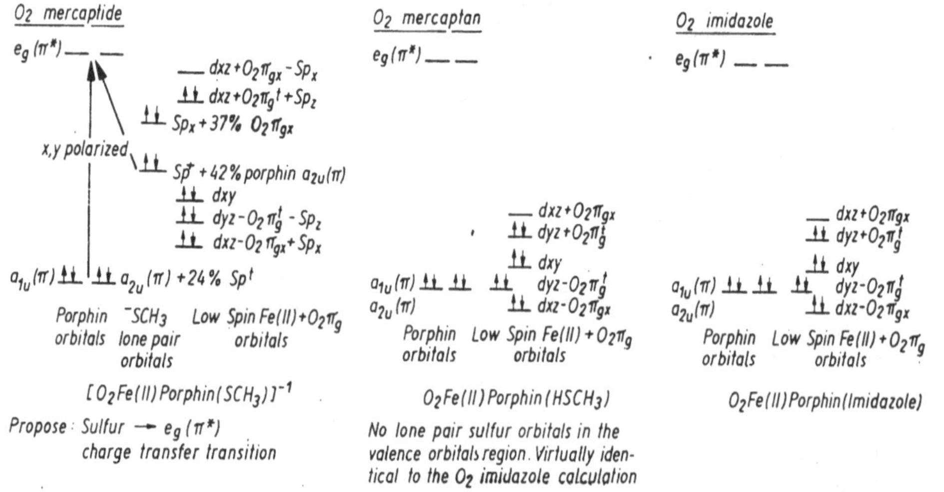

Fig. 7. Results of IEHT calculations for O_2—P–450 in comparison to oxyhemoglobin (HANSON et al., 1977). The used fifth heme iron ligand in the calculations was mercaptide (left molecular orbital scheme), mercaptane (middle scheme) and imidazole (right scheme)

spectral properties of models of oxy- and CO-complexes of hemoglobin and P–450 and of ferrous cytochrome b_5 performed with the π-INDO-CI method (JUNG, 1980) shows that the Q, the Soret and the UV band predominantly originate from the molecular orbital transitions $a_{1u}, a_{2u} \rightarrow e_g{}^*$ and $e_g(\pi_{ax}) \rightarrow b_{1u}^*$ (Figure 8):

"normal" spectral properties: CO-hemoglobin, oxyhemoglobin, oxy-P–450, cytochrome b_5

$$Q = a_{1u}, a_{2u} \rightarrow e_g{}^*$$
$$B = a_{1u}, a_{2u} \rightarrow e_g{}^*$$
$$UV = e_g(\pi_{ax}) \rightarrow b_{1u}^*$$

"anomalous" spectral properties: CO-P–450

$$Q = a_{1u}, a_{2u} \rightarrow e_g{}^*$$
$$B = 2/3\ (a_{1u}, a_{2u} \rightarrow e_g{}^*) + 1/3\ \left(e_g(\pi_{ax}) \rightarrow b_{1u}^*\right)$$
$$UV = 1/3\ (a_{1u}, a_{2u} \rightarrow e_g{}^*) + 2/3\ \left(e_g(\pi_{ax}) \rightarrow b_{1u}^*\right)$$

The strong interaction between the molecular orbital transitions $a_{1u}, a_{2u} \rightarrow e_g^*$ and $e_g(\pi_{ax}) \rightarrow b_{1u}^*$ in CO-P–450 is induced by the lower ionization potential of the negatively charged sulfur ligand and its higher π-orbital overlap with the iron-$d\pi$-orbitals compared to an imidazole ligand. The amount of this interaction is determined by the sulfur charge and the iron-sulfur distance. Only with negative sulfur charges less than "−0.6" and iron-sulfur distance between 1.9 and 2.3 Å the spectral anomaly of CO-P–450 can be simulated in this model (JUNG 1980). This distance is in agreement with Extended X-ray Absorption Fine Structure (EXAFS) studies of P–450 LM2 (ferric low-spin) yielding a value of 2.19 Å (CRAMER et al., 1978) and of CO-P–450 CAM revealing a value of 2.32 Å (DAWSON et al., 1978) and also with X-ray scattering measurements of a mercaptide CO model complex yielding 2.35 Å (CARON et al., 1979). The partial negative charge of the sulfur ligand may be explained by a hydrogen bond between the mercaptane ligand and a nitrogen base of an amino acid in the vicinity of the fifth iron ligand (JUNG, 1980). Recently, the first all-valence calculation of the excited states of CO-P–450, oxy-P–450 and CO-hemoglobin was published by LOEW and ROHMER (1980) using an INDO-CI method adapted for metal complexes (ZERNER et al., 1980). According to this calculation the Soret band of CO-hemoglobin should be produced by four intense electronic transitions with approximately equal os-

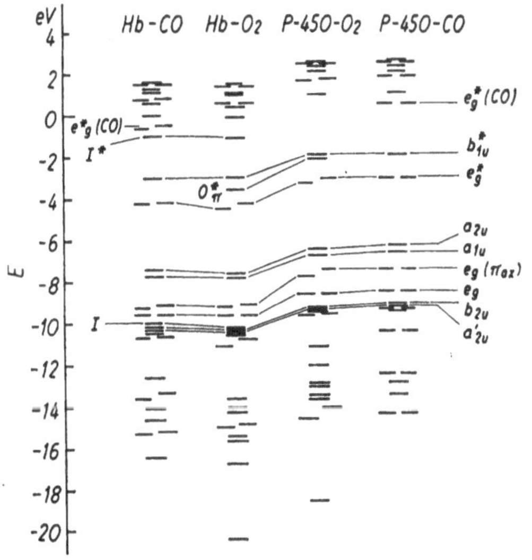

Fig. 8. Molecular orbital schemes for the dioxygen and carboxy complexes of P–450 and hemoglobin obtained with the π-INDO method (JUNG, 1980). The molecular orbitals a_{2u}, b_{2u}, a_{1u}, a_{2u} and b_{1u}^* are totally localized in the porphyrin ligand. O_π^* is the antibonding molecular orbital of the O_2 ligand. I and I* are pure imidazole molecular orbitals. e_g^*(CO) is the degenerate antibonding molecular orbital of the CO ligand. $e_g(\pi_{ax})$ is localized in the axial π-system for 50 to 80%

cillator strengths resulting from a strong mixing of the molecular orbital transitions a_{1u}, $a_{2u} \rightarrow e_g^*$, $d_{x^2-y^2} \rightarrow e_g^*$ and $d\pi \rightarrow d_{xy}^*$. These four transitions should split in CO-P-450 due to the participation of sulfur $3p_z$ in the a_{2u} molecular orbital and of the appearance of an energetically high-lying molecular orbital with 50% sulfur $3p_z$ character. The results of this calculation, however, are in contrast to MCD measurements which show a well resolved A-term for the Soret band of CO-hemoglobin which is in agreement with only one degenerate electronic transition (see 2.1.3.).

2.1.2. Circular Dichroism Spectra

Hemoproteins exhibit optically active absorption bands in different spectral regions. The optical activity of the peptide band is predominantly due to the helix content of the protein which correlates to the magnitude of the *CD* band (MOFFIT, 1956; BLAUER, 1974). In the near ultraviolet region the absorption bands of the aromatic amino acid residues as well as of the heme chromophore are optically active and intensive *CD* bands of the heme chromophore are also observed in the visible region.

As shown in 2.1.1. the relatively high symmetry of D_{4h} can be assumed for the heme complex to the first approximation which implies energetically degenerate electronic transitions. That argument agrees with the lack of optical activity of isolated hemin (RUCKPAUL, et al., 1970a). Even the binding of porphyrin to globin induces optical activity of heme bands. Peripheral porphyrin substituents and also the introduction of the iron into the porphyrin have only modifying influence (RUCKPAUL et al., 1970b). The appearance of optical activity of the heme bands in the protein bound heme indicates the perturbation of the D_{4h} symmetry of the iron porphyrin complex reflected in the disappearance of the degeneracy of the $\pi-\pi^*$ transitions. Symmetry perturbations are caused by the asymmetry of the protein in the vicinity of the heme chromophore. For example, in myoglobin the origin for the rotational strength of the Soret *CD* signal is the coupling of the Soret transitions with excited states of aromatic side chains (HSU and WOODY, 1971). The nature as well as the geometrical orientation of the aromatic amino acid residues with respect to the heme plane are important factors to determine the magnitude, the shape and the sign of these heme *CD* signals.

In hemoproteins both positive and negative signs of the Soret *CD* band are found. Hemoglobin and myoglobin (RUCKPAUL et al., 1970c) and cytochrome c (URRY, 1965; MYER et al., 1966), for example, exhibit positive *CD* bands whereas Chironomus hemoglobin (RUCKPAUL, et al., 1969), Lamprey hemoglobin (LAMPE et al., 1972; SUGITA et al., 1968), leghemoglobin (ELLFOLK et al., 1975; NICOLA et al., 1975) catalase (ULLMER and VALLEE, 1965) and also P-450 CAM (PETERSON, 1971) as well as P-450 LM (GUENGERICH et al., 1975; REIN et al., 1976; CHIANG and COON, 1979) show a negative Soret *CD* band (Fig. 9–11). The reason for this different sign of the Soret *CD* band can be seen in a different influence of the protein on the x and y polarized transition components of the heme. The magnitudes of the two individual components depend on the orientation of the heme transition in the heme plane but the total rotational strength is independent of the direction of polarization.

Quantitative determination of the rotational strength of the Soret B_x and B_y components were calculated for Sperm whale myoglobin and horse hemoglobin with 0.3 and 0.2 DBM, respectively (HSU and WOODY, 1971) and for the negative Soret *CD* band of Chironomus hemoglobin CT III with −0.2 DBM (FLEISCHHAUER and WOLLMER, 1972). These authors concluded that the main contribution in the sign inversion in Chironomus hemoglobin as compared to mammalian hemoglobin and myoglobin results from the coupling of the Soret transition with $\pi-\pi^*$ transitions mainly of phenylalanyl residues localized near the heme pocket of Chirnomus hemoglobin CT III

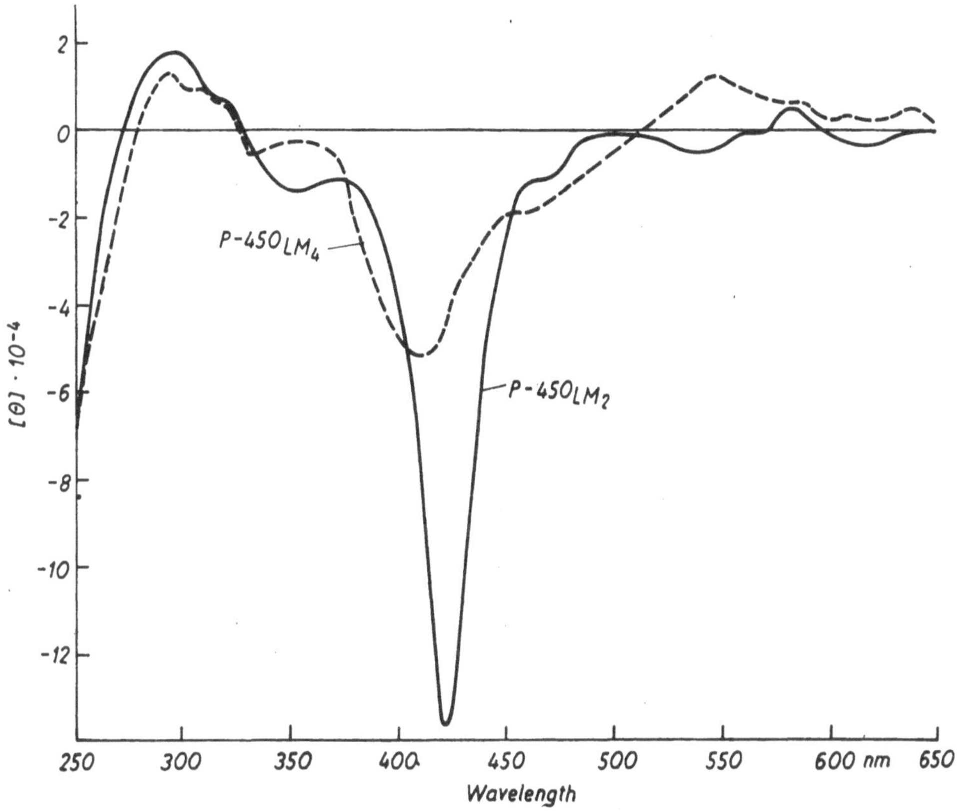

Fig. 9. CD spectra of oxidized P–450 LM2 and P–450 LM4, taken from CHIANG and COON (1979); reproduced with permission from Academic Press, Inc.

(FLEISCHHAUER and WOLLMER, 1972). A relatively high content of phenylalanine is present in P–450 CAM as well as in P–450 LM (YASUNOBU et al., 1980) which could explain the negative Soret Cotton effect.

There is experimental evidence that hemoproteins possessing a relatively open heme pocket such as Chironomus hemoglobin (HUBER et al., 1969), leghemoglobin (VUK-PAVLOVIC et al., 1976), Lamprey hemoglobin (LAMPE et al., 1972) exhibit Soret *CD* bands with negative sign. From the negative Soret *CD* bands of P–450 (Fig. 9—11) a relatively open heme pocket is indicated which is in agreement with proton magnetic relaxation studies of the enzyme (RUCKPAUL et al., 1976) as well as with infrared studies of the CO complex of P–450 (BÖHM et al., 1976). Functional requirements of the enzyme to bind relatively large molecules and probably also the reductase near the prosthetic group support this assumption.

The negative Soret *CD* band of P–450 CAM at 410 nm is shifted to 388 nm by the binding of camphor (PETERSON, 1971) which corresponds with the shift of the Soret absorption band in consequence of the change of the spin equilibrium from the low spin to the high spin state. In contrast to this behaviour P–450 LM shows no significant change of the Soret *CD* spectrum at binding of type I substrates, e.g. benzphetamine, which is in agreement with a smaller shift of the spin equilibrium compared with P–450 CAM (REIN et al., 1976; CHIANG and COON, 1979). This result may reflect the known low substrate specifity of P–450 LM,

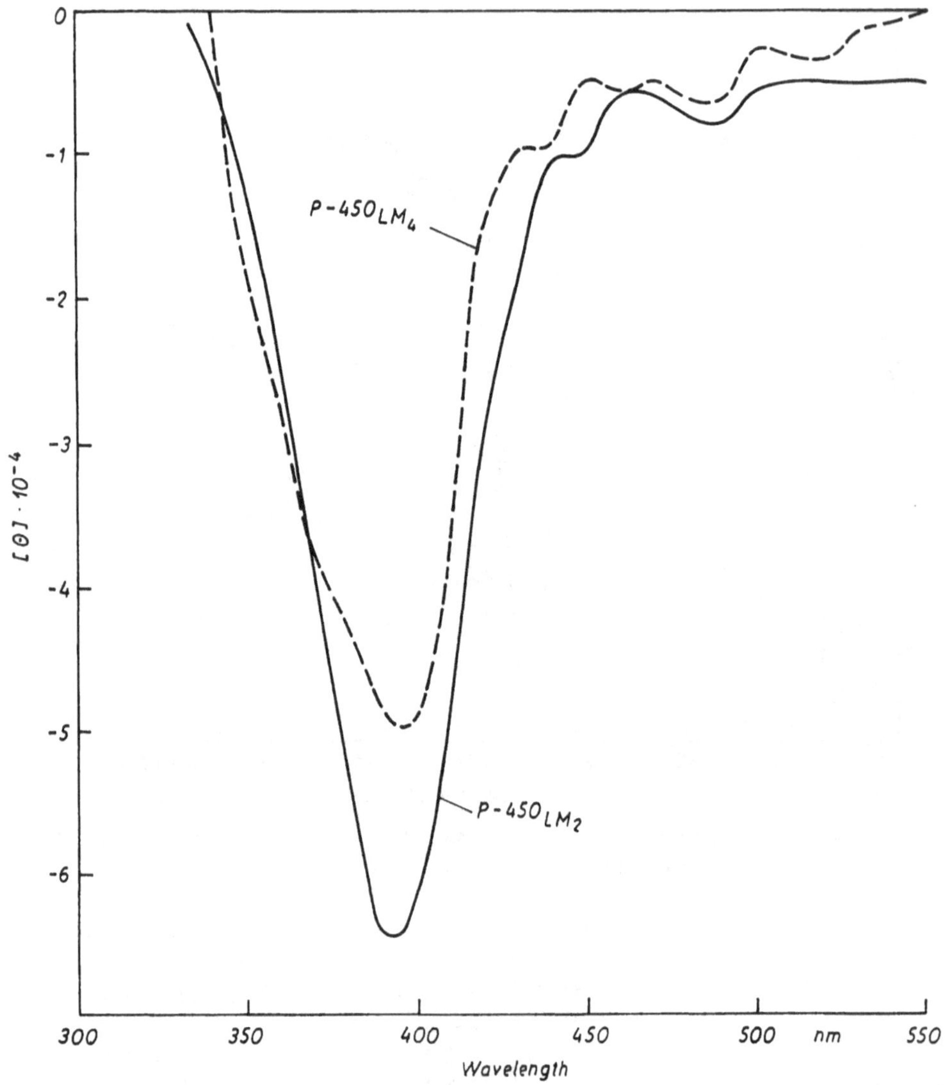

Fig. 10. CD spectra of reduced P–450 LM2 and P–450 LM4, taken from CHIANG and COON (1979); reproduced with permission from Academic Press, Inc.

i.e. the PB induced isozyme P–450 LM2. Type II substrates, however, significantly influence the the shape of the Soret CD band of P–450 LM meaning that interactions between the heme chromophore and the protein are disturbed by these substrates.

A significant change of the *CD* spectrum of P–450 is caused by the reduction of the heme iron (with dithionite) forming a relatively broad negative *CD* band with lower intensity as compared with the ferric form and a maximum at about 400 nm (REIN et al., 1976; CHIANG and COON, 1979) (Fig. 10). This result can be interpreted as a conformational change of the protein in the vicinity of the heme caused by the disruption of the sixth heme iron ligand involved in the reduction process.

The binding of CO to the ferrous form leads to a negative Soret *CD* band with

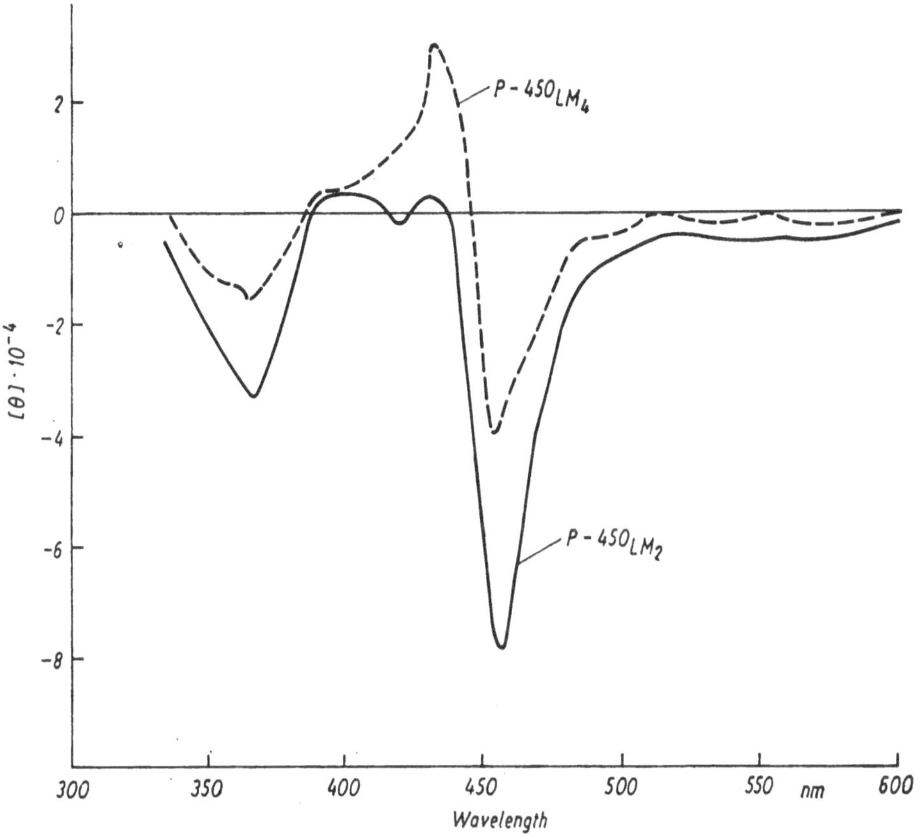

Fig. 11. CD spectra of CO complexes of reduced P–450 LM2 and P–450 LM4, taken from Chiang and Coon (1979); reproduced with permission from Academic Press, Inc.

a maximum at 454 nm and to a further intensive *CD* band at about 360 nm (Fig. 11). The magnitude of the 360 nm band which corresponds to the intense *UV* band in the absorption spectrum was shown to be equal to the intensity of the Soret *CD* band in the partially purified form (Rein et al., 1976) but half to that in the CO complexes of P–450 LM2 and P–450 LM4 (Chiang and Coon, 1979).

Complexes of reduced P–450 with alkylisocyanides show two *CD* bands in the Soret region at 428 nm and 455 nm (Rein et al., 1976) corresponding to two absorption bands described by Imai and Sato (1966) and Jefcoate et al. (1970).

The *CD* bands of the heme transitions disappear upon the conversion of P–450 to the denatured form, P–420 (Chiang and Coon, 1979; Shimizu et al., 1979). The loss of the *CD* spectrum is not due to a release of the heme group from the protein. Therefore, the heme in P–420 is bound at a symmetric environment, which could be caused by a loosening of van der Waals bonds between heme and protein (Shimizu et al., 1979).

Shimizu et al. (1979) estimated the α-helix content of P–450 LM from rabbit liver between 40% and 50% without significant differences between phenobarbital- and methylcholanthrene-induced P–450. Lower values are 37% and 25% which are determined for the purified forms P–450 LM2 and P–450 LM4, respectively (Chiang and Coon, 1979). In the presence of phospholipid (dilauroylglyceryl-3-phosphorylcholine) the α-helix

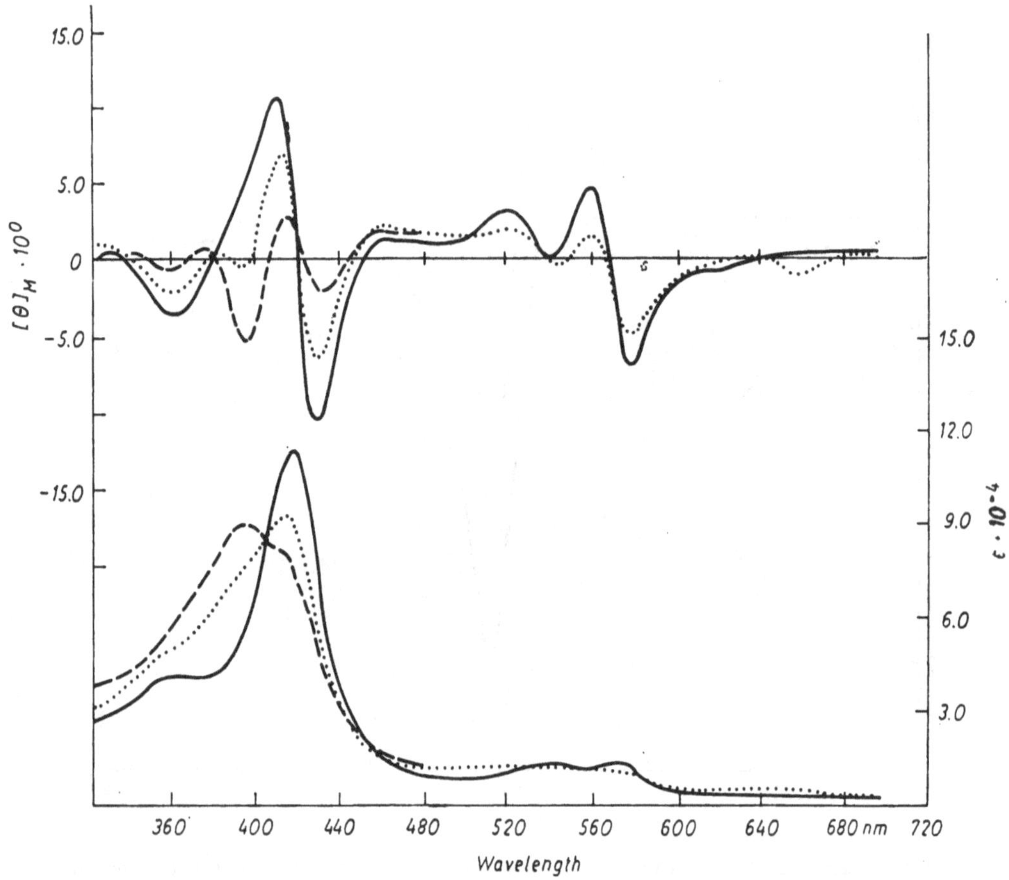

Fig. 12. MCD (top) and absorption (bottom) spectra of oxidized phenobarbital induced P–450 LM2 (———), phenobarbital induced P–450 LM3:4 (....) and β-naphthoflavone induced P–450 LM4 (– – – –), taken from DAWSON et al., (1978); reproduced with permission from Amer. Chem. Soc.

content is enhanced up to 44% and 49% in P–450 LM2 and P–450 LM4, respectively.

2.1.3. *Magnetic Circular Dichroism Spectra*

Several MCD studies of P–450 and of P–450 model complexes were performed because by this method detailed information about the symmetry of the excited and ground electronic states of the heme chromophore can be obtained from which conclusions about the molecular structure of the heme complex may be derived.

MCD measurements of partially purified P–450 from liver of phenobarbital-induced rabbits were carried out by SHIMIZU et al. (1975). DAWSON et al. (1974) measured the MCD spectra of microsomal suspensions from liver of phenobarbital- und 3-MC-induced rats and also of the pure isoenzymes LM2, LM3 : 4 and LM4 of rabbit liver P–450 (DAWSON et al., 1978). COLLMAN et al. (1976a) investigated highly purified P–450 from rabbit liver. P–450 CAM was studied by VICKERY et al. (1975 and 1976a) and by DAWSON and CRAMER (1978). All measurements result in well resolved *A*-terms of the Soret and *Q* bands of ferric P–450 in the low-spin state (Fig. 12 and 13).

A theoretical analysis of MCD spectra carried out by BUCKINGHAM and STEPHENS

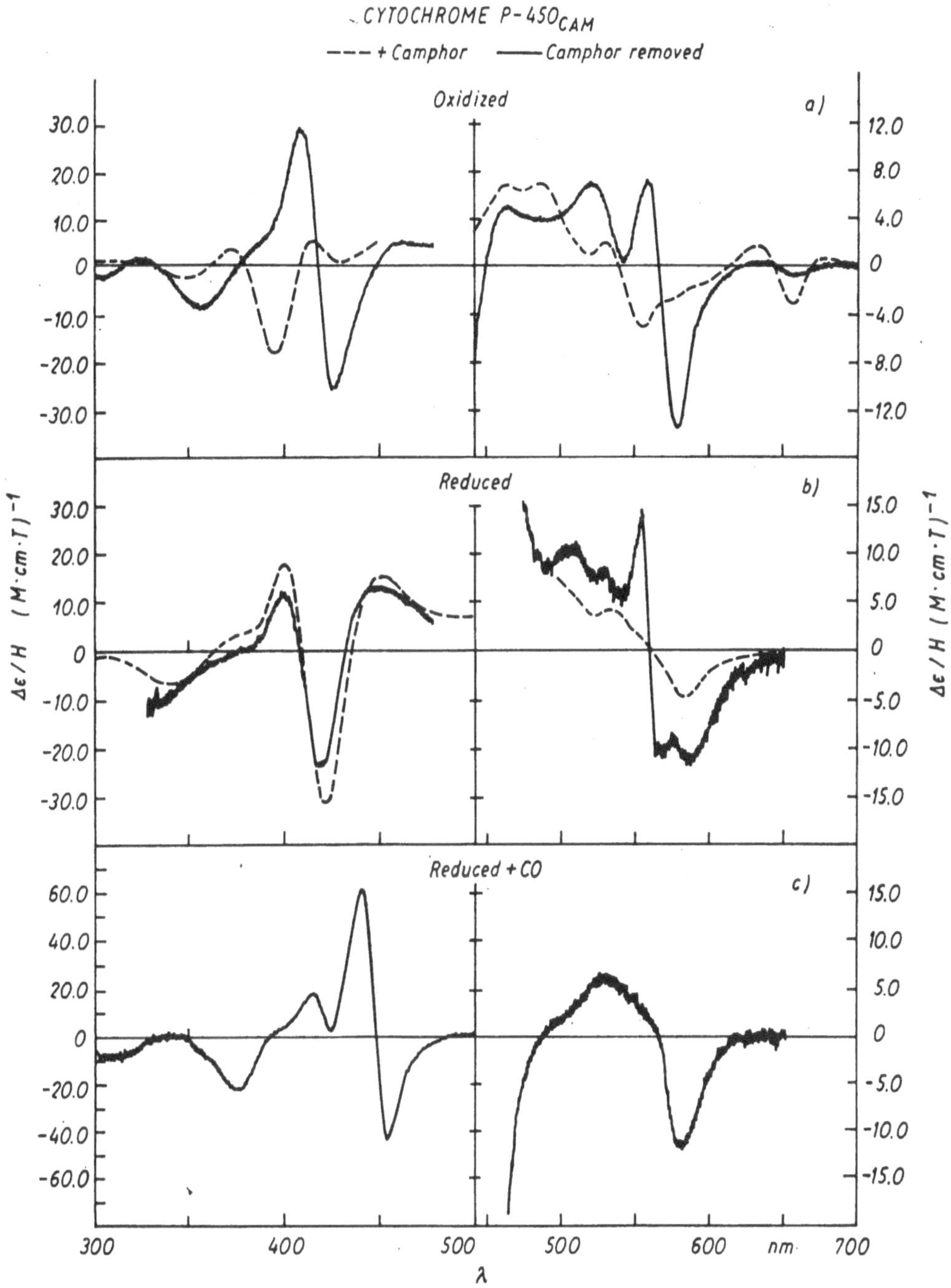

Fig. 13. MCD spectra of P–450 CAM in the presence of an excess of *D*-camphor (– – –) and in the absence of camphor (———), recorded at room temperature, taken from VICKERY et al. (1975); reproduced with permission from Elsevier Biomedical

(1966) yielded three terms in the equation for the molar ellipticity, showing characteristic line shapes. An *A*-term is expected if the electronic ground state or the electronic excited state or both states are symmetry-originated energetically degenerate. The *C*-term resulting from a degenerate electronic ground state shows a temperature dependence in consequence of the temperature dependent population of the split ground state levels in the magnetic field. The *B*-term is produced by electric and magnetic interaction between the excited state of the electronic transitions and the other excited states and the ground state of the molecule. The *B*- and *C*-term show a Gaussian line shape but the *A*-term exhibits a derivative line shape of a Gaussian (STEPHENS et al., 1966).

From symmetry reasons A-terms for the Soret- and *Q*-bands of the porphyrin complexes of the four-fold symmetry are expected. Diamagnetic porphyrin complexes have not degenerate ground state, and therefore should not show *C*-terms.

A temperature dependence of the MCD signals of the ferric low spin P–450 was found by SHIMIZU et al. (1975) by which a participation of *C*-terms originating from the degeneration of the ground state was indicated. The ferric high spin P–450 (P–450 CAM + camphor and the isozyme P–450 LM4) exhibit MCD curves which were interpreted as a superposition of *A*- and *C*-terms (Fig. 12 and 13). The results obtained for the ferric P–450 are generally in accordance with that for other hemoproteins in the ferric form (BRIAT et al., 1972; REIN et al., 1973, 1975b; TREU, 1975; KOBAYASHI et al., 1976; LIVSHITZ et al., 1976; SHARONOV et al., 1976; VICKERY et al., 1976a; MINEYEV et al., 1978).

A theoretical explanation of the MCD spectra of the hemoproteins in the ferric low spin state was given by MINEYEV et al. (1978) who showed that the *C*-terms may result from interaction between the iron orbital and spin momentum with the orbital and spin momentum of the *Q*- and *B*-(Soret) state of the porphyrin ligand.

In contrast to the ferric form the ferrous P–450 shows a MCD spectrum (Fig. 14) (SHIMIZU et al., 1975; VICKERY et al., 1975; DAWSON et al., 1978) in the Soret region which is qualitatively different from the spectrum of deoxymyoglobin (VICKERY et al., 1976a) and deoxyhemoglobin (TREU, 1975; SHARONOV et al., 1976). A large peak in the Soret region is observed for deoxymyoglobin, whereas reduced P–450 shows a strong trough which reflects different electronic properties of these hemoproteins, likely caused by different fifth axial heme iron ligands.

Measurements of the temperature dependence of the MCD spectrum of reduced P–450 indicate a participation of *C*-terms (SHIMIZU et al., 1975) (Fig. 15). Deoxyhemoglobin shows a similar temperature behaviour which was explained by TREU (1975) with spin-orbit coupling.

The CO complex of P–450 was investigated with the MCD method by DAWSON et al. (1978) for the isozymes P–450 LM2 and P–450 LM4, by SHIMIZU et al. (1975) for solubilized P–450 from rabbit liver and by DOLINGER et al. (1974) and VICKERY et al. (1975 and 1976a) for P–450 CAM. COLLMAN et al. (1976a) published a comparative study of the MCD spectra of highly purified P–450 from rabbits and P–450 model complexes with a mercaptide ligand bound to the heme iron. In accordance with the theoretical expectation all these CO complexes show well resolved *A*-terms for the *Q*- and Soret bands (Fig. 13 and 16). The *A*-term of the Soret band shows a shoulder at 420 nm. The energy difference between the *A*-term crossover point and this shoulder is 1590 cm^{-1} which is very similar to the difference of 1560 cm^{-1} observed for carboxymyoglobin (VICKERY et al., 1976a) and the P–420 LM2 carboxy complex (DAWSON et al., 1978) and indicates a vibronic structure of the Soret band. Contradictory to the experimental (HANSON et al., 1976) and the theoretical (JUNG, 1980) findings of its *x*,*y*-polari-

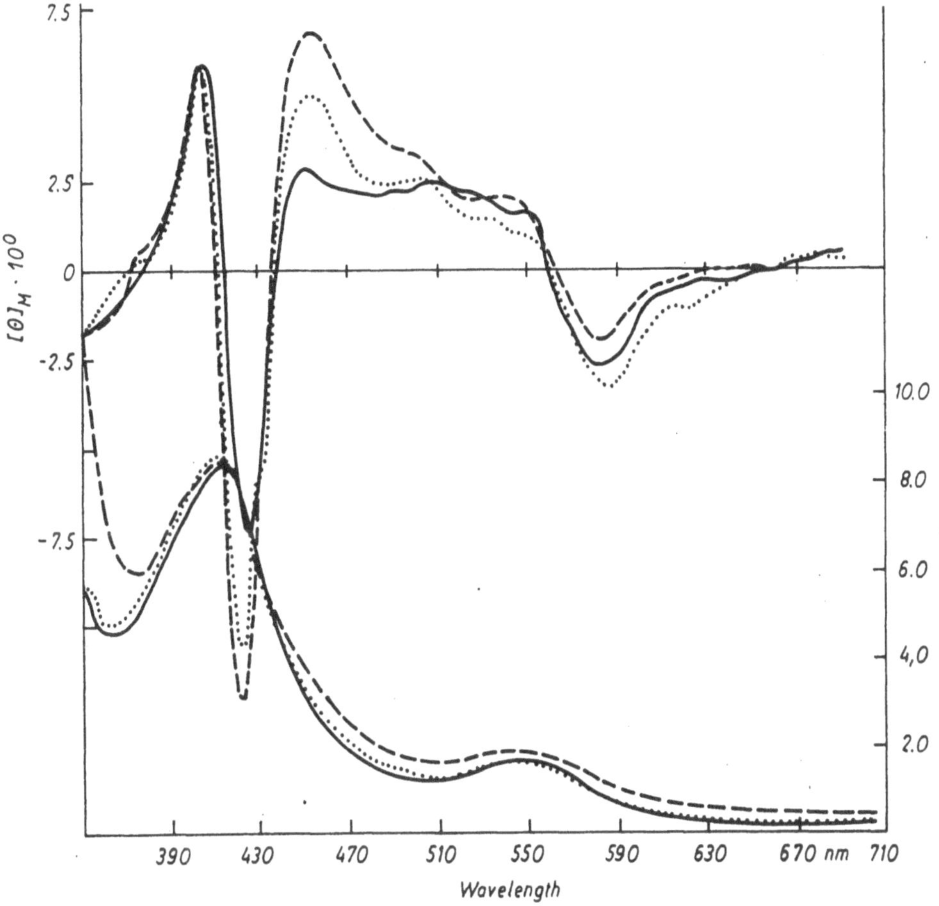

Fig. 14. MCD (top) and absorption (bottom) spectra of reduced phenobarbital induced P–450 LM2 (———), phenobarbital induced P–450 LM3:4 (....) and β-naphthoflavone induced P–450 LM4 (-----), taken from DAWSON et al. (1978); reproduced with permission from Amer. Chem. Soc.

zation the *UV* band exhibits a MCD signal different from an *A*-term which should be expected for *x,y*-polarized transitions. This deviation from the theoretical expectation may be explained by a superposition of *A*- and *B*-terms or by splitting of the degenerated components of the *UV* band. Because of its charge transfer character the *UV* band is very sensitive to symmetry distortions. For example inequivalence of p_x and p_y orbitals of the suggested axial thiolate ligation of the heme iron could be the reason for such distortion (see 2.1.1.).

In contrast to the theoretical expectation which excludes *C*-term for diamagnetic complexes, in CO-P–450 a weak temperature dependence of the MCD signal in the Soret region was found (SHIMIZU et al., 1975) (Fig. 15). Such temperature dependence revealing a sharpening of the *A*-term was also observed for the CO complex of myoglobin (VICKERY et al., 1976b) and may be explained by a temperature-dependent sharpening of the absorption band. Contrary to the CO complex of myoglobin the low-temperature MCD spectrum of CO-P–450 exhibits a new peak at 441 nm. Such new band, however, is not indicated in the low-

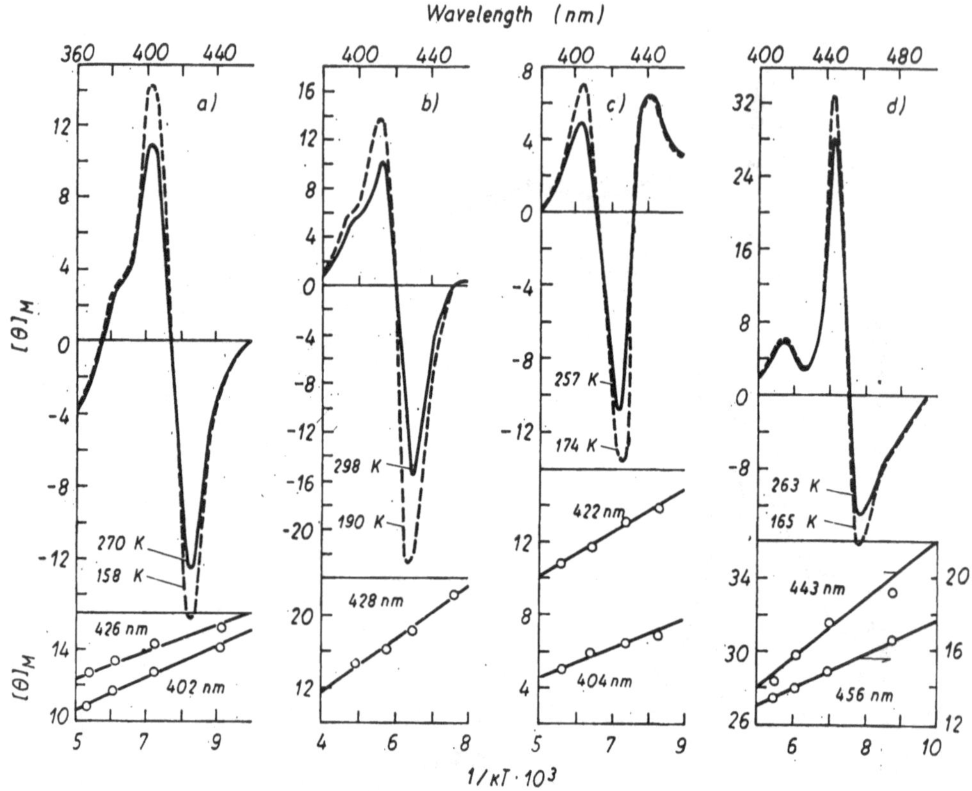

Fig. 15. Effect of temperature on the Soret MCD spectra of oxidized phenobarbital induced P–450 LM (*a*), metmyoglobin-N_3 (*b*), reduced phenobarbital induced P–450 LM (*c*) and its CO complex (*d*), taken from Shimizu et al. (1975); reproduced with permission from Amer. Chem. Soc.

temperature absorption spectrum. The origin of this temperature dependence of the MCD spectrum is not yet clear.

Until now MCD measurements of the dioxygen complex of P–450 were carried out only for P–450 CAM (Dawson et al., 1976). As shown in Fig. 17, the MCD spectrum of O_2-P–450 CAM complex is very different from that of oxymyoglobin. This difference may be explained by several differences in the charge transfer character of the electronic transitions in the Soret and *UV*- band region and by the larger split of the *x*- and *y*-polarized components of the Soret transition in O_2-P–450 (630 cm^{-1}) compared to oxyhemoglobin (100 cm^{-1}) (Jung, 1980).

Many efforts were made to simulate absorption and MCD spectra of P–450 by means of model complexes in order to get information about the nature of the fifth heme iron ligand. The MCD spectrum of oxidized camphor bound P–450 CAM could be very well simulated by the model complex Fe(III)-protoporphyrin IX dimethylester with $(SC_6H_4\text{—}p\text{—}NO_2)^-$ as axial ligand (Dawson et al., 1976) supporting the cysteinate-iron binding in P–450. Collman et al. (1976a) studied model complexes of CO-P–450 with MCD and compared the data with the CO compound of the natural enzyme. Almost identical MCD and absorption spectra of ferrous porphyrins with sodium methyl mercaptide and CO in benzene as compared to those of CO·P–450 also indicate a mercaptide of a cysteine in the fifth coordination position of the

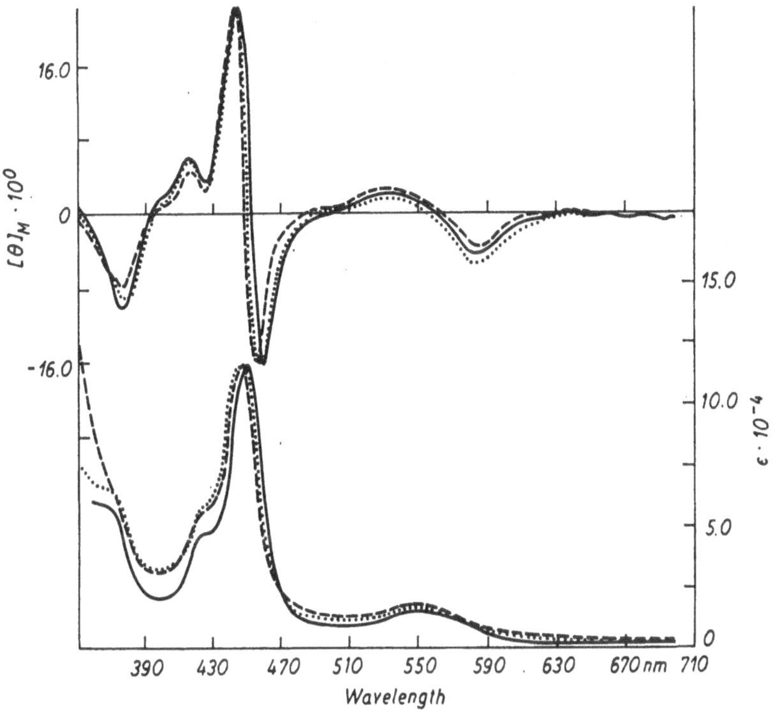

Fig. 16. MCD (top) and absorption (bottom) spectra of CO complexes of phenobarbital induced P–450 LM2 (———), phenobarbital induced P–450 LM3:4 (....) and β-naphthoflavone induced P–450 LM4 (–––), taken from DAWSON et al. (1978); reproduced with permission from Amer. Chem. Soc.

heme iron in P–450. This conclusion of a cysteinate-iron binding in P–450 is in accordance with results obtained by other methods like EPR, Resonance Raman, Mößbauer and Extended X-ray Absorption Fine Structure Spectroscopy (see 2.1.4 and 2.2).

Summarizing the MCD results it can be concluded that the fourfold symmetry of the iron porphyrin complex in ferric and ferrous low spin P–450 is only slightly disturbed. This fact is reflected in the appearance of clearly resolved A-terms of the Soret and Q-bands. This conclusion seems to be in contradiction with the *CD* measurements which reveal a perturbation of the fourfold symmetry (see 2.1.2.). Considering, however, that the MCD intensity of heme transitions is about 10^4 times larger than the *CD* intensity, it is not surprising that perturbations of symmetry indicated by the *CD* signals are not detectable with the MCD method. RISTAU and REIN (1976) have shown that a 100 cm^{-1} splitting of degenerate energy levels is still sufficient to produce A-terms.

Comparing the MCD spectra of P–450 with those of hemoglobin or myoglobin, ferrous P–450 shows some peculiarities. These peculiarities can be explained by the supposed mercaptide ligand in the fifth heme iron coordination site which induces a larger π-back-donation from the axial π-system to the porphyrin-π-system than the imidazole ligand in hemoglobin or myoglobin (JUNG, 1980). The assumption of a higher π-back-donation to the porphyrin-π-system in the reduced P–450 is experimentally supported by Resonance Raman measurements (see 2.1.4.).

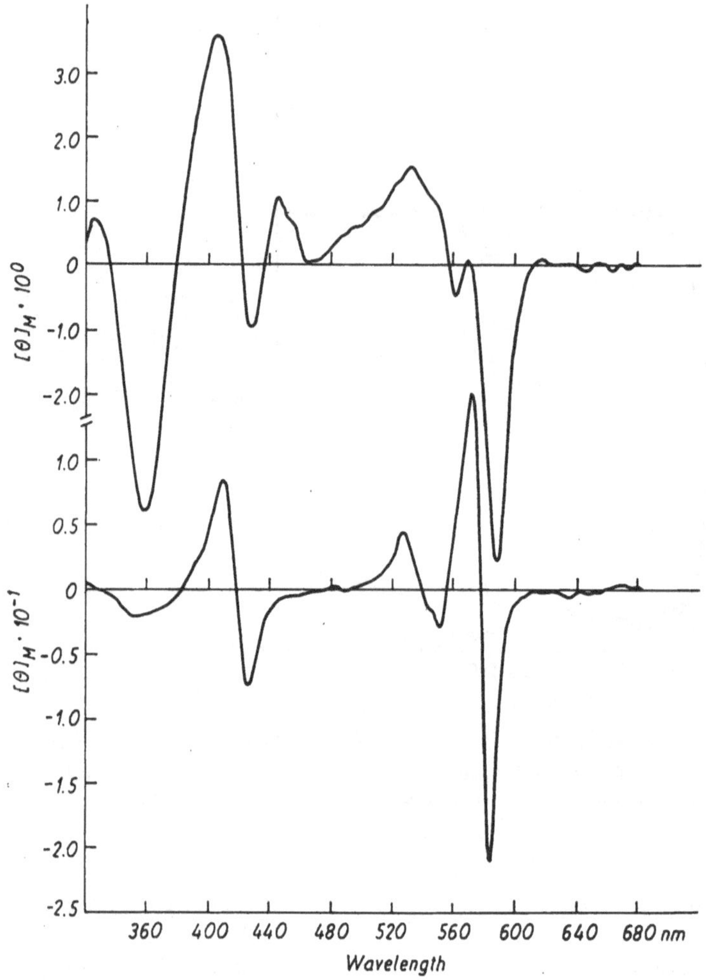

Fig. 17. MCD spectra of O_2-P–450 CAM and oxymyoglobin, taken from Dawson and Cramer (1978); reproduced with permission from Elsevier Biomedical

2.1.4. Resonance Raman Spectra

Compared with Raman spectroscopy the Resonance Raman method offers the advantage of an intensity enhancement of certain Raman lines if a compound is irradiated by a laser beam with a frequency near the excitation energy of an electronic transition. In this case the equilibrium conformation of the molecule is distorted along the normal coordinate of the Raman line during transition from the electronic ground state to the electronic excited state (Hirakawa and Tsuboi, 1975).

There are two important considerations with respect to Resonance Raman spectra: (i) The peak positions are a property of the vibrational states of the electronic ground state of the compound. (ii) The peak intensities are a property of the electronic ground and excited states. According to these two considerations the Resonance Raman investigations of hemoproteins and other porphyrin systems have been classified. The frequency dependent Resonance Raman intensity exhibits two kinds of scattering mechanism corresponding to the so-called *A*- and *B*-terms (Albrecht, 1961) in the Raman tensor expansion. A-terms involve vibrational coupling to a single electronic transition. Only totally symmetric

Table 2: Resonance Raman peak positions of P–450 compared with other hemoproteins (cm^{-1})

Compound	Oxidation state marker		Spin state marker		Oxidation and spin state marker		References
	A(*p*)	*B*(*dp*)	*C*(*ap*)	*D*(*p*)	*E*(*p*)	*F*(*dp*)	
Ferric low spin							
P–450_{CAM}	1372	—	—	—	1502	1635	Champion et al., 1978
P–450_{LM} (PB induced)	1371	—	—	—	—	—	Ozaki et al., 1978
chloroperoxidase-CN^-	1373	—	—	—	1503	1633	Remba et al., 1979
cytochrome *c*	1374	1562	1582	1582	1502	1636	Spiro and Strekas, 1974a
myoglobin-CN^-	1374	1564	1588	1583	1508	1642	Spiro and Strekas, 1974a
Ferric high spin							
P–450_{CAM}	1368 1372	—	—	—	1488 1502	1623	Champion et al., 1978
P–448 (PB-induced)	1371	—	—	—	—	1629	Ozaki et al., 1978
P–448 (MC-induced)	1371	—	—	—	—	1630	Ozaki et al., 1978
chloroperoxidase	1369	—	—	—	1490	1627	Remba et al., 1979
cytochrome *c′*	1372	—	—	—	1500	1637	Spiro and Strekas, 1974b
myoglobin-F^-	1373	1565	1555	1565	1482	1608	Spiro and Strekas, 1974a
horseradish peroxidase	1375	—	—	—	1500	1630	Rakshit and Spiro, 1974
P–420_{CAM}	1372	—	—	—	—	1627	Champion et al., 1978
Ferrous high spin							
P–450_{CAM}	1344	—	—	—	1466	1601	Champion et al., 1978
P–450_{LM} (PB-induced)	1341	—	—	—	—	—	Ozaki et al., 1978
P–448_{LM} (PB-induced)	1347	—	—	—	—	—	Ozaki et al., 1978
P–448_{LM} (MC-induced)	1346	—	—	—	—	—	Ozaki et al., 1978
chloroperoxidase	1348	—	—	—	1470	1612	Remba et al., 1979
cytochrome *c′*	1355	—	—	—	1475	1609	Spiro and Strekas, 1974b
horseradish peroxidase	1358	—	—	—	1472	1605	Rakshit and Strekas, 1974
deoxyhemoglobin	1358	1546	1552	1565	1473	1607	Champion et al., 1978
P–420_{CAM}	1360	—	—	—	—	—	Champion et al., 1978
P–420_{LM} (PB-induced)	1359	—	—	—	—	—	Ozaki et al., 1978
Ferrous low spin							
CO–P–450_{CAM}	1368	—	—	—	—	—	Champion et al., 1978
CO–P–450_{LM} (PB-induced)	1365	—	—	—	—	—	Ozaki et al., 1978
CO–P–448_{LM} (PB-induced)	1365	—	—	—	—	—	Ozaki et al., 1978

Table 2 (continued)

Compound	Oxidation state marker		Spin state marker		Oxidation and spin state marker		References
	$A(p)$	$B(dp)$	$C(ap)$	$D(p)$	$E(p)$	$F(dp)$	
$CO-P-448_{LM}$	1365	—	—	—	—	—	OZAKI et al., 1978
CO-chloroperoxidase	1370	—	—	—	—	—	SPIRO and STREKAS, 1974a
CO-hemoglobin	1372	—	—	—	—	—	RIMAI et al., 1975
$CO-P-420_{LM}$ (PB-induced)	1371	—	—	—	—	—	OZAKI et al., 1978
oxyhemoglobin	1377	1564	1586	1582	1506	1640	SPIRO and STREKAS, 1974a
cytochrome *c*	1362	1548	1584	1594	1483	1620	SPIRO and STREKAS, 1974a

PB = phenobarbital; MC = 3-methylcholanthrene; p = polarized; dp = depolarized; ap = anomalously polarized

vibrations (A_{1g}) can contribute to this term. B-terms include vibronic mixing of two excited electronic states. For the B-terms vibrations of such symmetry are allowed which are contained in the direct product representation of the two electronic transitions (A_{1g}, A_{2g}, B_{1g} and B_{2g}). The symmetry of the vibrations which correspond to Raman lines can be deduced from polarization measurements of the scattered light. A_{1g} is polarized, B_{1g} and B_{2g} are depolarized and A_{2g} is anomalously polarized (SPIRO, 1975). Vibrations of other symmetries of the D_{4h} group, for example E_u, are IR active (OGOSHI et al., 1972).

Most Resonance Raman investigations of hemoproteins including P-450 were performed in order to get informations about the ground state properties meaning the charge distribution and the interaction between the iron and the axial ligand and the porphyrin ring. Three classes of Raman lines related with the oxidation and spin state of the metal have been demonstrated (Table 2) (SPIRO, 1975). The oxidation state marker A lies at about 1375 cm^{-1} and 1360 cm^{-1} for Fe(III)- and Fe(II)-complexes, respectively, and its amplitude is enhanced by excitation into the Soret band. The spin state marker C is enhanced by exciting into the Q-band and shows values at 1555 cm^{-1} and 1585 cm^{-1} for the high and low spin state, respectively. A assignment of these markers to special vibrations was carried out using a normal coordinate analysis (STEIN et al., 1975); it is suggested that the oxidation state marker A corresponds to the totally symmetrical breathing vibration (A_{1g}) of the peripheral cycle of the porphyrin. However, experiments with ^{15}N enriched metal octaethylporphyrin (KITAGAWA et al., 1977, 1978) and a new calculation (ABE et al., 1976, 1978) indicate that the oxidation state marker A is basically a breathing mode of the pyrrol rings with an appreciable contribution from the C—N symmetric vibration, associated with the inphase displacement of the four pyrrol nitrogen atoms toward the metal ion. The spin state marker C corresponds to the A_{2g} stretch vibration of the methene bridges (STEIN et al., 1975). The Raman lines corresponding to the metal ligand stretch vibrations are observed in the low frequency region between 200 and 700 cm^{-1} which are especially enhanced by excitation into the Soret band (KITAGAWA et al., 1976) (Table 3). A significant deviation of the oxidation state marker A from the normal value of 1360 cm^{-1} for Fe(II)-

Table 3: Low-frequency vibrations in resonance Raman spectra (cm^{-1})

Stretching mode	Fe—$N_{pyrrole}$	Fe—5th ligand	Fe—6th ligand	References
myoglobin (III)				
$-F^-$	252	416	462	Desbois et al., 1979
$-H_2O$	249	411	—	Desbois et al., 1979
$-OH^-$	255	413	490	Desbois et al., 1979
myoglobin (II)				
deoxy*	222	408	—	Desbois et al., 1979
		220	—	Kitagawa et al., 1979
$-O_2$	218	409	577	Desbois et al., 1979
$-NO$	218	409	—	Desbois et al., 1979
hemoglobin-NO	—	—	533	Stong et al., 1980
cytochrome b_5 reduced	—	277	277	Kitagawa et al., 1979
Fe(II)-protoporphyrin IX-diimidazole complex	—	299	299	Verma and Bernstein, 1974
Fe(III)-protoporphyrin IX-diimidazole complex	—	276	276	Verma and Bernstein, 1974

* There are contradicting assignments of the low-frequency mode at 220 cm^{-1}: Desbois et al. (1979) assigned this line to the Fe-$N_{pyrrole}$ stretching mode, Kitagawa et al. (1979), however, assigned this line to the Fe—N (His F8) stretching mode.

complexes is observed for oxyhemoglobin and carboxyhemoglobin. The frequencies of 1377 cm^{-1} and 1372 cm^{-1} for oxyhemoglobin and carboxyhemoglobin, respectively, are similar to those of iron(III) complexes and support the Weiss' model (Fe^{3+}—O_2^-) (Weiss, 1964). The higher value for oxyhemoglobin is explicable by the stronger electron-withdrawing of the dioxygen ligand as compared with the CO ligand, resulting in a lower negative net charge of the porphyrin ligand in oxyhemoglobin. Quantum chemical calculations (π-INDO method) of the charge distribution in heme iron complexes with different axial ligands (Jung, 1980) show that the negative net charge of the porphyrin ligand correlates with the experimental frequency of the oxidation state marker *A* (Figure 18). Therefore, the low frequencies of this marker found at 1365 cm^{-1} and 1368 cm^{-1} for rabbit CO-P-450 (Ozaki et al., 1978) and CO-P-450 CAM (Champion et al. 1978), respectively, corresponds to a higher negative net charge of the

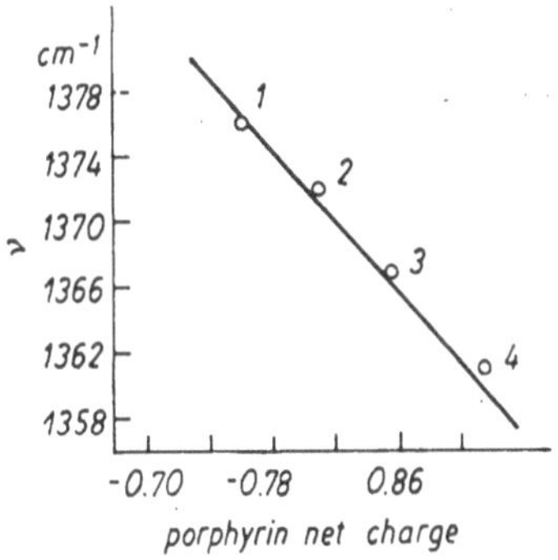

Fig. 18. Correlation between the π-INDO calculated net charge of the porphyrin ligand (Jung, 1980) and the experimental frequency of the Resonance Raman "oxidation state marker" vibration. The experimental values are taken for 1, oxyhemoglobin from Spiro and Strekas (1974a); for 2, CO-hemoglobin from Rimai et al. (1975); for 3, CO-P-450 LM from Ozaki et al. (1978); and for 4, cytochrome b_5 from Ozaki et al. (1976a)

Table 4: Infrared spectral parameters for ferrous complexes of cytochromes P-450, P-448, P-420, hemoglobin and model complexes

Cytochrome, species, pretreatment; models	ν_{CO}, cm^{-1}	$\Delta\nu_{1/2}$, cm^{-1}		ε_M, $M^{-1}cm^{-1} \cdot 10^{-3}$		References
		H_2O	D_2O	H_2O	D_2O	
P-450 LM, rabbit (PB)	1949	15–16	14	1.9	1.67	Böhm et al., 1976
P-420 LM, rabbit (PB)	1966	~20				
P-448 LM, rabbit (3-MC)	1954 1961* (shoulder)	~15	13–14	1.9	1.74	Böhm et al., 1979
P-420 LM, rabbit (3-MC)	1972**					
P-450 LM, rat (PB)	1948; 1948***	~25***	~20	—	2.15	
P-420 LM, rat (PB)	1966					
P-448 LM, rat (3-MC)	1952; 1954***	~30***	~21	—	1.77	
P-420 LM, rat (3-MC)	1970					
P-450 CAM, Pseud. putida (camphor-bound)	1940	13		2.3		O'Keeffe et al., 1978a
P-450 CAM, Pseud. putida (camphor-free)	1963, 1942	11–12, 19–21				
hemoglobin A	1951	8		3.7 ± 0.2		Yoshikawa et al., 1977
denatured hemoglobin A	1966	20				
N-methylimidazole-heme solvent:						
CCl_4 (D = 0)****	1980	18				Maxwell and Caughey, 1976
$CHCl_3$ (D = 1.1)	1969	29				
$(ClCH_2)_2$ (D = 1.7)	1959	26				
1-methylimidazole (D = 3.7)	1954	24				
Fe^{II}(TpivPP)-N-methylimidazole solvent: benzene	1964 1969					Collman et al., 1976a Collman et al., 1976b
Fe^{II}(DEPP)-N-methylimidazole	1970					Wang et al., 1958; Rougee and Brault, 1975
Fe^{II}(PPDME)-imidazolate solvent: dimethylsulfoxide	1935					Mincey and Traylor, 1979
Fe^{II}(TpivPP)-n-C_3H_7SH solvent: benzene	1970					Collman et al., 1976a
Fe^{II}(TpivPP)-Na SCH_3 solvent: benzene	1945					

Table 4 (continued 1)

Cytochrome, species, pretreatment; models	ν_{CO}, cm^{-1}	$\Delta\nu_{1/2}$, cm^{-1}		ε_M, $M^{-1}cm^{-1}\cdot10^{-3}$		References
		H_2O	D_2O	H_2O	D_2O	
Fe^{II}-heme-mercaptide	1923					CHANG et al., 1976a
Fe^{II}-heme-mercaptide solvent: chlorobenzene	1920					CARON et al., 1979

* The shoulder at 1961 cm^{-1} (H_2O) is clearly recognized as band in D_2O. ** The band at 1972 cm^{-1} of the thermally denatured P–448 from rabbit appears at the beginning of the P–420 formation. At higher levels of the thermally induced denaturation two bands appear at about 1970 and 1966 cm^{-1} indicating two different P–420 forms. *** Values of cytochrome P–450/P–448 microsomes measured from O'KEEFFE et al., 1978a. **** Dipole moment of the solvent in Debey. Abbreviations: (PB) = phenobarbital induced; (3-MC) = 3-methylcholanthrene induced; TpivPP = mesotetrapivalamidophenylporphyrin; DEPP = diethylprotoporphyrin; PPDME = protoporphyrindimethylester; Na SCH$_3$ = sodium methylmercaptide-Crown-ether complex.

porphyrin in P–450 in comparison with hemoglobin (Table 4). The reason for the higher negative charge of the porphyrin in P–450 can be seen in the postulated mercaptide as the 5th heme iron ligand. This ligand induces a stronger π-back-donation from the iron to the porphyrin than imidazole. The conversion of CO-P–450 to CO-P–420 is connected with a shift of the oxidation state marker to 1371 cm^{-1} (OZAKI et al., 1978) suggesting a substitution of the mercaptide ligand by the imidazole of a neighbouring histidine (STERN et al., 1973; REIN et al., 1977b) or by a breakdwon of the hydrogen bond between the iron-coordinated mercaptide and a nitrogen base forming a mercaptane as suggested from quantum chemical calculations (JUNG, 1980).

So far, no Resonance Raman measurements for O_2-P-450 are reported. From the mentioned correlation between the quantum chemically calculated porphyrin net charges and the experimental frequencies of the oxidation state marker a similar value to that of CO-hemoglobin is expected (see Fig. 18).

Reduced high-spin P–450 shows a similarly strongly lowered frequency for the oxidation state marker in the range between 1341–1347 cm^{-1} (Table 2 and Fig. 19). The relative shift to 1358 cm^{-1} for deoxyhemoglobin are apparently caused by different axial ligands of the 5-coordinated heme iron complexes, namely imidazole in hemoglobin and probably a mercaptide in P–450.

The conversion of ferrous high spin P–450 into the denatured form P–420 in the absence of carbon monoxide yields a change of the oxidation state maker to a normal value of 1360 cm^{-1} for P–420 CAM (CHAMPION et al., 1978) and 1359 cm^{-1} for P–420 from phenobarbital-induced rabbit liver (OZAKI et al., 1978).

In contrast to the ferrous state, ferric low-spin P–450 exhibits a value for the oxidation state marker similar to that of ferric hemoglobin and myoglobin (P–450 CAM without camphor: 1373 cm^{-1} (OZAKI et al., 1976a) and 1372 cm^{-1} (CHAMPION et al., 1978); highly purified P–450 from phenobarbital-induced rabbit liver: 1371 cm^{-1} (OZAKI et al., 1978)). The frequency of the oxidation state marker of the camphor bound P–450 CAM depends strongly on the wavelength of the exciting laser beam. The apparent shift of the marker from 1368 to 1372 cm^{-1} if the laser excitation is moved from the

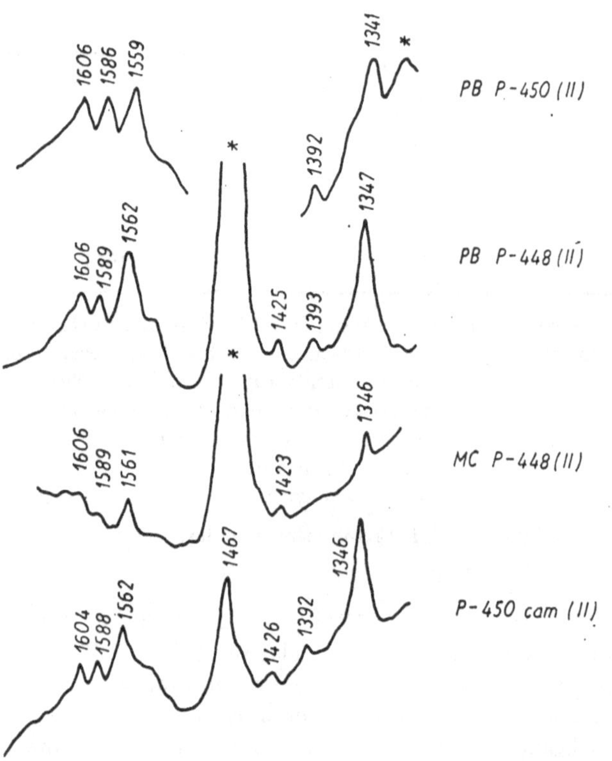

Fig. 19. Resonance Raman spectra of reduced P–450, taken from OZAKI et al. (1978). PB P–450 (II) and PB P–448 (II), purified reduced forms from phenobarbital induced rabbit liver; MC P–448 (II), purified reduced form from 3-MC induced rabbit liver. The intensive RAMAN line seen around 1465 cm^{-1} (*) is due to glycerol present in the solvent; reproduced with permission from Amer. Chem. Soc.

the high-energy side of the Soret band to the low-energy side therefore reflects a superposition of the high-spin spectrum with the spectrum of a small part of low-spin-P–450 CAM (CHAMPION et al., 1978). Additionally it is interesting to note that ferric myoglobin shows a similar change of the marker from 1370 to 1373 cm^{-1} if the spin state is changed from high to low by an increase of the pH-value from 7 to 11.6 (OZAKI et al., 1976b). A selective intensity enhancement by exciting into the red-side of the Soret band of P–450 CAM (camphor-bound) is also observed for the low-frequency mode of A_{1g} symmetry at 676 cm^{-1} (Fig. 20) (CHAMPION et al., 1978). On excitation into the high-energy side of the Soret band a Raman line at 351 cm^{-1} appears and the 676 cm^{-1} mode vanishes. The simultaneous selective enhancement of the intensity of oxidation state marker and the low-frequency Raman lines indicates a strong participation of charge-transfer between the axial π-system and the porphyrin-π-system during the excitation into the Soret band.

2.1.5. Infrared Studies of the Carbonyl Complex of Cytochrome P–450

It is well known that the binding of carbon monoxide to the heme iron is accompanied by a decrease of the frequency of the C—O stretch vibration from 2143 cm^{-1} for the unbound CO (COTTON and WILKINSON, 1967) to 1920—1970 cm^{-1} for the

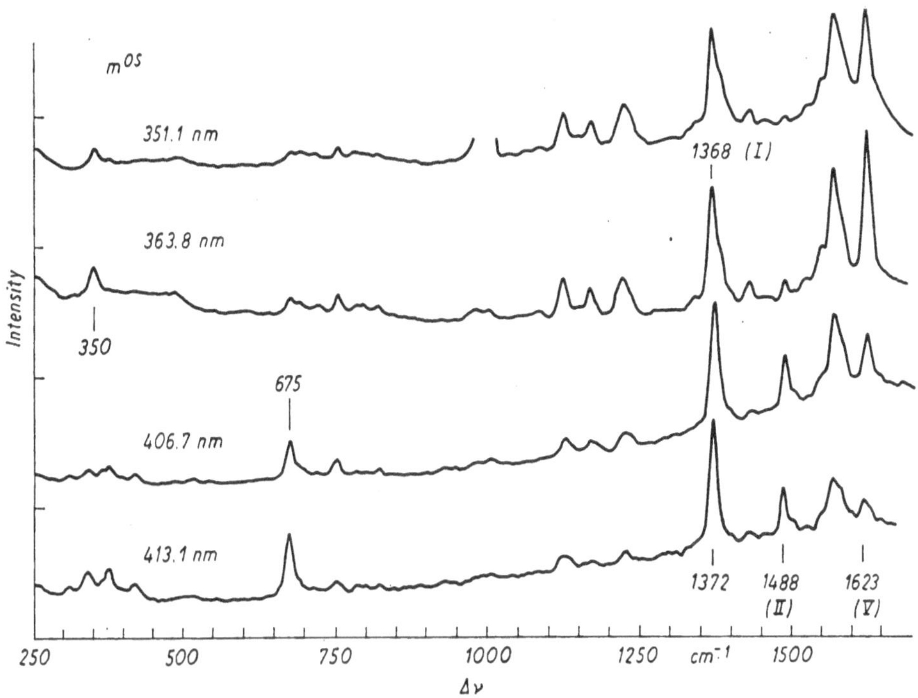

Fig. 20. Resonance Raman spectra of camphor bound oxidized P–450 CAM showing the dependence of Raman lines on the frequence of the exciting laser beam, taken from CHAMPION et al. (1978); reproduced with permission from Amer. Chem. Soc.

bound CO. The decrease of this vibration frequency indicates a loosening of the C—O-bond caused by a decrease of the C—O π-bond order. Two factors determine the degree of this loosening: (i) the nature of the axial iron ligand trans to CO and (ii) the nature of the environment of the heme-bound CO. For isolated heme it was observed that increase in negative charge of the fifth heme iron ligand produces an enhancement of the π-backdonation from the iron to the CO ligand which is reflected in a decrease of the frequency for the CO stretch vibration and of the π bond order of CO (Table 4). So, the substitution of the imidazole ligand by imidazolate shifts the frequency from 1964 cm^{-1} (COLLMAN et al., 1976a) to 1935 cm^{-1} (MINCEY and TRAYLOR, 1979). Deprotonation of the mercaptane ligand to mercaptide alters the frequency from 1970 cm^{-1} to 1945 cm^{-1} (COLLMAN et al., 1976). CHANG et al. (1976b) found a value of 1923 cm^{-1} for the mercaptide-heme-CO complex. It is important to note that the mercaptide ligand induces a lower frequency than the imidazole ligand (Table 4). The mercaptide-heme-CO complex is regarded as model for the prosthetic group of CO-P–450 derived from many studies using optical, MCD, and Mössbauer methods. However, in contrast to this low frequency for the model complex the CO stretch vibration frequency of the carbonyl complex of P–450 is not noticable changed as compared with the carbonyl complex of hemoglobin showing a value at 1951 cm^{-1} (ALBEN and CAUGHEY, 1968; YOSHIKAWA et al., 1977). The frequency for CO-P–450 LM from phenobarbital-induced rabbit liver is only 2 cm^{-1} lower and for CO-P–450 LM from 3 MC-induced rabbit liver is even 3 cm^{-1} higher than the value of CO-hemoglobin

(Table 4) (Böhm et al., 1979). Different CO stretch vibration frequencies are not only observed with CO-P–450 complexes from differently induced material but also if different species are compared. The frequencies of CO-complexes of 3-MC induced P–450 LM from rabbits is centered at 1954 cm^{-1} but from rat liver at 1952 cm^{-1} (Böhm et al., 1979). CO-P–450 CAM in the absence and presence of camphor exhibits a stretch vibration frequency of 1963 cm^{-1} and 1940 cm^{-1}, respectively (O'Keeffe et al., 1978a). These results show that infrared spectra of CO complexes of P–450 reflect very sensitively the environment of the heme and indicate different proteins in dependence on inducer and species.

The denaturation of phenobarbital-induced rabbit and rat P–450 LM, i.e. the conversion to P–420 LM produced by heating or by deoxycholate (Böhm et al., 1976; Rein et al., 1977b) is accompanied by an increase of the stretch vibration frequency to 1966 cm^{-1} which is identical with that of acid-denatured carboxyhemoglobin (Caughey et al., 1973). The frequency of the denaturated form is practically the same as that of CO-heme with imidazole coordinated as 5th ligand (Collman et al., 1976a). This finding indicates a linear binding structure of the CO molecule in denatured P–450 as it was found by X-ray measurements for a model of the prosthetic group — the CO-complex of the "Picket fence" porphyrin (Collman et al., 1975a). From these similarities a substitution of the mercaptide ligand in CO-P–450 by the imidazole residue of a histidine during the conversion to CO-P–420 is suggested (Rein et al., 1977b). In contrast to the CO complex of phenobarbital-induced P–420 LM, the 3-MC-induced CO-P–450 LM exhibits a higher frequency at 1970 cm^{-1} which is identical with the mercaptan model complex, suggesting a mercaptan ligand trans to CO at this denatured P–450.

A comparison with the already mentioned model complexes and CO-hemoglobin may be useful to explain the observed value of the stretch vibration frequency for CO-P–450. The lower frequency for CO-hemoglobin in comparison to its model complexes indicates a loosening of the CO bond. This loosening is caused by deviation of the Fe—C—O arrangement from the linear structure induced by interaction between CO and the protein. A bent Fe—C—O structure in CO-hemoglobin is proved by X-ray studies (Heidner et al., 1976). Assuming mercaptide sulfur as fifth ligand of the heme iron in P–450 and linear Fe—C—O arrangement a value for the vibration frequency in CO-P–450 of 1923—1945 cm^{-1} should be observed similar to the model complexes. A deviation from the linear Fe—C—O structure should reveal a further lowering of the frequency. This expected lower frequency, however, is not observed in the experiment which yields values between 1949 cm^{-1} and 1963 cm^{-1}. A weaker binding of the CO molecule to the heme iron in P–450 compared with hemoglobin and an almost linear Fe—C—O arrangement in P–450 as well may explain this apparent contradiction. The last conclusion is supported by the large line width of the CO band of CO-P–450 being about 13—21 cm^{-1} (see Table 4) which is very similar to denatured CO complexes of P–450 and hemoglobin and very different from native CO-hemoglobin, showing a value of only 8 cm^{-1} (Alben and Caughey, 1968). This small line width indicates a rigid environment of the bound CO in hemoglobin. The large line width of the CO band observed at CO-P–450, however, suggests a flexible protein or solvent environment of the CO ligand in P–450.

2.2. Magnetic Properties

2.2.1. Electron Paramagnetic Resonance Spectra, Axial Ligands and Models

The EPR spectrum of ferric P–450 was first described in 1962 (HASHIMOTO et al., 1962). The unusual g-values near 1.9, 2.2 and 2.4 found in rabbit liver microsomes differ from those of other metalloproteins and were suggested to belong to an unknown iron-containing protein called Fe_x. Later the identity of the Fe_x with ferric P–450 in its low spin state could be demonstrated (MURAKAMI and MASON, 1967). Since that time these g-values have been used as characteristic EPR spectral property of ferric P–450. They have been found also in non-mammalian organisms such as bacteria (GUNSALUS, 1968), plants (RICH et al., 1975), fishes (CHEVION et al., 1977a), and birds (GHAZARIAN et al., 1978) (Table 5).

In 1970, TSAI et al. discovered in the EPR spectrum of the substrate-bound soluble P–450 CAM the signals of the high spin iron with g-values near 8, 3.8, and 1.8. The corresponding signals of microsomal P–450 LM of rabbit pretreated with 3-MC were found in the same year (PEISACH and BLUMBERG, 1970), those of rats and mice in 1973 (PEISACH et al., 1973; STERN et al., 1973; NEBERT and KON, 1973). The g-values of the ferric high spin P–450 indicate a considerably larger deviation from the axial ligand symmetry than it is known from other hemoproteins (PEISACH et al., 1971) (Fig. 21).

Among the mammalian P–450 LM preparations those from rabbits pretreated with polycyclic hydrocarbons show especially intensive high spin signals. The successful isolation of the isoenzymes P–450 LM2 and P–450 LM4 in the laboratory of Coon (HAUGEN et al., 1975; HAUGEN and COON, 1976) proved the high spin signals in the EPR spectrum of rabbit liver microsomes to be caused by P–450 LM4 the heme iron of which is predominantly in a permanent high spin state ((75%) (HAUGEN and COON, 1976)).

Earlier EPR studies of P–450 were mainly performed with microsomes due to difficulties in solubilizing membrane bound P–450. EPR studies in microsomal P–450, however, provide only rough g-values because numerous isozymes of P–450 lead to an inhomogeneous broadening of the EPR signal. Therefore crystal field analysis based on highly accurate g-values seem to be questionable in the case of microsomal P–450. Such attempts were made in order to elucidate the nature of the sixth heme iron ligand of P–450 culminating in the paper of CHEVION et al. (1977b) where the imidazole group of a histidine residue has been proposed to be the sixth ligand. EPR investigations on the microsomal fraction are, on the other hand, well applicable for the determination of the spin state of the heme iron and of paramagnetic contaminations of the preparation, i.e. other hemoproteins and non-heme iron proteins. The latter show an absorption line near $g = 4.2$ (BEINERT and SANDS, 1960). After longer storage of samples containing P–450 at room temperature or after adding hydrogen peroxide to P–450 this signal is also formed, probably due to an oxidative porphyrin cleavage (FRIEDRICH et al., 1979).

Effect of Substrates

The shift of the spin equilibrium of the heme iron of ferric P–450 towards the high spin state at binding of type I substrates can be well observed by optical measurements but can be hardly detected in the EPR spectrum. WATERMAN et al. (1973) found a decrease of the low spin signal after adding cyclohexane to the microsomal sample but did not prove the corresponding increase of the high spin signal. Such an increase has been observed after binding of SKF 525-A

Table 5. *g*-values of cytochrome P–450 from various sources in the presence of several ligands

Low Spin

Species/Tissue*	Inducer**	Ligand	*g*-Values			References
Rat-LM	none	none	2.425	2.254	1.915	STERN et al., 1973
Rat-LM	PB	none	2.417	2.255	1.917	STERN et al., 1973
Rat-LM	MC	none	2.417	2.248	1.912	STERN et al., 1973
Rat-LM	PB	Metyrapone	2.47	2.26	1.90	GRASDALEN et al., 1975
Rat-LM	PB	Aniline	2.482	2.260	1.901	PEISACH & BLUMBERG, 1970
Rat-LM	PB	n-Octanethiol	2.44	2.26	1.91	ULLRICH et al., 1975
Rat-LM	PB	Diphenylphosphine	2.49	2.28	1.88	ULLRICH et al., 1975
Rat-LM	PB	Pentamethylenesulfide	2.48	2.26	1.90	ULLRICH et al., 1975
Rat-LM	PB	Aniline	2.45	2.24	1.90	CHEVION et al., 1977 b
Rat-LM	PB	Cyanide	2.51	2.29	1.83	CHEVION et al., 1977b
Rabbit-LM	PB	none	2.424	2.240	1.923	KUMAKI et al., 1978
Rabbit-LM	BNF	none	2.429	2.250	1.915	KUMAKI et al., 1978
Rabbit-LM	MC	none	2.428	2.252	1.917	KUMAKI et al., 1978
Rabbit-LM	none	none	2.41	2.25	1.91	ICHIKAWA & YAMANO, 1968
Rabbit-LM	none	Ethylisocanide	2.43	2.29	1.87	ICHIKAWA & YAMANO, 1968
Rabbit-LM	PB	Cyanide	2.54	2.28	1.83	REIN et al., 1977a
Mouse-LM	PB	none	2.413	2.245	1.918	CHEVION et al., 1977 b
Guinea pig-LM	PB	none	2.422	2.251	1.923	KUMAKI et al., 1978
Guinea pig-LM	MC	none	2.414	2.251	1.923	KUMAKI et al., 1978
Tulip bulb microsomes		none	2.40	2.25	1.91	RICH et al., 1975
Stenotomus versicolor-LM		none	2.411	2.249	1.920	CHEVION et al., 1977 a
P–450 LM2	PB	none	2.43	2.25	1.93	HAUGEN & COON, 1976

High Spin

Source/Tissue*	Inducer**	*g*-Values			Reference
Rat-LM	MC	7.9	3.7	≈1.7	STERN et al., 1973
Rabbit-LM	none	8.00	3.70	1.7	KUMAKI et al., 1978
Rabbit-LM	PB	7.94	3.78	1.74	KUMAKI et al., 1978
Rabbit-LM	BNF	7.96	3.76	1.74	KUMAKI et al., 1978
Rabbit-LM	MC	7.95	3.76	1.74	KUMAKI et al., 1978
Guinea pig-LM	none	7.95	≈3.8	≈1.7	KUMAKI et al., 1978
Guinea pig-LM	PB	7.95	≈3.8	≈1.7	KUMAKI et al., 1978
Guinea pig-LM	BNF	7.93	3.78	≈1.7	KUMAKI et al., 1978
Guinea pig-LM	MC	7.92	3.80	1.74	KUMAKI et al., 1978
Tulip bulb microsomes		8.3	?	?	RICH et al., 1975
P–450 CAM + camphor		8	4	1.8	TSAI et al., 1970
P–450 LM4	BNF	8.36	3.84	?	HAUGEN & COON, 1976

* LM = Liver microsomes

** PB = phenobarbital, BNF = β-naphthoflavone, MC = 3-methylcholanthrene

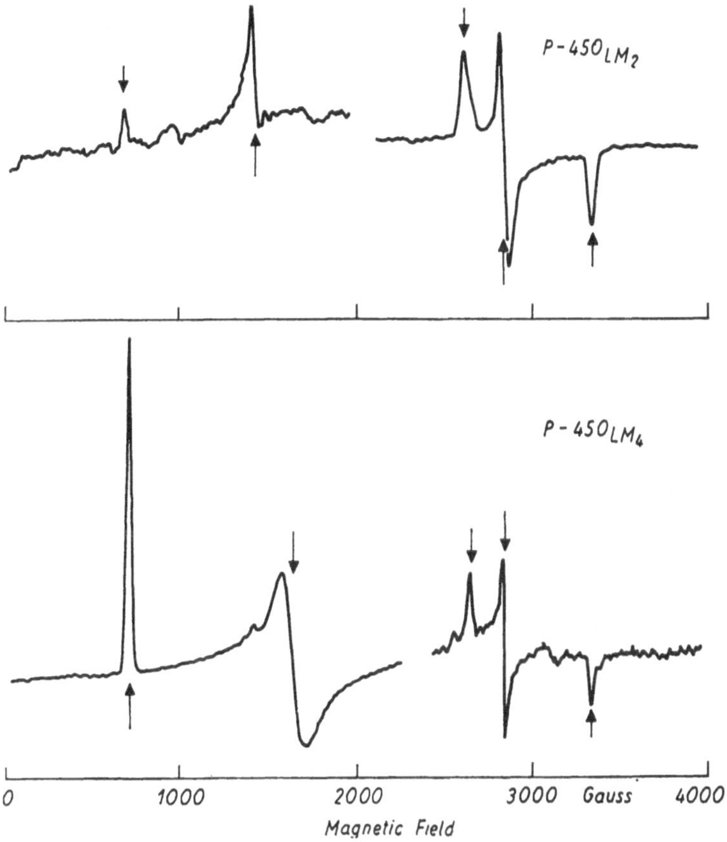

Fig. 21. EPR spectra of P-450 LM2 and P-450 LM4. Temperature, 7 K and 20 K for the low field and high field region, respectively. g-values see Tab. 5. P-450 LM2 was 0.15 mM in 0.03 M phosphate buffer, pH 7.4, containing 20% glycerol and 1.0 mM EDTA and P-450 LM4 was 0.13 mM in a similar mixture containing 0.1 M phosphate, taken from HAUGEN and COON (1976); reproduced with permission from the Amer. Soc. Biol. Chem., Inc.

to P-450 in rat liver microsomes (GRASDALEN et al., 1975). KUMAKI et al. (1978) report also an increase of the high spin signal at $g = 8$ connected with a decrease of the low spin signal with increasing amounts of cyclohexane at rat and mouse liver microsomes correlated with an increase of the amplitude of the optical difference spectrum. But other authors did not find enhanced high spin signals in the presence of type I substrates (EBEL et al., 1978). This failure was explained by substrate dissociation from P-450 caused by its decreased affinity at low temperatures typical for type I substrates (MISSELWITZ et al., 1976, 1977). Using a freeze-quenching technique to avoid substrate dissociation, EBEL et al. (1978) observed increased high spin signals in the presence of hexobarbital. The lack of high spin signals may also be explained by the existence of a temperature dependent spin equilibrium of ferric P-450 favouring the low spin state with lower temperatures (REIN et al., 1977c).

In contrast to type I substrates, type II substrates known to cause low spin shifts of the P-450 heme iron lead to a disappearance of the high spin signals and to a change of the low spin signals, characterized by an increased distance between the two outer low spin peaks (see Table 5).

Because type II substrates belong to nitrogeneous bases known as heme iron ligands it has been assumed that type II substrates act as heme iron ligands also in P–450 (ORRENIUS et al., 1972). This assumption includes the displacement of the sixth ligand of ferric P–450 which is hitherto unknown. ULLRICH et al. (1975a) investigated the binding of other ligands containing heteroatoms such as P, S, and O to P–450. These ligands lead to small changes of the EPR spectrum in a similar manner as nitrogen-containing ligands (see Table 5).

Axial ligands of P–450 Heme Iron

The axial ligands of the heme iron influence both strength and symmetry of the ligand field in the heme complex which is reflected in the EPR spectrum. The unusual g-values of P–450 in the low spin state (1.9, 2.2, 2.4) and in the high spin state (8, 3.8, 1.7) suggest the chelation of the heme iron by at least one amino acid residue not present in other hemoproteins. For example, the heme complexes of methemoglobin and metmyoglobin show in the high spin state a nearly complete axial ligand symmetry characterized by $g_{\perp} = 6$ and $g_{\parallel} = 2$ (GRIFFITH, 1956). Earlier studies on hemoproteins show only small deviations from the axial ligand symmetry, i.e. small splittings of the resonance absorption at $g = 6$. A wellknown example for a hemoprotein showing such a splitting is bovine liver catalase with $g = 6.3$, 5.3 and 1.92 (REIN et al., 1968) containing also a fraction with stronger rhombicity with $g = 6.9$ and 5.1 (PEISACH et al., 1971). However, the splitting is much larger in P–450 indicating an axial ligand with strong π-interaction with the d_{yz}- or d_{xz}-orbital of the heme iron (RISTAU, 1980).

Based on g-values of low spin ferric heme iron, Blumberg and Peisach developed a theoretical method for classifying the hemoproteins with respect to their ligand field parameters (BLUMBERG and PEISACH, 1971; PEISACH et al., 1973). The ligand field parameters of all P–450 forms investigated are very similar forming the so-called P-region in the ligand field diagram which contains also the mercaptide complexes of hemoglobin. Therefore the ligand field analysis of the EPR spectrum confirms the assumption that one of the axial ligands (the fifth) of ferric heme in P–450 is a mercapto sulfur (see below). This method has also been used in order to elucidate the nature of the sixth heme iron ligand of P–450 (CHEVION et al., 1977b; see below).

Due to alteration of the low spin EPR spectrum after addition of p-chloromercuribenzoate (MURAKAMI and MASON, 1967) the sulfhydryl group of cysteine was first proposed as fifth axial ligand of iron in P–450. The EPR spectrum caused by this sulfhydryl reagent is characterized by a strong absorption at $g = 6$ typical for high spin hemoproteins. Later studies on model complexes such as hemin derivatives (RÖDER and BAYER, 1969; RUF and WENDE, 1977; RUF et al., 1979; ULLRICH et al., 1979) and metmyoglobin in the presence of mercapto compounds (JEFCOATE and GAYLOR, 1969; HILL et al., 1970) confirmed this assumption. The EPR spectra of these model compounds show absorptions near $g = 1.9$, 2.2, and 2.4 which are typical for low spin P–450.

The EPR spectrum of high spin P–450 with g-values at 8, 3.8, and 1.7 has also been simulated using iron porphyrin-mercapto complexes (COLLMAN et al., 1975b; TANG et al., 1976). Examples for these model compounds are listed in Table 6. The model compounds have in common an at least partially deprotonated mercapto sulfur as heme iron ligand which seems to be essential for simulating the unique spectral properties of P–450, i.e. EPR, optical absorption and X-ray spectra.

Whereas corresponding results of many research groups justify the assumption

Table 6. Selected model compounds showing g-values similar to P–450 (the used solvents are not regarded in this table)

Low Spin

Compounds	g-Values			References
Hemin chloride, Dodecylmercaptane	2.355	2.23	1.956	RÖDER & BAYER, 1969
Hemin chloride, Cysteinemethylester	2.405	2.28	1.928	RÖDER & BAYER, 1969
Hemin chloride, Thiophenole	2.46	2.27	1.913	RÖDER & BAYER, 1969
Metmyoglobin, H_2S	2.4	2.3	1.91	HILL et al., 1970
Metmyoglobin, Methylmercaptane	2.38	2.25	1.94	HILL et al., 1970
Metmyoglobin, Ethylmercaptane	2.44	2.25	1.925	HILL et al., 1970
Methemoglobin A, pH > 10	2.40	2.25	1.90	REIN et al., 1975a
FePPIXDME, Thiophenole	2.385	2.262	1.936	RUF & WENDE, 1977
FePPIXDME, p-Nitrothiophenole	2.405	2.274	1.925	RUF & WENDE, 1977
FePPIXDME, p-Nitrothiophenole	2.36	2.26	1.94	TANG et al., 1976
FeOEP, Thiophenole	2.34	2.25	1.96	TANG et al., 1976
High Spin				
FePPIXDME, p-Chlorothiophenole	7.2	4.8	1.9	TANG et al., 1976
FeOEP, Thiophenole	7.2	4.7	1.9	TANG et al., 1976
FeTPP, Thiophenole	8.26	3.38	≈1.65	SATO et al., 1977

Abbreviations:
PPIXDME = Protoporphyrin IX dimethylester;
OEP = Octaethylporphyrin; TPP = Tetraphenylporphyrin

of a mercapto sulfur from a cysteine residue as the fifth heme iron ligand in P–450 contradictory suggestions exist about the sixth ligand. The reason is that the assumed sulfur ligand mainly contributes to the ligand field of the heme iron thus determining the essential heme-specific spectral properties of P–450. The sixth ligand, therefore, will produce only small changes in the EPR spectra (PEISACH et al., 1973; STERN et al., 1973).

Based on a comparative ligand field analysis of EPR spectra performed by BLUMBERG and PEISACH (1971) the imidazole group from a histidine residue was proposed to be the sixth ligand of the P–450 heme iron (CHEVION et al., 1977b). However, this method requires very accurate g-value determinations for certain calculations of the ligand field parameters which is not provided by microsomal P–450 due to inhomogeneous broadening of the EPR absorptions by the P–450 isozymes. The assumption of a histidine as sixth ligand therefore needs further experimental support.

ULLRICH et al. (1979) concluded from EPR investigations as well as from optical properties of model complexes that the sixth ligand must be a "hard" one favouring a hydroxyl group from water, serine or threonine. Based on investigations of the pH dependence of the optical spectra of P–450 the hydroxyl group of a tyrosine residue is suggested (RUCKPAUL et al., 1980).

Results from proton magnetic relaxation studies dealing with the sixth heme iron ligand of P–450 is discussed in chapter 5. Further information on the nature of the axial heme iron ligands comes from the EXAFS method.

The extended X-ray absorption fine structure approach to determine the nature and distances of ligands of a metal complex is a relatively new development. The method

Table 7. Table of distances of the iron-nitrogen (porphyrin), iron-carbon (porphyrin C_α) and iron-sulfur bonds in Angström evaluated by the EXAFS-method. The mean deviation of the Fe—N distances are 0.007 and of the other distances 0.025 Angström

Compound	Fe—N	Fe—C	Fe—S	References
ferric high spin				
chloroperoxidase	2.05	3.09	2.30	Cramer et al., 1978
P–450_{CAM} + camphor	2.05	3.06	2.35	Dawson et al., 1980
ferric low spin				
P–450_{LM2}	2.00	3.07	2.19	Cramer et al., 1978
P–450_{CAM}	2.02	3.05	2.24	Dawson et al., 1978
ferrous high spin				
P–450_{CAM}	2.10	3.12	2.43	Dawson et al., 1978
ferrous low spin				
P–450_{CAM}—CO	2.02	3.10	2.32	Dawson et al., 1978

uses the information of the modulation of the X-ray absorption curve caused by the K-shell (K-edge). This fine structure is attributable to the electron scattering effect of the atoms surrounding the X-ray absorbing atom.

Cramer et al. (1978) have examined by means of this method a set of iron porphyrins with crystallographically known structures. The results were then compared with EXAFS-spectra obtained from chloroperoxidase, P–450 LM2 and P–450 CAM (Dawson et al., 1980). In all three enzymes a sulfur atom was well established as the 5th iron ligand. The length of the iron sulfur bound was first quantitated in some P–450 species. The question for the nature of the 6th iron ligand could not be clearly answered. From the amplitude of the EXAFS nitrogen peak it was concluded that the 6th ligand may be a nitrogen. The results are summarized in Table 7.

2.2.2. Ligand Field Theory and Application to the High Spin and Low Spin State of Cytochrome P–450

The magnetic properties and particular the spin state of the prosthetic group of hemoproteins are controlled by the behaviour of the electrons in the $3d$ orbitals of the heme iron. Although these orbitals interact with the π and σ orbitals of the porphyrin and the axial ligands it is assumed in the ligand field theory that the resulting covalencies are small. Therefore for an approximate evaluation of the magnetic properties this delocalization of the d-electrons may be neglected. The ligands are considered only as carriers of charges influencing the energy of the d-electrons by their resulting electrostatic (ligand) field. The ligand field theory is widely used to evaluate g-factors of EPR experiments. In most cases the ligand field theory is a satisfactory model in spite of the neglected covalencies of the iron ligand bonds. Because EPR signals are observed only from paramagnetic ferric hemoproteins* this survey is limited to the evaluation of the g-factors of these hemoproteins.

With respect to the spin state the 5 d-electrons of a ferric ion can be arranged in three different ways: (i) all electrons are unpaired, i.e. high spin state $S = 5/2$ (sextet terms),

* High spin ferrous hemoproteins are pratically EPR silent because the nonmagnetic ground state is only influenced by the magnetic field in second order.

Table 8. The split of some free ion terms in a cubic ligand field. The numbers in the last column means the spatial degeneracy of each cubic term

terms in free ion	orbital degeneracy	terms in a cubic ligand field	spatial degeneracy
S	1	A_1	1
P	3	T_1	3
D	5	E, T_2	2, 3
F	7	A_2, T_1, T_2	1, 3, 3
G	9	A_1, E, T_1, T_2	1, 2, 3, 3
H	11	E, T_1, T_1, T_2	2, 3, 3, 3
I	13	A_1, A_2, E, T_1, T_2, T_2	1, 1, 2, 3, 3, 3

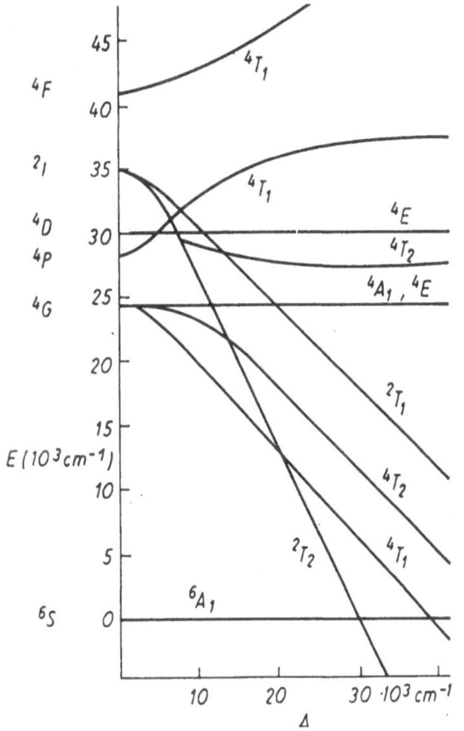

Fig. 22. Energy level diagram of a d^5 electron system in dependence on cubic ligand field strength. Only 10 cubic terms are shown. The low spin term $^2T_2$ crosses the high spin term $^6A_1$ at about 30,000 cm^{-1}; based on data from Tanabe and Sugano, 1954.

(ii) 3 electrons are unpaired, i.e. intermediate spin state S = 3/2 (quartet terms), (iii) 1 electron is unpaired, i.e. low spin state $S = 1/2$ (double terms).

With respect to the energy the five *d*-electrons can be arranged in many terms. The number of terms depends on the symmetry of the ligand field originating from the iron ligands. In a cubic ligand field there exist 43 terms designated by capital letters which represent the spatial degeneracy and by superscripts which denote the spin degeneracy. In Table 8 the splitting of seven terms of a free ion with different spatial degeneracy into the corresponding cubic terms is shown. The spin degeneracy in the cubic terms, however, is the same as in the parent free ion term. The high spin term is therefore designated $^6A_1$, the lowest low spin term $^2T_2$ and the lowest term with intermediate spin $^4T_1$. The spin state of the energetically lowest term determines the magnetic properties of a heme complex because at room or lower temperature no excited term is markedly populated.

The rough symmetry of an irregular octahedral heme complex corresponds to a cubic one. Therefore it is useful to calculate the energy levels (terms) of the d^5 system in such ligand field symmetry as first performed by Tanabe and Sugano (1954) (see Fig. 22). One result from these calculations is very important: while for small ligand field strength the high spin term $^6A_1$ is the lowest one at higher field strength the low spin term $^2T_2$ is the ground state. In the region near the cross-over point of both terms a thermal transition between them is possible which can be regarded as reason for the existence of thermal high spin/low spin equilibria in hemoproteins. For the transition in principle two mechanism are conceivable: (i) a thermal population of the excited term at constant ligand field strength or (ii) a thermal change of the ligand field strength induced by thermal variation of the metal-ligand bond length. As shown later, the analysis of the EPR data of P–450 in the high spin state favours the second possibility at least for this case. In a cubic ligand field only the high spin or the low spin term can form the ground term. But in ligand fields with strong tetragonal (axial) distortions theoretical calculations show that also the quartet term $^4T_1$ can form the ground state (Harris, 1968). In hemoproteins, however, no example for such a ground state with intermediate spin (quartet term) is known. Therefore it can be concluded that the cubic Tanabe-Sugano-diagram is approximately valid for hemoproteins. This means too, that in a heme complex axial and rhombic distortions are generally small leading Griffith (1956) to the hypothesis that hemoproteins exhibiting intermediate spin values are thermal mixtures of the pure high spin and low spin form. This prediction was experimentally confirmed by different methods (magnetic susceptibility, EPR, optical spectroscopy) for several hemoproteins (George et al., 1959; Beetlestone and George, 1964; Rein and Ristau, 1965; Iizuka and Kotani, 1968; Iizuka and Yonetani, 1970; and some others).

The concept of the ligand field theory

By the following remarks only a short introduction in the ligand field theory and its usefulness in the characterization of the heme complex will be given, for further details see Weissbluth (1974). The energies of the five d-electron orbitals of the iron are degenerated at the free ion but split by the influence of a ligand field. Character and magnitude of the splitting depends on the strength as well as on the symmetry of the ligand field. The ligand field theory calculates the splitting of the d-electron orbitals (one electron energies) on the electrostatic field of the ligands (Fig. 23).

In principle the strength of the ligand field can be calculated from the knowledge of charge and the bond length of the ligands to the iron but experience shows that such calculations are too inaccurate. Therefore the task of the ligand field theory is to extract the ligand field parameters from measured g-values or values of magnetic susceptibilities. Ligand field parameters obtained on such basis allow a comprehensive description of the magnetic properties of a transition metal complex. Additionally the effective symmetry of the ligand field is characterized by the ligand field parameters. The effective ligand field symmetry defines a ligand field which causes the same energetic splitting of the iron d-orbitals which occur by interaction of the d-electrons with the whole environment of the iron not only caused by electrostatic forces. Because the rigid porphyrin skeleton is tetragonal (axial) symmetry the observed rhombic distortion of the heme complex reflects the unequivalence of the d_{xz} or d_{yz} iron orbital caused by axial ligands (Ingram, 1972).

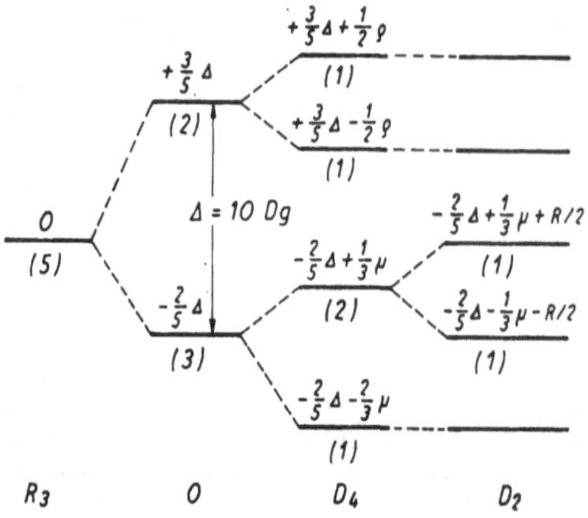

Fig. 23. One-electron energies ("orbital energies") of the five d-orbitals for cubis (Δ), axial (μ, δ) and rhombic (R) ligand field symmetries.

Low spin hemoproteins ($S = 1/2$) can be often described with sufficient accuracy by use of the low spin term alone without consideration of the spin orbit interaction with other terms. The ligand field theory then is used to calculate the split of the threefold degenerate $^2T_2$ term caused by ligand field components lower than cubic and by spin orbit coupling. This 1-term model (Griffith, 1957) leads to explicit formulas for the g-factors related to the linear combination coefficients which define the wave function of the new ground state in the presence of the real ligand field including spin orbit coupling. In this 1-term model a correction factor (k) for partial covalence according to Stephens (1953) is often introduced.

In the frame of STEPHENS' model k has to be $\leqq 1$. The following equations are obtained.

$$g_z = 2\big((1+k)\,A^2 - B^2 + (1-k)\,C^2\big) \quad [1]$$

$$g_x = 2\big(B^2 + 2AC + \sqrt{2}\,kB(A+C)\big) \quad [2]$$

$$g_y = 2\big(B^2 - 2AC + \sqrt{2}\,kB(A-C)\big) \quad [3]$$

The Stephens-factor k is defined by the relation for the interaction of a wave function with a magnetic field (Zeeman effect).

$$H_m = \beta \cdot H \cdot (k \cdot \hat{l} + 2\hat{s}) \quad [4]$$

The nonlinear equations (1—3) must be solved with the normalization condition $A^2 + B^2 + C^2 = 1$. This can be done taking the four equations as a system for optimization with respect to the 3 coefficients A, B, C and the Stephens-factor k. Thereafter the ligand field parameters can be calculated by means of the equations

$$R = C\,\frac{2A + \sqrt{2}B}{A^2 - C^2} \quad [5]$$

and

$$\mu = 0.5 + \frac{A}{2}\left(\frac{\sqrt{2}}{B} - \frac{R}{C}\right). \quad [6]$$

The description of the magnetic properties of *high spin hemoproteins* ($S = 5/2$) by use of a 1-term model is not possible because the high spin term $^6A_1$ itself does not split either by spin orbit coupling nor by ligand field directly. Therefore always an isotropic g-factor 2 is expected in all high spin complexes. On the contrary, the observed g-factors are of high anisotropy often with $g_\perp = 6$ and $g_\parallel = 2$. The usual magnetic description of the high spin complexes is based on spin Hamiltonian treatment. The spin Hamiltonian operator acts only on the spin part of the wave functions. In most cases the use of a truncated operator is sufficient. That means that only two parameters, namely the axial parameter D and the rhombic parameter E enter the spin Hamiltonian. However in the case of P–450, which shows a very strong rhombic distortion, the influence of the cubic parameter α becomes noticeable.

$$\hat{H} = D\left(\hat{S}_z^2 - \frac{35}{12}\right) + E\,(\hat{S}_x^2 - \hat{S}_y^2) + \frac{\alpha}{6}\left(\hat{S}_z^4 + \hat{S}_x^4 + \hat{S}_y^4 - \frac{707}{16}\right) + g_{\text{eff}} \cdot \vec{H} \cdot \hat{S} \quad [7]$$

For g_{eff} in the Zeeman-term the value 2 can be taken in a good approximation. From this spin Hamiltonian the following equations for the g-factors and subterm energies can be derived.

$$g_z = \left\{2[(6-3\eta+w)(2-15\eta+w) - 45\eta^2]^2 - 108\left(\frac{E}{D}\right)^2(6-8\eta+w)^2 + 100\left(\frac{E}{D}\right)^2(2-24\eta+w)^2\right\}\Big/ N^2 \quad [8]$$

$$g_{(x,y)} = \left\{6[(6-3\eta+w)(2-15\eta+w) - 45\eta^2]^2 \mp 48\left(\frac{E}{D}\right)(6-8\eta+w) \times[(6-3\eta+w)(2-15\eta+w) - 45\eta^2] + 120\left(\frac{E}{D}\right)^2(6-8\eta+w) \times(2-24\eta+w)\right\}\Big/ N^2 \quad [9]$$

$$w^3 + w^2(8-18\eta) + w\left(12 - 96\eta - 28\left(\frac{E}{D}\right)^2\right) - \left(\frac{E}{D}\right)^2(128-384\eta) = 0 \quad [10]$$

$$N^2 = [(6 - 3\eta + w)(2 - 15\eta + w) - 45\eta^2]^2 + 10\left(\frac{E}{D}\right)^2 (2 - 24\eta + w)^2 + 18\left(\frac{E}{D}\right)^2 (6 - 8\eta + w)^2 \quad [11]$$

In these equations the symbols w, η, N have the following meaning: $w = -\varepsilon/D - 8/3 + \alpha/D$, a reduced term energy; $\eta = \alpha/(6D)$, the reduced cubic Hamiltonian parameter α; N = the normalisation factor. The parameter ratios E/D and α/D can be calculated from these equations based on experimental g-values. In practice these parameter ratios can be evaluated taking the g-factor equations [8, 9] as an optimization problem that means the calculation of the minimum of the following sum.

$$(g_z^{\text{exp}} - g_z^{\text{theory}})^2 + (g_x^{\text{exp}} - g_x^{\text{theory}})^2 + (g_y^{\text{exp}} - g_y^{\text{theory}})^2 = \text{minimum} \quad [12]$$

For small rhombic distortions $E/D < 0.01$ the g-factor relations simplifies to

$$g_z = 2 \quad \text{and} \quad [13]$$

$$g_{(x,y)} = 6 \mp 24\left(\frac{E}{D}\right)(1 + 10\alpha/(9D)) \quad [14]$$

A better approximation is obtained by setting both w and η in equations [8, 9] to zero which yields (HERRICK, 1976)

$$g_z = \left(2 - 218\left(\frac{E}{D}\right)^2 \Big/ 9\right) \Big/ M \quad [15]$$

and

$$g_{(x,y)} = \left(6 \mp 24\left(\frac{E}{D}\right) + 10\left(\frac{E}{D}\right)^2\right) \Big/ M \quad [16]$$

with

$$M = 1 + 43\left(\frac{E}{D}\right)^2 \Big/ 9. \quad [17]$$

From equations [13, 14] it follows that for small values of the ratio E/D the parameter ratio α/D cannot be determined. Especially, the sign of the parameter ratio α/D is important to determine the electronic ground state. According to SATO et al. (1976) the high spin state is the ground state for positive α/D.

Advantages of a 3-term model

A common description of a low spin and a high spin complex with the same formalism is possible only by use of a multi term ligand field model. The minimal model which fulfils this demand is a 3-term model (RISTAU, 1980, 1981). In this model besides the low spin and the high spin term the lowest intermediate spin term $^4T_1$ is included coupling with both mentioned terms (see Fig. 22). The multi term model is further advantageous by avoiding a Stephens-factor k larger than one at the calculation of the low spin ligand field parameter. Such abnormal k-values especially evaluated from the g-values of P-450 LM2 using the normal procedure of low spin complexes indicate the influence of spin orbit coupling with excited terms. This coupling, however, is considered in the 3-term model.

At least by the 3-term model it is possible to describe the magnetic properties of hemoproteins that are called quantum mixed spin species of the spin states 5/2 with 3/2 or 1/2 with 3/2. Such quantum mixed spin states were postulated by MALTEMPO and MOSS (1976) to explain the unusual magnetic behaviour of cytochrome c′ and some peroxidases. This type of quantum spin mixing is caused by spin orbit interaction of neighbouring terms. From this it can be derived that a small quantum spin mixing is always present but becomes dominant if the energetic difference between the interacting term is sufficiently small.

The 3-term model contains 4 ligand field parameter, one cubic, two axial and one rhombic. This leads to some difficulties in the use of this model, because it is not possible to determine these 4 parameter unequivocally from 3 experimental g-values. But for P-450 CAM in the presence of camphor the zero field splitting of the high spin term in the three subterms is additionally measured (HERRICK and STAPLETON, 1976). Therefore the ligand field parameters of the high spin compounds can be calculated if it is assumed that all P–450 high spin complexes with very similar g-values have also very similar zero field splittings. In the low spin P–450 compounds the influence of the axial parameter ϱ on the calculated g-factors is very small. Therefore the ratio of the two axial parameters ϱ/μ can be taken as equal to the high spin P–450 compounds.

Calculation of the ligand field parameters from P–450 compounds

P–450 LM4

Experimental g-values: $g_z = 1.74$, $g_x = 3.77$, $g_y = 7.94$ (SATO et al., 1977).

The analysis of these g-factors with the spin Hamiltonian treatment (equations 8, 9) yields the following results: $E/D = 0.092$ and $\alpha/D = 0.017$.

The 3-term model yields the following results:

cubic ligand field parameter Δ	$51.75\ \lambda$
axial ligand field parameter μ	$7.0\ \lambda$
axial ligand field parameter ϱ	$8.4\ \lambda$
rhombic ligand field parameter R	$4.35\ \lambda$
zero field splitting $\pm 1/2 \rightarrow \pm 3/2$	$6.9\ cm^{-1}$
zero field splitting $\pm 1/2 \rightarrow \pm 5/2$	$20.5\ cm^{-1}$

The experimental values for the zero field splitting are 6.6 and 20.7 cm^{-1}.

P–450 LM2

Experimental g-values: $g_z = 2.45$, $g_x = 2.26$, $g_y = 1.93$ (RISTAU, 1980).

The 3-term model yields:

cubic ligand field parameter Δ	$58.9\ \lambda$
axial ligand field parameter μ	$5.2\ \lambda$
axial ligand field parameter ϱ	$6.2\ \lambda$
rhombic ligand field parameter R	$5.0\ \lambda$

The parameter ratio ϱ/μ was set to 1.2 as determined in the high spin compound P–450 LM4.

Contrary the usual 1-term analysis yields the following parameters:

axial parameter	μ	$6.69\ \lambda$
rhombic parameter	R	$5.85\ \lambda$
Stephens-factor	k	1.31

The spin orbit coupling constant λ was taken to be 430 cm^{-1}.

Conclusions drawn from the ligand field calculations

The most important result of the ligand field analysis is the answer to the question whether the high spin term of P–450 is a ground term or not. From the analysis of the g-values of P–450 LM4 it must be concluded (Hamiltonian parameter α/D positive) that the high spin term is the lowest term, respectively the ground term in the sense of ligand field theory. In consideration of the excistence of a thermal spin equilibrium in P–450 (SLIGAR, 1976; REIN et al., 1977a, c) the high spin state is not an excited state which is more populated with rising temperature by nearly fixed ligand field parameters or geometry of the heme complex, respectively, but is caused by a thermal change of the geometry coupled with a change of the ligand field strength. In the thermodynamical sense the high spin state is of course an excited one but at the transition of the low spin to the high spin state the energy is used for the geometric and conformational change of the heme and protein, respectively. The heme geometry depends on the properties of the two axial iron ligands (MASHIKO et al., 1978). In the high spin state one strong and one weak bound axial ligand forces the heme iron out of the porphyrin plane in direction to the strong field ligand. In ferric high spin hemoproteins the iron can be 6-coordinated with one short and one long iron ligand bond like methemoglobin or 5-coordinated like a P–450 model complex (TANG et al., 1976) where the weak ligand is lost. One cannot differentiate between these two possibilities. But kinetic measurements of the high spin/low spin transition in P–450 CAM (without camphor) show slow kinetics which was correlated with the dissociation rate of the sixth iron ligand (SLIGAR, 1976). In P–450 LM2 such slow kinetics was not found (RISTAU et al., 1979), meaning that the heme iron in the high spin state of P–450 LM2 may be mainly 6-coordinated.

Because the spin equilibrium in P–450 is a geometric one, the coupling of the spin state with the protein conformation can be easily understood. Both axial heme iron ligands are side chains of amino acids of the protein. Therefore a geometrical change of the heme is rigidly coupled with a conformational change of the protein predominantly in the vicinity of the heme. A thermodynamic model of the high spin/low spin equilibrium should therefore include two conformations of the enzyme one contains high spin and the other low spin iron.

2.2.3. ENDOR Studies of Cytochrome P–450

Compared with the usual EPR the electron nuclear double resonance technique (ENDOR) is a powerful method with enhanced resolution of hyperfine interactions of the unpaired electrons of the central metal ion with the spin of paramagnetic nuclei of ligands in transition metal complexes. Therefore in hemoproteins the ENDOR lines of the four equally bound pyrrol nitrogens of the porphyrin (^{14}N, $S = 1$) and also from the axial ligands of the heme iron if they possess a magnetic nucleus are expected. LO BRUTTO et al. (1980) have investigated P–450 CAM in the presence (high spin) and in the absence of camphor (low spin). In the low spin state ENDOR lines of protons were obtained which disappear by use of deuterated buffer. In the high spin complex, on the other hand, such exchangeable proton signals were not observed. In principle these protons could arise from H_2O, from amino acid side chains of the form RNH_2, ROH or $RCONH_2$, or conceivably from an exchangeable proton bound to a cysteinyl sulfur. The δ-N proton of histidine is an unlikely candidate for the observed splitting of 8—9 MHz; this proton would be ~ 5Å from the heme iron and has a dipolar coupling of only ≈ 1.3 MHz. The lack of the exchangeable proton signal in the high spin state suggest that the proton of the cysteinyl sulfur is not the reason for the observed signal in the low spin state because also in the high spin state the sulfur should be bound to the iron. Therefore it must be assumed that the observed ENDOR signal results from protons of the 6th ligand which could be a water (MARIČIČ et al., 1977; REIN and RISTAU, 1978), the hydroxyl group of a tyrosine (RUCKPAUL et al., 1980) or possibly further amino acids with above mentioned residues.

2.2.4. Linear Electric Field Effect in EPR and Electron Spin Echo Measurements

The EPR parameter of a sample can often be modified by applying an electric field accross it. Thereby induced frequency shifts are small, but have been detected in several cases for single-crystal materials (ROITSIN, 1972). By the method the odd portion of the crystal-field potential is detected. For a site with inversion symmetry, for example D_{4h} symmetry, no linear electric field effect will be there. Small deviations from pure inversion symmetry will result in electric field-induced g-factor shifts. Detailed studies of low symmetry sites in single crystal samples have been performed by MIMS and GILLEN (1967) who measured the shifts for various orientations of electric and magnetic fields, E and H_0, respectively. PEISACH and MIMS (1973) have adapted the method which was originally used for crystals also for frozen solutions commonly used in EPR studies of biological materials. They studied some low spin hemoproteins and heme complexes including models and P–450 itself (MIMS and PEISACH, 1976; PEISACH and MIMS, 1973; PEISACH and MIMS, 1977).

In low spin P–450 of rat liver, the presence of an electric field shift indicates a deviation from inversion symmetry (D_{4h}) of the heme complex and is compatible with two axial ligands attached to the heme iron are not the same. From comparison with P–450 model complexes (mercaptide as 5th ligand and different non-mercaptide ligands on the 6th coordination site of the heme iron) it is suggested that in P–450 the non-mercaptide ligand to the heme has a crystal field contribution larger than imidazole and smaller than guanidine (PEISACH and MIMS, 1973).

PEISACH et al. (1979) have also observed the nuclear modulation patterns in the envelope of electron spin echoes for various low spin paramagnetic heme proteins including P–450 CAM. This work was done using a three-pulse EPR technique (stimulated spin echo method). From the results it was concluded that the low spin ferric heme iron of P–450 is coordinated to a nitrogenous ligand, probably imidazole. Furthermore, it was demonstrated that the remote ^{14}N of the imidazole ligand (i.e. not the iron coordinated one) in an imidazole-heme-NO complex is coupled with the electron spin of the iron differently than in myoglobin nitroxide, demonstrating the direct effect of the protein on metal ligand bonding.

2.2.5. Magnetic Susceptibility

Magnetic susceptibility measurements of hemoproteins provides information on the spin state of the heme iron. A thermo — magnetic study may reflect a thermal equilibrium between the low spin and the high spin state. The temperature dependence of the magnetic susceptibility can be described by use of CURIE's law.

$$\chi = \frac{C}{T} = \frac{N_L \mu^2}{3kT} \qquad [18]$$

$$\chi = \frac{0.125 n^2}{T} \text{ [erg ml}^{-1}\text{ G}^{-2}] \qquad [19]$$

C = Curie-constant
μ = magnetic moment
k = Boltzmann constant
n = number of Bohr magnetons (B.M.)

If a plot of χ versus $1/T$ displays a straight line, then the magnetic moment of a paramagnetic material is independent of the temperature. Any deviation from a straight line may be interpreted as an effect of a thermal equilibrium existing between different magnetic forms.

Up to now studies of the magnetic susceptibility of P–450 have been performed only with P–450 CAM. PETERSON (1971) investigated the ferric form in the presence of camphor in the temperature range from −20 °C down to −179 °C. A constant magnetic moment of 5.22 B.M. was found which corresponds to 76% high spin if a pure spin susceptibility of 1.73 B.M. and 5.92 B.M. for the low spin and the high spin form, respectively, is assumed for the calculation. The lacking temperature dependence of the high spin/low spin ratio led PETERSON (1971) to the conclusion that in P–450 CAM in the presence of camphor a thermal spin equilibrium does not exist. But it must be considered that the spin equilibrium could be frozen in over the temperature range studied (MESSANA et al., 1978) because spin transitions of hemoproteins are coupled with conformational changes.

The magnetic susceptibility of ferrous P–450 CAM was measured by CHAMPION et al. (1975a) in the temperature range from 1.7 up to 4.5 K. The heme iron in reduced P–450 CAM is like in deoxyhemoglobin in the high spin (S = 2) state. A strong deviation from Curie's law at the measured very low temperature range is observed. This magnetical behaviour is explicable by the splitting of the 5-fold degenerate high spin term in three sub-terms by the spin orbit coupling and the axial ligand field whereby the population of the three excited subterms are temperature dependent.

One possibility to describe this zerofield splitting of the ground state exists by use of the spin Hamiltonian formalism.

$$\hat{H}_{sp} = D(S_Z^2 - 2) + \beta\check{S}g \cdot \vec{H} \qquad [20]$$

D, coefficient describing the zero field splitting
H, applied magnetic field
g, anisotropic Zeeman interaction

For the magnetic susceptibility the following equation can be derived: (valid for low magnetic fields $H \to 0$)

$$\chi = \frac{N_L\beta^2[(8xe^{-2x} + 2xe^x)\, g_z^2 - g_{x,y}^2(8/3e^{-2x} + 28/3e^x - 12e^{2x}]}{3D(2e^{-x} + 2e^x + e^{2x})} \qquad [21]$$

$$x = \frac{D}{kT} \qquad [22]$$

For very low temperatures the magnetic susceptibility is limited to the value:

$$\chi = \frac{4g_{x,y}^2 N_L\, \beta^2}{D} \qquad [23]$$

$$g_{x,y} \approx 2[1 + D/\lambda] \qquad [24]$$

$g_{z,x,y}$ = electronic g-values

By use of this mathematical formalism from the experimental data the zero field splitting D was fitted between 10 and 17 cm^{-1} for ferrous P–450 CAM (Champion et al., 1975a). From Mößbauer experiments a similar large value of $D = 13.9\ cm^{-1}$ was found by the same authors (Champion et al., 1975b). The large zero field splitting coefficient of ferrous P–450 CAM contrasts markedly to the value of $D = 5.2\ cm^{-1}$ found for deoxyhemoglobin and deoxymyoglobin (Nakano et al., 1971, 1972). Champion et al. (1975b) explain the low values of D of these hemoproteins by a small energetic difference of the ground state to the next higher spin quintet. That means that the ground state is not well isolated from the nearest excited state and therefore the spin Hamiltonian formalism used by Nakano et al. (1971, 1972) is not an adequate approach in this case. On the other hand, a description of the magnetic data from ferrous P–450 CAM in terms of a spin Hamiltonian is probably true and therefore it is unlikely that the zerofield splitting parameters of deoxyhemoglobin can be directly compared with those found for ferrous P–450.

From the high spin state measured for ferrous P–450 it may be concluded that the heme iron is 5-coordinated in analogy to deoxyhemoglobin and deoxymyoglobin in which this coordination state is established by X-ray analysis (Fermi, 1975) and other methods (Fabry and Reich, 1966).

2.2.6. Mößbauer Spectroscopy

Mößbauer spectra reflect the properties of the electronic structure of a metal complex. This type of spectroscopy is limited to a relatively small number of "Mößbauer active nucleus". Due to the presence of the stable Mößbauer isotope ^{57}Fe to an amount of 2.2% in all natural iron compounds it is possible to study hemoproteins by this method. But in several cases ^{57}Fe enriched samples are necessary to get sufficient sensitivity. Therefore until now only the P–450 CAM was investigated extensively because bacterial P–450 can be easily enriched by growth of *Pseudomonas putida* in a medium containing iron enriched with ^{57}Fe.

To discuss the Mößbauer data, some remarks about the interpretation of Mößbauer spectra are useful. In general these spectra are characterized by three types of interactions between the active nucleus and its electronic environment: isomer shift, quadrupole splitting and magnetic hyperfine interaction.

(*i*) *Isomer shift*

Experimentally the isomer shift is defined as the position of the center of the Mößbauer absorption spectra with respect to relative velocity of the source to an iron metal at 300 K. The magnitude of the isomer shift depends on the *s*-electron density at the nucleus. For ^{57}Fe a larger *s*-electron density of the sample compared with the source yields a negative isomer shift and vice versa. An additional effect is caused by *p*- and *d*-electrons. Owing to the shielding of the nucleus by these electrons the *s*-electron density on the nucleus of the sample may be diminished compared with the nucleus in the source which results in a more positive isomer shift. This behaviour is reflected in high spin ferrous hemoproteins which, due to one more *d*-electron compared with ferric hemoproteins, exhibit a more positive isomer shift. The CO and oxy compounds of ferrous hemoproteins show an isomer shift comparable with those of ferric low spin hemoproteins (see Table 9). This finding is compatible with the Weiss-model of the iron-oxygen bond in oxyhemoglobin. WEISS (1964) suggested that this bond might be ionic between a ferric ion and a superoxide ion ($Fe^{3+}O_2^-$). But the ferric type isomer shift of oxyhemoglobin can also be explained by a strong π-back donation from the iron to the oxygen.

(ii) Quadrupole splitting

The iron isotope ^{57}Fe shows a quadrupole moment in the excited state which causes a splitting of the excited state into two states in the presence of an electric field gradient. Therefore two absorption peaks are present in a Mößbauer spectrum of ^{57}Fe. The energy splitting of the two lines depends on the quadrupole moment at the nucleus and on the magnitude of the electric field gradient of the nucleus. From the magnitude of the quadrupole splitting the hemoproteins can be divided into two groups.

a. Ferrous hemoproteins and related model complexes

The largest quadrupole splittings are shown by oxy compounds of hemoproteins including model complexes with a quadrupole splitting of 2.1–2.5 mm/s, followed by other diamagnetic 6-coordinated heme complexes such as ferro chromogens, ferrous cytochrome *c* and dipyridin heme complexes with values between 1.1–1.3 mm/s (see Table 9). The smallest quadrupole splittings of diamagnetic compounds are shown by the carboxy complexes with values at 0.32–0.37 mm/s. The ferrous high spin hemoproteins yield similar high values as the oxy compounds.

b. Ferric hemoproteins and related model complexes

The quadrupole splitting of ferric hemoproteins depends on the spin state. Whereas low spin compounds show splittings between 1.4–2.82 mm/s the splitting of the high spin compounds is smaller (in the range of 0.5–1 mm/s). Based on this difference SHARROCK et al. (1973, 1976) has determined the high spin/low spin ratio of P–450 CAM in the presence of camphor in dependence on temperature. The high spin content varies from 45% at 4.2 *K* up to 77% at 200 K.

(iii) Magnetic hyperfine splitting

The ^{57}Fe nucleus possesses the spin of $S = 1/2$ in the ground state but the spin $S = 3/2$ in the excited state. For axial symmetry the excited state splits into four subterms in a magnetic field. Because a quadrupole moment is also present in the nucleus ^{57}Fe the 4-fold spin degeneracy is lifted and split into two quadrupole terms, each of them possessing twofold degeneracy. The ground state itself splits into two terms so that 6 transitions are possible. A more complicated situation is observed for lower symmetry (WEISSBLUTH, 1974).

b. Paramagnetic compounds

The interaction between the nuclear spin and the electron spin leads to a complicated form of the Mößbauer spectra. The interaction depends very strongly on the magnitude of the electronic spin lattice relaxation time. If the electronic relaxation time is short as compared with the duration of the larmor frequency cycle of the nucleus the influence of the electron-produced magnetic field on the nucleus vanishes (case a). But in the reverse case (b), if the electronic relaxation rate is low as compared with the larmor frequency of the nucleus, the electrons produce an additional magnetic field at the nucleus. This additional magnetic field is a superposition of all magnetic states which are occupied by the electrons by thermal energy. On the other hand, the influence of the magnetic field of the nucleus on the electrons is often negligible because this field is only weak ($\approx$ 10 Gauss).

At ferric heme complexes, which have a half

Table 9. Mößbauer parameters of hemoproteins

Compound	Temperature K	Quadrupole splitting mm/s	Isomer shift mm/s	References
	diamagnetic Fe(II)-complexes			
Oxyhemoglobin	1.2	2.24	0.24	1
Oxymyoglobin	77	2.27	0.22	2
Fe(N-Me-imid)TpivPP-O_2	4.2	2.106	0.28	8
Fe(THT) TpivPP-O_2	4.2	2.29	0.31	8
O_2—P-450 CAM + camphor	4.2	2.15	0.31	4
CO-hemoglobin	4.2	0.36	0.26	1
CO-myoglobin	4.2	0.363	0.266	3
CO—P-450 CAM+camphor	4.2	0.32	0.29	4
CO-Chloroperoxydase	4.2	0.52	0.29	10
Cytochrome c	4.2	1.17	0.45	5
Fe-protoporphyrin-dipyridine	77	1.21	0.45	7
Fe-octaethylporphyrin-dipyridine	4.2	1.13	0.56	7
	paramagnetic Fe(II)-complexes (high spin)			
Hemoglobin	4.2	2.4	0.91	1
Myoglobin	77	2.24	0.86	2
P-450 CAM	4.2	2.46	0.83	4
Chloroperoxidase	4.2	2.5	0.85	6
Horseradish peroxidase	4.2	2.7	0.9	6
	low spin Fe(III)-complexes			
Hemoglobin—CN^-	195	1.39	0.17	1
—N_3^-	195	2.3	0.15	1
—OH^-	77	1.9	0.2	1
P-450 CAM	130	2.82	0.33	4
+ phenylimidazole	200	2.81	0.32	4
+ camphor	210	2.75	0.31	4
Chloroperoxidase	4.2	2.9		
	high spin Fe(III)-complexes			
Hemoglobin—F^-		2.0	0.9	11
—H_2O	195	2.0	0.2	1
P-450 CAM				
+ camphor	210	0.78	0.34	4
+ camphor +putidaredoxin	197	0.99	0.39	4
Chloroperoxidaseiodid	1.5	0.65	—	10
Fe-protoporphyrin				
-p-nitrophenolat	4.2	0.76	0.33	9
-ethanolate	4.2	0.46	0.32	9
-monoimidazolechloride	4.2	0.78	0.26	9

[1]LANG and MARSHALL (1966); [2]GONSER et al. (1974); [3]TRAUTWEIN et al. (1974); [4]SHARROCK et al. (1976); [5]LANG et al. (1968); [6]CHAMPION et al. (1975c); [7]DOLPHIN et al. (1976); [8]SPARTALIAN et al. (1975); [9]TANG and KOCH (1976); [10]CHAMPION et al. (1973); [11]WEISSBLUTH, (1974)

Abbreviations: N-Me-imid = N-methylimidazole; Tpivpp = tetra(pivalamidophenyl)-porphyrin; THT = tetrahydrothiophen.

integral spin, the Kramers degeneracy is lifted by an external magnetic field and by $H \gg 10$ Gauss the influence of the nuclear field on the electronic state is prevented. The ferrous high spin hemoproteins, however, do not show spontaneous magnetic hyperfine interaction. In the presence of a strong external magnetic field the spins of the electrons are partially aligned especially at low temperatures, and a magnetic hyperfine interaction occurs.

The electronic relaxation times of the ferric complexes are longer than those of the ferrous forms. Therefore magnetic hyperfine interaction can be obtained (case *b*). The theoretical interpretation of the spectra is easier and the resolution is better in the presence of an external magnetic field. But in most cases resolved magnetic hyperfine interaction is observed also in ferric hemoproteins only at low temperatures. The fluoride complex of methemoglobin is an exception which shows paramagnetic hyperfine interaction already at 200 K (WEISSBLUTH, 1974).

Mößbauer investigations on P–450 CAM show the parameters of the oxy- and carboxy compound of P–450 CAM to be comparable with those of other hemoproteins (see Table 9). The small isomer shift of both compounds suggests a strong charge transfer (π-back donation) from the iron to oxygen and CO, respectively. The Mößbauer spectrum of ferrous P–450 CAM exhibits a magnetic hyperfine interaction different from other ferrous hemoproteins like deoxyhemoglobin or horseradish peroxidase but is similar with those of chloroperoxidase. This similarity implies similar internal fields for both reduced hemoproteins. The g-values calculated from Mößbauer data shows Table 10. The g-factors of the oxidized species obtained from EPR spectra are included. The similarity of the g-values of both hemoproteins in the reduced as well as in the oxidized state suggests a mercaptide as 5th heme iron ligand to be present at this position in chloroperoxidase, too. The Mößbauer spectrum of ferrous P–450 CAM is not changed by camphor. Ferric P–450 CAM exists as equilibrium between the high and low spin state. Both states are well separated in the Mößbauer spectrum by different quadrupole splitting. By the addition of putidaredoxin to the P–450 CAM-camphor complex the quadrupol splitting is enlarged and the high spin portion diminished.

Table 10. g-values of P–450 CAM and chloroperoxidase calculated from Mößbauer data for the reduced state and experimentally obtained from EPR measurements for the oxidized high spin state (CHAMPION et al., 1975b,c; HOLLENBERG et al., 1980)

enzyme	oxydation state	g-values		
P–450$_{CAM}$	reduced	2.24	2.32	2.0
chloroperoxidase		2.24	2.32	2.0
P–450$_{CAM}$	oxidized	≈ 8	≈ 4	1.8
chloroperoxidase		7.4	4.3	1.9

Because it is laborious to enrich P–450 from animals with ^{57}Fe only few investigations have been performed with hepatic P–450. MAY and HOLTZMAN (1970) have recorded the Mößbauer spectrum of lyophilized liver microsomes from phenobarbital induced rats. Unfortunately the lyophilization causes partially denaturation of P–450 to high spin P–420. The obtained Mößbauer data are predominantly caused by this species with quadrupole splitting of 0.64 mm/s and an isomer shift of 0.64 mm/s.

2.2.7. Nitric Oxide, a Spin Label Probe to Study the Oxygen Binding Site

As several other hemoproteins, P–450 can bind not only oxygen but also carbon monoxide and nitric oxide. Complexes with these ligands have been used to study the electronic structure and stereochemical properties of the heme in P–450 (EBEL et al., 1975; O'KEEFFE et al., 1978b) (see also 2.1.5.) although both of these ligands inhibit the enzyme function. Nitric oxide forms complexes with ferric as well as with ferrous P–450.

Whereas hemoglobin and myoglobin do not form stable ferric NO complexes (Kon, 1968; Kon and Kataoka, 1969) horseradish peroxidase, cytochrome c peroxidase and P–450 form stable spin-paired diamagnetic complexes (Yonetani et al., 1972; O'Keeffe et al., 1978b). On the other hand, at the binding of nitric oxide to ferrous heme iron paramagnetic complexes are formed the properties of which can be studied by the EPR method (Rein et al. 1964). The paramagnetism is caused by the unpaired electron of the nitric oxide radical which is not paired at the iron binding because all *d*-electrons of the ferrous iron in the heme complex are already paired. Therefore nitric oxide can be seen as a unique spin label ligand and the nitric oxide complexes of hemoproteins were used as a model for the diamagnetic oxygen complex (Rein et al., 1972; Trittelvitz et al., 1972; Maxwell and Caughey, 1976; O'Keeffe et al., 1978b). The EPR spectra were always measured at low temperatures (77 K or lower) and are characterized by a signal centered around $g = 2$ having rhombic symmetry. In dependence on the property of the protein and its conformational state a well resolved hyperfine structure in the g_z-region is observed. Comparative measurements of nitric oxide heme complexes using ^{14}NO or ^{15}NO have shown that the hyperfine structure is caused by the interaction of the unpaired electron with the nucleus spin of the nitrogen in the NO group (Kon, 1975). Moreover, also in dependence on the conformation of the protein at certain hemoproteins a superhyperfine structure is observed caused by the electron-spin nucleus-spin interaction of the unpaired electron of the heme bound nitric oxide with the nucleus spin of the axial ligand trans to the nitric oxide group. This splitting of each hyperfine line into 3 lines is observed at nitric oxide complexes of hemoglobin (Kon and Kataoka, 1969; Chien, 1969), myoglobin (Dickinson and Chien, 1971), horseradish peroxidase and cytochrome c peroxidase (Yonetani et al., 1972) but not in the respective complexes of catalase, lactoperoxidase (Yonetani et al., 1972) and P–450 (O'Keeffe et al., 1978b). The exhibition of a superhyperfine structure requires a nucleus spin of the trans ligand. Despite the presence of such a ligand the superhyperfine structure is not observed in all cases, e.g. at isolated α, β, γ subunits and α, β hybrids of hemoglobin and mutant hemoglobins (Henry and Banerjee, 1973; Trittelvitz et al., 1975). The reason for lacking superhyperfine structure although the trans ligand possesses a nucleus spin could be (i) a very quick electron spin relaxation and/or (ii) a specific protein conformation which causes a weak binding of the trans ligand to the heme iron. On the other hand, an observable superhyperfine structure means a strong sigma-bond of the trans ligand to the heme iron.

Lacking superhyperfine structure in the EPR spectrum of NO-P–450 agrees with the postulated thiol as the 5th heme iron ligand because sulfur possesses no nucleus spin. But other trans ligands also those with a nucleus spin are not ruled out considering the above remarks.

Studies of NO derivatives of different hemoglobins have shown two types of hyperfine spectra characterized by different hyperfine splitting constants (Trittelvitz et al., 1975). The type I splitting constants amounts to 18 G(^{14}NO) or 23 G(^{15}NO) whereas the splitting constants in the type II spectra are 23 G(^{14}NO) or 32 G(^{15}NO), respectively. In the presence of detergents perturbing the native structure of the protein (Kon, 1968) and also in the presence of allosteric effectors (Rein et al., 1972) NO-hemoglobin exhibits type I spectra. Kon (1975) has shown that such hyperfine splitting (16–18 G) will be observed at 5-coordinated heme model NO complexes whereas 6-coordinated heme complexes yields hyperfine splittings of 20–23 G. From these results it was concluded that type I spectra generally may produced from 5-coordinated hemoproteins or at least by hemoproteins with a strong distorted (elongated) bond between the iron and the 5th ligand. In type II spectra, on the other hand, a strong binding of the trans ligand is expressed.

The observable hyperfine structure in the g_z-region of NO-P–450 LM and NO-P–450 CAM with splitting constants of 20 G (^{14}NO), see Table 11 (O'Keeffe et al., 1978b; Schlosnagle, 1974) suggest a 6-coordinated heme complex existing at the native NO derivatives of P–450. Contrarily, a splitting constant of 16 G (^{14}NO) for denatured P–450, i.e.

Table 11. Electron spin resonance parameters of nitric oxide complexes of cytochrome P–450

Complex	g-values			Hyperfine coupling constant	No. of lines	References
	g_x	g_z	g_y	Ag_z		
P–450$_{LM}$* · ^{14}NO (+ hexobarbital)	2.068	2.008	1.978	20	3	O'KEEFFE et al., 1978b
P–420$_{LM}$* · ^{14}NO	2.074	2.015	2.074	16	3	
P–420$_{LM}$* · ^{15}NO	2.072	2.015	2.072	22	2	
P–450$_{CAM}$ · ^{14}NO (± camphor)	2.073	2.009	1.976	19.2	3	
P–450$_{CAM}$ · ^{14}NO	2.08	2.004	1.97	20	3	SCHLOSNAGLE et al., 1974
P–450$_{CAM}$ · ^{15}NO (+ camphor)	2.074	2.009	1.975	27.4	2	O'KEEFFE et al., 1978b
P–420$_{CAM}$ · ^{14}NO	2.072	2.015	2.072	16	3	
P–420$_{CAM}$ · ^{14}NO	2.08	2.01	2.08	16	3	SCHLOSNAGLE et al., 1974

* Microsomes from phenobarbital induced male albino rats

for NO-P–420 is observed (see Table 11) being independent of the agent by which P–420 is produced. From the splitting constant of 16 G it can be derived that apparently the 5th ligand of NO-P–420, i.e. the thiol group is disrupted without binding of another ligand as it is discussed for CO-P–420 (STERN et al., 1973; REIN et al., 1977b). Preliminary optical studies of NO-P–420 LM2 show a strong diminution of the Soret aborption band as compared with the native state (unpublished results of the authors). Therefore it is suggested that in NO-P–420 not only the 5th heme iron ligand is disrupted but also the specific heme protein interaction is destroyed. Additionally, it could be found that CO-P–420 cannot be produced by gassing of NO-P–420 with CO supporting the 5-coordinated state of the heme iron in NO-P–420.

The native liver ferrous NO complexes are instable and denaturate to the P–420 complexes within a few minutes also in the presence of type I substrates which stabilize the native complex to some extent. On the other hand, the ferric NO complexes of P–450 are stable in the absence of oxygen. The ferric NO complex of P–450 CAM exhibits a stability of several hours at 25 °C between pH 7 and 8 (O'KEEFFE et al., 1978). Moreover, also the ferrous NO complex of the bacterial P–450 CAM is stable in the absence of oxygen. Different stability of NO complexes of P–450 from different species shows that the bonding interaction between the iron and the axial ligand trans to NO is sensitive to the protein moiety. But a comparison of the optical (see Table 12) and magnetic properties of the NO complexes of P–450 LM and P–450 CAM suggest very similar heme electronic and magnetic environments. From changes in the splitting of g_x and g_y signals when ^{14}NO is replaced with ^{15}NO it was concluded that at NO complexes of both species significant electron spin density is delocalized into the porphyrin nitrogens (O'KEEFFE et al., 1978). It can be suggested that the NO group in P–450 as in hemoglobin (CHIEN, 1969) and in model complexes (Nitrosyl-tetraphenyl-porphinato-1-methylimidazole-iron II) (SCHEIDT and PICIULO, 1976) is bound angularly. Because a d_π-p_π-electron back-

Table 12. Optical absorbance properties of nitric oxide complexes of cytochrome P–450 (O'KEEFFE et al., 1978b)

	Soret region		visible region	
	λ_{max}, nm	ε, $mM^{-1}cm^{-1}$	λ_{max}, nm	ε, $mM^{-1}cm^{-1}$
ferric P–450$_{CAM}$ · NO (± camphor)	430; 432*	103	α: 541 β: 571	15.0, 12.3
ferric P–450$_{LM}^{**}$ · NO (+ substrate***)	436*			
ferrous P–450$_{CAM}$ · NO (± camphor)	438, 440*	75.7	557	12.8
ferrous P–450$_{LM}$ · NO (+ substrate***)	444*			
ferrous Mb NO	420	127	548, 579	11.3, 10.1

* Maximum in the difference spectrum, the NO complex versus the NO free form
** Microsomes from phenobarbital induced male albino rats
*** The used type I substrate was 5α-androstandione-3,17-dione
Mb = myoglobin

bonding is involved in the binding between the heme iron and NO a slight turn of the NO group around the heme axis will change the π-bond between the p_π orbitals of the NO group and the d_{xz} or d_{yz}-orbitals of the iron. Such changes which are sensitively expressed in the EPR spectra should occur by changes of the steric environment about the Fe—NO bond or at changes of the strength of the bonding between the trans axial ligand as it could be observed in NO hemoglobin by the influence of allosteric effectors (REIN et al., 1972; TRITTELVITZ et al., 1972). But lacking conformational changes in the g_z signal of NO-P–450 LM were found in the presence of type I substrates (hexobarbital) (O'KEEFFE et al., 1978). Therefore it must be concluded that by substrates (type I) the oxygen binding site is not influenced, which agrees with lacking effects of type I substrates on CD spectra of P–450 in the Soret region (REIN et al., 1976) (see also 2.1.2.).

2.2.8. *Studies of the Active Site with Spin Label Molecules*

Due to their high sensitivity to the local environment several spin labeled substrate analogues have been utilized for the analysis of stereochemical properties of the active site of P–450.

Spin labels are compounds with a nitroxide group the free radical structure of which is stereochemically protected in general by four methyl groups being in the immediate neighbourhood of the NO. This arrangement results from two tertiary carbon atoms directly bound to the nitrogen of the NO which carry the methyl groups. Furthermore the spin label molecule has to bear reactive functional groups which can selectively bind to reactive groups of the protein or can react as a substrate analogue of an enzyme. The EPR spectrum of the nitroxide free radical is characterized by three hyperfine lines of equal intensity around $g = 2$ caused by the interaction of the unpaired electron with the magnetic ^{14}N nucleus ($I = 1$; $2I + 1 = 3$). The hyperfine coupling constant and the g-values depend on the orientation of the NO group in the magnetic field. This magnetic anisotropy is averaged almost to zero at rapid tumbling of the molecule, for example, being in water or other nonviscous solvents. If, however, the viscosity of the solution is increased or the label molecule is rigidly bound to a protein molecule, the rotational motion of the NO group decreases and the spectra become increasingly asymmetric, often designated as a powder or polycrystalline spectrum. The strongly immobilized spin label forms an asymmetric spectrum which is characterized by a line broadening accompanied with enlarged hyperfine splitting constants in comparison with the spectrum of the free label molecule. If additionally a magnetic dipole interaction of the spin label molecule with the iron spin occurs as it is possible in hemoproteins, a further broadening takes place and it may be difficult to measure the concentration of the bound spin label.

RUF and NASTAINCZYK (1976) and GRIFFIN et al. (1974) used the spin labeled inhibitor metyrapone which was bound to microsomal P–450 and P–450 CAM, respectively. A rotational correlation time of 0.05 μs for the bound label was estimated by RUF and NASTAINCZYK (1976). This value is characteristic for a residual motion of the spin label molecule which was ascribed by the authors to a motion of the P–450 molecule in the microsomal membrane. No indication for a rapid anisotropic motion was observed in the EPR spectrum.

The use of a spin labeled isocyanide (4-isocyanide-1-oxyl-2,2,6,6-tetramethyl-piperidine) in P–450 studies leads to apparently controversial results. On the one hand, REICHMAN et al. (1972) assigned a signal to an only weakly immobilized spin label which is not broadened by addition of $K_3[Fe(CN)_6]$ and concluded that the spin label is hydrophobically bound to the enzyme far from the heme iron. Because REICHMAN et al. (1972) used microsomes for their study it can be suggested that the observed signal of the weakly immobilized spin label is due to the label being in the membrane. This suggestion is experimentally supported by comparative studies on microsomal and solubilized P–450 showing that in the presence of microsomes the portion of a hydrophobic spin label, not bound to the enzyme, is weakly immobilized (REIN et al., 1980). This result evidences the presence of hydrophobic label molecules in the membrane and may be true in general for hydrophobic molecules in vivo, too. A signal originating from a strongly immobilized spin labeled isocyanide was not detected by REICHMAN et al. (1972). On the other hand, RUF et al. (1974) have shown by a very strong broadening of the signal that the portion of the label tightly bound to the enzyme is abolished. Comparative studies of two spin labeled isocyanides (see Fig. 24) with different chain lengths between the isocyanide and spin label group (distance between both groups 6 Å and 10 Å, respectively) show that the spin labeled isocyanide with the short chain identical with the spin labeled isocyanide used by REICHMAN et al. (1972) and by RUF et al. (1974) indeed exhibits a strong magnetic interaction of the spin label group with the heme iron (PIRRWITZ et al., 1979a). This could be shown in the EPR spectrum of the label as well as in that of the heme iron. The spin label with the long chain, however, exhibits no magnetic interaction with the heme iron as shown in the appearance of a strongly immobilized part in the EPR spectrum and binds with a 1 : 1 stoichiometry to the electrophoretically homogeneous fraction P–450 LM2 (Table 13). The determination of one binding site with high affinity for this label leads to the conclusion that the label is bound to the active site of the enzyme.

It is well known that isocyanide derivatives which are bound to the ferrous iron in different hemoproteins are also bound to ferric iron in P–450 (GRIFFIN and PETERSON, 1971). The reason of this fact may be seen in better π-electron back-donor properties of the heme iron in P–450 caused by the axial mercaptide group which is assumed to be bound to the 5th coordination site of the heme iron. Different distances between the NO group and the heme iron exist necessarily at the bound labels with different chain lengths. Therefore the spin labeled isocyanide with the short chain forms a strong dipolar interaction between the paramagnetic heme iron and the spin

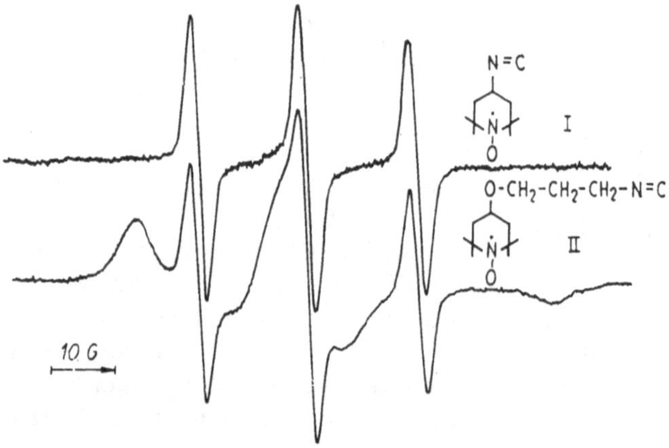

Fig. 24. EPR spectra of spin labeled alkylisocyanides without (I) as well as with (II) chain between the label and the isocyanide group in the presence of partially purified P–450 LM from phenobarbital induced rabbit liver (P–450: $1.98 \cdot 10^{-4}$ M and spin label: $5 \cdot 10^{-4}$ M). EPR spectrum of II shows portions of free and bound label. In comparison to II in the EPR spectrum of I only the free label molecule is shown. The EPR spectrum of the bound label I is broadened probably caused by magnetic interaction between the spin label and the iron spin, taken from PIRRWITZ et al. (1979)

Table 13. EPR determined apparent dissociation constants and binding sites for complexes of P–450_{LM} with spin labeled (SL) compounds (PIRRWITZ et al., 1979b)

Compound	P–450	K_{s1} [mM]	n_1	K_{s2} [mM]	n_2
SL-n-alkylamine $(CH_2)_{10}$	solub.*	0.015	0.02	1.85	5.8
SL-isocyanide I	solub.*	0.005	0.35	0.22	2.8
SL-isocyanide II	solub.*	0.002	0.34	0.78	3.9
SL-isocyanide II	LM2	0.016 ± 0.01	0.8 ± 0.13	—	—

* solub. = partially purified P–450_{LM} from rabbit liver

label group resulting in an almost completely broadened signal of the bound label. On the other hand, the isocyanide with the long chain cannot so strongly interact, consequently leading to a well pronounced signal of the bound label (see Fig. 24). The strong magnetic interaction between the label and iron spin may be explained by the direct interaction of the label group with the conjugated π-electron system of the porphyrin considering that spin density of the unpaired electron of the ferric heme iron is delocalized into the porphyrin orbitals (in myoglobin cyanide this delocalization was determined to be 25%, OGAWA et al., 1972). Such an idea agrees with results of spin labeled alkylamines with different alkyl chain lengths between the NO and amino group. Like unlabeled alkylamines these labeled compounds cause type II substrate difference spectra. The binding affinity of unlabeled alkylamines increases with longer alkyl chain lengths, the highest affinity shows octylamine (JEFCOATE et al., 1969). Spin labeled

Table 14. Optical determined apparent dissociation constants for complexes of P–450$_{LM}$ with spin labeled compounds. P–450, 4 · 10^{-6} M (solub./SL-n-alkylamines); P–450, 3.4 · 10^{-6} M (LM2/SL-n-alkylamines); P–450, 1.98 · 10^{-6} M (solub., LM2/SL-isocyanide); PIRRWITZ et al., 1979b

Compound	P–450	λ_{max} [nm]	λ_{min} [nm]	K_{s1} [mM]	K_{s2} [mM]	Distance$^*_{max}$ [Å]
SL-n-alkylamines						$>\dot{N}-O\cdots-NH_2$
$(CH_2)_1$	solub**	432	411	0.02	1.6	7
$(CH_2)_2$	solub.	433	411	0.11	3.9	8.5
$(CH_2)_2$	LM2	434	413	0.96	—	
$(CH_2)_7$	solub.	432	411	1.0	7.9	14
$(CH_2)_{10}$	solub.	432	411	0.09	1.6	18.5
$(CH_2)_{10}$	LM2	433	413	0.99	—	
SL-isocyanides***						$>\dot{N}-O\cdots-N{=}C$
I	solub.	433	411	0.002	0.15	6
I	LM2	433	412	0.008	—	
II	solub.	433	411	0.004	0.83	10
II	LM2	433	412	0.016	—	

* from Dreiding models; ** solub. = partially purified P–450$_{LM}$ from rabbit liver;
*** formula see Fig. 24

alkylamines exhibit an inverse behaviour. The alkylamine with the shortest chain shows the highest affinity to P–450 (PIRRWITZ et al., 1977) (see Tab. 14). The high affinity of this label should be caused by the label group being involved in the binding of this substrate. Considering that the amino group of alkylamines is bound to the heme iron (JEFCOATE et al., 1969; ULLRICH et al., 1975a) it is suggested that the pyrrolidine ring of the label group with the shortest chain (7 Å between the heme iron and the NO group) is included in the hydrophobic binding site for substrates. Because the porphyrin ring has a diameter of about 8 Å (HOARD, 1971) it can be inferred that the porphyrin ring is involved in the binding site not only of the labeled isocyanide and alkylamine with the short chain but also in the binding of substrates being hydroxylated.

Further information about the architecture of the active site of P–450 results from studies with a spin labeled type I substrate. The primary aromatic amine 1,1,3,3-tetraethyl-2-oxy-5-amino-isoindoline induces a type I substrate differences spectrum and binds with a high affinity of $K_s = 15\,\mu M$ to P–450 LM2 (REIN et al., 1980). The binding occurs with a 1 : 1 stoichiometry and identical binding constants were calculated from EPR as well as from optical titration experiments for the binding to the high affine binding site. Therefore it can be concluded that the spin labeled type I substrate analogue is bound to the active site of the enzyme. The spin labeled aminoisoindoline can be chemically regarded as an aniline derivate which are known to form type II spectra. Surprising therefore is the formation of a type I spectrum by the spin labeled amino isoindoline the reason of which can be seen in the stereochemical properties of the molecule. Due to the condensation of the aromatic benzene ring with the heterocyclic label group the size of the spin label significantly increases as compared with aniline. Moreover, this spin label molecule is characterized by 4 ethyl groups protecting the radical state instead of 4 methyl groups usually occupying these positions. The largest diameter of the spin label molecule in the plane of the 4 ethyl groups was quantitated from the Dreiding model for the distance between the two opposite ethyl groups with about 10 Å. As compared with aniline the diameter of the spin label is more than twofold increased. This difference of the diameters between both different molecules suggests a steric hindrance of the spin labeled amino isoindoline inhibiting the aromatic group to bind to the iron as a prerequisite for the formation of a type II substrate difference spectrum. Furthermore, from the Dreiding model a distance of 7 Å between the amino and NO group is estimated for the spin labeled amino isoindoline. But because the amino group of the spin labeled amino isoindoline does not contact the heme iron, consequently the distance between the NO group and the heme iron should be significantly larger than 7 Å. This assumption is indeed true because in the EPR spectrum of the spin labeled amino isoindoline a line broadening due to a dipole dipole interaction between the bound spin label and the heme iron is not observed. The inhibited binding of the amino group of the spin labeled amino isoindoline to the heme iron means that the heme pocket of P–450 is supposed to be larger than in other hemoproteins (REIN et al., 1976) but smaller than the largest diameter of the spin labeled amino isoindoline, i.e. 10 Å.

2.2.9. Nuclear Magnetic Resonance

High resolution proton magnetic resonance (NMR) spectroscopy of diamagnetic but also of paramagnetic macromolecules has recently been employed for investigating structure and structure-function relationship especially in metalloproteins.

Two effects enhance the value of NMR in the study of hemoproteins. (i) The ring-current shifts which can shift resonances from protons positioned above or below the plane of an aromatic ring. The aromatic porphyrin ring exhibits very large ring current shifts and the affected resonance peaks are shifted to a greater extent than can be realized by other aromatic rings. (ii) The very large paramagnetic hyperfine shift of heme-bound protons and side chains due to the electron paramagnetism of the heme iron which shift the resonance peaks outside of the normal diamagnetic range.

The first high resolution NMR study of P–450 CAM was made by KELLER et al. (1972). The down field side of proton NMR spectra of the two high spin species ferrous and ferric P–450 CAM in the presence of the 6 mM camphor was studied together with the low spin cyanide complex of ferric P–450 CAM. Several resonances with small intensities could be observed in the region above +10 ppm. These large downfield shifts are due to hyperfine interactions (case ii) with unpaired electrons and thus show that in the three forms of P–450 CAM mentioned above the heme group is in a paramagnetic state. The temperature dependence of the paramagnetic hyperfine shift corresponds to a CURIE-type behaviour.

MORISHIMA et al. (1974) and IIZUKA (1975) compared data of the ferric P–450 CAM cyanide complex with those of other ferric hemoprotein cyanide complexes and suggested that the signals at 23 and 14.7 ppm of P–450 CAM cyanide are due to methyl and presumably vinyl protons at the side chain of protoheme, respectively. The origin of the third peak at −12 ppm which did not obey CURIE's law with varying temperature is unknown at present. Additionally it was found that the hyperfine shift of the three lines is changed if the substrate camphor was omitted meaning that the conformation of the cyanide complex of P–450 CAM is changed by camphor.

BERZINIS and TRAYLOR (1979) provided evidence for the binding of a mercaptide to the heme iron in a P–450 model complex by NMR spectroscopy. The CO complex of ferroprotoheme-bis-dimethyl-amide in the presence of phenethyl mercaptide forms a typical CO-P–450 spectrum. The complex formation induces a strong chemical shift of the methylene protons of phenethyl mercaptide ion which has not been observed on the corresponding mercaptan complex. BERZINIS and TRAYLOR (1979) have also studied the ^{13}CO compound of their model complex. A large upfield shift of the ^{13}CO resonance was found after its binding. This upfield shift is larger than obtained from CO complexes of hemoglobin, myoglobin and heme complexes containing imidazole as 5th heme iron ligand (MOON et al., 1977). Therefore from the striking shift of the methylene proton resonances together with the typical optical spectrum of the phenethyl mercaptide CO heme complex it can be concluded that really the mercaptide sulfur is ligated to the heme iron at the 5th coordination position as it is assumed for P–450, too. PARMELY and GOFF (1980) investigated some high spin ferrous heme models of P–450 by the NMR method. NMR peaks of the heme iron bound butylmercaptide could be identified. The shift of the methylene protons of the coordinated butylmercaptide is with 250 ppm the largest one observed in NMR spectroscopy of iron porphyrins. This large value supports a very efficient spin delocalisation mechanism from the iron to the mercaptide which may be also true for ferrous P–450.

3. Reaction Cycle of Cytochrome P-450 — Desrciption of Intermediate Steps

3.1. *Thermodynamic of the Overall Reaction*

In hydroxylating reactions catalyzed by the monooxygenatic enzyme system with P–450 as terminal oxidase, according to equation [25] it was shown by isotopic experiments that only one atom of molecular oxygen is incorporated into the substrate (HAYAISHI, 1962; HAYANO et al., 1955; MASON, 1957).

$$\begin{aligned} &RH + O_2 + NADPH + H^+ \\ &\rightarrow ROH + H_2O + NADP^+ \end{aligned} \qquad [25]$$

Therefore in the overall reaction [25] two reactions which require high energy of activation are involved, the cleavage of the oxygen molecule as well as of the CH-bond of the substrate (RH). The monooxygenase system is able to reduce the activation energy by decomposing the overall reaction [25] into several intermediate steps. For understanding the catalytic power of the monooxygenase one has to consider the thermodynamics of the overall reaction (JUNG and RISTAU, 1978).

$$NADPH + H^+ \rightarrow H_2 + NADP^+ \quad \Delta G = 19.3 \text{ kJ/mole} \qquad [26]$$

$$^3O_{2(g)} \rightarrow 2O_{(g)} \quad \Delta G = 460.5 \text{ kJ/mole} \qquad [27]$$

$$H_{2(g)} + O_{(g)} \rightarrow H_2O_{(fluid)} \quad \Delta G = -470.1 \text{ kJ/mole} \qquad [28]$$

$$RH_{(fL)} + O_{(g)} \rightarrow ROH_{(fluid)} \quad \Delta G = -393.9 \text{ kJ/mole} \qquad [29]$$

$$\sum \Delta G = -384.2 \text{ kJ/mole}$$

(RH, e.g. cyclohexane)

Because of the negative free enthalpy reaction [25] proceeds spontaneously from the thermodynamic point of view. The reaction rate, however, is very slow caused by the two reaction steps which require high energies, the high dissociation enthalpies of the oxygen molecules (460.5 kJ/mole) and of the CH-bond in the substrate (about 418 kJ/mole) (HOFMANN, 1972). Neglecting its enzymatic character, a value of about 418 to 460 kJ/mole is expected for the activation energy of the overall reaction. For the P–450 catalyzed reaction, however, an activation energy only of about 38 to 71 kJ/mole was found (AKHREM et al., 1977). That means P–450 causes a decrease of the mentioned dissociation enthalpies. Because of lacking substrate specificity of P–450 the interaction between the enzyme and the substrate cannot be the essential effect of the catalytic action. Therefore a substrate activation as well as a substrate induced activation of the hydroxylation mechanism is unlikely. Contrary, the interaction between molecular oxygen and P–450 is a specific one and leads to the oxygen activation. From this point of view the overall reaction can be divided into two main parts, the formation of the activated oxygen species and the incorporation of the oxygen atom into the substrate. Even these two processes are in the centre of the discussion of the question: "How can P–450 catalyze the substrate hydroxylation?" Different opinions about these processes exist in the literature which will be discussed below.

3.2. Formation of the Activated Oxygen Species of Cytochrome P–450

3.2.1. Binding of Substrates to Cytochrome P–450

Contrary to other enzymes P–450 binds a variety of substrates with different chemical structures. The binding to the ferric enzyme is accompanied by changes of the optical absorption spectrum caused by an indirect or direct influence of the substrate on electronic transitions of the heme chromophore. These changes are relatively small but large enough for the titration of difference spectra in the Soret region. Depending on the nature of the difference spectra 3 types of substrates can be differentiated.

At the binding of type I substrates a trough at 417 nm and a maximum at 387 nm is produced. The binding of type II produces a maximum in the range of 420 nm to 430 nm and a minimum at about 400 nm. A further type of substrates are the so-called inverse or reversed type I substrates producing difference spectra with peaks at the same position as they are formed by type I substrates but maximum and minimum are reversed.

The substrate induced change in the absorption spectrum of P–450 is a blue or a red shift of the Soret band at binding of type I or type II substrates, respectively. An explanation of the change of the optical spectrum at substrate binding is given by the changes of the spin state of the heme iron. The two existing spin states of the ferric heme iron, the high spin and the low spin state, exhibit individual optical spectra (SMITH and WILLIAMS, 1968). The low spin and the high spin state of P–450 LM2 are characterized by absorption maxima at 417 nm, 545 nm, 575 nm and 387 nm, 500 nm, 630 nm, respectively (HAUGEN and COON, 1976). The formation of a positive maximum of the high spin Soret band (387 nm) and of a minimum for the low spin Soret band (417 nm) in the difference spectrum therefore means a spin state change from low to high. A reverse direction of the low spin and high spin band, respectively, means a shift from high to low. Additional changes of the position of the low spin and high spin band, respectively, especially produced by type II substrates have to be assigned to substrate specific effects.

A lot of work has been done to evaluate the binding sites for the three types of substrates. The situation seems to be rather complicated because the induced spectral type can depend (i) on the concentration of the substrate (SCHENKMAN et al., 1969) (ii) on the source of P–450 or (iii) on the inducer chosen (IMAI and SATO, 1966; SCHENKMAN et al., 1972). Partially the difficulties are due to the microsomes used in most studies. It is known that the microsomes contain isozymes with different concentrations depending on source and inducer. The substrate difference spectra therefore are the sum of all reacting isozymes which may bind the substrates with different affinities and with different spectral effects.

For type II substrates as various nitrogeneous bases (amines, anilines, pyridines, imidazoles) (SCHENKMAN et al., 1967; SCHENKMAN, 1970; IMAI and SATO, 1967; JEFCOATE et al., 1970) and isocyanides (GRIFFIN and PETERSON, 1971; PIRRWITZ et al., 1979) it seems to be clear that the heme iron represents the binding site (JEFCOATE et al., 1969). Some type II substrates (aniline) can displace carbon monoxide from ferrous P–450 (OMURA and SATO 1964) and cyanide from ferric P–450 (MATSUBARA and TOCHINO, 1976), clearly indicating an interaction with the heme iron. The binding of aniline to reduced P–450 can also be detected directly by the optical changes which occur by titration (IMAI and SATO, 1968). On the other hand, aniline is hydroxylated and therefore aniline must also bind at the type I binding site (SCHENKMAN, 1970). Pure type II substrates (iron ligands) are inhibitors of the enzymatic function (metyrapone) because they compete with oxygen.

For type I and reversed type I substrates the binding to two sites in dependence on the substrate concentration is assumed by SCHENKMAN et al. (1972). These authors found that hexobarbital and aminopyrine forms a type I spectrum at low concentrations, whereas at higher concentrations a reversed type I spectrum occurs. But this behaviour appears only by use of liver microsomes of benzopyrene-treated rats. Therefore this result can also be explained by different patterns of the isozymes caused by different inducers. Different isozymes can exhibit different substrate affinities and spectral types for the same substrate. Therefore using microsomes the integral spectral type can change with the concentration of the substrate because the saturation of the isozymes is obtained at different concentrations.

The occurrence of different spectral types (type I and reversed type I), also if the same binding site for both spectral types is assumed, may be explained by two conformations of the enzyme exhibiting different affinities for substrates. Such different affinities could be proved for the high spin and low spin conformation of the isolated isozyme P–450 LM2 (RISTAU et al., 1979) (see 3.2.3.). The existing equilibrium between the two conformations is shifted to the high spin conformation by type I substrates which have higher affinities to the high spin conformation. On the other hand, the conception of SCHENKMAN et al. (1972) assuming different binding sites for both types of substrates (type I and reversed type I) can also be explained with the two-conformation model. For the typical reversed type I substrate n-butanol VAN DEN BERG et al. (1979a, b) assumed the simultaneous binding to the heme iron (type II site) and to the type I binding site, but the optical effect of the type II binding site dominates. The binding of n-butanol to the heme iron was also assumed by YOSHIDA and KUMAOKA (1975) and by NEBERT et al. (1976). But the binding of n-butanol to the type I site is supported by the finding of TESCHKE et al. (1975) showing that n-butanol is metabolized via the alcohol-oxidizing system in which P–450 may be involved.

Summarizing these results it can be concluded that two binding sites for substrates exist. One binding site (type I) should be localized at the protein in the heme environment where the porphyrin could also be partly involved. The second site (type II) represents the heme iron which binds type II substrates directly at the sixth ligand position. Lipophilic compounds with a nitrogen, organic sulphides, thiols, selenides and oxygen ethers are to be considered as such typical ligands. The reason for the formation of inverse type I spectra can be seen in a favoured binding to the type I binding site of the low spin conformation but in some cases also the binding to the heme iron may be possible. Also the simultaneous binding to both binding sites could be possible.

At the binding of type I substrates a shift of the spin equilibrium to the high spin state is induced. But this conversion from the low to the high spin state occurs only partially (Rein et al., 1977a; Werringloer et al., 1979). This result observed at microsomes is differently interpreted. Waterman et al. (1973), Ebel et al. (1978), Werringloer et al. (1979) assumed that the isozyme which binds the substrate completely undergoes the spin conversion up to 100% high spin but isozymes which are not able to bind a substrate remain in the low spin state. To overcome these difficulties several different substrates were simultaneously added (Ebel et al., 1978), but again the conversion to the high spin state of the microsomal P–450 occurs only partially (70—80%). Another explanation assumes that substrate-saturated P–450 is not completely converted to the high spin state. This idea is supported by the temperature dependence of the substrate saturated P–450 LM2 which is larger than with substrate-free P–450 LM2 (Ristau et al., 1979). This result agrees with a thermal spin equilibrium existing also in the type I substrate saturated state of P–450 which is only possible if the spin conversion by substrates occurs incompletely. A different extent of the spin shift of P–450 LM2 caused by various type I substrates agrees also with an incomplete spin conversion (unpublished results of the authors). Some authors explain the temperature induced increase of the high spin content in microsomes by an increased binding of an endogeneous substrate at higher temperature (Ebel et al., 1978; Werringloer et al., 1979). In this case the spin shift caused by an exogeneous substrate should be diminished with higher temperature considering the hypothesis of complete spin conversion by substrates. Substrate titration at different temperatures, however, demonstrate that the saturation extinction increases with rising temperature (Ebel et al., 1978). The two-conformation model of P–450 explains the substrate I induced high spin shift by the preferential binding of type I substrates to the high spin form. A complete conversion to the high spin state by substrates requires a much higher affinity of the substrate to the high spin conformation as compared with the low spin conformation. However, it was found that the substrate affinity to the high spin conformation is only moderately increased as compared with the low spin conformation. Several attempts were made to measure the substrate induced spin conversion by means of EPR (Waterman et al., 1973; Kumaki et al., 1978; Ebel et al., 1978). The problems connected with this approach are discussed in 2.2.1.

3.2.2. Spin Equilibrium

The existence of a spin equilibrium in ferric hemoproteins between the high spin state with a total spin of $S = 5/2$ and the low spin state with a total spin of $S = 1/2$ is well known. It was found in certain methemoglobin and metmyoglobin derivatives (George et al., 1964) in peroxidase compounds (Tamura, 1971) as well as in cytochrome c peroxidase (Iizuka et al., 1968) and in compounds of beef liver catalase (Rein et al., 1968). A spin equilibrium could be detected also in P–450 from the bacteria *Pseudomonas putida* (Sligar, 1976) and in liver microsomal P–450 (Rein et al., 1977a, c).

A spin equilibrium can be proved by magnetic measurements at different temperatures. From the experimental point of view, however, it is difficult to analyse small changes of the spin state with the magnetic balance. EPR is a convenient approach to determine the spin state

$S = 5/2$ and $S = 1/2$ because both states exhibit specific spectra. But EPR is not suited to determine a spin equilibrium at room temperature. Because of very short electronic relaxation times of heme complexes ($T_{1e} \approx 10^{-10}$ s) at room temperature line broadening renders EPR signals unobservable. Due to a linkage of optical properties with the spin state also optical measurements are appropriate to determine the spin state of the iron. As demonstrated by SCHELER et al. (1957) and HAVEMANN and HABERDITZL (1958) the values of the magnetic susceptibility of certain methemoglobin derivatives are linearly correlated with the position of the Soret band. With increasing magnetic susceptibility the Soret band is shifted to shorter wavelengths, and vice versa, with decreasing magnetic susceptibility a red shift of the Soret band is observed. The shift results from the fact that the observed Soret band of a ferric hemoprotein possessing both spin states simultaneously is composed of two band with different positions: the high spin band at shorter wavelengths and the low spin band at longer wavelengths. The exact position of the Soret band of the respective spin form varies with the hemoprotein. At hemoglobin and myoglobin the position of the low spin and high spin band are relatively close together: the pure high spin compound methemoglobin fluoride exhibits the Soret band at 406 nm; the pure low spin compound methemoglobin cyanide at 418 nm. This hemoprotein therefore exhibits an overlapping of the high spin and low spin Soret band if both spins states exist in the respective complex. In dependence on the spin state the position of this composed Soret band will be found between the two extrema 406 and 418 nm. On the other hand, the low spin and the high spin Soret band at 417 nm and 387 nm, respectively, in P–450 exhibit a much greater distance in their positions. Therefore a correlation between the position of the Soret band and the spin state does not exist in P–450, rather both Soret bands are observed in the presence of both spin states. Shifts of the spin states are directly indicated by a change of the amplitudes of both bands.

3.2.3. Analysis and Experimental Data of the Spin Equilibrium of Cytochrome P–450 LM2

Based on theoretical considerations of the spin equilibrium of hemoproteins in the following quantitative data of P–450 LM2 (RISTAU et al., 1978, 1979) are described. In P–450 LM2 a thermal spin equilibrium in the absence as well as in the presence of a substrate is observed. Therefore two equations for these spin equilibria can be formulated

$$K_1 = \frac{[\text{P–450}_{\text{high}}]}{[\text{P–450}_{\text{low}}]}, \qquad [30]$$

$$K_2 = \frac{[\text{P–450} \cdot S_{\text{high}}]}{[\text{P–450} \cdot S_{\text{low}}]} \qquad [31]$$

Assuming that the substrate S can be bound to the low spin form as well as to the high spin form two equilibria for the binding of substrates are obtained.

$$K_3 = \frac{[\text{P–450}_{\text{low}}] \cdot [S]}{[\text{P–450} \cdot S_{\text{low}}]}, \qquad [32]$$

$$K_4 = \frac{[\text{P–450}_{\text{high}}] \cdot [S]}{[\text{P–450} \cdot S_{\text{high}}]} \qquad [33]$$

Under the condition that substrate binding only shifts the spin equilibrium and an additional substrate specific influence on the enzyme can be excluded spin and substrate equilibria can be described by the following scheme

$$\begin{array}{ccc} \text{P–450}_{\text{high}} & \xleftrightarrow[\pm S]{K_4} & \text{P–450} \cdot S_{\text{high}} \\ \updownarrow K_1 & & \updownarrow K_2 \\ \text{P–450}_{\text{low}} & \xleftrightarrow[\pm S]{K_3} & \text{P–450} \cdot S_{\text{low}} \end{array} \qquad [34]$$

The experimentally available substrate binding constant K_s is given by

$$K_s = \frac{[\text{P-450}_{\text{low}} + \text{P-450}_{\text{high}}] \cdot [S]}{[\text{P-450} \cdot S_{\text{low}} + \text{P-450} \cdot S_{\text{high}}]} \quad [35]$$

By this thermodynamic model the shift of the spin equilibrium by substrates can be explained assuming that for the low spin and the high spin form different affinities to the substrate exist.

The spin equilibria constants can be determined by the analysis of temperature difference spectra. But the exact determination of K_2 must consider full substrate saturation in the temperature range measured. Because this is not sure at any temperature, especially not at low temperatures corresponding to the predominantly hydrophobic bond, it is useful to titrate the substrate at different temperatures and to calculate K_2 from these data. The calculation of the spin equilibrium constant in the absence of a substrate K_1 follows from the quantitative formulation of the temperature difference spectrum

$$\Delta A = c\,(\alpha \text{ sample} - \alpha \text{ reference}) \cdot \Delta\varepsilon \quad [36]$$

α, high spin content; ΔA, difference in absorption; $\Delta\varepsilon$, molar difference in molar extinction; c, concentration of the enzyme.

In the following equations the high spin content is formulated as temperature dependent equilibrium constant

$$\alpha_{\text{sample}} = \frac{K_{1,T}}{1 + K_{1,T}} \quad [37]$$

$$\alpha_{\text{reference}} = \frac{K_{1,\vartheta}}{1 + K_{1,\vartheta}} \quad [38]$$

T, ϑ, temperature in the sample and reference cuvette, respectively.

The calculation of the dissociation constants of the high spin and low spin substrate complexes K_3 and K_4, respectively, are based on the spin and substrate equilibria. Under the condition that the optical spectrum depends only on the spin, excluding substrate specific changes from the thermodynamic model the following epuations can be derived.

$$\Delta A = \Delta A_\infty \frac{S}{K_s + S} \quad [39]$$

$$\Delta A_\infty = \frac{cK_1(K_4 - K_3)\,\Delta\varepsilon}{(1 + K_1)(K_4 + K_1K_3)} \quad [40]$$

$$K_s = \frac{(1 + K_1)\,K_3}{1 + K_2} = \frac{(1 + K_1)\,K_3K_4}{K_4 + K_1K_3} \quad [41]$$

ΔA_∞, molar extinction in the substrate difference spectrum under saturation conditions; K_s, substrate overall dissociation constant, this means including the dissociation of the substrate from the low spin as well as from the high spin form.

The dissociation constant K_s can be determined from the substrate difference spectra (equation [35]). The constants K_3 and K_4 can be calculated from equations [39]—[41] because these formulas contain the experimentally accessible parameters ΔA_∞ and $\Delta\varepsilon$.

But also the spin equilibrium constant K_2 can be calculated from these formulas and therefore a possibly imperfect direct determination caused by an insufficient saturation of the enzyme with the substrate can be proved. The explicit equations for K_2, K_3 or K_4 can be formulated.

$$K_2 = \frac{K_1 \cdot K_3}{K_4} \quad [42]$$

$$K_3 = \frac{K_s}{1 - \Delta A_\infty(1 + K_1)/\Delta\varepsilon} \quad [43]$$

$$K_4 = \frac{K_s}{1 + \Delta A_\infty(1 + K_1)/(K_1 \cdot \Delta\varepsilon)} \quad [44]$$

The calculated equilibria constants for P-450 LM2 are shown in Table 15. From the temperature dependence of the equilibria the thermodynamic parameters were calculated (Table 16). A comparison of K_3 and K_3 demonstrates that the substrate (benzphetamine) is bound about 5 times stronger to the high spin conformation. In agreement with the thermodynamic model this is the reason for the

Table 15. Various parameters characterizing the spin equilibrium and the substrate binding of P–450 LM2: (i) the apparent K_s-values of the binding of benzphetamine. (ii) spin equilibria without substrate, K_1 and (iii) in presence of substrate, K_2. (iv) dissociation constants of the binding of benzphetamine to the low spin form, K_3 and to the high spin form, K_4. (v) percentage of the high spin portion without and in the presence of benzphetamine

T (°C)	K_s (mM)	$\varepsilon\infty$ (mM)	K_1	K_2	K_3 (mM)	K_4 (mM)	high spin (%) P–450	P–450-S
11	0.63	7.7	0.043	0.23	0.73	0.15	4	17
13	0.49	10.0	0.049	0.26	0.59	0.10	5	19
16	0.34	9.1	0.059	0.31	0.41	0.088	6	24
20	0.28	12.4	0.076	0.39	0.37	0.07	7	28
24	0.20	14.3	0.10	0.48	0.28	0.053	9	33
28	0.18	14.4	0.12	0.59	0.25	0.054	11	37
32	0.135	16.3	0.16	0.72	0.20	0.043	14	42
37	0.125	17.3	0.21	0.91	0.20	0.045	17	48

Table 16. Thermodynamic constants of the four equilibria included in the thermodynamic mode for the interrelationship between substrate binding (benzphetamine) and spin equilibrium

Equilibria	K (20 °C)	ΔH (kJ/mole)	ΔS (J/mole/grad)	ΔG (kJ/mole)
$K_1 = \frac{[\text{P–450}_{\text{high}}]}{[\text{P–450}_{\text{low}}]}$	0.08	-44.3 ± 0.8	-130 ± 5	-6.3 ± 0.8
$K_2 = \frac{[\text{P–450} \cdot S_{\text{high}}]}{[\text{P–450}_{\text{low}}]}$	0.39	-38.0 ± 0.8	-121 ± 5	-2.5 ± 0.8
$K_3 = \frac{[\text{P–450}_{\text{low}}] \cdot [S]}{[\text{P–450} \cdot S_{\text{low}}]}$	0.37 mM	31.4 ± 2.1	171 ± 9	-18.8 ± 2.5
$K_4 = \frac{[\text{P–450}_{\text{high}}] \cdot [S]}{[\text{P–450}_{\text{high}}]}$	0.07 mM	26.3 ± 2.1	171 ± 9	-23.8 ± 2.5

experimentally observable shift of the spin equilibrium to the high spin state by type I substrates and also for the formation of type I substrate difference spectra (see equations [39]—[41]). From the thermodynamic parameters in Table 16 it can be seen that the binding of benzphetamine to P–450 is endothermic which is characteristic for type I substrates (MISSELWITZ et al., 1976, 1977). The calculated high entropy value of the binding reaction means a hydrophobic binding overcompensating the endothermic character of the reaction. This means that the binding of benzphetamine to P–450 is an entropy driven process and it can be assumed that this is true for type I substrates in general because the binding of other type I substrates show the same behaviour (MISSELWITZ et al., 1977).

The high entropy values of the spin equilibria of P–450 LM2 in the absence and in the presence of substrates is a characteristic phenomenon of spin equilibria (MORASSI et al., 1973; GOODWIN, 1976). This large value results from the large change in the degree of thermodynamic freedom during the spin transition of a metal complex inbedded in a protein and therefore the entropy is much greater than the pure spin change entropy ($R \cdot \ln 3 = 9.2$ J/mole/K). The excess part is named configuration entropy by König and Ritter (1974) and is much larger at protein metal complexes than in organic metal complexes. One reason for the configuration entropy in hemoproteins can be seen in the cooperative breaking of van der Waals heme protein contacts (Otsuka, 1970). From the statistical thermodynamic model described by Ikegami (1977) follows that entropy and enthalpy of the breaking of van der Waals contacts are proportional to the number of the heme protein contacts (N) according to

$$\Delta S = -\beta N \quad \text{and} \quad \Delta H = -\alpha N. \qquad [45]$$

The proportionality factors β and α denote chain entropy and contact energy respectively. For the chain entropy values between 8.4 and 16.7 JK^{-1} have been found in several proteins (Kanehisa and Ikegami, 1977). The contact energy yields 0.62 or 2.1 kJ/mole for aliphatic or aromatic contacts, respectively (Nemethy and Scheraga, 1962).

Based on these considerations the number of heme protein contacts involved in the spin transition of P–450 LM2 were estimated. The calculated entropy of $-144\ JK^{-1}$ exceeds the entropy value of $-9.2\ JK^{-1}$ due to the change of the spin state ($\Delta S = -R \ln 3 = -9.2\ JK^{-1}$) about 15 fold. Taking the entropy value for the spin transition of P–450 LM2 together with the probable value of the chain entropy of $-8.4\ JK^{-1}$ the number of heme protein contacts (equation 45) is estimated to be 15. With this number of van der Waals contacts which are cooperatively broken by the transition from the low spin to the high spin state and the mean value for the contact energy of 1.3 kJ/mole an enthalpy value of 18.8 kJ/mole was calculated for this process. To obtain the pure enthalpy for the spin transition comparable with organic metal complexes this value (18.8 kJ/mole) must be subtracted from the experimental value (Table 16) yielding an enthalpy of approximately 25.1 kJ/mole for P–450 LM2.

In agreement with the relatively small benzphetamine induced spin shift of P–450 LM2 the thermodynamic parameters of the spin equilibrium are only slightly influenced by the substrate used. The somewhat larger spin equilibrium constant in the presence of benzphetamine is mainly caused by a higher enthalpy value. This higher enthalpy value can be explained by the higher substrate binding enthalpy of the high spin form because the spin equilibrium constant K_2 according to the thermodynamic model does not exist independently of the other equilibrium constants but is determined by them. These results from the closed cycle of the thermodynamic model match the following relation

$$\Delta G_2 = \Delta G_1 + \Delta G_3 - \Delta G_4. \qquad [46]$$

The thermodynamic parameters of the substrate binding are typical for a hydrophobic binding. The high positive entropy value is explicable by the release of immobilized water from the binding site at the binding of the substrate. According to Nemethy and Scheraga (1962) 1 mole released water yields approximately 5.4 JK^{-1} (true especially for aliphatic surfaces). Using this value together with the binding entropy of benzphetamine to P–450 LM2 it can be estimated that approximately 33 water molecules are released at the binding of benzphetamine.

3.2.4. Importance of the Spin Equilibrium for the Reduction Rate

Optical and EPR studies show that ferric heme iron of P–450 LM2 as well as of P–450 CAM in the absence of substrates is predominantly in the low spin state. On the other hand, from measurements of the magnetic susceptibility (Champion et al., 1975a) it is known that ferrous P–450 CAM is in the high spin state which should be true also for ferrous P–450 LM2. From the statistic point of view the reduction of the ferric low spin state is unfavoured because this is a spin forbidden reaction. Only the reduction of the ferric high spin state occurs with constant total spin considering that the total spin of the reduced enzyme is $S = 2$.

$$\left.\begin{array}{l} Fe^{3+}(S = \pm 1/2) + e^-(S = \pm 1/2) \\ Fe^{3+}(S = \pm 5/2) + e^-(S = \pm 1/2) \end{array}\right\} \rightarrow Fe^{2+}(S = 2) \qquad [47]$$

The assumption of a spin dependent reduction reactivity of P–450 is supported by studies of the reduction rate of heme complexes (Kadish and Davis, 1973). Kadish and Davis (1973) could show an about 100 times faster reduction of the heme iron if the reduction is not accompanied by a spin state change. It can be postulated that the low spin state of P–450 is the inactive state of the enzyme because in this case the reduction, the second elementary step of the reaction cycle is inhibited. By this fact unnecessary consumption of NADPH and the formation of hydrogen peroxide is inhibited. Only after the binding of a substrate (type I) producing a shift of the spin equilibrium to the high spin state the reduction reaction is favoured and by this event the enzyme function is triggered. This proposal is supported by the well known fact that in the presence of type I substrates the reduction rate is enhanced (Diehl et al., 1970; Schenkman, 1972; Blanck et al., 1976; Matsubara et al., 1976; Björkhem, 1977; Imai et al., 1977; Sato et al., 1979). Moreover, for certain type I substrates a correlation between the amount of the spin shift and the reduction rate could be shown (Rein et al., 1979) supporting the hypothesis of triggering the second elementary step of the reaction cycle by the substrate induced high spin state. This hypothesis does not exclude that certain substrates can induce additional conformational changes of P–450 leading to a reduction favoured state as it was described by Holtzman et al. (1973).

Summarizing these considerations it can be concluded that P–450 responses to the binding of a variety of compounds with the uniform reaction of a spin state change, thereby triggering the process of oxygen activation. The broad substrate specifity excludes a substrate specific structure of the active site. Therefore the substrate binding energy need not necessarily be coupled with the hydroxylation mechanism of the substrate.

3.2.5. Relationship between Spin State and Redox Potential

The transition from the ferric to the ferrous state of P–450 is characterized by a redox potential which can be calculated according to the NERNST equation

$$E = E_0 + \frac{RT}{F} \ln \frac{c_{\text{oxidized state}}}{c_{\text{reduced state}}} \qquad [48]$$

The NERNST equation is only valid for electron transitions within the same chemical complex. If we use this equation for the determination of the redox potential we have to ask for the structure of the heme complex in the oxidized and the reduced state.

Structure of the ferrous heme complex

By measurements of the magnetic susceptibility (Champion et al., 1975a) as well as by Mößbauer studies (Sharrock et al., 1976) the reduced P–450 CAM has been shown to exist in the high spin state. From these studies it is concluded that the heme iron as in deoxyhemoglobin is in the 5-coordinated state. Moreover, similar spectral properties of reduced P–450 CAM (Peterson, 1971) and reduced P–450 LM (HAUGEN and COON, 1976) support the conclusions that P–450 LM in the reduced state is also 5-coordinated.

Structure of ferric heme complex

The structure of several ferric hemoproteins are studied by X-ray analysis (DEATHERAGE et al., 1976a, b). These results demonstrate that all low spin hemoproteins are 6-coordinated whereby the iron is roughly in the prophyrin plane. The iron of high spin ferric hemoproteins, on the other hand, is out of this plane directed to the strongly bound 5th ligand. High spin ferric hemoproteins are also usually 6-coordinated, e.g. methemoglobin and metmyoglobin and some complexes of these hemoproteins. However, some exceptions are observed. VUK-PAVLOVIĆ and SIDERER (1977) concluded from EPR studies of peroxidase using $H_2{}^{17}O$ a 5-coordinated state of the heme iron in this hemoprotein. Model complexes of high spin P–450 have been investigated by X-ray analysis by TANG et al. (1976) and were found to be 5-coordinated. SLIGAR (1976) concludes from kinetic experiments that the transition from the low spin to the high spin state in P–450 CAM is connected with the dissociation of the 6th heme iron ligand. On the other hand preliminary kinetic studies of the spin equilibrium of P–450 LM (RISTAU et al., 1979) show that this equilibrium cannot be connected with the dissociation of a heme iron ligand because the spin equilibrium relaxation time is shorter than 100 μs. Therefore the heme iron of high spin ferric P–450 LM should be most likely in the 6-coordinated state.

From the validity of the NERNST-equation it is necessary to conclude that in agreement with the 5-coordinated ferrous state the 5-coordinated ferric state is the actual precursor of the reduced state. On the other hand, from the fast spin equilibrium a small dissociation constant of the 6th heme iron ligand one has to expect because this apparently occurs to such a small extent that it is not detectable.

$$\text{P–450} - \text{6th ligand} \xrightarrow{K_D} \text{P–450} + \text{6th ligand} \qquad [49]$$

Based on the spin equilibrium data of P–450 LM2 (Table 15) (RISTAU et al., 1979), the redox potential of −330 mV for P–450 LM (GUENGERICH et al., 1975) and the assumption that the 5-coordinated ferric high spin state has the same redox potential as this compound of P–450 CAM namely −170 mV (SLIGAR and GUNSALUS, 1976) a dissociation constant of the 6th heme iron ligand of $K_D = 0.03$ at 20 °C by means of the NERNST-equation was calculated (REIN et al., 1979). This small dissociation constant corresponds well to the observed fast spin equilibrium of P–450 LM and means that only 3% of the ferric high spin iron are in the 5-coordinated state.

Controversial results of the redox potential of P–450 LM in the presence of substrates (SIES and KANDL, 1970; GUENGERICH et al., 1975) lead SLIGAR et al. (1979) to reinvestigate the problem. In the presence of several substrates they observed different shifts of the redox potential to more positive values which were correlated with the spin state. In general it can be stated that if any change of the redox potential of P–450 LM by substrates may occur this change is not so intensive as in the case of P–450 CAM being −340 mV without and −175 mV with camphor (SLIGAR and GUNSALUS, 1976). The small shift of the redox potential of P–450 LM by substrates agrees with the relatively small spin shift caused by these substrates. This leads to the assumption that substrate specific influences on the redox potential do not exist. Therefore the dissociation constant of the 6th heme iron ligand is not influenced by substrates. Such an assumption is supported by the following facts: (i) P-450 reacts with substrates of different chemical structure and therefore a specific enzyme-substrate interaction is absent (MISSELWITZ et al., 1975); (ii) weak enzyme-substrate interaction excludes strong conformational changes of the protein (MISSELWITZ et al., 1976); (iii) the very small change of the CD spectrum in the Soret region at the binding of type I substrates (REIN et al., 1976) also excludes a strong conformational change of the protein in the immediate environment of the heme.

Based on these characteristics of P-450 LM it can be concluded that in a first approximation a two state model is sufficient to describe the behaviour of ferric P-450. Both conformational states are differentiated by the high spin and the low spin state of the heme iron and characterized by different binding affinities for substrates. The binding of substrates (type I), however, has no remarkable effect on the structure of the two conformations. The same may be true for the dissociation of the loosely bound 6th heme iron ligand of the high spin conformation.

Based on the data of the spin equilibrium of P-450 LM2, its shift by benzphetamine and the calculated binding constants of benzphetamine to the low spin and high spin state, respectively (see Table 15) a thermodynamic model of the spin, ligation and redox state of P-450 LM was developed (see Fig. 25). The thermodynamic parameters of the model were calculated using the redox potential (E^0_{exp}) of −330 mV for P-450 LM (GUENGERICH et al., 1975) and a dissociation constant K_D of 0.3. Up to now not all parameters of the model are experimentally available but only 7 constants are necessary for a quantitative description. In the model the calculated free enthalpies of the reactions at 20 °C expressed in millivolts are shown. The simplification of the model results from lacking substrate specific conformational changes of the enzyme as described above. The calculated parameters of the model describe quantitatively the two first elementary steps of the reaction cycle of P-450: substrate binding and reduction. In these steps the 5-coordinated ferric high spin state is involved. This state has to be regarded as actual precursor of the reduced state. By the model a definite sequence of the reaction process is determined. Structural changes of the heme complex involved in the first two elementary steps are shown in Fig. 26.

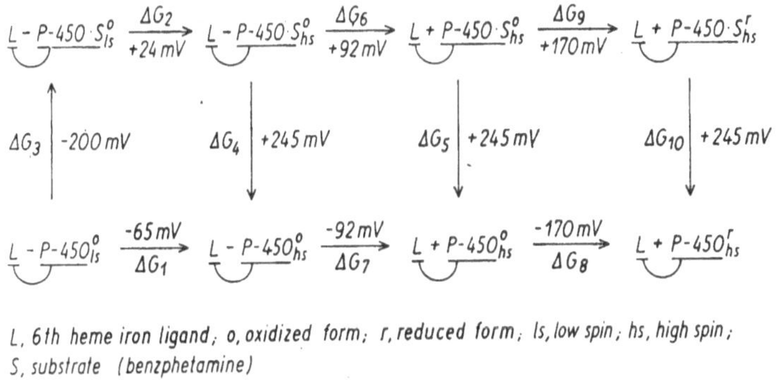

Fig. 25. Thermodynamic model of spin, redox and ligation states of the heme iron of P-450 LM2 considering the essential 5-coordinating high spin state of the reduced heme iron. The calculations are based on the data of P-450 LM2 (RISTAU et al., 1979): $K_1 = 0.08$; $K_2 = 0.39$; $K_6 = K_7 = 0.027$; $K_3 = 0.37$ mM; $K_4 = 0.07$ mM; $K_5 = K_{10} = 0.07$ mM; S = benzphetamine

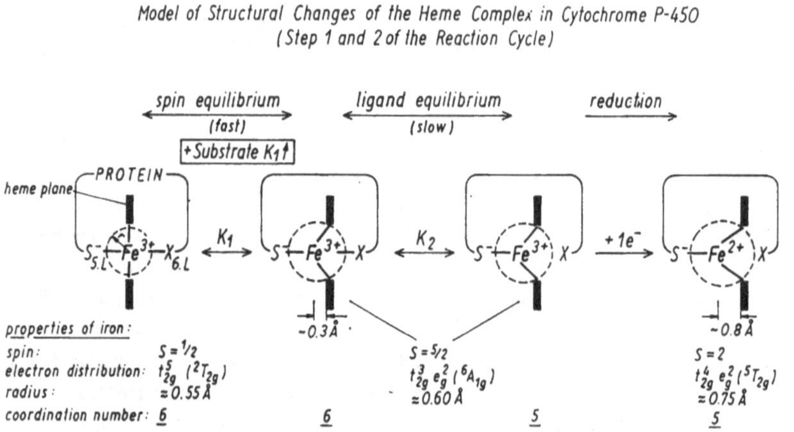

Fig. 26. Model of structural changes of the heme complex in P–450 occurring during the first two elementary steps of the reaction cycle, taken from REIN et al., 1979

3.2.6. Binding of Molecular Oxygen

As in oxyhemoglobin and oxymyoglobin the oxygen molecule is bound to the ferrous heme iron and a bent arrangement of the Fe—O—O structure can be assumed also for oxy-P–450 (PHILLIPS, 1978). The binding of dioxygen to the heme iron induces a strong polarization of the O—O bond. By quantum chemical calculations (π-INDO) of the oxygen complex considering thiol and imidazole, respectively, as the fifth heme iron ligand the following charge distribution are obtained (JUNG, 1980).

model of oxy-P–450

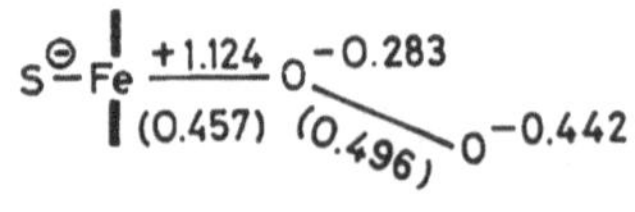

model of oxyhemoglobin

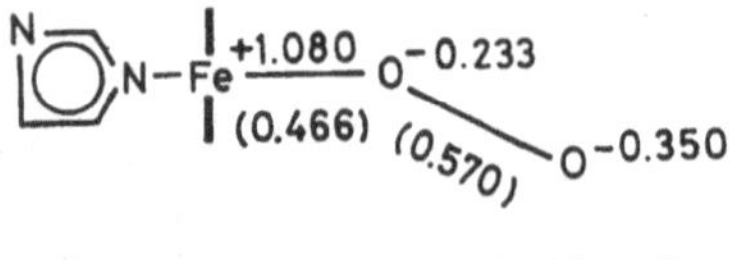

In comparison to imidazole the thiol sulfur induces a stronger π-backdonation from the iron to the oxygen molecule, reflected in a lower O—O π-bond order (values in parantheses) and a larger negative net charge of the oxygen molecule in P–450 (values at the atoms). The distant oxygen atom of the iron is stronger negative than the proximate one (Table 17). It can be assumed that a proton can easily attache to the strongly negative oxygen which could be important for the activation process of oxygen by P–450 (see below).

Oxy-P–450 is very unstable but stabilized at low temperature at about −30 °C in the presence of glycerol (BONFILS et al., 1980; MAUREL et al., 1980). The autoxidation of oxy-P–450 is connected with the formation of hydrogen peroxide (NORDBLOM and COON, 1977; ESTABROOK et al., 1979) which results from dismutation of superoxide formed by dissociation of the dioxygen complex of P–450 (KUTHAN et al., 1978).

3.2.7. Second Reduction of cytrochrome P–450

The first reaction steps of P–450 (substrate binding, first reduction, oxygen binding) in the catalytic cycle are relatively clarified whereas the mechanism of the following steps (second reduction, dioxygen splitting and substrate hydroxylation) is still under discussion. Until now, no

Table 17. Net charges and π-bond orders in oxygen complexes of hemoglobin and P–450 models before and after second reduction and the influence of a proton attached at the outer oxygen atom obtained with the π-INDO method (JUNG, 1980)

	Hemoglobin model: $N_{Im}-Fe^{II}-O_2$ A	B	C	D	P–450 model: $S^{-}-Fe^{II}-O_2$ A	B	C	D
Charges								
iron	1.080	1.043	0.907	0.667	1.124	0.984	1.362	0.922
imidazole	0.264	0.264	0.258	0.258	—	—	—	—
sulfur	—	—	—	—	−0.090	−0.110	0.051	−0.049
porphyrin	−0.761	−1.554	0.390	−0.293	−0.809	−1.544	−0.492	−0.819
inner-oxygen	−0.233	−0.300	−0.627	−0.702	−0.283	−0.317	−0.497	−0.627
outer-oxygen	−0.350	−0.453	—	—	−0.442	−0.513	—	—
outer-oxygen attached a proton	—	—	0.072	0.070	—	—	0.076	0.073
bond-orders								
Fe—O	0.466	0.501	0.231	0.324	0.457	0.473	0.269	0.435
O—O	0.570	0.487	0.028	0.016	0.496	0.444	−0.033	−0.054

$A = [-Fe^{II}-O_2]$; $B = [-Fe^{II}-O_2]^{-}$; $C = [-Fe^{II}-O_2H^{+}]$; $D = [-Fe^{II}-O_2H^{+}]^{-}$

spectroscopic measurements of oxy-P–450 after the second reduction could be carried out which allow some conclusions about the electronic structure of this reduced complex. From quantum chemical calculations (JUNG, 1980), however, some estimations about the changes of charge distributions and bond orders can be made (Table 17). The calculations show that after reduction of the oxy-P–450 model complex without previous attachement of a proton to the distant oxygen atom, the extra electron is localized predominantly in the porphyrin-π-system (74%) and only to 10% and 14% at the dioxygen and the iron, respectively (complex B in Table 17). Only when a proton is attached to the distant oxygen atom, then the extra electron is localized to 13% at the dioxygen, to 44% at the iron and only to 33% in the porphyrin-π-system (complex *D* in Tables 17 and 18). In this case the O—O bond is strongly weakened reflecting an antibonding π-bond order of −0.054. The charge changes at the sulfur ligand are small compared with other parts of the heme complex (max. 10%). The influence of the fifth

Table 18. Distribution of an extra electron to oxygen complexes of hemoglobin and P-450 models (%) (data derived from Table 17)

	hemoglobin model B	D	P–450 model B	D
iron	4	24	14	44
imidazole	0	0	—	—
sulfur	—	—	2	10
porphyrin	79	68	74	33
O_2	17	—	10	—
O_2H^{+}	—	8	—	13
summarized for axial system	21	32	26	67

$B = [X-Fe^{II}-O_2]^{-}$;

$D = [X-Fe^{II}-O_2H^{+}]^{-}$; $X = -N_{Im}, -S^{-}$

heme iron ligand on the distribution of the extra electron can be estimated by comparison of the calculated model complexes for oxy-hemoglobin and oxy-P–450. For both complexes a similar tendency in the changes of the charge distribution and π-bond orders after the attachement of a proton are obtained (Table 17). For the proton-attached oxy-P–450 model complex, however, the extra electron is localized mainly in the axial-π-system (67%) contrary to oxyhemoglobin showing a localization predominantly in the porphyrin (68%) (Table 18). From these quantum of chemical calculations it seems that the porphyrin-π-system can serve as charge buffer which is filled or emptied depending on the nature of the fifth heme iron ligand. This behaviour may be an important property to understand the different catalytic function of hemoproteins.

3.3. Substrate Hydroxylation

P–450 LM catalyzes a remarkable variety of reaction types with a multitude of substrates. The distinct purified forms of liver P–450 from rat, mouse, rabbit and human, recently reviewed by Lu and West (1980), show somewhat different but overlapping substrate specificity. Although the binding site of the substrates is not known in detail from the altered heme absorption induced by substrate binding it is assumed that the binding site is localized at the protein near the prosthetic group (Rein and Ristau, 1978). The following consideration will focus on the most important reaction types of substrate hydroxylation.

3.3.1. Hydroxylation of Aromatic Compounds

Substituted aromatic compounds are hydroxylated at all C-atoms but to different amounts. For example, the following product distribution was found: aniline (85% para, 15% ortho and 0% meta position), acetanilid (94% para, 5% ortho and 1% meta position) (Ullrich et al., 1968), toluene (28% para, 59% ortho and 13% meta position (Ullrich, 1969). Aromatic compounds with electron attracting substituents are not hydroxylated (Ullrich and Staudinger, 1971). Boyland and Wiltshire (1953) have found that P–450 converts naphthalene into the 1,2-dihydrodiol form. This is reasonable by an intermediate epoxide formation (Jerina et al., 1968). The so-called NIH-shift which means an intramolecular migration or shift of the group displaced by hydroxyl to an adjacent position on the aromatic ring (Guroff et al., 1967) is often observed during the hydroxylation of aromatic compounds catalyzed by P–450. An explanation for this finding could be an epoxide intermediate (Ullrich and Staudinger, 1971) followed by an rearrangement to the stable hydroxylated aromatic compound. The appearance of the NIH-shift seems less dependent on the enzyme but more on the substrate (Guroff et al., 1967).

3.3.2. Hydroxylation of Compounds with Isolated Double Bonds

Compounds with isolated carbon double bonds are epoxidized by the attack of the active oxygen (Ullrich and Staudinger, 1971; Jerina et al., 1970). A NIH-shift could be also observed at this type of reaction (Byington and Leibman, 1965). Cyclohexene is transformed into the epoxide (cyclohexene-oxid) and the allylalcohol (2-cyclohexene-1-ol) in approximately equal proportion (White et al., 1979).

3.3.3. Hydroxylation of Saturated Compounds

Frommer et al. (1970) have studied the hydroxylation of many saturated hydrocarbons. It was found that these compounds are hydroxylated mainly at the tertiary CH-bond and to a smaller extent at the secondary or primary CH-bond. For example, for 2-methylbutane a distribution of 87.3% tertiary, 11.9% secondary and 0.8% primary hydroxylation product were found. The corresponding distribution for methylcyclohexane exhibits 50.2%, 49.3% and 0.5% (ULLRICH and STAUDINGER, 1971). Recent studies on the hydroxylation of methylcyclohexane catalyzed by P–450 LM2 yielded a similar ratio of tertiary : secondary : primary alcohols of 1.25 : 1 : 0.072 (WHITE et al., 1979). Caused by steric effects (DENO et al., 1977) the long-chain fatty acids are hydroxylated predominantly at the primary and secondary (ω- and ($\omega -1$)-) position (PREISS and BLOCH, 1964). Steroids are hydroxylated at the secondary CH-bond with retention of configuration or at the methyl groups (HAYANO, 1962; BJÖRKHEM, 1977). A kinetic isotopic effect (k_H/K_D) in the range of 1—4 was found depending on the position of the hydroxylated carbon atom (BJÖRKHEM, 1977). Norbornane is preferentially hydroxylated in the exo-CH-bond. A large intramolecular isotopic effect (k_H/K_D) of about 11 is shown if the exo-hydrogen atoms are replaced by deuterium atoms. 14—18% of norbornane are transformed into the alcohol without retention of configuration (WHITE et al., 1979). A similar large intramolecular isotopic effect of about 11 was observed for 1,3-diphenylpropane-1,1-D_2 (HJELMELAND et al., 1977) .It seems that a large intramolecular isotopic effect occurs, if two exactly equivalent CH and CD bonds are compared in one substrate molecule (FOSTER et al., 1974; HJELMELAND et al., 1977).

3.3.4. Dealkylation

The dealkylation can be regarded as a special case of hydroxylation of saturated compounds. A heteroatom in the vicinity of CH-bond causes a destabilization of these CH-bonds by its polarizing effect. The consequence is a preferred attack on these CH-bonds by the active oxygen (BRODIE et al., 1958). The formed semiacetal is unstable and undergoes heterolytic cleavage yielding the corresponding aldehyde. RENSON et al. (1965) have shown that the oxygen atom of the formed aldehyde comes from molecular oxygen.

3.4. Shunt Mechanism

P–450 is able to catalyze the hydroxylation of organic substrates at the expense of alkyl hydroperoxides and some other oxidants as peroxyacids, iodosobenzene, iodosobenzene diacetate, sodium periodate, sodium chlorite and pyridine N-oxide (KADLUBAR et al., 1973; HRYCAY et al., 1975; LICHTENBERGER et al., 1976; GUSTAFSSON and BERGMANN, 1976; GUSTAFSSON et al., 1976 and 1979). The peroxide-supported hydroxylation reaction first observed by KADLUBAR et al. (1973) take place without NADPH reduction of the heme and without the uptake of molecular oxygen i.e. without the first reaction steps of the P–450 reaction cycle. Therefore this reaction type is often called the shunt mechanism. The actual reaction performed may be aliphatic and aromatic hydroxylation, N- or O-dealkylation and alkene epoxidation.

3.5. Proposals about the Mechanism of Insertion of an Oxygen Atom into the Substrate

Fig. 27 summarizes different opinions about the kind of splitting of dioxygen (heterolytic or homolytic) and the C—H (with or without a radical intermediate) bonds and about the kind of insertion of the oxygen atom into the substrate.

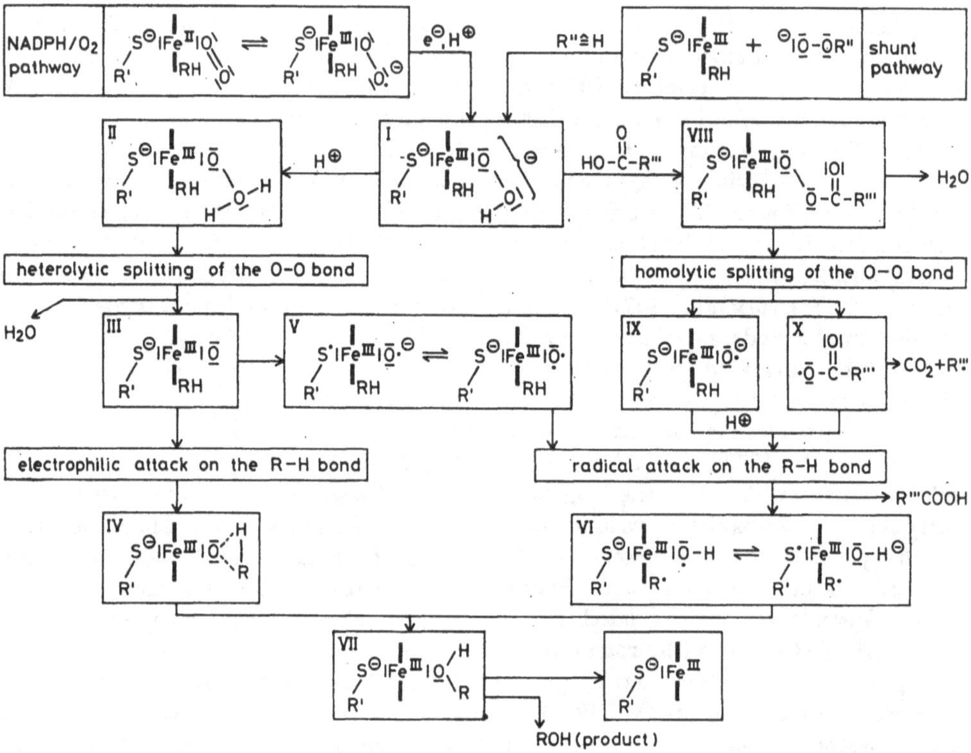

Fig. 27. Summary of the most important opinions (ULLRICH and STAUDINGER, 1971; WHITE and COON, 1980; WHITE et al., 1980) about the kind of the O—O bond splitting and of the attack of the substrate by the active oxygen species. Fe symbolized the iron-protoporphyrin IX complex. RH = substrate; R′ = cysteine residue; R″ = alkyl residue or cumole; R‴ = hydrocarbon residue of the carbon acid. The coordinating bond between the axial ligands and the iron ion is demonstrated by the electron pair from the ligand. The scheme has two starting points — the NADPH/O_2 pathway on the left upper side and the shunt pathway on the right upper side. The alkyl residue R″ is assumed equal H in order to unify the following steps common for the NADPH/O_2 and the shunt pathway

3.5.1. *Heterolytic Splitting of the O—O Bond Non-radicalic Splitting of the C—H Bond*

The most important opinions (ULLRICH and STAUDINGER, 1971; WHITE and COON, 1980; WHITE et al., 1980) about the molecular mechanism of the P–450 dependent hydroxylation start from the assumption that the NADPH/oxygen dependent pathway leads to an intermediate complex

$$S|^{\ominus}Fe^{3+}|\overline{O}-OH^{\ominus}$$

(complex I in Fig. 27). According to ULLRICH and STAUDINGER (1971) complex I should be transformed into complex III

$$S|^{\ominus}Fe^{3+}|\overline{O}$$

by a heterolytic splitting of the O—O bond of complex I or complex II accompanied by the formation of OH^- or water, respectively. This conclusion was drawn from extensive studies on the reaction type of the hydroxylation of the substrates in comparison to model reactions. Such model reactions are the substrate hydroxylation by trifluoro-

Table 19. Reactions of active oxygen species in comparison to monooxygenase reactions, taken from ULLRICH and STAUDINGER (1971)

active oxygen	source	hydroxylation		epoxi-dation	migration of substituents
		aromatic	aliphatic		
H_2O_2	3% solution	–	–	–	–
$HO_2^{\cdot}$	H_2O_2 + Ce(IV)	–	–	–	–
$OH^{\cdot}$	H_2O_2 + Fe(II) EDTA	+ electrophilic	only tertiary CH-bond	–	–
OH^+	trifluoro-per-acetic acid	+ electrophilic	+ selective	+	+
O	$N_2O + h \cdot \nu$ (gas phase)	–	+ unselective	–	–
[O]	N-oxides + $h \cdot \nu$	+ electrophilic	+ selective	+	+
monooxy-genase		+ electrophilic	+ selective	+	+

peroxyacetic acid (CHAMBERS et al., 1959; ULLRICH et al., 1968) and by "N-oxides + $h \cdot \nu$" (STREITH et al., 1967; IGETA et al., 1968; TSUCHIYA et al., 1969) both of which show the same reaction characteristics as the P-450 catalyzed hydroxylation, i.e. electrophilic hydroxylation, epoxidation, migration of substituents (NIH-shift) in aromatic or olefinic substrates and the selective hydroxylation of aliphatic and alicyclic substrates (Table 19). In both model reactions an intermediately formed oxene oxygen (|O̲; ^{1}S state) is the reactive component (BECKER, 1965; ULLRICH and STAUDINGER, 1971).

Hydroxylation of substrate R—C—H by trifluoroperoxyacetic acid

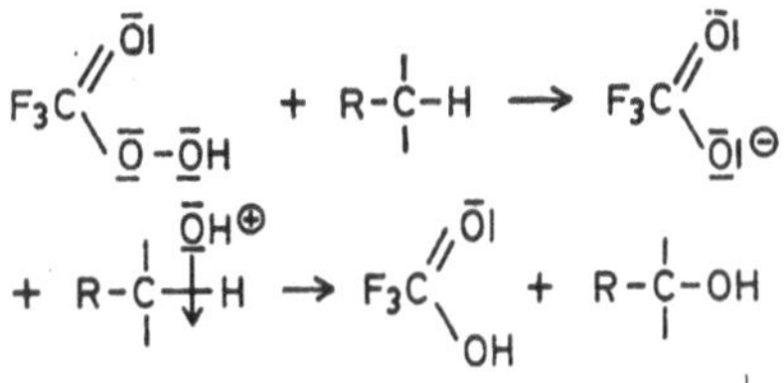

[50]

Hydroxylation of substrate R—C—H by N-oxides and light

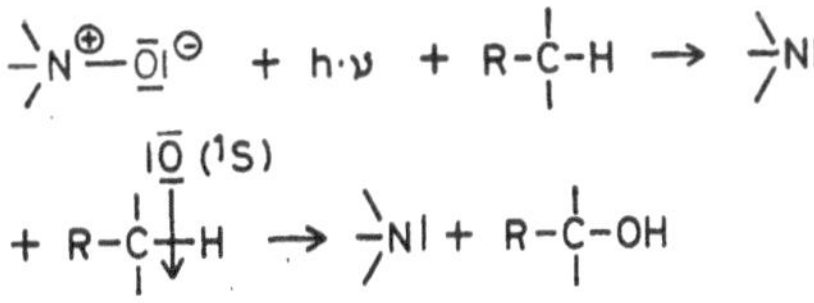

[51]

Other model studies with substituted perbenzoic acids as oxygen donor also indicate the electrophilic character of the hydroxylation of alicyclic hydrocarbons (MÜLLER and SCHNEIDER, 1979). The assumption of an iron bound "oxene" oxygen as formulated in complex III which electrophilically attacks the substrate is further supported by the P-450 catalyzed formation of the N-oxide from N,N-dimethylaniline (HLAVICA and HÜLSMANN, 1979).

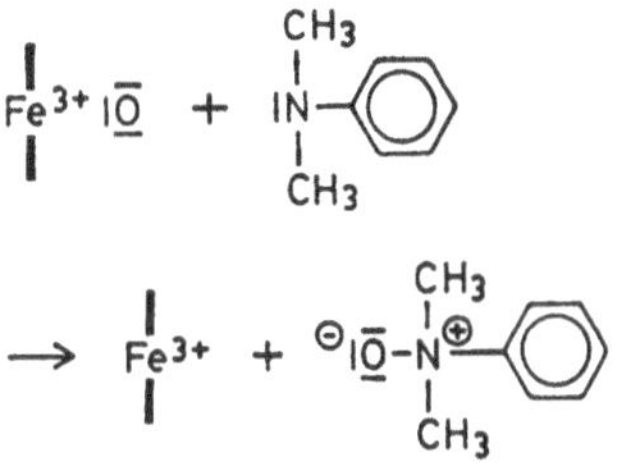

[52]

Furthermore, such a mechanism is also assumed for the P-450 catalyzed hydroxylation of steroids (GUSTAFSSON et al., 1979) and the O-dealkylation of 7-ethoxycoumarin (LICHTENBERGER et al., 1976) in the presence of iodosylbenzene.

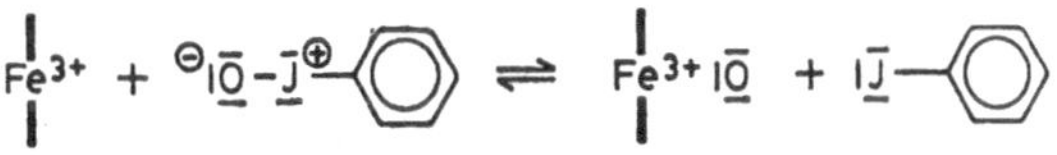

[53]

Table 20. Net charges and optimized CH-bond distances, calculated according to the CNDO/2 method considering the Mulliken population analysis (JUNG, unpublished)

	ethylene	aniline*	2-methylbutane*
C	−0.0815	−0.055 (ortho) −0.008 (meta) −0.030 (para)	−0.068 (primary) −0.004 (secundary) 0.044 (tertiary)
H	0.0407	0.025 (ortho) 0.017 (meta) 0.017 (para)	0.018 (primary) 0.003 (secundary) −0.007 (tertiary)
R(C−H) (Å)	—	—	1.121 (primary) 1.127 (secundary) 1.133 (tertiary)

* values averaged about all atoms or bonds in equivalent positions (ortho, para, primary, secundary, respectively)

Recent studies carried out by BURKA et al. (1980) have shown the reversibility of this reaction.

The function of iodosylbenzene and other agents as $NaIO_4$ and $NaClO_2$ as oxygen sources (HRYCAY et al., 1975; GUSTAFSSON et al., 1976; GUSTAFSSON and BERGMANN, 1976) supports not only the "oxene" oxygen as the active oxygen species but gives also indications that from the dioxygen complex of P–450 in the NADPH/O_2 pathway the directly iron bound oxygen atom is transferred to the substrate.

The designation of the iron bound oxene as electrophilic agent does not mean that the net charge of the oxygen atom is positive. Quantum chemical calculations (LOEW et al., 1977a) by means of the IEHT-method based on a $S = 3/2$ ground state configuration for the $(Fe-O)^{3+}$ heme complex yield a net charge of the oxygen atom of −0.31 and low-energy virtual orbitals which contain considerable contributions of the oxygen atom. The low-energy virtual orbitals can serve as incipient electron acceptors. From this result a "covalently controlled" pathway of the electrophilic attack of the $(Fe-O)^{3+}$ species should be favoured in comparison to a "charge controlled" pathway (LOEW et al., 1977b). In this connection an interesting fact is that the quantum chemical calculations also show a negative oxygen atom and low-energy virtual orbitals for trifluoroperoxyacetic acid and chromyl chloride (LOEW et al., 1977b) which both exhibit P–450 similar reactions. The "covalently controlled" pathway is in agreement with the electrophilic hydroxylation and the epoxidation of aromatics and olefines. Substrate radicals should not be formed during such hydroxylation reaction.

The reaction can be written in the following sequence (BECKER, 1965; LOEW et al., 1980):

i) aromatics

(equation [54] pp. 235)

The preferred attack of the ortho and para position showing a more negative net charge than the meta one (Table 20) indicates the electrophilic mechanism.

ii) olefines

(equation, [55] pp. 235)

For aliphatic substrates two types of reactions may be discussed:

i) *A "concerted" reaction*

The lone-pair of the oxene species polarizes the CH-bond (complex IV in Fig. 27). Such electronic "push-pull"-mechanisms are often discussed in organic chemistry (BECKER, 1965).

(equation [56], pp. 235)

This mechanism is in agreement with the retention of configuration observed for most alicyclic substrates and also

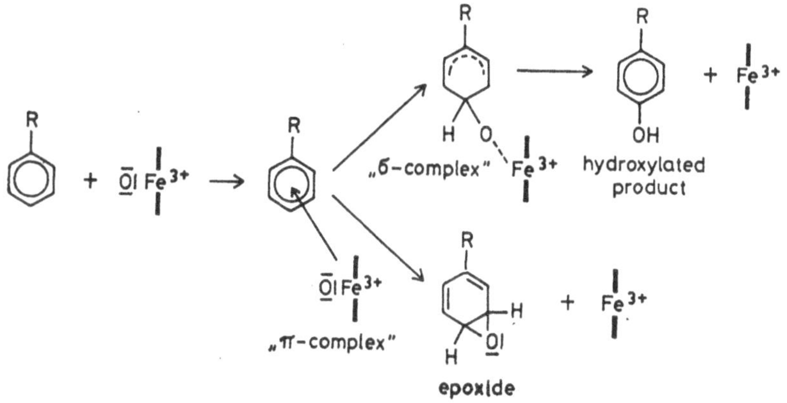

[54]

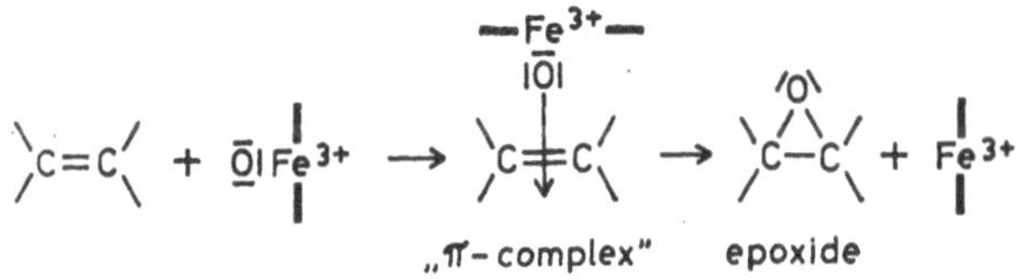

[55]

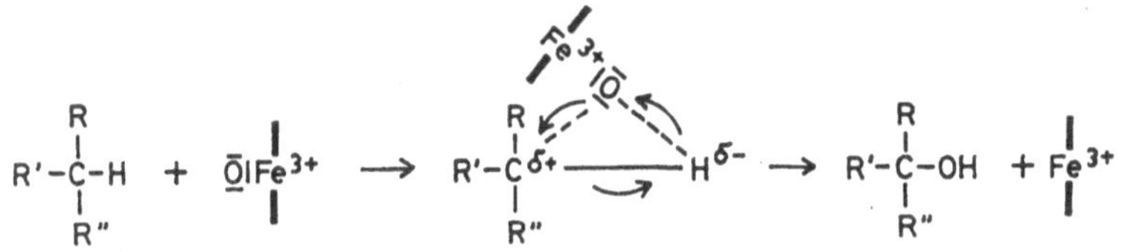

[56]

with the preferred hydroxylation at the tertiary CH-bond in aliphatic substrates because in this position the CH-bond is already weakly polarized in the correct direction and the CH-bond distance is somewhat expanded in comparison to a primary CH-bond (Table 20).

ii) *The "attachment-rearrangement" mechanism*

From quantum chemical calculations according to the MINDO/3 method the energy profile for the pathway of the following reaction was determined (PUDZIANOWSKI and LOEW, 1980):

$$CH_4 + |\overline{O}\ (^1S) \longrightarrow CH_3OH \qquad (57)$$

These calculations show that methane reacts with the oxene oxygen $|\overline{O}\ (^1S)$ to methanol via an oxygen-methane-attachment complex $O \cdots H{-}CH_3$. Methanol is formed by the rearrangement of this attachment complex. A true hydride abstraction to form OH^- and CH_3^+ is not indicated by these results. Assuming this "attachment-rearrangement" mechanism then the retention of configuration of the substrate during the hydroxylation is expected.

3.5.2. Heterolytic Splitting of the O—O Bond — Radicalic Splitting of the C—H Bond

In contrast to the observed retention of configuration during the hydroxylation of alicyclic substrates (BJÖRKHEM, 1977; HANANO, 1962) WHITE et al. (1979) found for the hydroxylation of exo-tetradeuterated norbornane, that 14—18% of the exo-deuterium were converted into the endo-position. The authors interpreted this observation as indication of an intermediate norbornane radical (complex VI in Fig. 27) produced by hydrogen abstraction using the complex V as starting radical. The recombination of complex VI with the substrate radical should yield the substrate alcohol, still bound to the iron (complex VII) (WHITE and COON, 1980). The idea of a radicalic cleavage of the substrate CH-bond was generalized by WHITE and COON (1980) for all hydroxylation reactions catalyzed by P-450. The preferred hydroxylation of the tertiary C—H bond in methylcyclohexane (WHITE et al., 1979) and the observed large intramolecular isotopic effect k_H/k_D for the hydroxylation of norbornane and substrates with equivalent C—H and C—D bond (HJELMELAND et al., 1977; FOSTER et al., 1974) were taken as further indications for a radicalic split of the substrate C—H bond.

The following three facts seem to argue against a radicalic mechanism. (i) The tertiary carbon atom is not only a preferred position for radical stabilization but also for a positive charge (carbenium ion) or at least for a partial positive charge (BECKER, 1965); (ii) A large intramolecular isotopic effect is not sufficient evidence for a radicalic split of the C—H bond but only an indication that the split of the C—H bond (radicalic, ionic or rearrangement including C—H opening) is an elementary step. The origin of a kinetic isotopic effect k_H/k_D is the lowered zero-point energy of the C—D bond in the lowest vibration state, caused by the higher mass of deuterium compared with hydrogen. If the oxene C—H attachment complex proposed by PUDZIANOWSKI and LOEW (1980) is also valid for P-450 catalyzed reactions, then the rearrangement of this complex into alcohol may produce such a large intramolecular isotopic effect; (iii) Epoxidation and NIH-shift are not explicable by a radicalic mechanism. Quantum chemical calculations according to the MINDO/3 method (PUDZIANOWSKI and LOEW, 1980) show that the theoretically possible attack of the oxygen radical ($O(^3P$ state)) to ethylene does not lead to the epoxide but yields an asymmetric diradical. The activation barrier for this radical reaction is about 14 kcal/mole higher than that for the attack of an oxygen atom in the singlet state ($O(^1S)$) yielding the epoxide as product.

3.5.3. *Homolytic Splitting of the O—O Bond — Radicalic Splitting of the C—H Bond*

In contrast to the previously discussed heterolytic splitting of the O—O bond recent studies by WHITE et al. (1980) lead to the assumption of a homolytic splitting. This assumption results from investigations of the P-450 catalyzed hydroxylation of several substrates using peroxyphenylacetic acid as oxygen donor. The binding of peroxyphenylacetic acid to the oxidized P-450 yields complex VIII in Fig. 27 if R''' represents $-CH_2-C_6H_5$. The homolytic splitting of the O—O bond in complex VIII is followed by the formation of the complex

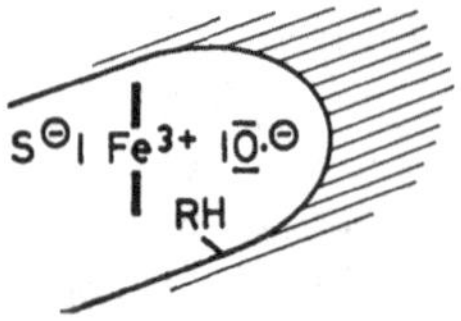

(complex IX) and the phenylacetoxy radical (complex *X* in Fig. 27). This radical should abstract the hydrogen atom from the substrate leading to the substrate radical (complex VI in Fig. 27) and the carbon acid. After proton attachment to complex IX the substrate radical recombines with complex VI producing the iron bound substrate alcohol (complex VII). The competitive reaction to this substrate hydroxylation is the decarboxylation of the phenylacetoxylradical generating the benzyl radical. In the presence of a proton the benzyl radical and complex IX can recombine to benzylalcohol, which was also detected (WHITE et al., 1980). Investigations with water enriched in ^{18}O bound in the benzylalcohol, indicating that no $OH/^{18}OH$ or $OH^-/^{18}OH^-$ exchange in complex IX or complex VI, respectively, has taken place. If indeed an intermediate benzyl radical is formed then a hydrogen abstraction from the substrate caused by the benzylradical is expected yielding toluene. Toluene, however, was not detected (WHITE et al., 1980). The P–450 catalyzed hydroxylation using peroxyphenylacetic acid is regarded as model pathway for the $NADPH/O_2$ supported hydroxylation called as "quasi-Fenton" mechanism. This assumption is supported by recent work by SLIGAR et al. (1980) showing that one oxygen atom of $^{18}O_2$ is incorporated into a fatty acid acting as effector molecule. This finding is explained by the acylation of the iron-bound dioxygen (complex I in Fig. 27) resulting from $NADPH/O_2$ pathway which leads to an iron-bound peroxyacid (complex VIII), which should be also protein-bound in the normal native O_2 activation by P–450 (WHITE and COON, 1980; WHITE et al., 1980).

3.5.4. *Common Reaction Intermediates of the NADPH/Dioxygen and of the Shunt Pathway*

The shunt pathway of the P–450 catalyzed hydroxylation reaction is often used to model the $NADPH/O_2$ pathway and to compare the enzymatic functions of different hemoproteins, e.g. peroxidase. Two questions are derived from the experimental analysis of the product pattern and from the spectral properties of the reaction intermediates of the shunt pathway compared with the $NADPH/O_2$ supported reaction. (i) Does the $NADPH/O_2$ and the shunt pathway of the P–450 catalyzed hydroxylation reaction exhibit common reaction intermediates, so that the shunt pathway can be regarded as a real cut of the $NADPH/O_2$ pathway? (ii) Is the system "P–450 + organic hydroperoxide" equivalent to the system "peroxydase + organic hydroperoxide"? These two questions cannot unambiguously be answered until now. But some experimental findings should be discussed here.

To (i). If the $NADPH/O_2$ as well as the shunt pathway lead to the same oxygen intermediate than the same substrate product pattern should be expected irrespective of the oxygen source used. HRYCAY et al. (1976) and GUSTAFSSON et al. (1979) have shown that androstenedione is hydroxylated by P–450 using several different organic hydroperoxides ($NaIO_4$, $NaClO_2$ and iodosylbenzene derivates) as oxygen source. Although the same positions of androstenedione are attached during the reaction supported by different oxygen agents as well as by $NADPH/O_2$, the quantitative product pattern, however, depends on the nature of the oxygen source used. Two possible interpretations of this finding are noted here. (α) In the $NADPH/O_2$ and in the shunt pathway different oxygen intermediates are formed which attack the positions of the substrate with different efficiency. (β) A common oxygen intermediate is formed in the $NADPH/O_2$ and the shunt pathway. However, fragments which are formed after splitting of the oxygen atom from the oxygen agent used, block certain positions of the substrate. The qualitatively similar spectral behaviour of the intermediates formed by cumenehydroperoxide (HRYCAY et al., 1976) and (RAHIMTULA et al., 1974) and $NaIO_4$ (HRYCAY et al., 1976) seems to support the second interpretation.

BLAKE and COON (1980) studied the spectral intermediate formed during the interaction of P–450 LM2 with different substituted aromatic hydroperoxides and peroxy acids. Two intermediates (P–450 LM2 + oxidant $\rightleftharpoons$ C $\rightleftharpoons$ D) could be detected by stopped flow measurements. Complex D shows a difference spectrum (complex D minus oxidized P–450 LM2) which is qualitatively similar to those formed by cumenehydroperoxide and $NaIO_4$, i.e. a trough at about 420 nm and a peak at about 435 nm. A small dependence of the difference spectrum on the oxygen agent and its substituted derivatives was found indicating the influence of the split organic fragments of the peroxides to complex D.

A generalization of this interpretation of all P–450 catalyzed reaction is rendered more difficult by many controversial experimental findings. For example CAPDEVILA et al. (1980) have reported that the cumenehydroperoxide supported P–450 catalyzed hydroxylation of benzo(a)pyrene predominantly leads to the quinones. In contrast the NADPH/O_2 supported reaction predominantly yields phenoles and diols, resulting from the hydrolysis of the epoxides. Recently, RENNEBERG et al. (1981) could show that the H_2O_2 supported P–450 catalyzed hydroxylation of benzo(a)pyrene yields about 50% of the total products formed as diols and phenols, which is more similar to the 70% for the NADPH/O_2 than to the 6% for the cumenehydroperoxide supported reaction. A detailed investigation of the reactions following the P–450 catalyzed hydroxylation has shown that the quinones are formed by a one-electron oxydation of the phenols in the presence of cumenehydroperoxide (CAPDEVILA et al., 1980) and H_2O_2 (RENNEBERG et al., 1981). NAGATA et al. (1974) and LESKO et al. (1975) demonstrated that the oxidation of phenoles of benzo(a)pyrene leads to an intermediate oxo-radical, which is detectable by EPR-measurements.

The experimental findings therefore support the assumption that reactions of different types may coexist in P–450 catalyzed hydroxylation of substrates. Depending on the nature of the oxygen agent and on the substrate used the initial electrophilic attack of the active oxygen species on the substrate is followed by several non-radicalic and radicalic processes. The qualitative and quantitative product pattern is essentially determined by these following processes. Therefore, unambiguous conclusions about the mechanism of the initial reaction of the active oxygen with the substrate from the product pattern can be drawn only by exclusion of the following processes.

To (ii). HRYCAY and O'BRIEN (1972, 1973, 1974) have shown that NADPH and NADH are oxidized by organic peroxides catalyzed by P–450. This so-called peroxidase-activity of P–450 raised the question for a common active oxygen intermediate for peroxidase and P–450. Derived from studies using different oxygen agents and model systems a $(Fe{-}O)^{3+}$ structure for the active oxygen species of P–450 was concluded (see above) which should be similar to the peroxidase compound I (SCHONBAUM and LO, 1972; DUNFORD and STILLMAN, 1976). MÖSSBAUER studies (MOSS et al., 1969; MAEDA and MORITA, 1967), experimental (IIZUKA et al., 1968) and theoretical (RISTAU, 1975) investigations of the magnetic susceptibility and EPR measurements (YONETANI et al., 1966) indicate that the iron of compound I exists in the Fe(IV) state, and that a free radical is present which may be localized near the active site in an aromatic amino acid residue of the protein (YONETANI et al., 1966) or in the porphyrin (SCHONBAUM and LO, 1972). The physico-chemical properties correspond with an electronic spin state of $S = 3/2$ originating from the Fe(IV) state ($S = 1$) and the free radical ($S = 1/2$) (LOEW et al., 1977a; DOLPHIN et al., 1971). Because of its free-radical character the peroxidase compound I is able to one-electron-oxidations of substrates (AH_2) (DUNFORD and STILLMAN, 1976), yielding intermediate substrate radicals.

Peroxidase:

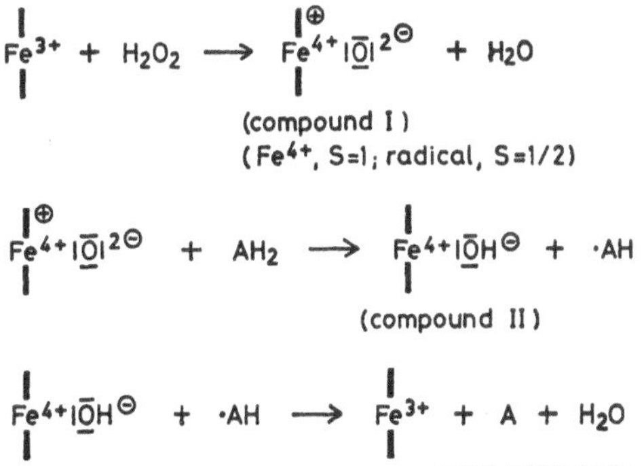

[58] [59] [60]

The electron octer of the iron-bound oxygen atom should hinder on electrophilic attack of the substrate by peroxidase compound I.

P-450:

In contrast to peroxidase compound I the experimental findings discussed above suggest an "oxene" oxygen bound to the iron in the P–450 $(Fe-O)^{3+}$ species which electrophilically attacks the substrate. There are no indications that a free radical like in peroxidase compound I exists in P–450 $(Fe-O)^{3+}$. Therefore, an electronic spin state of $S = 1/2$ originating from the Fe(III), may be assumed for P–450 $(Fe-O)^{3+}$.

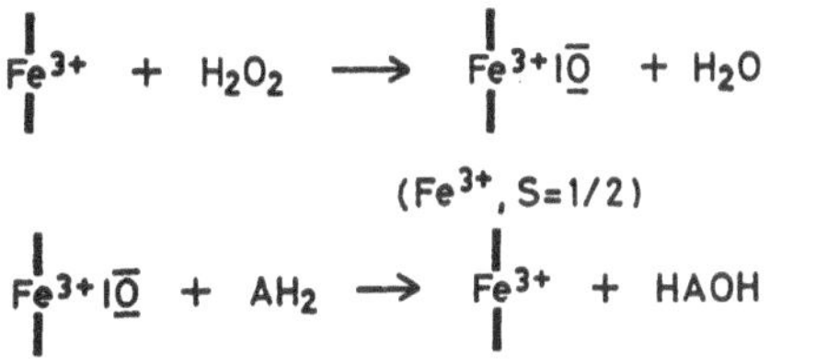

[61] [62]

Loew et al. (1977b) assumed a $S = 3/2$ ground state of peroxidase compound I as well as of P–450 $(Fe-O)^{3+}$ species for the quantum chemical calculations according to the IEHT method. Assuming, however, a $S = 1/2$ ground state for P–450 $(Fe-O)^{3+}$ the conclusion of a "covalently-controlled" pathway of the electrophilic attack on the substrate by the active oxygen is not restricted but rather supported because an empty energetically low-lying orbital of the active oxygen species enforces the electrophilic property.

The different electronic states for the $(Fe-O)^{3+}$ species of peroxidase and P–450 may be caused by different fifth iron ligands (imidazole for peroxidase (Brill and Sandberg, 1968) and mercaptide for P–450 (Bayer et al., 1969)) and by different protein environment of the prosthetic group. Consequently differences in the electronic state of the Fe—O species should lead to different molecular mechanism in substrate transformations catalyzed by peroxidase and P–450.

Such a conclusion is in agreement with EPR stopped flow measurements of the demethylation of aminopyrine catalyzed by horseradish peroxidase and P–450 using H_2O_2 as oxygen source (Renneberg, unpublished results). Whereas the horseradish peroxidase catalyzed reaction yields intermediate substrate radicals, no radicals could be detected in the P–450 catalyzed reaction. Under special conditions (high concentrations of H_2O_2 (50 mM) or use of cumenehydroperoxide as oxygen supporting agent) Griffin et al. (1980) were able to detect aminopyrine radicals by electronic absorption spectroscopy. The formation of an intermediate substrate radical seems to be specific for such an aromatic substrate like aminopyrine which easily forms moderately stable radicals compared with aliphatic compounds and for organic peroxides like cumenehydroperoxide which more easily split in radicals than H_2O_2. This interpretation is supported by the finding of Griffin et al. (1980) that no radicals are detected during the demethylation of benzphetamine.

The presented experimental findings and suggestions show that peroxidase compound I and P–450 $(Fe-O)^{3+}$ species seem to be formally equivalent but at nearer view a difference in the electron distribution may be detected. This difference seems to occur between P–450 $(Fe-O)^{3+}$ and cytochrome c peroxidase compound I, which possesses two separated no-interacting spin systems (Fe(IV) $S = 1$ and a radical of an aromatic amino acid residue $S = 1/2$ (Yonetani et al., 1966). In contrast, in horseradish peroxidase compound I the free radical ($S = 1/2$) is very close to the Fe(IV) ($S = 1$) (AASA et al., 1975; DUNFORD and STILLMAN, 1976 and HERMAN, 1980), possibly in the porphyrin π-system (LOEW et al., 1976). In this case, assuming a possible thermal spin state equilibrium between $S = 3/2$ and $S = 1/2$ for horseradish peroxidase compound I was well as for P-450 $(Fe-O)^{3+}$, some similarities in functional properties may be expected.

The different proposals for the mechanism of the P-450 catalyzed hydroxylation of substrates show that no final conclusion about the mechanism can be drawn up to now. The question about a common reaction pathway for all types of substrates is still open. But, it seems that the initial reaction between the active oxygen species and the substrate is an electrophilic attack of the substrate. For substrates easily forming stable radicals, a competitive radicalic attack may be possible. The shunt pathway of the P-450 catalyzed hydroxylation of substrates seems to simulate the NADPH/O_2 pathway only, if following reactions of the oxidant (peroxides etc.) or its fragments with the substrate or the products can be neglected.

References

AASA, R., T. VÄNNGÅRD, and H. B. DUNFORD (1975): Biochim. Biophys. Acta **391**, 259 to 264

ABE, M., T. KITAGAWA, and Y. KYOGOKU (1976): Chem. Lett. 249–252

ABE, M., T. KITAGAWA, and Y. KYOGOKU (1978): J. Chem. Phys. **69**, 4526–4531

AKHREM, A. A., D. I. METELITZA, S. N. BIELSKI, P. A. KISELEV, M. E. SHURKO, and S. A. USANOV (1977): Croat. Chem. Acta **49**, 223–235

ALBEN, J. O., and W. S. CAUGHEY (1968): Biochemistry **7**, 175–183

ALBRECHT, A. C. (1961): J. Chem. Phys. **34**, 1476–1484

ANTONINI, E. and M. BRUNORI (1971): Hemoglobin and Myoglobin in their Reactions with Ligands, North-Holland Publ. Comp., Amsterdam, London

BALNY, C., P. DEBEY, and P. DOUZOU (1976): FEBS Lett. **69**, 236–239

BAYER, E., H. A. O. HILL, A. RODER, and R. J. P. WILLIAMS (1969): Chem. Commun. 109

BECKER, H. (1965): Einführung in die Elektronentheorie organisch-chemischer Reaktionen, VEB Deutscher Verlag der Wissenschaften, Berlin

BEETLESTONE, J., and P. GEORGE (1964): Biochemistry **3**, 707–714

BEINERT, H., and R. H. SANDS (1960): Biochem. Biophys. Res. Commun. **3**, 41–46

BERG, A. P. van den, J. NOORDHOEK, and E. KOOPMAN-KOOL (1979a): Biochem. Pharmacol. **28**, 31–36

BERG, A. P. van den, J. NOORDHOEK, and E. KOOPMAN-KOOL (1979b): Biochem. Pharmacol. **28**, 37–41

BERZINIS, A. P., and T. G. TRAYLOR (1979): Biochem. Biophys. Res. Commun. **87**, 229 to 235

BJÖRKHEM, I. (1977): Pharmacol. Ther. A **1**, 327–348

BLAKE, R. C., and M. J. COON (1980): J. Biol. Chem. **255**, 4100–4111

BLANCK, J., G. SMETTAN, G.-R. JÄNIG, and K. RUCKPAUL (1976): Stud. Biophys. **57**, 105 to 108

BLAUER, G. (1974): Structure and Bonding **18**, 69–129

BLUMBERG, W. E., and J. PEISACH (1971) in: Probes of Structure and Function of Macromolecules and Membranes. Vol. II. Probes of Enzymes and Hemoproteins. (B. CHANCE, T. YONETANI, and A. S. MILDVAN eds.) Academic Press, New York, London, p. 215–229

BÖHM, S., H. REIN, G. BUTSCHAK, G. SCHEUNIG, H. BILLWITZ, and K. RUCKPAUL (1979): Acta Biol. Med. Germ. **38**, 249–255

BÖHM, S., H. REIN, G.-R. JÄNIG, and K. RUCKPAUL (1976): Acta Biol. Med. Germ. **35**, K27–K32

BONFILS, C., P. DEBEY, and P. MAUREL (1979): Biochem. Biophys. Res. Commun. **88**, 1301 to 1307

BONFILS, C., C. BALNY, P. DOUZOU, and P. MAUREL (1980) in: Biochemistry, Biophysics and Regulation of Cytochrome P-450. (GUSTAFSSON, J.-A., J. CARLSTEDT-DUKE, A. MODE and J. RAFTER) Elsevier/North-Holland, Amstedam, pp. 559–564

BOYLAND, E., and G. H. WILTSHIRE (1953): Biochem. J. **53**, 424–426

Briat, B., D. Berger, and M. Leliboux (1972): J. Chem. Phys. 57, 5606–5607

Brill, A. S., and H. E. Sundberg (1968): Biochemistry 7, 4254–4260

Brodie, B. B., J. R. Gillette, and B. N. La Du (1958): Ann. Rev. Biochem. 27, 427 to 454

Buchler, J. W. (1975) in: Porphyrins and Metalloporphyrins. (K. M. Smith ed.), Elsevier Sci. Publ. Comp., Amsterdam, Oxford, New York, pp. 157–231

Buckingham, A. D., and P. J. Stephens (1966): Ann. Rev. Phys. Chem. 17, 399–432

Burka, L. T., A. Thorsen, and F. P. Guengerich (1980): J. Amer. Chem. Soc. 102, 7615–7616

Byington, K. H., and K. C. Leibman (1965): Molec. Pharmacol. 1, 247–254

Capdevila, J., R. W. Estabrook, and R. A. Prough (1980): Arch. Biochem. Biophys. 200, 186–195

Caron, C. C., A. Mitschler, G. Revière, L. Ricard, M. Schappacher, and R. Weiss (1979): J. Amer. Chem. Soc. 101, 7401–7402

Caughey, W. S., C. H. Barlow, D. H. O'Keeffe, and M. C. O'Toole (1973): Ann. N.Y. Acad. Sci. 206, 296–309

Chambers, R. D., P. Goggin, and W. K. R. Musgrave (1959): J. Chem. Soc. 1804–1807

Champion, P. M., E. Münk, P. G. Debrunner, P. F. Hollenberger, and L. P. Hager (1973): Biochemistry 12, 426–435

Champion, P. M., E. Münk, P. G. Debrunner, T. H. Moss, J. D. Lipscomb, and I. C. Gunsalus (1975a): Biochem. Biophys. Acta 376, 579–582

Champion, P. M., J. D. Lipscomb, E. Münk, P. G. Debrunner, and I. C. Gunsalus (1975b): Biochemistry 14, 4151–4158

Champion, P. M., R. Chiang, E. Münk, P. Debrunner, and L. P. Hager (1975c): Biochemistry 14, 4159–4166

Champion, P. M., I. C. Gunsalus, and G. C. Wagner (1978): J. Amer. Chem. Soc. 100, 3743–3751

Chang, C. K., and D. Dolphin (1976a): Proc. Natl. Acad. Sci. USA 73, 3338–3342

Chang, C. K., and D. Dolphin (1976b): J. Amer. Chem. Soc. 98, 1607–1609

Chevion, M., J. J. Stegeman, J. Peisach, and W. E. Blumberg (1977a): Life Sci. 20, 895–900

Chevion, M., J. Peisach, and W. E. Blumberg (1977b): J. Biol. Chem. 252, 3637–3645

Chiang, Y. L., and M. J. Coon (1979): Arch. Biochem. Biophys. 195, 178–187

Chien, J. C. W. (1969): J. Chem. Phys. 51, 4220 to 4227

Collman, J. P., and T. N. Sorrell (1975): J. Amer. Chem. Soc. 97, 4133–4134

Collman, J. P., R. R. Gagne, C. A. Reed, I. R. Halbert, G. Lang, and W. T. Robinson (1975a): J. Amer. Chem. Soc. 97, 1427–1439

Collman, J. P., T. N. Sorrell, and B. M. Hoffman (1975b): J. Amer. Chem. Soc. 97, 913–914

Collman, J. P., T. N. Sorrell, J. H. Dawson, J. R. Trudell, E. Bunnenberg, and C. Djerassi (1976a): Proc. Natl. Acad. Sci. USA 73, 6–10

Collman, J. P., J. I., Brauman, T. R. Halpert, and K. S. Suslick, (1976b) Proc. Natl. Acad. Sci. USA 73, 3333–3337

Cotton, F. A., and G. Wilkinson (1967): Anorganische Chemie. VEB Deutscher Verlag der Wissenschaften, Berlin

Cramer, S. P., J. H. Dawson, K. O. Hodgson, and L. P. Hager (1978): J. Amer. Chem. Soc. 100, 7282–7290

Dawson, J. H., P. M. Dolinger, J. R. Trudell, G. Barth, R. E. Linder, E. Bunnenberg, and C. Djerassi (1974): Proc. Natl. Acad. Sci. USA 71, 4594–4597

Dawson, J. H., R. H. Holm, J. R. Trudell, G. Barth, R. E. Linder, E. Bunnenberg, C. Djerassi, and S. C. Tang (1976): J. Amer. Chem. Soc. 98, 3707–3709

Dawson, J. H., and S. P. Cramer (1978): FEBS Lett. 88, 127–130

Dawson, J. H., J. R. Trudell, R. E. Linder, G. Barth, E. Bunnenberg, and C. Djerassi (1978): Biochemistry 17, 33–42

Dawson, J. H., L. A. Andersson, I. M. Davis, and J. E. Hahn (1980) in: Developments in Biochemistry, Vol. 13, Biochemistry, Biophysics and Regulation of Cytochrome P-450, (Gustafsson, J. A., J. Carlstedt-Duke, A. Mode, and J. Rafter eds.) Elsevier/North-Holland, Amsterdam, p. 565 to 572

Deatherage, J. F., R. S. Loe, Ch. M. Anderson, and K. Moffat (1976a): J. Molec. Biol. 104, 687–706

Deatherage, J. F., R. S. Loe, and K. Moffat (1976b): J. Molec. Biol. 104, 723 to 728

Deno, N. C., E. J. Jedzimak, L. A. Messer,

and E. S. Tomeszko (1977) in: Bioorganic Chemistry, Vol. 1, (Tamelen, E. E. V., ed.) Academic Press, New York, p. 79–92

Desbois, A., M. Lutz, and R. Banerjee (1979): Biochemistry 18, 1510–1518

Dickinson, L. Ch., and J. C. W. Chien (1971): J. Amer. Chem. Soc. 93, 5036–5040

Diehl, H., J. Schädelin, and V. Ullrich (1979): Hoppe-Seylers Z. Physiol. Chem. 351, 1359–1371

Dolinger, P. M., M. Kielczewski, J. R. Trudell, G. Barth, R. E. Linder, E. Bunnenberg, and C. Djerassi (1974): Proc. Natl. Acad. Sci. USA 71, 399–403

Dolphin, D., A. Forman, D. C. Borg, J. Fajer, and R.-H. Felton (1971): Proc. Natl. Acad. Sci. USA 68, 614–618

Dolphin, D., J. R. Sams, T. Bik Tsin, and K. L. Wong (1976): J. Amer. Chem. Soc. 98, 6970–6975

Dunford, H. B., and J. S. Stillman (1976): Coord. Chem. Rev. 19, 187–251

Ebel, R. E., D. H. O'Keeffe, and J. A. Peterson (1975): FEBS Lett. 55, 198 to 201

Ebel, R. E., D. H. O'Keeffe, and J. A. Peterson (1978): J. Biol. Chem. 253, 3888 to 3897

Eisenstein, L., P. Debey, and P. Douzou (1977): Biochem. Biophys. Res. Commun. 77, 1377–1383

Ellfolk, N. and G. Sievers (1975): Biochim. Biophys. Acta 405, 213–227

Estabrook, R. W., S. Kawano, J. Werringloer, H. Kuthan, H. Tsuji, H. Graf, and V. Ullrich (1979): Acta Biol. Med. Germ. 38, 423–434

Fabry, T. L., and H. A. Reich (1966): Biochem. Biophys. Res. Commun. 22, 700–703

Fermi, G. (1975): J. Molec. Biol. 97, 237–256

Fleischhauer, J., and A. Wollmer (1972): Z. Naturforsch. 27b, 530–532

Foster, A. B., M. Jarman, J. D. Stevens, P. Thomas, and J. H. Westwood (1974): Chem.-Biol. Interact. 9, 327–340

Franklin, M. R. (1972): Molec. Pharmacol. 8, 711–721

Friedrich, J., G. Butschak, G. Scheunig, O. Ristau, H. Rein, K. Ruckpaul, and G. Smettan (1979): Acta Biol. Med. Germ. 38, 207–216

Frommer, U., V. Ullrich, and H. J. Staudinger (1970): Hoppe-Seylers Z. Physiol. Chem. 351, 903–912

George, P., J. Beetlestone, and J. S. Griffith (1959) in: Hematin Enzyme Symposium, Canberra, Part 1. (Falk, J. E., R. Lemberg, and R. K. Morton eds.) Pergamon Press, London, p. 105

George, P., J. Beetlestone, and J. S. Griffith (1964): Rev. Modern Phys. 36, 441–458

Ghacarian, J. G., J. L. Weber, and R. C. Sealy (1978): FEBS Lett. 86, 213–218

Gonser, U., Y. Maeda, F. Trautwein, F. Parak, and F. Formanek (1974): Z. Naturforsch. 296, 241–244

Goodwin, H. A. (1976): Coord. Chem. Rev. 18, 293–325

Gouterman, M., F. P. Schwarz, P. D. Smith, and D. Dolphin (1973): J. Chem. Phys. 59, 676–690

Gouterman, M., L. K. Hanson, G. E. Khalil, J. W. Buchler, K. Rohbock, and D. Dolphin (1975): J. Amer. Chem. Soc. 97, 3142 to 3149

Grasdalen, H., D. Bäckström, L. E. Ericksson, and A. Ehrenberg (1975): FEBS Lett. 60, 294–299

Griffin, B. W., and J. A. Peterson (1971): Arch. Biochem. Biophys. 145, 220–229

Griffin, B. W., S. M. Smith, and J. A. Peterson (1974): Arch. Biochem. Biophys. 160, 323–332

Griffin, B. W., C. Marth, Y. Yasukochi, and B. S. S. Masters (1980): Arch. Biochem. Biophys. 205, 543–553

Griffith, J. S. (1956): Proc. Roy. Soc. (London) A 235, 23–36

Griffith, J. S. (1957): Nature 180, 30–31

Gunsalus, I. C. (1968): Hoppe-Seylers Z. Physiol. Chem. 349, 1610–1613

Gunsalus, I. C., and S. G. Sligar (1978): "Advances of Enzymology and Related Areas of Molecular Biology, Vol. 47, 1–44

Guengerich, F. P., D. P. Ballou, and M. J. Coon (1975): J. Biol. Chem. 250, 7405 to 7414

Gurinovitch, G. P., A. N. Sevchenko, and K. N. Solovev (1963): Usp. Fiz. Nauk 79, 173–234

Guroff, G., J. W. Daly, D. M. Jerina, J. Renson, B. Witkop, and S. Udenfriend (1967): Science 157, 1524–1530

Gustafsson, J. A., and J. Bergman (1976): FEBS Lett. 70, 276–280

Gustafsson, J. A., E. G. Hrycay, and L. Ernster (1976): Arch. Biochem. Biophys. 174, 440–453

GUSTAFSSON, J. A., L. RONDAHL, and J. BERGMAN (1979): Biochemistry 18, 865–870
HANSON, L. K., W. A. EATON, S. G. SLIGAR, I. C. GUNSALUS, M. GOUTERMAN, and C. R. CONNELL (1976): J. Amer. Chem. Soc. 98, 2672–2674
HANSON, L. K., S. G. SLIGAR, and I. C. GUNSALUS (1977): Croat. Chim. Acta 49, 237–250
HARRIS, G. (1968): Theor. Chim. Acta 10, 119 to 154
HASHIMOTO, Y., T. YAMANO, and H. S. MASON (1962): J. Biol. Chem. 237, 3843–3844
HAUGEN, D. A., T. A. van der HOEVEN, and M. J. COON (1975): J. Biol. Chem. 250, 3567 to 3570
HAUGEN, D. A., and M. J. COON (1976): J. Biol. Chem. 251, 7929–7939
HAVEMANN, R., and W. HABERDITZL (1958): Z. Phys. Chem. 209, 135–160
HAYANO, M., M. C. LINDBERG, R. I. DORFMAN, J. E. H. HANCOCK, and E. W. von DOERING (1955): Arch. Biochem. Biophys. 59, 529 to 532
HAYANO, M. (1962): Oxygenases, Hayaishi, O. Ed., Academic Press, New York, p. 182
HAYAISHI, O. (1962): Oxygenases, Academic Press, New York
HEIDNER, E. J., R. S. LADNER, and M. F. PERUTZ (1976): J. Molec. Biol. 104, 707–722
HENRY, Y., and R. BANERJEE (1973): J. Molec. Biol. 73, 469–482
HERRICK, R. C. (1976): Ph. D. Thesis, University of Illinois, Urbana, USA
HERRICK, R. C., and H. J. STAPLETON (1976): J. Chem. Phys. 65, 4786–4790
HILL, H. A. O., A. RÖDER, and R. J. P. WILLIAMS (1970): Naturwissenschaften 52, 69 to 72
HIRAKAWA, A. Y., and M. TSUBOI (1975): Science 188, 359–361
HJELMELAND, L. M., L. ARONOW, and J. R. TRUDELL (1977): Biochem. Biophys. Res. Commun. 76, 541–549
HLAVICA, P., and S. HULSMANN (1979): Biochem. J. 182, 109–116
HOARD, J. L. (1971): Science 174, 1295–1302
HOFMANN, E. (1972): Enzyme und energieliefernde Stoffwechselreaktionen, Part II, WTB-Reihe, Akademie Verlag, Berlin
HOLLENBERG, P. F., L. P. HAGER, W. E. BLUMBERG, and J. PEISACH (1980): J. Biol. Chem. 255, 4801–4807
HOLTZMAN, J. L., and B. H. RUMAK (1973): Biochemistry 12, 2309–2313
HRYCAY, E. G., and P. J. O'BRIEN (1972): Arch. Biochem. Biophys. 153, 480–494
HRYCAY, E. G., and P. J. O'BRIEN (1973): Arch. Biochem. Biophys. 157, 7–22
HRYCAY, E. G., and P. J. O'BRIEN (1974): Arch. Biochem. Biophys. 160, 230–245
HRYCAY, E. G., J. A. GUSTAFSSON, M. INGELMAN-SUNDBERG, and L. ERNSTER (1975): Biochem. Biophys. Res. Commun. 66, 209 to 216
HRYCAY, E. G., J. Å. GUSTAFSSON, M. INGELMAN-SUNDBERG, and L. ERNSTER (1976): Europ. J. Biochem. 61, 43–52
HSU, M.-C., and R. W. WOODY (1971): J. Amer. Chem. Soc. 93, 3515–3525
HUBER, R., O. EPP, and H. FORMANEK (1969): Naturwissenschaften 56, 362–367
ICHIKAWA, Y., and T. YAMANO (1967): Biochim. Biophys. Acta 131, 490–497
ICHIKAWA, Y., and T. YAMANO (1968): Biochim. Biophys. Acta 153, 753–765
IGETA, H., T. TSUCHIYA, M. YAMAD, and H. ARAI (1968): Chem. Pharm. Bull. 16, 767 to 769
IIZUKA, T., and M. KOTANI (1968): Biochim. Biophys. Acta 181, 275–286
IIZUKA, T., M. KOTANI, and T. YONETANI (1968): Biochim. Biophys. Acta 167, 257 to 267
IIZUKA, T., and T. YONETANI (1970): Adv. Biophys. 1, 157–182
IIZUKA, T. (1975): Seikagaku (Japanese) 47, 1061–1090
IKEGAMI, A. (1977): Biophys. Chem. 6, 117 to 130
IMAI, Y., and R. SATO (1966): Biochem. Biophys. Res. Commun. 22, 620–626
IMAI, Y., and R. SATO (1967): J. Biochem. (Tokyo) 62, 239–249
IMAI, Y., and R. SATO (1968): J. Biochem. (Tokyo) 64, 147–159
IMAI, Y., R. SATO, and T. IYANAGI (1977): J. Biochem. (Tokyo) 82, 1237–1246
INGRAM, D. J. E. (1972) in: Proceedings of the International Conference on Magnetic Resonance in Biological Research, Cagliari. (C. FRANCONI ed.) Gordon and Breach, New York, pp. 41–66
JEFCOATE, C. R. E., and F. C. GAYLOR (1969): Biochemistry 8, 3464–3472
JEFCOATE, C. R. E., J. L. GAYLOR, and R. L. CALABRESE (1969): Biochemistry 8, 3455 to 3463

JEFCOATE, C. R. E., R. L. CALABRESE, and J. L. GAYLOR (1970): Molec. Pharmacol. **6**, 391–401

JERINA, D. M., J. W. DALY, B. WITKOP, P. ZALTZMAN-NIRENBERG, and S. UDENFRIEND (1968): J. Amer. Chem. Soc. **90**, 6525 to 6527

JERINA, D. M., J. W. DALY, B. WITKOP, P. ZALTZMAN-NIRENBERG, and S. UDENFRIEND (1970): Biochemistry **9**, 147–156

JUNG, C., and O. RISTAU (1977): Chem. Phys. Lett. **49**, 103–108

JUNG, C., and O. RISTAU (1978): Pharmazie **33**, 329–331

JUNG, C. (1979): Chem. Phys. Lett. **61**, 630 to 636

JUNG, C., J. FRIEDRICH, and O. RISTAU (1979): Acta Biol. Med. Germ. **38**, 363–377

JUNG, C. (1980): Doctoral Thesis, Acad. Sci. GDR, Berlin

KADISH, K. M., and D. G. DAVIS (1973): Ann. N.Y. Acad. Sci. **206**, 495–503

KADLUBAR, F. F., K. C. MORTON, and D. M. ZIEGLER (1973): Biochem. Biophys. Res. Commun. **54**, 1255–1261

KANEHISA, M., and A. IKEGAMI (1977): Biophys. Chem. **6**, 131–149

KELLER, R. M., K. WÜRTHRICH, and P. G. DEBRUNNER (1972): Proc. Natl. Acad. Sci. USA **69**, 2073–2075

KITAGAWA, T., M. ABE, Y. KYOGOKU, H. OGOSHI, E. WATANABE and Z. YOSHIDA (1976): J. Phys. Chem. 80, 1181–1186

KITAGAWA, T., M. ABE, Y. KYOGOKU, H. OGOSHI, H. SUGIMOTO, and Z. YOSHIDA (1977): Chem. Phys. Lett. **48**, 55–58

KITAGAWA, T., M. ABE, and H. OGOSHI (1978): J. Chem. Phys. **69**, 4516–4525

KITAGAWA, T., K. NAGAI, and M. TSUBAKI (1979): FEBS Lett. **104**, 376–378

KNOP, J. V., and A. KNOP (1970a): Z. Naturforsch. **25a**, 1720–1725

KNOP, J. V., and A. KNOP (1970b): Z. Naturforsch. **25a**, 1726–1728

KOBAYASHI, H., T. HIGUCHI, and K. EGUCHI (1976): Bull. Chem. Soc. Japan **49**, 457–463

KÖNIG, E., and G. RITTER (1974) in: Mössbauer Effect Methodology, Vol. **9**. Gruverman, I. J., C. W. SEIDEL, and (D. K. DIETERLY, eds.) Plenum Press, New York, p. 3

KON, H. (1968): J. Biol. Chem. **243**, 4350 to 4357

KON, H., and N. KATAOKA (1969): Biochemistry 8, 4757–4762

KON, H. (1975): Biochim. Biophys. Acta **379**, 103–113

KUMAKI, K., M. SATO, H. KON, and D. W. NEBERT (1978): J. Biol. Chem. **253**, 1048 to 1058

KUTHAN, H., H. TSUJI, H. GRAF, V. ULLRICH, J. WERRINGLOER, and R. W. ESTABROOK (1978): FEBS Lett. **91**, 343–345

LAMPE, J., H. REIN, and W. SCHELER (1972): FEBS Lett. **23**, 282–284

LANG, G., and W. MARSHALL (1966): Proc. Phys. Soc. (London) **87**, 3–34

LANG, G., D. HERBERT, and T. YONETANI (1968): J. Chem. Phys. **49**, 944–950

LESKO, S., W. CASPRY, R. LORENTZEN, and P. O. P. Ts'o (1975): Biochemistry **14**, 3978 to 3984

LICHTENBERGER, F. W., W. NASTAINZYK, and V. ULLRICH (1976): Biochem. Biophys. Res. Commun. **70**, 939–946

LIVSHITZ, M. A., A. M. ARUTYUNYAN, and YU. A. SHARONOV (1976): J. Chem. Phys. **64**, 1276–1280

LOBRUTTO, R., CH. P. SCHOLES, G. C. WAGNER, I. C. GUNSALUS, and P. G. DEBRUNNER (1980): J. Amer. Chem. Soc. **102**, 1167 to 1170

LOEW, G. H., C. J. KERT, L. M. HJELMELAND, and R. F. KIRCHNER (1977a): J. Amer. Chem, Soc. **99**, 3534–3536

LOEW, G. H., L. M. HJELMELAND, and R. F. KIRCHNER (1977b): Int. J. Quant. Chem.: Quant. Biol. Symp. **4**, 225–244

LOEW, G. H., and Z. S. HERMAN (1980): J. Amer. Chem. Soc. **102**, 6173–6174

LOEW, G. H., and M.-M. ROHMER (1980): J. Amer. Chem. Soc. **102**, 3655–3657

LOEW, G. H., M. POULSEN, J. FERRELL, and D. CHAET (1980): Chem.-Biol. Interact. **31**, 319–340

LU, A. Y., and S. B. WEST (1980): Pharmacol. Rev. **31**, 277–295

MAEDA, Y., and Y. MORITA (1967): Biochem. Biophys. Res. Commun. **29**, 680–685

MAINES, M. D., and M. W. ANDERS (1973): Arch. Biochem. Biophys. **159**, 201–205

MALTEMPO, M. M., and T. H. MOSS (1976): Quart. Rev. Biophys. **9**, 181–215

MARIČIĆ, S., S. VUK-PAVLOVIĆ, B. BENKO, J. POROK, H. REIN, G.-R. JÄNIG, and K. RUCKPAUL (1977): Croat. Chem. Acta **49**, 323–331

MASHIKO, T., M. E. KASTNER, K. SPATABLIAN,

W. R. SCHEIDT, and CH. A. REED (1978): J. Amer. Chem. Soc. **100**, 6354–6362
MASON, H. S. (1957) in: Advances in Enzymology, Vol. **19** (F. F. NORD ed.) Chider Science Publ. Inc., New York, pp. 79–233
MATSUBARA, T., J. BARON, L. L. PETERSON, and J. A. PETERSON (1976): Arch. Biochem. Biophys. **172**, 463–469
MATSUBARA, T., and Y. TOCHINO (1976): Arch. Biochem. Biophys. **172**, 696–705
MAUREL, P., C. BONFILS, P. DEBEY, and C. BALNY (1980) in: Microsomes, Drug Oxidations and Chemical Carcinogenesis (M. J. COON, A. H. CONNEY, R. W. ESTABROOK, H. V. GELBOIN, J. R. GILLETTE, and P. J. O'BRIEN eds.) Academic Press, New York, p. 359 to 362
MAXWELL, J. C., and W. S. CAUGHEY (1976): Biochemistry **15**, 388–396
MAY, L., and J. L. HOLTZMAN (1970): Biochem. Biophys. Res. Commun. **39**, 296 to 300
MENDELSOHN, R., S. SUNDER, and H. J. BERNSTEIN (1975): J. Raman Spectroscop. **3**, 303–312
MESSANA, C., M. CERDONIO, P. SHENKIN, R. W. NOBLE, G. FERMI, R. N. PERUTZ, and M. F. PERUTZ (1978): Biochemistry **17**, 3652 to 3662
MIMS, W. B., and R. GILLEN (1967): J. Chem. Phys. **47**, 3518–3532
MIMS, W. B., and J. PEISACH (1976): J. Chem. Phys. **64**, 1074–1091
MINCEY, T., and T. G. TRAYLOR (1979): J. Amer. Chem. Soc. **101**, 765–766
MINEYEV, A. P., and YU. A. SHARONOV (1978): Theor. Chim. Acta **49**, 295–307
MISSELWITZ, R., G.-R. JÄNIG, H. REIN, E. BUDER, D. ZIRWER, and K. RUCKPAUL (1975): Acta Biol. Med. Germ. **34**, 1755 to 1766
MISSELWITZ, R., G.-R. JÄNIG, H. REIN, E. BUDER, D. ZIRWER, and K. RUCKPAUL (1976): Acta Biol. Med. Germ. **35**, K19–K25
MISSELWITZ, R., G.-R. JÄNIG, H. REIN, E. BUDER, D. ZIRWER, and K. RUCKPAUL (1977): Acta Biol. Med. Germ. **36**, K35 to K41
MOFFITT, W. (1956): J. Chem. Phys. **25**, 467 to 478
MOON, R. B., K. DILL, and J. H. RICHARDS (1977): Biochemistry **16**, 221–228
MORASSI, R., I. BERTINE, and L. SACCONI (1973): Coord. Chem. Rev. **11**, 343–402
MORISHIMA, I., T. IIZUKA, and Y. ISHIMURA (1974): Seikagaku (Japanese) **46**, 535–541
MOSS, T. H., A. EHRENBERG, and A. J. BEARDEN (1969): Biochemistry **8**, 4159–4162
MÜLLER, W., and H.-J. SCHNEIDER (1969): Angew. Chem. **91**, 438–439
MURAKAMI, K., and H. S. MASON (1967): J. Biol. Chem. **242**, 1102–1110
MYER, Y. P., A. J. MURPHY, and H. A. HARBURY (1966): J. Biol. Chem. **241**, 5370–5374
NAGATA, C., Y. TAGASHIRA, and H. KODAMA (1974) in: Chemical Carcinogenesis (P. O. P. Ts'o, and J. A. DI PAOLO eds.) Dekker, New York, part A, pp. 87–111
NAKANO, N., J. OTSUKA, and A. TASAKI (1971): Biochim. Biophys. Acta **236**, 222 to 233
NAKANO, N., J. OTSUKA, and A. TASAKI (1972): Biochim. Biophys. Acta **278**, 355 to 371
NEBERT, D. W., and H. KON (1973): J. Biol. Chem. **248**, 169–178
NEBERT, D. W., K. KUMAKI, M. SATO, and H. KON (1976): Hoppe-Seylers Z. Physiol. Chem. **357**, 1044–1045
NEMETHY, G., and H. A. SCHERAGA (1962): J. Phys. Chem. **66**, 1773–1789
NICOLA, N. A., E. MINASIAN, C. A. APPLEBY, and S. J. LEACH (1975): Biochemistry **14**, 5141–5149
NORDBLOM, G. D., and M. J. COON (1977): Arch. Biochem. Biophys. **180**, 343–347
OGAWA, S., R. G. SHULMAN, and T. YAMANE (1972): J. Molec. Biol. **70**, 291–300
OGOSHI, H., Y. SAITO, and K. NAKAMOTO (1972): J. Chem. Phys. **57**, 4194–4202
O'KEEFFE, D. H., R. E. EBEL, J. A. PETERSON, J. C. MAXWELL, and W. S. CAUGHEY (1978a): Biochemistry **17**, 5845–5852
O'KEEFFE, D. H., R. E. EBEL, and J. A. PETERSON (1978b): J. Biol. Chem. **253**, 3509 to 3516
OMURA, T., and R. SATO (1964): J. Biol. Chem. **239**, 2370–2378
ORRENIUS, S., B. J. WILSON, C. von BAHR, and J. B. SCHENKMAN (1972) in: Biological Hydroxylation Reactions (G. S. BOYD, and R. M. S. SMELLIE eds.) Academic Press, London and New York, pp. 55–77
OTSUKA, J. (1970): Biochim. Biophys. Acta **214**, 233–235
OZAKI, Y., T. KITAGAWA, Y. KYOGOKU, H. SHIMADA, Y. IIZUKA, and Y. ISHIMURA (1976a): J. Biochem. Tokyo **80**, 1447–1451

Ozaki, Y., T. Kitagawa, and Y. Kyogoku, (1976b): FEBS Lett. **62**, 369–372

Ozaki, Y., T. Kitagawa, Y. Kyogoku, Y. Imai, C. Hashimoto-Yutsudo, and R. Sato (1978): Biochemistry **17**, 5826–5831

Parmely, R. C., and H. M. Goff (1980): J. Inorg. Biochem. **12**, 269–280

Peisach, J., and W. E. Blumberg (1970): Proc. Natl. Acad. Sci. USA **67**, 172–179

Peisach, J., W. E. Blumberg, S. Ogawa, E. A. Rachmilewitz, and R. Oltzik (1971): J. Biol. Chem. **246**, 3342–3355

Peisach, J., and W. B. Mims (1973): Proc. Natl. Acad. Sci. USA **70**, 2979–2982

Peisach, J., J. O. Stern, and W. E. Blumberg (1973): Drug. Metab. Dispos. **1**, 45–61

Peisach, J., and W. B. Mims (1977): Biochemistry **16**, 2795–2799

Peisach, J., W. B. Mims, and J. L. Davis (1979): J. Biol. Chem. **254**, 12379–12389

Peterson, J. A. (1971): Arch. Biochem. Biophys. **144**, 678–693

Phillips, S. E. V. (1978): Nature **273**, 247–248

Pirrwitz, J., G. Lassmann, H. Rein, O. Ristau, G.-R. Jänig, and K. Ruckpaul (1977): FEBS Lett. **83**, 15–18

Pirrwitz, J., H. Rein, G. Lassmann, G.-R. Jänig, S. Pečar, and K. Ruckpaul (1979a): FEBS Lett. **101**, 195–200

Pirrwitz, J., G. Lassmann, H. Rein, G.-R. Jänig, S. Pečar, and K. Ruckpaul (1979b): Acta Biol. Med. Germ. **38**, 235–247

Plus, R., and M. Lutz (1974): Spectroscop. Lett. **7**, 73–84

Preiss, B., and K. Bloch (1964): J. Biol. Chem. **239**, 85–88

Pudzianowski, A. T., and G. H. Loew (1980): J. Amer. Chem. Soc. **102**, 5443–5449

Rahimtula, A. D., P. J. O'Brien, E. G. Hrycay, J. A. Peterson, and R. W. Estabrook (1974): Biochem. Biophys. Res. Commun. **60**, 695–702

Rakshit, G., and T. G. Spiro (1974): Biochemistry **13**, 5317–5323

Reichman, L. M., B. Annaev, and E. G. Rozantzev (1972): Biochim. Biophys. Acta **263**, 41–51

Rein, H., O. Ristau, and F. Jung (1964): Folia Haematol. **82**, 191–205

Rein, H., and O. Ristau (1965): Biochim. Biophys. Acta **94**, 516–524

Rein, H., O. Ristau, F. Hackenberger, and F. Jung (1968): Biochim. Biophys. Acta **167**, 538–546

Rein, H., O. Ristau, and W. Scheler (1972): FEBS Lett. **24**, 24–26

Rein, H., K. Ruckpaul, and W. Haberditzl (1973): Chem. Phys. Lett. **20**, 71–76

Rein, H., O. Ristau, and K. Ruckpaul (1975a): Biochim. Biophys. Acta **393**, 373 to 378

Rein, H., K. Ruckpaul, O. Ristau, and W. Haberditzl (1975b): Acta Biol. Med. Germ. **34**, 325–344

Rein, H., G.-R. Jänig, W. Winkler, and K. Ruckpaul (1976): Acta Biol. Med. Germ. **35**, K41–K50

Rein, H., O. Ristau, J. Friedrich, G.-R. Jänig, and K. Ruckpaul (1977a): Croat. Chem. Acta **49**, 251–261

Rein, H., S. Böhm, G.-R. Jänig, and K. Ruckpaul (1977b): Croat. Chem. Acta **49**, 333–338

Rein, H., O. Ristau, J. Friedrich, G.-R. Jänig, and K. Ruckpaul (1977c): FEBS Lett. **75**, 19–22

Rein, H., and O. Ristau (1978): Pharmazie **33**, 325–328

Rein, H. (1979): Discussion to Ullrich et al., 1979 Acta biol. med. germ. **38**, 287–297

Rein, H., O. Ristau, R. Misselwitz, E. Buder, and K. Ruckpaul (1979): Acta Biol. Med. Germ. **38**, 187–200

Rein, H., J. Pirrwitz, J. Friedrich, O. Ristau, G.-R. Jänig, and K. Ruckpaul (1980) in: Biochemistry, Biophysics and Regulation of Cytochrome P-450 (Gustafsson, J.-A., J. Carlstedt-Duke, A. Mode, and J. Rafter, eds.) Elsevier/North-Holland, Amsterdam, New York, Oxford, p. 577 to 581

Remba, R. D., P. M. Champion, D. B. Fitchen, R. Chiang, and L. P. Hager (1979): Biochemistry **18**, 2280–2290

Renneberg, R., J. Capdevila, N. Chacos, R. W. Estabrook, and R. A. Prough (1981): Biochem. Pharmacol. **30**, 843–848

Renson, J., W. Weissback, and S. Udenfriend (1965): Molec. Pharmacol. **1**, 145 to 148

Rich, P. R., R. Cammack, and D. S. Bendall (1975): Europ. J. Biochem. **59**, 281–286

Rimai, L., I. Salmeen, and D. H. Petering (1975): Biochemistry **14**, 378–382

Ristau, O. (1975) in: Struktur und Funktion der Erythrozyten. VII. Internationales Symposium, (S. Rapoport, and F. Jung, eds.) Akademie Verlag, Berlin, pp. 159–163

Ristau, O., and H. Rein (1976): Acta Biol. Med. Germ. **35**, 1613–1624

Ristau, O., H. Rein, G.-R. Jänig, and K. Ruckpaul (1978): Biochim. Biophys. Acta **536**, 226–234

Ristau, O., H. Rein, S. Greschner, G.-R. Jänig, and K. Ruckpaul (1979): Acta Biol. Med. Germ. **38**, 177–185

Ristau, O. (1980): Acta Biol. Med. Germ. **39**, 71–81

Ristau, O. (1981): Period. Biol. **83**, 39–49

Röder, A., and E. Bayer (1969): Europ. J. Biochem. **11**, 89–92

Roitsin, A. B. (1972): Sov. Phys. Usp. **14**, 766 to 782

Rougee, M. and D. Brault (1975): Biochemistry **14**, 4100–4106

Ruckpaul, K., H. Rein, and F. Jung (1969): Acta Biol. Med. Germ. **23**, K1–K4

Ruckpaul, K., H. Rein, O. Ristau, and F. Jung (1970a): Biochim. Biophys. Acta **221**, 9–19

Ruckpaul, K., H. Rein, and F. Jung (1970b): Naturwissenschaften **57**, 131–132

Ruckpaul, K., H. Rein, and F. Jung (1970c): Acta Biol. Med. Germ. **24**, 445–454

Ruckpaul, K., S. Maričić, G.-R. Jänig, B. Benko, S. Vuk-Pavlović, and H. Rein (1976): Croat. Chem. Acta **48**, 69–86

Ruckpaul, K., H. Rein, J. Blanck, O. Ristau, and M. J. Coon (1980) in: Biochemistry, Biophysics and Regulation of Cytochrome P-450 (J.-A. Gustafsson, J. Carlstedt-Duke, A. Mode, and J. Rafter, eds.) Elsevier/North-Holland, Amsterdam, New York, Oxford, pp. 539–549

Ruf, H. H., W. Nastainczyk, and V. Ullrich (1974) in: 6th International Conference on Magnetic Resonance in Biological Systems, Kandersteg, Switzerland. Abstr. C14

Ruf, H. H., and W. Nastainczyk (1976): Europ. J. Biochem. **66**, 139–146

Ruf, H. H., and P. Wende (1977): J. Amer. Chem. Soc. **99**, 5499–5500

Ruf, H. H., P. Wende, and V. Ullrich (1979): J. Inorg. Biochem. **11**, 189–204

Sato, M., A. S. Rispin, and H. Kon (1976): Chem. Phys. **18**, 211–224

Sato, M., H. Kon, K. Kumaki, and D. W. Nebert (1977): Biochim. Biophys. Acta **498**, 403–421

Sato, R., Y. Imai, and H. Tanaguchi (1979) in: The Induction of Drug Metabolism (R. W. E. Estabrook, E. Lindenlaub, eds.) Schattauer, Stuttgart, New York 1979, pp. 213–223

Schaffer, A. M., and M. Gouterman (1970): Theor. Chim. Acta **18**, 1–13

Scheidt, W. R., and P. L. Piciulo (1976): J. Amer. Chem. Soc. **98**, 1913–1919

Scheler, W., G. Schoffa, and F. Jung (1957): Biochem. Z. **329**, 232–246

Schenkman, J. B., H. Remmer, and R. W. Estabrook (1967): Molec. Pharmacol. **3**, 113 to 123

Schenkman, J. B., H. Greim, M. Zange, and H. Remmer (1969): Biochim. Biophys. Acta **171**, 23–31

Schenkman, J. B. (1970): Biochemistry **9**, 2081–2091

Schenkman, J. B. (1972): Molec. Pharmacol. **8**, 178–188

Schenkman, J. B., D. L. Cinti, S. Orrenius, P. Moldeus, and R. Kraschnitz (1972): Biochemistry **11**, 4243–4251

Schlosnagle, D. C., J. C. M. Tsibris, and I. C. Gunsalus (1974): Fed. Proc. **33**, 1370

Schonbaum, G. R., and S. Lo (1972): J. Biol. Chem. **247**, 3353–3360

Sharonov, Y. A., N. A. Sharonova, and B. P. Atanasov (1976): Biochim. Biophys. Acta **434**, 440–451

Sharrock, M., E. Münk, P. G. Debrunner, V. Marshall, J. D. Lipscomb, and I. C. Gunsalus (1973): Biochemistry **12**, 258–265

Sharrock, M., P. G. Debrunner, C. Schulz, J. D. Lipscomb, V. Marshall, and I. C. Gunsalus (1976): Biochim. Biophys. Acta **420**, 8–26

Shimizu, T., T. Nozawa, M. Hatano, Y. Imai, and R. Sato (1975): Biochemistry **14**, 4172–4178

Shimizu, T., T. Nozawa, M. Hatano, H. Satake, Y. Imai, C. Hashimoto, and R. Sato (1979): Biochim. Biophys. Acta **579**, 122–133

Sies, H., and M. Kandel (1970): FEBS Lett. **9**, 205–208

Sligar, S. G. (1976): Biochemistry **15**, 5399 to 5406

Sligar, S. G., and I. C. Gunsalus (1976): Proc. Natl. Acad. Sci. USA **73**, 1078–1082

Sligar, S. G., D. L. Cinti, G. G. Gibson, and J. B. Schenkman (1979): Biochem. Biophys. Res. Commun. **90**, 925–932

Sligar, S. G., K. A. Kennedy, and D. C. Pearson (1980): Proc. Natl. Acad. Sci. USA **77**, 1240–1244

SMITH, D. W., and R. J. P. WILLIAMS (1968): Biochem. J. **110**, 297–301

SPARTALIAN, K., G. LANG, J. P. COLLMAN, R. R. GAGNE, and CH. A. REED (1975): J. Chem. Phys. **63**, 5375–5382

SPIRO, T. G., and T. C. STREKAS (1974a): J. Amer. Chem. Soc. **96**, 338–345

SPIRO, T. G., and T. C. STREKAS (1974b): Biochem. Biophys. Acta **351**, 237–245

SPIRO, T. G. (1975): Proc. Roy. Soc. London A **345**, 89–105

STEIN, P., J. M. BURKE, and T. G. SPIRO (1975): J. Amer. Chem. Soc. **97**, 2304 to 2305

STEPHENS, K. W. H. (1953): Proc. Roy. Soc. London A **219**, 542–555

STEPHENS, P. J., W. SUETAKA, and P. N. SCHATZ (1966): J. Chem. Phys. **44**, 4592 to 4602

STERN, J. O., J. PEISACH, W. E. BLUMBERG, A. Y. H. LU, and W. LEVIN (1973): Arch. Biochem. Biophys. **156**, 404–413

STERN, J. O., and J. PEISACH (1974): J. Biol. Chem. **249**, 7495–7498

STONG, J. D., J. M. BURGE, P. DALY, P. WRIGHT, and T. G. SPIRO (1980): J. Amer. Chem. Soc. **102**, 5815–5819

STREITH, J., B. DANNER, and C. SIGWALT (1967): Chem. Commun. 979–980

SUGITA, Y., Y. DOHI, and Y. YONEYAMA (1968): Biochem. Biophys. Res. Commun. **31** 447–456

TAMURA, M. (1971): Biochim. Biophys. Acta **243**, 239–248

TANABE, J., and S. SUGANO (1954): J. Phys. Soc. (Japan) **9**, 766–779

TANG, S. C., S. KOCH, G. C. PAPAEFTHYMIOU, S. FONER, R. B. FRANKEL, J. A. IBERS, and R. H. HOLM (1976): J. Amer. Chem. Soc. **98**, 2414–2434

TESCHKE, R., Y. HASUMURA, and C. S. LIEBER (1975): J. Biol. Chem. **250**, 7397–7404

TRAUTWEIN, A., Y. MAEDA, F. E. HARRIS, and H. FORMANEK (1974): Theor. Chim. Acta **36**, 67–76

TREU, J. I., and J. J. HOPFIELD (1975): J. Chem. Phys. **63**, 613–623

TRITTELVITZ, E., H. SICK, and K. GERSONDE (1972): Europ. J. Biochem. **31**, 578–584

TRITTELVITZ, E., K. GERSONDE, and K. H. WINTERHALTER (1975): Europ. J. Biochem. **51**, 33–42

TSAI, R., C. A. YU, I. C. GUNSALUS, J. PEISACH, W. E. BLUMBERG, W. H. ORME-JOHNSON, and H. BEINERT (1970): Proc. Natl. Acad. Sci. USA **66**, 1157–1163

TSUCHIYA, T., H. ARAI, and H. IGETA (1969): Tetrahed. Lett. **32**, 2747–2750

ULLMER, D. D., and B. L. VALLEE (1965): Adv. Enzymol. **27**, 37–104

ULLRICH, V., J. WOLF, E. AMADORI, and H.-J. STAUDINGER (1968): Hoppe-Seylers Z. Physiol. Chem. **349**, 85–94

ULLRICH, V. (1969): Z. Naturforsch. **24**b, 699 to 704

ULLRICH, V., and H.-J. STAUDINGER (1971) in: Handbuch der experimentellen Pharmakologie, Vol. **28**/2 (B. B. BRODIE, J. R. GILLETTE, and H. S. ACKERMAN, eds.) Springer, Berlin, pp. 251–263

ULLRICH, V., W. NASTAINCZYK, and H. H. RUF (1975): Biochem. Soc. Transact. **3**, 803 to 806

ULLRICH, V., H. H. RUF, and P. WENDE (1977): Croat. Chem. Acta **49**, 213–222

ULLRICH, V. (1979): Topics in Current Chemistry, Part Biochemistry, **83**, 67–104

ULLRICH, V., H. SAKURA, and H. H. RUF (1979): Acta Biol. Med. Germ. **38**, 287–297

URRY, D. W. (1965): Proc. Natl. Acad. Sci. USA **54**, 640–648

VERMA, A. L., and H. J. BERNSTEIN (1974): J. Chem. Phys. **61**, 2560–2565

VICKERY, L., A. SALMON, and K. SAUER (1975): Biochim. Biophys. Acta **386**, 87 to 98

VICKERY, L., T. NOZAWA, and K. SAUER (1976a): J. Amer. Chem..Soc. **98**, 351–357

VICKERY, L., T. NOZAWA, and K. SAUER (1976b): J. Amer. Chem. Soc. **98**, 343–350

VUK-PAVLOVIĆ, S., B. BENKO, S. MARIČIĆ, G. LAHAJNAR, I. P. KURANOVA, and B. K. VAINSHTEIN (1976): Int. J. Pept. Prot. Res. **8**, 427–434

VUK-PAVLOVIĆ, S., and Y. SIDERER (1977): Biochem. Biophys. Res. Commun. **79**, 885 to 889

WANG, J. H., A. NAKAHARA, and E. E. FLEISCHER, (1958): J. Amer. Chem. Soc. **80**, 1109 to 1113

WATERMAN, M., V. ULLRICH, and R. W. ESTABROOK (1973): Arch. Biochem. Biophys. **155**, 355–360

WEISS, C., H. KOBAYASHI, and M. GOUTERMAN (1965): J. Molec. Spectroscop. **16**, 415 to 450

WEISS, J. J. (1964): Nature **202**, 83–84

WEISSBLUTH, M. (1974): Hemoglobin, Cooperativity and Electronic Properties (Kleinzeller, A. G. F. SPRINGER, and H. G. WITTMANN, eds.) Springer, Berlin, Heidelberg, New York, p. 159

WERRINGLOER, J., S. KAWANO, and R. W. ESTABROOK (1979): Acta Biol. Med. Germ. **38**, 163–175

WHITE, R. E., J. T. GROVES, and G. A. McCLUSKY (1979): Acta Biol. Med. Germ. **38**, 475–482

WHITE, R. E., and M. J. COON (1980): Ann. Rev. Biochem. **49**, 315–356

WHITE, R. E., S. G. SLIGAR, and M. J. COON (1980): J. Biol. Chem. **255**, 11108–11111

WILLIAMS, R. J. P. (1956): Chem. Rev. **56**, 299 to 328

YAMAZAKI, H., and K. J. COETANOVIC (1964): J. Chem. Phys. **41**, 3703–3710

YASUNOBU, K. T., L. G. ARMES, J. W. CRABB, M. HANIU, I. C. GUNSALUS, and M. J. COON (1980) in: Biochemistry, Biophysics and Regulation of Cytochrome P-450. (J.-Å. GUSTAFSSON, J. CARLSTEDT-DUKE, A. MODE, and J. RAFTER, eds.) Elsevier/North-Holland, Amsterdam, New York, Oxford, pp. 49–56

YONETANI, T., H. SCHLEYER, and A. EHRENBERG (1966): J. Biol. Chem. **241**, 3240–3243

YONETANI, T., H. YAMAMOTO, J. E. ERMAN, J. S. LEIGH, Jr., and G. H. REED (1972): J. Biol. Chem. **247**, 2447–2455

YOSHIDA, Y., and H. KUMAOKA (1975): J. Biochem. **78**, 455–468

YOSHIKAWA, S., M. G. CHOC, M. C. O'TOOLE, and W. S. CAUGHEY (1977): J. Biol. Chem. **252**, 5498–5508

YU, C. A., and I. C. GUNSALUS (1974): J. Biol. Chem. **249**, 102–106

ZERNER, M. C., G. H. LOEW, R. F. KIRCHNER, and U. T. MUELLER-WESTERHOFF (1980): J. Amer. Chem. Soc. **102**, 589–599

Acknowledgement

The authors are very grateful to Dr. P. DEBRUNNER from the Dept. of Physics of the University of Illinois, Urbana, USA, for helpful discussions about the subjekt reviewed in this chapter.

Chapter 5

Structural Studies of Monooxygenases by Solvent-Proton Magnetic Relaxation

S. Maričić

Contents

1. Preface

The primary purpose of this review is to bring the method of solvent-proton magnetic relaxation (S-PMR) to the attention of haemoprotein-experts, in particular those elucidating the structure/function relationship of monooxygenases. The method itself has come out of age with serious problems in the interpretation of the relaxation effects from protein-bound metal ions, as reflected in a recent critical survey by Burton et al. (1979). The survey left aside completely the application of S-PMR to studies of haemoproteins, thus following the suit of other earlier reviews, although S-PMR studies of proteins with bound metal ions had actually begun with haemoprotein solutions. Whereas the studies of oxygen-carriers by S-PMR may not have proved very exciting, the application of the method to monooxygenase enzyme systems seems to warrant now a review of the state of the art.

The NMR-expert will probably find here a few of the well known problems and a couple of new ones. The presentation is, however, deliberately of a qualitative nature in order to serve the above stated primary purpose. For the enzymologist interested in monooxygenases this article is hoped to offer a reasonable balance between the information on the S-PMR method as applied to haemoproteins (Section 2), without the technicalities, and the results obtained thus far about the macromolecular suprastructures of microsomes (Section 3) and about the neighbourhood of the haem (Section 4).

The reader interested mainly in the microsomes may well read directly Section 3 because the S-PMR results are discussed therein only in terms of relative structural changes of the microsomal constituents. However, Section 4 might require reading of Section 2 beforehand. The latter is hoped to serve as a guide in recognizing the potential and the limitations of the S-PMR method especially from the point of view of circumventing some of the problems for its fuller application to haemoprotein studies.

2. Solvent-Proton Magnetic Relaxation (S-PMR) Technique for Solutions of Haemoproteins

The high resolution mode of NMR has proved most useful in studies of proteins whose spatial, static (X-ray), structures had been known, so that assignement of a multitude of resonance peaks was less difficult. On the other hand, the less we know about the three-dimensional structure of the protein bearing paramagnetic centre(s), the more useful solvent-proton magnetic relaxation may be in learning about the structure and dynamics of the protein domain around the paramagnetic centre. Still, a certain degree of confidence in S-PMR for haemoproteins had to be achieved first by studies of haemoglobin(s) and myoglobin(s) whose spatial structures were known.

No three-dimensional structure of any monooxygenase is yet (as of December 1980) known. From this structural point of view the status of the field in the recent past was much like the studies of haemoglobin were in the fifties, when the first looks were cast on haemoglobin by S-PMR (Davidson and Gold, 1957; Kon and Davidson, 1959; Wishnia, 1960; Lumry et al., 1961; Scheler, 1963). However, at that time the very model of "simple" inorganic paramagnetic aquoions in NMR relaxation measurements was being developed (Solomon, 1955; Bloembergen, 1957; McConnell, 1958; Bernheim et al., 1959; Swift and Connick, 1962; Pfeifer, 1963; Luz and Meiboom, 1964). Two years after Luz and Meiboom (1964) verified experimentally the characteristic temperature dependence of the S-PMR rates for aqueous methanol Co- and Ni-solutions, a similar behaviour was observed for Fe^{3+}-haemoglobin and -myoglobin (Maričić et al., 1966). For a full elucidation of the relaxation mechanism and, hence, to obtain maximum information, a frequency dispersion study by S-PMR is necessary (Koenig et al., 1973).

In a strive to gain structural information from S-PMR of protein solutions the classical formula derived by Solomon (1955) for inorganic paramagnetic aquoions has frequently been used without sufficient caution called for by the much more complicated metal-protein complexes. Dwek (1973) has enumerated six conditions which ought to be fulfilled in this respect. There is hardly any metal-protein solution for which this could be claimed thus far.

In order to derive any structural information about the relation of the paramagnetic centre, its surrounding protein matrix and the solvent, one must first derive the purely paramagnetic contribution (R_{pmg}) to the overall solution rates. This implies only the direct dipole-dipole interaction between the unpaired electron spin of the metal ion and the proton spins of the solvent. It has become common to measure R_{pmg} for haemoproteins (Maričić et al., 1966) by subtracting the rates of the diamagnetic-(CO-bound)-protein solution. However, there are two assumptions implicit in this procedure.

Firstly, it is assumed that the relaxation rate within the protein solution consists only of R_{pmg} plus the diamagnetic contribution, and, secondly, that the latter is identical to that obtained from the independent blank measurement with CO-bound haemoprotein. It was only recently that the attention has again been drawn to the phenomenon of cross-relaxation (Edzes and Samulski, 1978; Eisenstadt et al., 1978; Koenig et al., 1978). It is envisaged that through this relaxation mechanism there is a very efficient relaxation "flow" from the paramagnetic centre by spin-diffusion (ABRAGAM, 1961) within the protein, i.e. indirectly to the solvent protons other than those in immediate dipolar interaction with the paramagnetic centre from which R_{pmg} originates. Hence, the total relaxation rate of a solution containing paramagnetic protein (i.e. protein with a paramagnetic centre) comprises an additional paramagnetic (cross-relaxation) component, which is absent in the blank measurement. All the R_{pmg} values derived after deduction of the diamagnetic blank rates are therefore apparently overestimated.

Whereas unequivocal structural data are difficult to obtain from S-PMR measurements on haemoproteins, the dynamic side of the method is very useful in that it, also, reflects to a certain degree the protein structure around the paramagnetic centre. To this end, then, the experimentally simpler way by S-PMR temperature dependence studies may suffice without the frequency dispersion measurements. An approach using stereochemical probes in temperature studies will be described here. In essence, it is an experimental method based on comparisons, so that most of uncertainties due to unwarranted applications of theory owing to the complexities of the metal-protein systems, can be circumvented.

2.1. The temperature dependence of S–PMR

This section provides for the general basis necessary in the later discussion of the results from the literature. Many important detail will be left out. For them, the reader is referred to the original papers quoted in the introduction of this section and to several specialists' reviews (DWEK, 1973; for several other see also BURTON et al., 1979). The longitudinal (spin-lattice) relaxation time, T_1, will be mostly dealt with here.

The total relaxation rate of a paramagnetic haemoprotein solution, $(1/T_1)_{tot} = R_{1,tot}/s^{-1}$, is composite entity:

$$R_{1,tot} = R_1^{I} + R_1^{II} + R_1^{+} + R_0 \quad (1)$$

R_0 is the contribution from relaxation mechanism(s) within the solvent (water, mainly) itself, which is very small and for most practical purposes may be neglected, as it does not influence the ensuing discussion of the $R_{1,tot}$-temperature dependence.

R_1^{I} is the dominant part at higher temperatures when the frequency of proton exchange between the site closest to the paramagnetic metal ion and bulk solvent is very fast. While $R_0 \sim 1$, $R_1^{I} = 10^2 - 10^6\ s^{-1}$. R_1^{I} is regarded as the "first-coordination sphere" relaxation, a term taken over from the model of metal aquoions. In case of metaloproteins the exchanging entity (protons of H_2O molecules) may not necessarily be directly coordinated to the paramagnetic metal ion and yet it may be its closest neighbour for which SOLOMON's (1955) equation does define the rate $1/T_{1M}$.

The temperature dependence of $R_{1,tot}$ stems from the thermally activated mechanism of proton exchange between the site closest to the metal ion, and bulk solvent, and from additional temperature dependent parameters, as follows:

$$R_1^{I} = \frac{n}{[H](T_{1M} + \tau_M)} \quad (2)$$

where n = number of exchanging protons; [H] = proton concentration in the solution, with the Arrhenius relation for the proton exchange rate

$$1/\tau_M = A \exp(-E_a/RT). \quad (3)$$

R_1^{II} used to be taken, again reflecting the metal aquo-ions model, as the "outer-sphere" relaxation. Actually it, too, is due to proton fast-exchange, only in this case from the outer coordination sphere of the metal ion. The proton-to-metal distance is related to this relaxation rate, R_1^{II}, as d^{-3}, whereas that for the first coordination sphere appears in the corresponding relations as d^{-6} with R_1^{I} (in fact $1/T_{1M}$). However, in view of the recently emerging picture about the cross-relaxation contribution, R_1^{+}, it seems impossible to disentangle experimentally R_1^{II} from R_1^{+}. This is not a serious drawback with regard to determining d because, as it has already been pointed out (PIFAT et al., 1973), it is a meaningless parameter for haemoproteins owing to lack of bulk-solvent spherical symmetry around the haem-iron. The more important question arises as to how to obtain R_1^{I} from $R_{1,tot}$, which we shall discuss later. Experimental evidence thus far indicates that R_1^{II} and R_1^{+} have similar temperature dependence due to the correlation time, τ_c, for the dipole-dipole interactions.

Fig. 1 depicts the $R_{1,tot}$ temperature dependence in an Arrhenius plot. At lower temperatures (region II) proton exchange between the metal site and bulk solvent is very slow. Few solvent molecules (with their protons) in a time unit take advantage of the powerful magnetic dipole of the haem-iron to experience much faster proton magnetic relaxation than in bulk solvent. Hence an inefficient overall relaxation mechanism and therefore low $R_{1,tot}$ values. The latter is due only to $R_1^{II} + R_1^{+}(+R_0)$. Were it possibly, R_1^{+} should be measured as an appropriate blank. It represents in our notation both (i) the so-called diamagnetic contribution from the magnetic interaction of protons from the protein with those in solvent, $R_{1,dia}$, and (ii) the part induced by spin

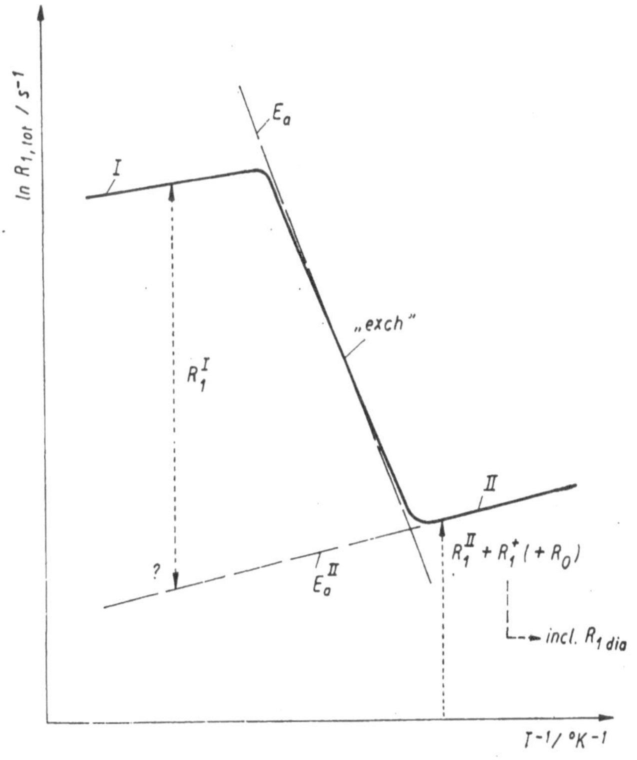

Fig. 1. A schematic Arrhenius plot of the solvent-proton magnetic relaxation rates in dependence on temperature for solutions of haemoproteins

diffusion from the paramagnetic haem-iron, through the protein. The two contributions are inseparable, because, in a sense, the whole protein molecule shares the "advantage" of being host to the haem-iron, acting thus as a spatially spread-out paramagnetic "centre". It is desirable to obtain R_1^{II} because it does represent a semiquantitative assessment of the haem-iron accessibility for solvent protons outside their closest possible approach within the "first coordination sphere". R_1^{II} originates from direct, though dipole-dipole interaction at longer range, between Fe^{3+} and H_2O protons from the "outer-sphere" solvent.

On raising the temperature, the exchange frequency from the sphere of closest approach to Fe^{3+} (with radius, r) increases, so that in a unit time more solvent protons experience the Fe^{3+} paramagnetic influence. Their complete relaxation is accomplished during the encounter because the time they spend at the metal site, τ_M is longer than their (proton) relaxation time at this site, T_{1M}. For the "exch"-region in Fig. 1, $R_{1,tot}$ is proportional to $1/\tau_M$ (eq. 3) because τ_M is much longer than T_{1M} throughout this temperature region (eq. 2). τ_M is usually on order of 10^{-2} to 10^{-4} s, but could be shorter.

At temperatures causing $\tau_M \ll T_{1M}$, i.e. for region I in Fig. 1, $R_{1,tot}$ is proportional to $1/T_{1M}$ (eq. 2). The relaxation mechanism for the whole solution is very efficient, because protons can only partially relax during their encounter ($\tau_M \ll T_{1M}$) with the powerful paramagnetic Fe^{3+} at short distance. On their return to bulk solvent the partially relaxed protons induce additional relaxation in other protons.

The limiting conditions for the three temperature regions in Fig. 1 may be formulated (BENKO, 1976) as follows:

$$1/\tau_M \geqq \frac{[H]E_a^{II}}{nE_a}(R_1^{II} + R_1^{+}) \qquad (4)$$

defines the onset of the "exch" region. For exchange of one molecule of water ($n = 2$) from the sphere of closest approach, $1/\tau_M = 5.5(R_1^{II} + R_1^{+})$, as $E_a^{II}/E_a \sim 1/10$.

Neglecting the value of $(R_1^{II} + R_1^{+})$ at higher temperatures to a first approximation, $R_{1,tot}$ attains the maximum value for

$$T_{1M}/\tau_M = E_a/E_a^{II} \simeq 10 \qquad (5)$$

Only when R_1^{I} is obtained from this temperature region (I in Fig. 1) one may, in principle, apply the Solomon-Bloembergen theory (SOLOMON, 1955; BLOEMBERGEN, 1957) to calculate r, the closest distance between the exchanging proton(s) and Fe^{3+} of the haem. This procedure assumes the physical reality of the proton exchange in haemoprotein solutions. It has indeed been proved for two inorganic aquo-ions (Co^{2+} and Ni^{2+}) by LUZ and MEIBOOM (1964). For several metaloprotein solutions the selfconsistency of the parameters derived by applying Solomon-Bloembergen theory left little doubt about the exchange mechanism. However, DERZHANSKI et al. (1970) have put forward an alternative interpretation of the observed temperature dependence of $R_{1,tot}$ for ferrimyoglobin solutions.

2.2. *The physical reality of proton exchange mechanism in haemoprotein solutions*

The "exch"-region (Fig. 1) observed for ferrimyoglobin solutions (MARIČIĆ et al., 1966) was attributed by DERZHANSKI et al. (1970) to a temperature induced conformational change of the haem surrounding, so that the paramagnetic haem-iron moves closer to the protein surface on raising the temperature. As R_1^{II} is proportional with d^{-3}, the outer-sphere relaxation may explain the whole curve in Fig. 1, if the distance, d, between Fe^{3+} and water molecules outside the "first-coordination-sphere" becomes shorter. DERZHANSKI et al. (1970) proposed this to be the case due to a temperature controlled conformational equilibrium of A and B states. State A for region II, and state B for region I (in Fig. 1) with $d_A > d_B$. Hence, there is no need to invoke proton exchange for ferrimyoglobin solutions, or, for that matter, in any metaloprotein solution because a similar conformational equilibrium cannot be excluded *a priori*. Moreover, the observed "exch"-temperature relaxation behaviour cannot be taken *per se* as proof of proton exchange mechanism for metaloproteins.

The decisive experiment would then be to measure $R_{1,\text{tot}}$ in the absence of solvent protons other than the undissociable protons on a marker solvent molecule larger than the water molecule. If the conformational hypothesis is correct the same temperature dependence ("exch" in Fig. 1) ought to be observed irrespective of the size of the marker molecule, because the hypothesis is based on magnetic interaction of solvent protons outside the protein, with the "buried" Fe^{3+} physically inaccessible to them. On the other hand, if proton exchange does take place in ordinary aqueous solutions, the temperature dependence of $R_{1,\text{tot}}$ should depend on the size of the marker molecule as it may not approach Fe^{3+} to the same site as water protons. Using deuterated aqueous solutions with deuterated methanol — $CH_3(O^2H)$, 1,2,ethandiol — $C_2H_4(O^2H)_2$ and 1,2,3,propanetriol — $C_3H_5(O^2H)_3$,VUK-PAVLOVIĆ et al. (1974) were able to show for several ferrihaemoprotein solutions that proton exchange relaxation mechanism is indeed a physical reality. Fig. 2 illustrates how the shape of the $R_{1,\text{tot}}$ temperature dependence may be varied at will by "titration" with free protons in solution.

In this figure a "temperature-window shift" is examplified for two regions of Fig. 1. Because of the well defined shapes of the marker molecules the term "Stereochemical S-PMR" will be used hereinafter for brevity when referring to this technique.

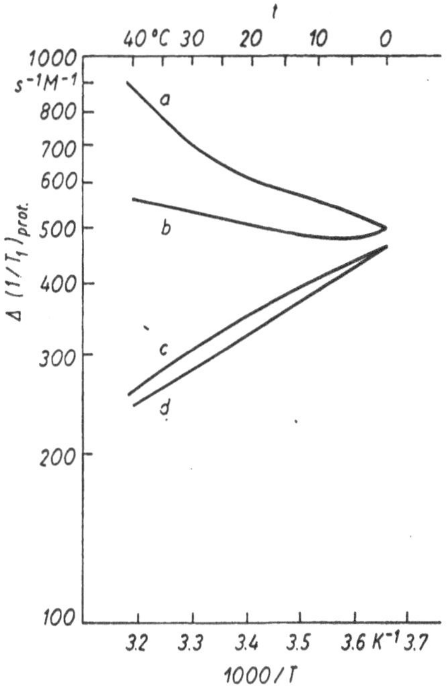

Fig. 2. Arrhenius plot of S-PMR for human (A) H_2O-ferrihaemoglobin

(a) 35% $C_2H_4(OH)_2$ in H_2O
(b) 35% $C_2H_4(OH)_2$ + 10% H_2O in D_2O
(c) 35% $C_2H_4(OH)_2$ in D_2O
(d) 35% $C_2H_4(OD)_2$ in D_2O

2.3. Stereochemical S-PMR

There are three aspects of this experimental approach towards structural studies of proteins bearing paramagnetic (metal-ion) centres. The first two are of a methodological interest, whereas the third gives its scope and, generally, that of S-PMR, in structural studies of haemoprotein aqueous solutions.

(i) *Proof of proton-exchange* mechanism — as described above. It will be implicit in presenting the scope of the technique (see iii), and will not be dealt with further.

(ii) *Determination of* R_1^{I}. In Fig. 1

this parameter is depicted as questionable because of the uncertainty in extrapolating the value $(R_1^{II} + R_1^+)$ to higher temperatures, especially when the "temperature-window" allows only relaxation type I to be observed. The smaller the value of $R_{1,tot}$, the more important $(R_1^{II} + R_1^+)$ is. In several cases the latter value was found to be 25–50% of $R_{1,tot}$. It should be mentioned in passing, that although the proton-exchange has been found real for haemoproteins one must not exclude any effects due to the conformation-controlled changes in relaxation dependent on temperature. Without the latter one is expecting a straight line corresponding to the extrapolated one in Fig. 1. In the cases encountered thus far a downward curvature was observed indicating that if there is any conformational effect it is due to a decrease in haem-iron accessibility, not to the shortening of d.

Though not dramatic, there are alterations of the haem accessibility in haemoglobin and myoglobin dependent on ethanediol concentration in the aqueous solution (Vuk-Pavlović et al., 1977). As these effects turned out to be quite different for the two haemoproteins, the best procedure to account for $(R_1^{II} + R_1^+)$ is to measure $R_{1,tot}$ in presence of the same amount of the marker alcohol in H_2O as for $(R_1^{II} + R_1^+)$ in ^2H_2O. One is then left with two unavoidable differences of the ^1H_2O and ^2H_2O solvents. The first concerns the possible differences in protein hydration, and hence the relaxation rates (included in R_1^+), while the second has to do with different translational and rotational motion of the larger alcohol molecules compared to the molecule of water. The latter would be difficult to estimate quantitatively, whereas the first seems to be negligible according to measurements of Mildvan et al. (1971) compared to those of Vuk-Pavlović et al. (1974).

Soluble monooxygenases are usually prepared in aqueous glycerol solutions which makes it convenient to use the stereochemical S-PMR method. On the other hand, if there happens to be no protein cleft surrounding the haem, the method cannot be applied to determine $(R_1^{II} + R_1^+)$, and, hence, R_1^I. Namely, in such a case even the molecule of glycerol may not be obstructed in getting as close to the haem iron as water molecules do. However, for the monooxygenases other (substrate or inhibitor) molecules may be used instead (see later).

(iii) *The scope of (stereochemical) S-PMR*

a) Allosteric alteration of haem-iron accessibility, which may be relevant to interactions of monooxygenases with their substrates.

Binding of inositolhexaphosphate to ferrihaemoglobin doubles the proton-exchange rate (region "exch" in Fig. 1 obtained in the usual S-PMR measurements) and increases also two to threefold the $(R_1^{II} + R_1^+)$-value as measured by stereochemical S-PMR (Pifat et al., 1974). This has been interpreted as an increase in haem-iron accessibility both in the sense of the protein dynamics ("exch") and its (static) structure (region II). This was convincingly confirmed (Benko et al., 1977) by stereochemical S-PMR using methanol (CH_3O^2H) as the marker molecule. The "exch"-region was observed on addition of inositol hexaphosphate, in distinction when region II was found without this allosteric effector. Hence, binding of it far away from the haem-iron induces an "opening" of the protein cleft comprising the haem. This enables molecules of methanol to exchange between the interior of the cleft and bulk solvent, whereas they could hardly approach the outer-cleft entrance in the absence of the effector. In the same paper a gradual impedence of the solvent dynamics inside the haem-crevice was demonstrated by stereochemical S-PMR on increasing the alkalinity of the aquo- and fluoro-ferrihaemoglobin solutions. In this case the dissociation of protons from the protein surface induces an alteration of protein conformation or/and its dynamics around the haem. The stereochemical S-PMR method has thus proved quite sensitive to detect allosteric effects.

b) Temperature-window "shift"

The full curve represented schematically in Fig. 1 was observed by LUZ and MEIBOOM (1964) for Ni^{2+}- and Co^{2+}-aqueous methanol solutions within some 150 centigrades. For aqueous protein solutions the temperature span is hardly 40 degrees above zero centigrade. Whereas the upper limit cannot be extended because of denaturation, one would expect to widen the temperature range by supercooling the aqueous alcohol solutions. However, cooling down to $-20\,°C$ did not alter the relaxation mechanism (VUK-PAVLOVIĆ et al., 1974).

Thus, the temperature range being always the same, from 0° to 40 °C, the relaxation mechanism observed by S-PMR, i.e. the various segments of the curve in Fig. 1, will be unique for a given haemoprotein structure. Twelve different haem-iron accessibilities were observed, and the corresponding haemoproteins could be arranged in a series of ever increasing accessibility (BENKO and MARIČIĆ, 1978). The term accessibility is defined loosely because of lack of a clear-cut spatial parameter which could be derived from S-PMR. In fact the outer-sphere relaxation (region II in Fig. 1), when observable, is closest to a structural interpretation related to accessibility (PIFAT et al., 1973). It has also been argued that the distance, d, of closest proton approach within the inner sphere is not sensitive to conformational characteristics. Rather, it is the proton exchange rate, $1/\tau_M$ (eqs. 2 and 3), which is much more dependent on protein conformation and its dynamics (VUK-PAVLOVIĆ, 1976). The mentioned series of haem-iron accessibilities in haemoproteins was constructed taking mainly into account these qualitative arguments which emerge from an analysis of the S-PMR temperature dependence.

However, the stereochemical S-PMR enables a more quantitative way of defining the haem accessibility — by a "shift" of the available temperature-window. The method has not yet been put to full use in this respect, but a good example is a study of *Lupinus luteus* leghaemoglobin (VUK-PAVLOVIĆ et al., 1976), in which $CH_3(O^2H)$, $C_2H_4(O^2H)_2$ and $C_3H_5(O^2H)_3$ were used as marker molecules. The temperature range was, of course, same as usual, but the observed segment of the $R_{1,tot}$-Arrhenius plot (Fig. 1) varied according to the size of the marker molecule. Among the oxygen carriers studied so far legHb (in solution) has a most open haem-crevice, but, still, a crevice. It will be shown later here that some monooxygenases extend this "structural spectrum" even further.

It is possible presently to distinguish four main threshold sizes of increasing haem accessibility by defining that each corresponds to the marker molecule size which has the temperature-window "open" only for region II in Fig. 1, i.e. when the outer sphere relaxation is observed:

- for H_2O region II is evident from the very low relaxation rates, usually 1/10 those of region I, like in the case of the tightest haem-pocket from *Glycera dibranchiata* haemoglobin III;
- for $CH_3(O^2H)$, which "shifts" the temperature-window of human aquoferrihaemoglobin in ^1H_2O from "exch" region to region II;
- for $C_2H_4(O^2H)_2$ with a shift to (mostly) region II even from I;
- for $C_3H_5(O^2H)_3$ which for fluoroferrilegHb with I in ^1H_2O yields a full II, and also II for aquoferrilegHb from its quasi-I in ^1H_2O with $C_3H_5(OH)_3$.

Intermediate cases can be envisaged when a "shift" of the temperature window is observed from I to "exch", rather than to II.

It has been suggested (Benko, 1976) that stereochemical S-PMR technique could be improved considerably by recording the partially relaxed high-resolution PMR spectra from deuterated haemoprotein solutions containing only small amount of $^{1}H_2O$ together with the several aliphatic alcohol markers. It will become apparent later here that other markers for monooxygenases could be used in the same manner. With appropriate signal-enhancement techniques the protein perturbation due to the presence of alcohols or other marker molecules could be minimized by lowering their concentration.

3. Structural Aspects of Microsomes

We shall address ourselves in this Section to problems of supramolecular structure arrangements within the endoplasmic reticulum as amenable to experimental studies in the form of microsomes. The primary interest is to find out to what extent the S-PMR method can give useful information. Though in a sense very specific, as it is based on the presence of the paramagnetic haem-iron in monooxygenases, the system under study is of such complexity, that the problems to be encountered are not few. On the other hand microsomes are closest to the natural environment for a large group of monooxygenases so that the study of their structure/function relationship cannot avoid to tackle microsomes.

Although the focus is on the S-PMR results, other techniques which have been used in delineating the microsomal structures will have to be taken into account inasmuch they deal with seemingly same effects as does S-PMR. In discussing the results of the latter mainly the phenomenology will be presented for cases where relative comparisons are valid without going into quantitative details, so that, as pointed out in the Preface, the reading of this Section ought to be independent of the others (for those with special interest limited only to microsomes).

3.1. Protein(m^0) — membrane interrelationship in microsomes (Arrhenius discontinuities; reconstitution)

Ruckpaul (1978) has summarized diverse experimental findings of Arrhenius plots with discontinuities in the range 17° to 24 °C. The experiments quoted therein were performed preponderately with *rat* microsomes, their constituents and the reconstituted systems. They fall into two main groups with seemingly controversial results, let alone conclusions and models derived therefrom. Ruckpaul's (1978) rationalization is that one group of studies involved diffusion controlled steps (in various enzyme-m^0 reaction kinetics) and should therefore reveal the lipid phase transitions, whereas the second type of reactions ($m^0 \rightarrow m^r$ reduction) do not.

As to whether the observed Arrhenius breaks are due to any temperature-induced alterations of the membrane structure is still debatable. To resolve it, many techniques have been in use: electron spin resonance (ESR); fluorescent dye spectroscopy (FDS); calorimetry (CAL); freeze-etch electron microscopy (FEM); nuclear magnetic resonance.

The ensuing picture will depend on whether the parameter measured reflects local or bulk properties. With ESR/FDS it is important just where and to what extent the label-molecule is fixed on the protein, while CAL and FEM give an overall information on the bulk state. In distinction to ESR and FDS the NMR method does not require introduction of extraneous label molecules. It may give both local and bulk information.

Blazyk and Steim (1972) and Stier and Sackmann (1973) demonstrated that the bulk of the microsomal lipid phase does not show any thermally induced phase-transitions above 5 °C, a conclusion reached most recently also by Sligar and Briggs (1979).

In a very comprehensive study with regard to the techniques used Wunderlich et al. (1975) confirmed a discontinuity at 17 °C for microsomes from *Tetrahymena puriformis*, which, however, though a unicellular eukaryote, may well differ in its endoplasmic reticulum structure from the mammalian. The latter authors, too, do not claim any bulk lipid liquid-crystalline → crystalline phase-transition.

In point of fact a consensus appears to emerge as to at least two lipid environments: a bulk one (without any "phase transition" above 5 °C) and another, in a tighter interaction with the protein (m^0). Whereas Stier and Sackmann (1973) ascribe a crystalline → liquid-phase transition at 32 °C to this "rigid halo" (less than 20% of total phospholipids), Wunderlich et al. (1975) ascribe the 17 °C-thermotropic alterations to a clustering of "rigid" liquid-crystalline lipid environments which exclude membrane-interacting proteins.

In all of the above mentioned cases various extrinsic molecular probes and/or substrates had to be used. The question is whether any kind of structural discontinuity with temperature could be observed in the microsomes *per se*, i.e. with the enzyme in the resting state and without any external perturbation of the structure as a whole?

In this respect the results obtained with PB-induced rat hepatic m^0_{450} microsomes using the S-PMR method are confirmative (Ruckpaul et al., 1976). Similar measurements for rabbit m^0 and with another inducer (Maričić et al., 1979) were not detailed enough to warrant a more general conclusion on the appearence of the transition-temperature, though, it is also likely for rabbit m^0_{450} microsomes. Anyway, as most other observations from the literature are about rat hepatic m^0_{450} microsomes, the S-PMR results might deserve a comment here, too.

Firstly, the discontinuity in the Arrhenius plots for the S-PMR rates was observed for the reduced, CO-saturated samples containing therefore $m^{r,CO}_{450}$ in the microsomes. These PMR rate enhancements are caused by interactions of solvent (mainly water) protons with the macromolecular constituents (i.e. mainly with their protons) of the microsomes. (The paramagnetic influence of the haem-iron from m^0 in the original microsomes was excluded by the mentioned conversion into the diamagnetic state.) An overall abrupt change in the solvation structure of macromolecules and its dynamics between 13° and 20 °C was thus deduced from S-PMR data. It should be pointed out that the solvent proton magnetic relaxation measurements record the information gathered at short distance between the protons within the quasisolid (microsomes) — liquid (water) interfaces, information which is subsequently spread very efficiently to the bulk of solvent by fast diffusion of water molecules. Hence, any local structural and dynamical alterations sensed by S-PMR may not necessarily be detected by other techniques (like CAL, FDS, ESR, for instance).

Secondly, the Arrhenius-discontinuities were observed also for the paramagnetic contribution (by m^0_{450}) to the S-PMR rates. In other words, the nearest environment of the m^0-haem-iron senses whatever (structural) alteration is induced in m^0 at the temperature threshold. Even more important is the fact that this observation was valid for the original microsomes as well as for the crudely solubilized sample. The latter was in fact in the form of large ($\sim 10^6$ daltons) micelles (BEHLKE et al., 1979) because of the retained lipid, but with an estimated much smaller number of lipid molecules per molecule of m^0, than there is in the microsomes. The absolute paramagnetic S-PMR rates were only about half those of original microsomes, but the discontinuity in the crudely solubilized sample was the same. On the other hand, only partial reconstitution of the system was sufficient to reproduce the microsomal m^0 S-PMR rates (partially purified NADPH-dependent-m^0-reductase, rd, was mixed with the solubilized m^0 without making up for the rest of the missing lipids from the original microsomes). Thus it is not clear whether (i) the lipids carried along by the solubilized m^0 and *rd* were sufficient to recover the native state of m^0 as characterized by S-PMR rates, or (ii) that the very presence of rd was also necessary to reestablish the proper m^0 — lipid — *rd* structural assembly. Studies of GIBSON et al. (1979) and SCHENKMAN et al. (1979) show that the m^0-spin-equilibrium is controlled by the lipids (and, according to WILLS, 1979, by free fatty acids). This strongly suggests that the reconstitution effects observed by S-PMR (RUCKPAUL et al., 1976) are due to protein-lipid (i.e. case (i) above) rather than m^0-*rd* interaction.

In both rat and rabbit solubilized m^0_{450} the paramagnetic S-PMR rates are about half those maesured on corresponding microsomes (RUCKPAUL et al., 1976, MARIČIĆ et al., 1979). The low-spin content of solubilized m^0_{450} was found in several laboratories to be 80—90% at 20 °C. GIBSON et al. (1979) found equal amounts of high- and low-spin forms in microsomes. However, GRASDALEN et al. (1978) report that microsomal m^0_{450} is also 80% low-spin. On the other hand, they estimated from S-PMR titration with Type I and II substrates that the low-spin paramagnetic rate is 800 s^{-1} M^{-1} and the high-spin one 3000 to 4000 s^{-1} M^{-1}. With these estimates and GIBSON'S et al. (1979) increase in high-spin content of microsomal vs. solubilized m^0_{450} the observed S-PMR rate difference between the two (RUCKPAUL et al., 1976, MARIČIĆ et al., 1979) could be accounted for, at least semiquantitatively.

One may thus conclude that only a small part of the microsomal lipid (+ free fatty acids?) controlls fully the native spin equilibrium. A molecular field calculation (MARČELJA, 1976, 1977) suggests that interactions between protein molecules in membranes do not extend over more than two intervening lipid molecules. This short-range protein-lipid interaction acts obviously upon the m^0-haem-iron via the protein structure. If the protein-lipid (+ free fatty acids?) interaction changes, so will the spin-state equilibrium.

GRASDALEN et al. (1978) suggested that the Arrhenius-plot discontinuity in the S-PMR rates is simply a consequence of the spin equilibrium alteration. However, their differential light-absorption data clearly show that the high-to-low-spin ratio is constant (for rat microsomes) up to 15—17 °C. Above this temperature the low-spin fraction falls off sharply with temperature. (In their Fig. 2 GRASDALEN et al., 1978, make use of this independently determined temperature dependence of high/low spin ratio to disentangle the composite S-PMR data into their high- and low-spin part. Both are drawn as falling off above 20 °C whereas one would expect the high-spin relaxation contribution to increase.) There is no such discontinuity for the solubilized (rabbit) m^0_{LM2} (RISTAU et al., 1979), the high-spin content increasing with temperature monotonically. Hence, in microsomes it is not solely the intrinsic temperature-controlled mechanism of spin equilibrium in the haem-iron which is reflected in the Arrhenius plots of S-PMR rates. This is definitely corroborated by the fact that at temperatures above the discontinuity the S-PMR rates fall off whereas the enhancement of high-spin content would require quite opposite an effect — an increase in the paramagnetically induced rates. Finally, no spin-equilibrium change with temperature could explain the break in the S-PMR rates for the diamagnetic $m_{450}^{r,CO}$.

The preceding discussion leads, on balance, to the conclusion that (i) there is a temperature induced alteration in the short-range m_{450}^{0}-lipid assemblies around 20 °C, which, in turn, causes (ii) a conformational change of m_{450}^{0} restricting the haem accessibility.

3.2. *Location of m^0 in microsomes with regard to cytoplasmic vs. cisternal liquid of microsomal suspensions*

The molar paramagnetic enhancement of S-PMR rates in homogeneous macromolecular solutions is based on the (total) concentration of paramagnetic centres. When such rates are derived for inhomogeneous suspensions, like microsomes, it is questionable whether all the solvent protons whose relaxation rate is being measured are under equal influence of the paramagnetic centres (m^0 haem-iron ions). With regard specifically to microsomal suspensions there are two interrelated questions: (i) is the proton magnetic relaxation in the cisternal liquid equal to that within the cytoplasmic liquid, and (ii) are the monooxygenase molecules equally distributed on the cytoplasmic and cisternal surface (i.e. on the external and internal membrane surface)? It is assumed, of course, that the m^0-haems are accessible to the liquid(s) of the two compartments, a point we shall come back to at the end of this subsection.

S-PMR studies of rat (Ruckpaul et al., 1976) and rabbit (unpublished results from the author's laboratory, 1976—1978) hepatic microsomes indicate rather convincingly that the cisternal liquid behaves as do highly immobilized "structured" water molecules:

(a) When homogeneous aqueous solutions are dialysed against ^2H_2O the S-PMR rates diminish owing to the nuclear magnetic dilution (protons vs. deuterons). The effect was observed, as expected, with the solubilized m_{450}^{0} sample (Ruckpaul et al., 1976). However, such an effect was small for microsomal suspensions, and was recorded, again as expected, only during the initial replacement of some 20% of the ^1H_2O content by ^2H_2O (during the first 10 minutes of dialysis). On further dialysis until equilibrium with large excess of ^2H_2O, the total S-PMR rate increased, i.e. it almost doubled, from the initial value of the ^1H_2O suspensions. The equilibrium content of solvent protons after dialysis was still 18—20% of the initial amount.

These results are characteristic of water restricted motionally by macromolecular structures. In this case it could be ascribed only to the cisternal lumen within the microsomes, and the implicit conclusion is thus that water molecules do not exchange between the cisternal and cytoplasmic compartments, because the observed effect is possible only if the ^1H_2O molecules present initially in the lumen were still there after the ^2H_2O dialysis.

(b) Immobilized water, either encaged, or within hydration layers, freezes gradually well below zero centigrades (Kuntz and Kauzmann, 1974). Such was the case with the rabbit hepatic microsomes in ^1H_2O when the percent "unfrozen" protons was determined by following the total magnetization (M_0 at 200 µs after the 180° pulse) in dependence on temperature. At −2 °C there was 17% protons unfrozen. 10% were unfrozen at −7 °C with a subsequent milder dependence on cooling (8% at −12 °C). These results (unpublished, from author's laboratory) may be rationalized as follows:

the first part of unfrozen water (down to $-7\,°C$) may be due to the external hydration, whereas the rest of it could only be that encaged within the cisternal cavities.

In all likelyhood, therefore, the microsomal membrane is impermeable for H_2O molecules. Strange as it seems, this finding is not at variance with the most detailed study of microsomal structures published by NILSSON et al. (1973). Using a variety of experimental methods, but not S-PMR, these authors found that rat liver microsomes are permeable to uncharged molecules of up to at least 600 daltons but that they are impermeable for charged substances (> 90 daltons) and for macromolecules.

The water molecule with its large dipole moment and (hence) the very probable structuring as indicated by the S-PMR experiments, especially in view of the polar surface vs. apolar body of the microsomal membrane, may find it, indeed, impossible to reach the cisternal core of the microsomes. This is corroborated by nuclear magnetic relaxation studies on model compounds using $^1H_2O/^2H_2O$: water does not penetrate into the apolar cavity of Triton X-100/dimyristoyl-phosphatidylcholine micelles (RIBEIRO and DENNIS, 1975 — see therein quoted similar evidence from other studies).

The second question posed at the beginning of this section concerned the location of m^0 within the microsomal membrane. Having summarized here the evidence that the cisternal liquid, though perhaps less than 10% of the liquid in microsomal pellets (packed at 100000 g), does behave extraordinarily, with much faster relaxation rates of protons than in the solvent outside, it becomes important to know to what extent the m^0's are in contact with either of the two liquid compartments.

RUCKPAUL et al. (1976) described the experiment aimed at answering this question. Deuterated microsomes were prepared by using ^2H_2O instead of ^1H_2O at all stages of the preparative procedure. The sample contained 3% residual (water) protons. GRASDALEN et al. (1978) prepared the microsomes (abbrev. here as "D_2O") similarly with a final (water) proton content of less than 5%. The more sensitive experimental equipment enabled the latter authors to measure the S-PMR rates with these residual (water) protons, i.e. in magnetically extremely diluted solvent with regard to the protons. RUCKPAUL et al. (1976), on the other hand, had to add ordinary water to enable measurements of required precision but also with the specific purpose of that particular experiment (abbrev. here as "$D_2O + H_2O$").

One would have expected, on the basis of the suggested difference between the cisternal and cytoplasmic liquid that GRASDALEN et al. (1978) ought to have observed two-exponential magnetic relaxation (free-induction) decays, from the protons in the cisternal water, but this is not discussed by the authors. In the other study, "$D_2O + H_2O$", the addition of ^1H_2O well in access to that assumed to have been encaged inside the microsomes during their preparation, provided for the observation of paramagnetic relaxation rates due to the m^0-haem irons exposed on the surface of the microsomes. The R_{pmg}-values at room temperature are identical from both studies ($1500\ s^{-1}M^{-1}$) within the experimental uncertainty. However, the temperature dependence of R_{pmg} differs in the two experiments. The proton content in the "$D_2O + H_2O$" study was just at about the threshold when the two-exponential relaxation showed up in another case (BRNJAS-KRALJEVIĆ et al., 1977). The "D_2O" study was done, as mentioned, under extreme magnetic dilution. Under those conditions the relaxation behaviour may be very complicated due to the so-called cross-relaxation between the very efficient spin-diffusion controlled relaxation of the protons from the quasi-solid phase in the membrane and those in the liquid (see here Section I). On the other hand, just because of this very probable relaxation-effects-transmission the measured relaxation rates may be expected to be averaged out efficiently throughout the whole microsomal sample.

A further important point must be given due to consideration: are the diamagnetic contributions to the total S-PMR rates in the original (m^0) and reduced ($m^{r,CO}$) microsomes identical for the magnetically diluted systems ("D_2O"), as has usually been taken for granted for the H_2O-suspensions or solutions? Even for the latter the same question must be posed. In other words, the liquid proton relaxation rates have thus far been taken to be a sum of independent contributions from the haem-iron and the interface protons interaction (hydration). However, the latter appears to be under control of the paramagnetic centre from the quasi-solid phase as well (see Section I here). Hence, the interface-induced relaxation may not necessarily be equal in m^0- and $m^{r,CO}$-microsomes. The difference is probably dependent on the extent of magnetic dilution ($^2H/^1H$). In conclusion, until careful studies disentangle these effects quantitatively, we may assume that whatever difference there is between the "$D_2O + H_2O$" and "D_2O" data, it is due to the different magnetic dilution.

However, the latter qualification is not likely to be very important when comparing the "$D_2O + H_2O$" experiment with the normal one ("H_2O"), both reported in the same study (RUCKPAUL et al., 1976). The two experiments differ in their respective Arrhenius plots as well as in the R_{pmg}-values. The latter are lower for the "$D_2O + H_2O$" experiment at all temperatures. Above the transition (break-point discontinuity in the Arrhenius-plot) temperature of 20 °C R_{pmg} for "$D_2O + H_2O$" is 75% of that for "H_2O", both calculated using total m^0-content of the microsomes. These S-PMR rate differences are yet an independent confirmation that the cytoplasmic side of the microsomes is not identical with the cisternal one. The important conclusion is, however, that at least 75% of the m^0-haem irons are accessible to the cytoplasmic liquid. (For convenience, all studies of microsomes by S-PMR are based on the total m^0-concentration because the error is judged to be small. Nevertheless, the calculated R_{pmg}-values may be thus underestimated.)

Strictly speaking, the S-PMR data prove only that the major portion of the R_{pmg}-value is provided by the m^0-haem irons from the microsomal surface, but it is impossible to prove solely by S-PMR that all of the m^0 located at the surface are effective in the relaxation mechanism, i.e. whether all the haem clefts are equally accessible to solvent molecules. However, very large R_{pmg}-values make it quite improbable that part of the surface haems are inaccessible to water molecules. That m^0-haems are directly communicating with the liquid in microsomal suspensions has been shown in studies with completely different experimental techniques, like antibody binding (OMURA et al., 1980) and proteolytic digestion (DE PIERRE and ERNSTER, 1980).

4. Soluble Monooxygenases

The soluble bacterial monooxygenase from *Pseudomonas putida*, m_{pp-450}, has been regarded in numerous studies as a biological model for monooxygenases in general. Whereas there is indeed in many respects a "profound similarity" (DEBRUNNER et al., 1978) of m_{pp-450} with mammalian monooxygenases, we shall discuss here results which clearly indicate that the haem environments in monooxygenases are species unique. However, as the mammalian monooxygenases ougth to be removed from their natural environment when solubilized, the question arises as to the possible artefacts brought in this process with regard to the immediate neighbourhood of the haems.

4.1. The m^0-haem-iron accessibility, from microsomes to solubilized m^0

In liver microsomes the monooxygenase protein is in intimate contact with the lipids and other protein components of the whole monooxygenase system. On solubilization, depending on the degree of purity attained, some of the lipids are still retained, while the proteins other than m^0 are more or less removed. Ultracentrifugation studies (Behlke et al., 1979) show that the apparent molecular weight of solubilized monooxygenase preparations ranges from about one million for the lower purity, to some 300000 daltons for the highest purity attainable, that of LM2 fraction (for the nomenclature of the purified fractions see Coon et al., 1977). It was also shown in the same study that the solubilized samples could be titrated with nonionic detergents to attain an equilibrium molecular weight of m^0 which corresponds to the monomer unit of some 50000 daltons. The question therefore may be raised as to whether the structuring of the solubilized samples into micelles influences the accessibility of the haem iron. In other words, is it also true of these solubilized samples that the haems are mostly all equally exposed to the solvent as they appear to be in microsomes (Ruckpaul et al., 1976; see Section II here). An affirmative answer was obtained (Maričić et al., 1979) by a S-PMR titration of two LM2 preparations from phenobarbital induced rabbit liver, using Triton N-101 and X-100. The final m^0/Triton ratio was similar to the one from the ultracentrifuge study when the monomers of m^0 were attained. The paramagnetic-iron contribution to the solvent-proton relaxation rates (R_{pmg}) was found to remain constant irrespective of the amount of Triton added. The inference drawn is that the accessibility of the haem-iron in the large LM2-micelles is not altered by the micelle formation. This is consistent with the notion that part of the monooxygenase protein molecule may be highly hydrophobic and embedded within the lipid bilayer of the microsomes, while the haem-part is protruding towards solvent. Likewise, in the micelles the hydrophobic parts may stick together in combination with the retained lipids while the haem-iron crevice would be freely accessible to the solvent, a conclusion of more general importance.

To sum up, the S-PMR evidence thus far suggests that (i) the haems are exposed to the solvent at the surface of the microsomes or of the micellar associates in the solubilized samples, and (ii) that, once solubilized, the degree and the way of association will not appreciably change the relaxation behaviour. Hence, whatever differences in relaxation are observed between species, the purification steps or the types of monooxygenases (m^0_{450} vs. m^0_{448}) they are taken to reflect truly the particular haem environment.

However, comparison of the S-PMR results obtained on microsomes with those from solubilized samples (Maričić et al., 1979) reveals that disruption of the native macromolecular organization in the endoplasmic reticulum does affect the accessibility of the haem iron: (a) for the m_{450} form the haem-iron becomes less accessible on solubilization, both for the rabbit and the rat liver m_{450}; (b) for the rabbit m_{448} (3-MC induced) the solubilization does not diminish the relaxation rates as much as in the former (a) case, the difference being very small.

The actual meaning of the term "accessibility" as used here cannot be defined in a clear-cut manner (see Section I). Suffice to say at this point that the above comparisons were made for the cases with the same S-PMR temperature region (I in Fig. 1), characteristic of fast-exchange relaxation mechanism. On the whole one may conclude that

- the S-PMR behaviours of rabbit and rat monooxygenases are more alike in microsomes than in the partially purified solubilized form, but
- the solubilization alters only for the rat 3-MC-induced m^0 the S-PMR temperature dependence by shifting the temperature-window (see Section I) from region I to I + "exch" (see Fig. 1).

Further comparisons of the m^0-haem immediate environments by S-PMR must and can therefore be made separately for microsomes and for solubilized m^0's.

4.2. The m^0-haem-iron accessibility series for several monooxygenases

4.2.1. m^0_{pp-450} and $m^0_{rab-450}$

These two monooxygenases can be compared by their respective S-PMR data, thanks (a) to two independent experimental studies of m^0_{pp-450} (Griffin and Peterson, 1975; Philson et al., 1979) and of $m^0_{rab-450}$ solubilized to various purities (Rein et al., 1976; Maričić et al., 1977; Maričić et al., 1979), and (b) because all these monooxygenases contain preponderately (more than 85%) low-spin haem-iron. Philson et al. (1979) argue convincingly that the observed S-PMR behaviour is by and large due to this low-spin paramagnetic iron ion.

There are significant differences in the S-PMR results for the two representative monooxygenases, i.e. for the bacterial (m_{pp-450}) and the mammalian ($m^0_{rab-450}$):

1) the relaxation type for $m^0_{rab-450}$ is that of fast-exchange only (temperature window I in Fig. 1), whereas m^0_{pp-450} has the exchange region in addition (temperature window for "exch" and I in Fig. 1);
2) the absolute value of R_{pmg} (for I) of m^0_{pp-450} is about one half that of $m^0_{rab-450}$.

These quantitative findings suggest that the bacterial enzyme could have indeed a haem "pocket" of medium accessibility.

However, any firm structural inference from (1) and (2) for the mammalian enzyme would be impossible although the S-PMR results might seem to suggest a more accessible haem-iron in $m^0_{rab-450}$. That this is indeed the case was shown (Rein et al., 1976) by the stereochemical *S-PMR* method making use of the methylene proton relaxation from the glycerol present in the deuterated solution (see Section I). The R_{pmg} -values for the glycerol molecules were somewhat lower at room temperature, but coincided with those for H_2O protons at lower temperatures. The result was definitely different from what one observes with oxygencarrier haemoproteins (Vuk-Pavlović et al., 1974, and Fig. 2 here) where the R_{pmg}-values for the marker-molecules larger than H_2O are an order of magnitude smaller than for H_2O protons. The latter fact is due, of course, to the inability of the larger molecules to approach the haem-iron as close as water molecules do, which, in turn, indicates a rather well defined "pocket" connecting the haem-iron with the solvent through the protein. It thus follows that no such "pocket" could be expected in the mammalian enzyme. That the protein structure surrounding the haem in $m_{rab-450}$ is very open was corroborated in the same paper by the observation that even the ("five-coordinated") reduced enzyme with its divalent high-spin iron, $m^r_{rab-450}$, induces well measurable R_{pmg}-rates in complete distinction with reduced haemoglobin, for which no such rates could be measured at all (FABRY and REICH, 1966).

The S-PMR results presently available to this reviewer reveal thus that the mammalian monooxygenase (induced by PB in rabbit liver) has a shallow protein cleft, if any, accomodating the haem. This is quite distinct from the bacterial enzyme (camphor-induced from *Pseudomonas putida*) which in all likelyhood has its haem embedded in a protein pocket. The stereochemical S-PMR method has not been applied (yet) to directly test the latter conclusion. On the face of the available evidence one would predict the method will corroborate the existence of a haem pocket in m^0_{pp-450}.

4.2.2. Mammalian m_{448} vs. m_{450}

In the preceding paragraph the comparison was explicit as to monooxygenases from evolutionary widely different species — bacteria vs. mammals. However, there is also an implicit difference in the degree of the enzymatic specificity of the two monooxygenases. There seems to be a consensus in the literature that, generally speaking, m_{pp-450} is very substrate-specific, whereas $m_{hep-450}$, considering even the multiplicity of its forms (COON et al., 1977), is a more versatile enzyme. It also appears that the aryl-hydroxylases (m_{448}) are substrate-specific irrespective of the species they originate from, resembling in this respect m_{pp-450}.

This functional aspect is, of course, of primary interest. It is therefore important to compare the two enzymes excluding wide species-differences, i.e. it is desirable to compare m_{450} and m_{448} from mammalian liver. Notwithstanding the paucity of S-PMR data and the difficulties in their interpretation owing to insufficiently pure isolation of m_{448} (MARIČIĆ et al., 1979), a structural pattern could be delineated from this experimental approach.

The 3-MC-induced m_{448} from rat is not identical with m_{448} from rabbit (liver). There is a more complete set of S-PMR data for the former, but the conclusions are not straightforward even in that case. They could be summarized as follows:

(i) for rat-m_{448} the S-PMR "temperature-window" shows the transition from exchange-controlled to fast-exchange relaxation mechanism ("exch" and I in Fig. 1 here), much like the behaviour of m_{pp-450}, but, however, with almost three times as large R_{pmg}-rates. (The latter finding will be considered in the next paragraph of this Section.) The authors ascribed the results to the existence of a relatively tight haem-pocket. A rather closed protein domain around haem in this case would be in agreement with the indicated exchange-controlled relaxation mechanism below room temperature, implying additional closure of the pocket. This temperature sensitivity of the protein folding would be consistent with the temperature sensitive spin-equilibrium and is rationalized by the interdependence of the protein structure and the magnetic state of the iron-ion (FRIEDRICH et al., 1979).

(ii) for rabbit-m_{448}, the inference about a very restricted accessibility of haem-iron for solvent protons is even more evident because the S-PMR temperature window depicts only region II in Fig. 1, i.e. the "outer-sphere" relaxation. The haem-iron appears to be thoroughly shielded from solvent molecules, which could be rationalized in two ways: (a) the inducer, 3-methylcholanthrene, binds firmly and blocks the sites available otherwise for a close encounter of the haem-iron with solvent molecules; (b) there is a completely different, closed protein cleft of rabbit m_{448} preventing the interaction of solvent protons with the haem-iron. FRIEDRICH et al. (1979) favour rather convincingly the latter alternative.

Their arguments are the lack of temperature-dependent spin equilibrium (in the absence of substrate) im $m_{rab-448}$, its inability to bind type I substrates, and the diminished rate of binding and reduction. They also add to the list the low S-PMR rate which is thus involved in a circular reasoning. However, a "frozen" protein-environment around the $m_{rab-450}$-haem (compatible, above all, with the lack of a temperature-dependent spin-equilibrium) seems indeed very probable. The similarity of S-PMR-data for $m_{rab-448}$ in microsomes and when solubilized (see the first paragraph in this Section) is in accord with this consideration.

We may conclude that, on balance, there is a profound structural difference between the substrate nonspecific, mammalian monooxygenases (from rat and rabbit liver), m_{450}, on the one hand, and bacterial (m_{pp-450}) and mammalian (rat and, especially, rabbit) m_{448}, on the other. The first group has a shallow protein-cleft accommodating the haem, whereas the second group of monooxygenases have their haems embedded in a relatively closed protein pocket. The latter pattern is more common in oxygen-carrier haemoproteins. For them a series of increasing haem-accessibility has been constructed on the basis of S-PMR data (Benko and Maričić, 1979). This series is reproduced here in Table 1 as a "scale" for a similar series of monooxygenases.

The qualitative S-PMR series of haem-accessibility in haemoproteins was constructed considering the type of the temperature dependence of the S-PMR rates, the absolute values of the paramagnetically induced rates, bearing in mind the intricate relationship between the stereochemical factors and the dynamics of the protein domain surrounding the haem (see Section I). The accessibility itself is defined loosely, meaning in the extremes an open haem cleft with free communication of the solvent directly with the haem-iron, or a very fastly breathing channel through which the solvent molecules may communicate with the interior of the haem "pocket". Various degrees of both the static openness and the dynamics within the protein, and their interplay may be envisaged to produce intermediate types of accessibilities.

There are two independent lines of evidence supporting the scale in Table 1. From an infrared spectroscopic study of carbon monoxide binding to ferrous monooxygenase forms, m^r, O'Keeffe et al. (1978) drew the conclusion "that the heme-carbonyl of the liver microsomal enzymes senses a less ordered nearest-neighbour

Table 1. Haem Accessibility Scale for Haemoproteins

Oxygencarriers H_2O-Fe^{3+}		Monooxygenases, m^0
	↑	m^0_{450}/rabbit liver (LM2)
	a	
Leghaemoglobin — (*Lupinus luteus*)	c	
	c	
Petromyzon marinus (II) and *Chironomus thummi* (III)	e s	m^0_{450}/rat liver
	s	
myoglobins (horse, seal, bovine)	i	
	b	m^0_{448}/rat liver
Lampetra fluviatilis and	i	m^0_{450}/*Pseudomonas putida*
Petromyzon marinus (III), IV, V) α-chains of HbA β-chains of HbA	l i t	
	y	
Aplysia and *Glycera dibranchiata* (III)	|	m^0_{448}/rabbit liver

environment than in the bacterial enzyme, perhaps partially the result of these enzymes having larger and more flexible substrate-binding sites". Whereas the latter statement refers to evolutionary distinct species, CHIANG and COON (1979) took the difference in the circular dichroism anisotropy value as evidence of a less open haem pocket in $m^0_{450,LM4}$ than in $m^0_{450,LM2}$, i.e. a structural distinction of enzymes from the biologically identical species.

4.2.3. *The sixth ligand to the iron of m^0-haem*

The most recent and very comprehensive review entitled "Oxygen activation by cytochrome P–450" comprises a chapter on "Structure of cytochrome P–450 at the heme locus", which is very relevant to this subsection (WHITE and COON, 1980). The authors presented well balanced pros and contras for the fifth and the sixth ligand of the haem-iron in monooxygenases. It seems that a consensus emerges as to the thiolate being the fifth ligand as originally proposed by MASON et al. (1965). In his critical review of preponderately indirect chemical data on model compounds ULLRICH (1979) comes to the same conclusion. The most convincing evidence from model compounds came, however, very recently, as pointed out by WHITE and COON (1980), from the work of TRAYLOR and MINCEY (1979) who prepared the model compound incorporating a thiolate anion in the absence of aromatic ligands or crown ethers. The latter had been used before, and had left the possibility of charge-transfer effects to explain the similarity of the spectra with m_0. Further direct support WHITE and COON (1980) see in the extended X-ray absorption fine structure spectra (EXFAS) of native $m^0_{450-LM2}$ reported by CRAMER et al. (1978). The latter data were suggestive of nitrogen as the sixth ligand, but were not conclusive.

On the other hand, ULLRICH (1979) summarized the available evidence in favour of a hydroxyl as the sixth ligand, without specifying its precise nature, but invoking also the S-PMR data and conclusions from the Thesis of the first author in PHILSON et al. (1979), which we shall discuss here in more detail. WHITE and COON (1980) were cautious as to specifying the sixth ligand, but they conclude that "a body of evidence comprising NMR work, model studies, and ligand replacement reactions supports the presence of an oxygen ligand, probably water". However, they have overstated the case for water by quoting GRIFFIN and PETERSON (1975) to have "provided the first physical evidence for water coordination, based on NMR relaxation rates of bulk water protons induced by the presence of the paramagnetic iron in P–450_{cam}". As discussed earlier (MARIČIĆ et al., 1977) and again analyzed later in this subsection, the S-PMR data quoted by WHITE and COON (1980) could not provide the physical evidence for water coordination. PHILSON et al. (1979) did provide physical evidence by S-PMR for proton(s) coordination, but they suggest water as one among the candidates for the sixth ligand.

The S-PMR method did look as predestined to solve this structural problem and it seems worthwhile to examine in this subsection to what extent this hope has been rendered true. Our survey will be limited to results obtained with the soluble(ized) forms of m^0 to avoid unnecessary complications in interpretation due to macromolecular interactions in microsomes.

4.2.4. *The correlation time, τ_c for the interaction of the iron and proton spins (magnetic dipoles)*

Because protons are most directly involved in the S-PMR method, effort was not spared to establish the distance of closest approach of proton(s) to the m^0-haem-iron as a first step in deciphering the sixth ligand. This distance, d, is related through K_{IS}, a constant comprising the physical parameters of the interacting spins, to the proton relaxation time, T_{1M}, of proton(s) within the "first coordination sphere", but in fast exchange with bulk solvent (eq. 6).

$$d = [K_{IS} T_{1M} f(\tau_c)]^{1/6} \qquad (6)$$

However, it should be born in mind that d is proportional to the arbitrarily chosen number of exchanging protons equidistant to haem-iron, from eq. (2), which for the fast-exchange case simplifies into

$$T_{1M} = \frac{n}{[H]\, R_1^{I}} \qquad (7)$$

The (correlation) time, τ_c, during which the magnetic dipoles (spins) of the paramagnetic iron and solvent-protons interact, is determined by the shortest of the three microdynamic processes, namely either by the time the proton species spends within the "first coordination sphere", or by the tumbling time of the protein-(paramagnetic)iron complex via which the iron-proton vectors fluctuate spatially and thereby influence the magnetic relaxation mechanism, or by the electron-spin relaxation time of iron, τ_s. It has been found that both for m^0_{pp} (PHILSON et al., 1979) and for m^0_{LM2} (MARIČIĆ et al., 1977, 1979) the correlation time in eq. (6) is, in fact, the low-spin (haem) iron-electron relaxation time, τ_s. (For the analytical form of $f(\tau_c)$ consult the last quoted papers).

There is only one direct method to determine τ_c (here τ_s), and that is the measurement of S-PMR rates in dependence on magnetic field and therefore at various frequencies. Table 2 has been compiled from such data.

The τ_s-value for m_{pp} in Table 2 has been extrapolated by the authors from

Table 2. The Iron-Electron Relaxation Times in Low-Spin Monooxygenases (at room temperature)

m^0	τ_s/s	source
P. putida (soluble)	$5.4 \times 10^{10} \pm$?	PHILSON et .al. (1979)
Rabbit liver (LM2-solubilized)	$1 \times 10^{-10} - 10^{-9}$	MARIČIĆ et. al. (1979)
Rat liver, m^0_{448} (solubilized, medium purity "LM4")	$(4.1 \pm 0.4) \times 10^{-11}$.	Ibid.

the directly measured value of 4.7×10^{-10} s at 34 °C. This temperature was necessary for the m^0_{pp} solution to attain the fast proton exchange (region I in Fig. 1) as required for making use of eq. (6). The obtained value is more reliable (though the error was not stated) than that for m^0_{LM2}, because in the latter case the field-strength was not sufficiently large to observe the dispersion step in the relaxation rates.

Irrespective of the particular use made of $\tau_s = 5 \times 10^{-10}$ s in these studies, such a long relaxation time for low-spin haemoproteins came as a surprise when compared with those for some ferric complexes of the corresponding oxygen carriers with $\tau_s \simeq 10^{-12}$ s, an observation still requiring a quantitative theoretical explanation.

4.2.5. The proton (s)-to-iron distance, d

This distance could be derived through eq. (6) once the fast exchange condition had been found valid, which is the case for m^0_{pp} and m^0_{LM2}. However, there are still two kinds of uncertainties in doing this calculation.

Firstly, it is the experimental uncertainty stemming from the cross-relaxation mechanism discussed here in Section I. The apparently inherent inability (but see below) to measure its contribution to R_{pmg} means that all the T_{1M}-values used thus far are smaller than the expected true ones. Hence, the correct distance (d in eq. 6) can only be longer than claimed in the literature.

Secondly, there are uncertainties in other parameters of eqs. (6) and (7). For one thing, the number, n, of protons taking part in the fast exchange from equidistant site(s) close to the paramagnetic centre is not known a priori. Besides, if the exchanging protons are not all equidistant to iron, eq. (7) must be replaced by an appropriate sum, and there will be several relations of the kind in eq. (6), with the corresponding T_{1M}'s derived from eq. (7). Obviously, under the assumption that there is only one proton site for fast exchange with bulk solvent ($n = 1$) one obtains the shortest possible distance (d) separating that site from the haem-iron. Further, the unknown direction(s) of the in-and-outgoing proton(s) with respect to the electron (g-tensor) anisotropy of the haem-iron adds to the overal uncertainty (for an illustration, see VUK-PAVLOVIĆ et al., 1976). For the low-spin iron of m^0_{pp} and m^0_{LM2} (g-values from 1.9 to 2.5) this affects the results less than for the high-spin case like in haemoglobin (g's equal 2 and 6) but the values of d may be uncertain to a tenth of an Ångstrom.

Notwithstanding all these limitations it is now possible to compare the values of d for m^0_{pp} (PHILSON et al., 1979) and m^0_{LM2} (MARIČIĆ et al., 1977), because of the similarities in the systems with regard to the enumerated uncertainties, and especially because of the firmly established electron-spin relaxation time ($\tau_s = 5 \times 10^{-10}$ s). However, the experimental procedure differed somewhat in the two cases: for m^0_{pp} the blank-measurement was done with metyrapone (2-methyl-1,2di-3-pyridyl-1-propanone) bound to the haem-iron of m^0_{pp}; in case of m^0_{LM2} the reduced, CO-bound form ($m^{r,CO}_{LM2}$) was used as a blank. GRIFFIN and PETERSON (1975), on the other hand, state: "In order to determine the diamagnetic contribution of protein to the experimental relaxation times, ... measurements were made on the diamagnetic carbon monoxide and metyrapone complexes of ferrous cytochrome P-450; sodium dithionite was utilized to reduce the enzyme and was also added to the buffer controls for this particular experiment".

The author of this review must apologize to the reader for this side-walk into the morass of experimental detail, but it seems unavoidable with complex systems for which answers to (unduly?) precise questions are being sought for.

Now, it is not clear whether PHILSON et al. (1979) reduced their sample while adding metyrapone. If it was not reduced, it is strange that, according to the authors: „... substantially the same reference values were obtained with the protein-free buffer or the complexes of m^0_{pp} with either pyridine, metyrapone, or CN^-". The haem-iron (Fe^{3+}) is almost fully converted into its low-spin state by metyrapone binding, and the paramagnetic Fe^{3+} in this complex (m^{0s}) is as much effective, if anything, as it is in m^0, which has been made use of in an ESR study (RUF and NASTAINCZYK, 1976).

Therefore, it may just well be that m^{0s} with metyrapone bound is the blank (proper) for S-PMR if, as it seems quite likely, it only blocks any close direct magnetic interaction of protons from bulk solvent with the low-spin haem-iron, allowing thereby the evaluation of true $R_1^+(+R_0) + R_1^{II}$ (see Fig. 1), i.e. the cross-relaxation rate contribution to the overall relaxation rate ($R_{1,tot}$ in eq. 1) originating from the purely diagmagnetic one (due to protein-solvent interaction) and from the long-range paramagnetic one (due to the paramagnetic $Fe^{3+}_{low\text{-}spin}$ centre).

We now return to comparing the d's (proton distances) for m^0_{pp} and m^0_{LM2} by assuming that the values derived for the former are experimentally more certain (because the cross-relaxation term has been apparently taken into account). As for the latter, one has to consider measurements with the purest LM2 fraction (MARIČIĆ et al., 1979), for which R_{pmg} was some 20% lower than those obtained with the partially purified sample (MARIČIĆ et al., 1977). Further, the more likely value for τ_s is 5×10^{-10} s, rather than 7×10^{-10} s as it had been assumed before. When these two corrections are taken into account in eq. (6) they roughly cancel. Hence, both for m^0_{pp} and for m^0_{LM2} the already suggested conclusions regarding the sixth ligand remain valid, with $d(m^0_{pp}) = 2.6$ Å and $r(m^0_{LM2}) \geqq 2.4$ Å if there is only one fastly exchanging proton (n = 1 in eq. 7).

The sign $\geqq$ in the latter indicates that for m^0_{LM2} d may even be somewhat longer because of the unaccounted cross-relaxation contribution, or it may be equal to 2.4 Å if this contribution is actually negligible. However, it ought to be stressed that the observed inequality $d_{pp} > d_{LM2}$ is very real: the actual measurements yielded $R^{pp}_{pmg} \simeq 1/2 R^{LM2}_{pmg}$, which, because of the 1/6-power dependence in eq. (6) results in the corresponding Δd derived therefrom. The cross-relaxation term certainly cannot account for the measured difference in R_{pmg}'s, so that $d_{pp} > d_{LM2}$ appears to be quite probable. Just because of the mentioned 1/6-power dependence any small relative difference in d, once it is close to 2 Å, is highly significant because it means very large differences in the actually measured R_{pmg}-rates.

We have not yet dealt here with the transverse, or spin-spin, proton magnetic relaxation rates ($1/T_2$). Suffice to say that while $1/T_1$ measures the rate of nuclear spin (de)magnetization along the laboratory magnetic field, $1/T_2$ is a similar, though not identical, rate of magnetization transverse to the magnetic field direction. It is also sensitive to the presence of paramagnetic centres, but, in distinction to the corresponding equation for $1/T_1$, the transverse relaxation time expression has an additive term due to the magnetic interaction of the proton spin with the electron spin through the chemical bond within which the nuclear magnetic spin (proton) is located. Thus, an experimentally measured rate $R_{2,pmg}$ ($1/T_{2M}$) exceeding ($1/T_{1M}$) occurs only if the fast-exchanging proton is connected to the paramagnetic centre through covalent bonds. Such was the observation by PHILSON et al. (1979), which is the first physical evidence that the exchanging proton(s) in m^0_{pp} is (are) directly coordinated to the haem-iron. Moreover, as the authors concluded, the histidyl proton as a sixth ligand may be definitely excluded because d for that case should be about 5 Å, far too long compared to the observed $d = 2.6$ Å. The additional very strong independent evidence in this respect (though for a frozen solution of m^0_{pp}, at 1.7 °K) came from electron-nuclear double resonance (ENDOR) experiments by LO BRUTTO et al. (1980). The elegance of the method is in that it is highly specific with regard to the haem-iron. A particular electron-spin frequency can be chosen and in this case it was that corresponding close to the z-component of the anisotropic low-spin ferric m^0-haem, with $g_z = 2.45$. Hence, the observed proton magnetic resonance spectrum actually measured under such conditions will be highly relevant to the direct interaction of the proton(s) with the m^0-iron, along the axis perpendicular to the haem, indeed if there is any such coupling it will be maximal. The existence of such coupled exchanging proton(s) was proved also by doing the ENDOR measurement in a frozen deuterated solution. The proton spectrum observable otherwise in ordinary aqueous solution was absent in the D_2O-solvent. The observed splitting of the proton magnetic resonance peak was in agreement with the S-PMR ($1/T_{2M}$)-data, which, again, excludes the proton from a histidyl as the sole contributor to this (hyperfine magnetic) interaction. However, in a footnote (17), the authors announce in the proton spectrum another, weak coupling, about one eight of the former. This may mean that at a distance of about 5 Å another proton (from a histidyl?) interacts with haem-iron electrons.

The final conclusions to be drawn from the S-PMR *and* ENDOR experiments with m^0_{pp} may be summarized as follows:

- histidyl is not the closest sixth (axial) ligand to the haem-iron, though it might well be within its "first coordination sphere";
- one or two protons exchanging fast, with a rate of about 1 MHz at 25 °C, from the site(s) closest to haem-iron (2.6 or 2.9 Å) may belong to a water molecule, or to amino acid side chains of the form RNH_2, ROH, or $RCONH_2$, or there may be an exchangeable proton bound to cysteinyl sulphur.

For m^0_{LM2} these conclusions may well be valid, too, but the crucial data, $(1/T_{2M})$ and ENDOR, are still lacking. However, one may predict on the basis of the present S-PMR $(1/T_{1M})$ data that the "first coordination sphere" of the m^0_{LM2} haem-iron on the sixth-ligand side differs from that in m^0_{pp}. Additional fast-exchanging protons in m^0_{LM2} can be envisaged in order to reconcile simply the m^0_{pp} S-PMR data with the fact that m^0_{LM2} has a much more accessible haem iron (see the preceding subsection). It is also possible that the sixth ligand is a water molecule firmly located at the sixth iron site by only one appropriate hydrogen bond to an amino-acid side-chain, while the remaining proton undergoes fast exchange. In addition, one or more protons, or a water molecule with double-proton exchange could be present at sites further away from haem-iron, but still within the "first coordination sphere" accommodated in the shallow protein cleft of m^0_{LM2}.

4.2.6. Substrate-m^0 interaction

There are two approaches to this line of study by proton magnetic relaxation.

One of them is within the bounds set up by the title of this review. One looks, in essence, for changes of the solvent-protons relaxation due to the interference of their interaction with haem-iron when the substrate or inhibitor molecule binds to m^0. That is, one makes use of the S-PMR approach in the presence of a binding molecule. In all the papers reporting about S-PMR results referred to herein, measurements with various substrate (or other) molecules are to be found. The interpretation, however, is not straightforward and a general comparison seems to be premature. There are namely difficulties in controlling fully such experiments. Firstly, there is the change of the haem-iron spin-equilibrium due to substrate binding, and its temperature dependence, a point dealt with in the papers by Grasdalen et al. (1978) and Philson et al. (1979). In a more general sense the same difficulty pertains especially to comparisons of results from microsomes, where one can think of m^0 in permanent interaction with the surrounding lipid (and other) molecules (see Section II).

The simpler way of following the (temperature dependence of) spin equilibrium is by differential visible spectrophotometry, but it would be a more direct way to do this by magnetic susceptibility measurements using high-resolution NMR. The method has been adapted for haemoproteins with special attention having been paid to the large susceptibility of the fully diamagnetic protein (Bartle et al., 1973), which is seldom taken properly into account in NMR measurements. Susceptibility data obtained with the same samples for S-PMR measurements would remove the present uncertainties.

Another experimental difficulty ensuing from changes in spin-state equilibria is the determination of the appropriate correlation times (i.e. τ_s's for high- and low-spin components with and without the substrate bound). Finally, a reliable common blank allowing for the cross-relaxation term must be defined.

The other approach is to monitor differential relaxation effects of protons (or, other magnetic nuclei, for that matter, like for instance ^{13}C) from specific moieties of substrates or other binding molecules. As distinct from the S-PMR method which reflects the specific interactions indirectly through the overall (averaged out) relaxations of the protons in bulk solvent, the other method monitors directly the specific interactions of the binding molecule with the protein. To the best of our knowledge there is thus far only one such example related to m^0_{LM2}, by NOVAK et al. (1977). It is, strictly speaking, outside the scope of the present review, but is very closely related to some of the questions raised.

NOVAK et al. (1977) adapted actually the general approach that has been in use from the early days of NMR penetration into biochemistry when for sensitivity reasons it was easier to study protein/drug interaction by high-resolution NMR spectra of the binding molecule. Nowadays it is possible to follow the relaxation parameters for each of the chemically equivalent-protons group on the ligand molecule directly, thanks to the sophistication of apparatus developed since. NOVAK et al. (1977) have measured R_{pmg} for the methyl and phenyl protons of xylidine (2,6-dimethylaniline) interacting with m^0 solubilized from liver microsomes of PB-treated rats. They encountered the difficulties discussed above (unknown spin state and τ_s), but proved that xylidine is in fast-exchange with m^0_{LM}, thus enabling estimation of d for both proton groups. It now seems more probable that xylidine/m^0_{LM} complex is mostly low-spin, which means $\tau_s \simeq 5 \times 10^{-10}$ s, and we conclude that out of the four groups of calculated d's given by NOVAK et al. (1977) the most probable is the pair $d_{meth} \simeq 4$ Å $\gtrsim d_{phe}$. This is an interesting result from the point of view of our general conclusion that m^0_{LM} has an extremely accessible haem-iron (see here the preceding subsection). Namely, NOVAK et al. (1977) found for xylidine in interaction with haemoglobin these distances to be more than twice as large. (The most interesting comparison would be, of course, with m^0_{pp}). Once again, quite independently, the conclusion that in m^0_{LM} the haem is accommodated within a shallow protein cleft is confirmed. In point of fact, the xylidine-PMR experiment is equivalent to the stereochemical S-PMR in its more sophisticated mode along the lines suggested by BENKO (1976) (see here the end of Section I). It would be very useful to extend xylidine-PMR measurements to the various m^0-forms.

5. General Comments

The usefulness of structural studies of enzymes is proportional to the relevance of the established facts in elucidating the enzyme function, and this Section is such an assessment.

The S-PMR work on microsomes has not been very rewarding. However, it has clearly demonstrated the species differences of haem-iron accessibility. The mechanisms controlling the protein conformation are obviously more complex and delicate in microsomes. Only a systematic study of partially recombined systems will clear the way to further direct studies of the native microsomes. One encouraging sign in this respect suggested thus far by S-PMR is that in all likelihood only a few lipid (and other?) molecules surrounding m^0 are important for its magnetic/conformational state. The other fact, that of the break-point in the S-PMR Arrhenius-plots for microsomal suspensions remains to be related to the enzyme-function, if any relatedness is to be expected at all. Finally, if the endoplasmic reticulum *in vivo* is impenetrable for H_2O molecules, as it has been concluded for microsomes by S-PMR, it would mean an enhanced solute-concentration gradient along the axis of the cisternal tubing. Could it be functionally irrelevant?

One of the more reliable outcomes of the S-PMR approach is a clear differentiation of the monooxygenases studied into two main structural classes correlated with the enzyme specificity. Thus, less specific monooxygenases seem to have a haem-iron more accessible for solvent molecules, and m^0_{450}, LM2 fraction, from rabbit liver has the most open protein haem-cleft (including for comparison all the oxygen-carrier haemoproteins studied thus far). In a review on hepatic porphyrias De Matteis (1978) invokes this S-PMR finding of an exposed haem-cleft to be relevant to the apoprotein/haem reconstitution.

While the established qualitative scale of the haem-accessibility (see the preceding Section III) may prove useful it would be desirable to establish more quantitative criteria. One way would be to use a common suitable marker-molecule as xylidine (see the end of Section III), perhaps making use of both proton and carbon-13 NMR partially relaxed spectra. The other way could be a test of the correlation between the temperature gradient in spin-equilibria and the "plasticity" of the protein. The assumption is that the more "frozen" a protein is, the less sensitive its spin-equilibrium to temperature. In other words, it is assumed that the haem-iron spin-state is interdependent with the protein conformation in a characteristic way. It is expected that the temperature gradient for the spin-equilibrium may turn out to be a quantitative parameter related to the haem-iron accessibility. The crucial role of the haem-iron spin-state coupling to the protein conformation sensitivity for substrate binding has been suggested independently by Sligar (1976) for m^0_{pp} and for m^0_{LM} by Rein et al. (1977) to be at the heart of the monooxygenase enzyme-mechanism.

The most definitive conclusion from S-PMR (and ENDOR) is that histidyl is excluded as candidate for the haem-iron sixth ligand, and that at least one, possibly two fast-exchanging protons in m^0_{pp} and m^0_{LM2} are present at the sixth Fe-site (and a few more nearby in m^0_{LM2}?). The "more aqueous" haem environment in m^0_{LM2} means that the microdissociation constants of any of the amino-acid side chains in the region between the active centre and the substrate binding-site(s) are expected to be more alike to the bulk constants than in m^0_{pp}. Whether this has anything to do with substrate specificity and the enzyme mechanism, remains to be seen. Further, if the Fe^{3+}-coordinated protons have to leave before dioxygen enters on reduction of Fe^{3+} to Fe^{2+} in the course of the enzyme reaction, what is the role of that immediate Fe^{3+}-environment in the whole mechanism? Could it be only a fortuitous play of selective evolutionary pressures which stabilized the Fe^{3+}-monooxygenases in a proper spin-state by a weak sixth ligand, because it could not be done away with the problem otherwise, while, in fact, the presence of the sixth ligand, per se, is not relevant directly to the enzyme mechanism?

References

Abragam, A. (1961): "Principles of Nuclear Magnetism", Oxford University Press, London

Bartle, K. D., B. J. Dale, D. W. Jones, and S. Maričić (1973): J. Magn. Reson. **12**, 286–295

Behlke, J., G.-R. Jänig, and D. Pfeil (1979): Acta biol. med. germ. **38**, 389–397

Benko, B. (1976): Ph. D. Thesis, University of Zagreb

BENKO, B., S. VUK-PAVLOVIĆ, and S. MARIČIĆ (1977): Biochim. Biophys. Acta 491, 457–468
BENKO, B., and S. MARIČIĆ (1978): Croat. Chem. Acta 51, 369–377
BERNHEIM, R. A., T. H. BROWN, H. S. GUTOWSKY, and D. E. WOESSNER (1959): J. Chem. Phys. 30, 950–956
BLAZYK, J., and J. M. STEIM (1972): Biochim. Biophys. Acta 266, 737–739
BLOEMBERGEN, N. (1957): J. Chem. Phys. 27, 572–573
BRNJAS-KRALJEVIĆ, J., S. MARIČIĆ, and V. BRAČIKA (1977): Biophys. Chem. 6, 191–200
BURTON, D. R., S. FORSÉN, G. KARLSTRÖM, and R. A. DWEK (1979): Progress in NMR Spectroscopy 13, 1–45.
CHIANG, Y.-L., and M. J. COON (1979): Arch. Biochem. Biophys. 195, 178–187
COON, M. J., R. E. WHITE, G. D. NORDBLOM, D. P. BALLOU, and F. P. GUENGERICH (1977): Croat. Chem. Acta 49, 163–177
CRAMER, S. P., J. H. DAWSON, K. O. HODGSON, and L. P. HAGER (1978): J. Amer. Chem. Soc. 100, 7282–7290
DAVIDSON, N., and R. GOLD (1957): Biochim. Biophys. Acta 26, 370–373
DEBRUNNER, P. G., I. C. GUNSALUS, S. G. SLIGAR, and G. S. WAGNER (1978): "Monooxygenase Hemoproteins: Cytochromes P–450" in: Metal Ions in Biological Systems. Vol. 7. Iron in Model and Natural Compounds (H. SIGEL ed.), New York, Marcel Dekker, Inc., pp. 241–275
DE MATTEIS, F. (1978): "Hepatic Porphyrias" in: Heme and Hemoproteins (F. DE MATTEIS and W. N. ALDRIDGE, eds.), Berlin, Springer-Verlag, pp. 129–155
DERZHANSKI, A., A. GEORGIEVA, K. KOTEV, and B. P. ATANASOV (1970): Biochim. Biophys. Acta 214, 83–93
DWEK, R. A. (1973): "Nuclear Magnetic Resonance in Biochemistry, Applications to Enzyme Systems", Oxford, Clarendon Press
EDZES, T. H., and E. T. SAMULSKI (1978): J. Magn. Reson. 31, 207–229
EISENSTADT, M., E. HSI, and R. G. BRYANT (1978): J. Magn. Reson. 29, 591–597
DE PIERRE, J. W., and L. ERNSTER (1980) in: Biochemistry, Biophysics and Regulation of Cytochrome P–450, GUSTAFSSON, J. Å., J. CARLSTEDT-DUKE, A. MODE, and J. RAFTER, eds. Elsevier-North-Holland Biomedical Press, pp. 431–443
FABRY, T. L., and REICH, H. A. (1966): Biochem. Biophys. Res. Commun. 22, 700–703
FRIEDRICH, J., G. BUTSCHAK, G. SCHEUNIG, O. RISTAU, H. REIN, K. RUCKPAUL, and G. SMETAN (1979): Acta Biol. Med. Germ. 38, 207–216
GIBSON, G. G., D. L. CINTI, J. B. SCHENKMAN, S. G. SLIGAR (1980) in: Biochemistry, Biophysics and Regulation of Cytochrome P–450, GUSTAFSSON, J. Å., J. CARLSTEDT-DUKE, A. MODE, and J. RAFTER, eds. Elsevier-North-Holland Biomedical Press, pp. 119–130
GRASDALEN, H., L. E. GÖRAN ERIKSSON, A. EHRENBERG, and D. BÄCKSTROM (1978): Biochim. Biophys. Acta 541, 521–527
GRIFFIN, B. W., and J. A. PETERSON (1975): J. Biol. Chem. 250, 6445–6451
KOENIG, S. H., R. D. BROWN, and C. F. BREWER (1973): Proc. Natl. Acad. Sci. USA 70, 475–479
KOENIG, S. H., R. G. BRYANT, K. HALLENGA, and G. S. JACOB (1978): Biochemistry 17, 4348–4358
KON, H., and N. DAVIDSON (1959): J. Mol. Biol. 1, 190–191
KUNTZ Jr. I. D., and W. KAUZMANN (1974): Adv. Prot. Chem. 28, 239–345
LO BRUTTO, R., C. P. SCHOLES, G. C. WAGNER, I. C. GUNSALUS, and P. G. DEBRUNNER (1980): J. Amer. Chem. Soc. 102, 1167–1170
LUMRY, R., H. MATSUMIYA, F. A. BOVEY, and A. KOWALSKY (1961): J. Phys. Chem. 65, 837–843
LUZ, Z., and S. MEIBOOM (1964): J. Chem. Phys. 40, 2686–2692
MARČELJA, S. (1976): Biochim. Biophys. Acta 455, 1–7
MARČELJA, S. (1977): Croat. Chem. Acta 49, 347–358
MARIČIĆ, S., A. RAVILLY, and A. S. MILDVAN (1966): "Proton Relaxation Measurements of Dissolved and Crystalline Methemoglobin and Metmyoglobin" in: Hemes and Hemoproteins, (B. CHANCE, D. W. ESTABROOK, and T. YONETANI, eds.) Academic Press, New York, London, pp. 157–170
MARIČIĆ, S., S. VUK-PAVLOVIĆ, B. BENKO, J. POROK, H. REIN, G.-R. JÄNIG, and K. RUCKPAUL (1977): Croat. Chem. Acta 49, 323–331
MARIČIĆ, S., B. BENKO, G. BUTSCHAK, G. SCHEUNIG, G.-R. JÄNIG, H. REIN, and K. RUCKPAUL (1969): Acta Biol. Med. Germ. 38, 217–233

MASON, H. S., J. C. NORTH, M. VANNESTE (1965): Fed. Proc. **24**, 1172–1180

MCCONNELL, H. M. (1958): J. Chem. Phys. **28**, 430–432

MILDVAN, A. S., N. M. RUMEN, and B. CHANCE (1971) in: Probes of Structure and Function of Macromolecules and membranes, Vol. II – Probes of Enzymes and Haemoproteins (B. CHANCE, T. YONETANI, eds.), Academic Press, New York, p. 205

NILSSON, R., E. PETERSON, and G. DALLNER (1973): J. Cell Biol. **56**, 762–776

NOVAK, R. F., I. M. KAPETANOVIĆ, and J. J. MYELAL (1977): Mol. Pharmacol. **13**, 15–30

O'KEEFFE, D. H., R. E. EBEL, J. A. PETERSON, J. C. MAXWELL, and W. S. CAUGHEY (1978): Biochemistry **17**, 5845–5852

OMURA, T., M. NOSHIRO, and N. HARADA (1980) in: Biochemistry, Biophysics and Regulation of Cytochrome P-450, GUSTAFSSON, J. Å., J. CARLSTEDT-DUKE, A. MODE, and J. RAFTER, eds. Elsevier-North-Holland Biomedical Press, pp. 445–453

PFEIFER, H. (1963): Biochim. Biophys. Acta **66**, 434–439

PHILSON, S. B., P. G. DEBRUNNER, P. G. SCHMIDT, and I. C. GUNSALUS (1979): J. Biol. Chem. **254**, 10173–10179

PIFAT, G., S. MARIČIĆ, and S. GRANDJA (1973): Biopolymers **12**, 905–920

PIFAT, G., B. BENKO, S. MARIČIĆ, and S. VUK-PAVLOVIĆ (1974): Croat. Chem. Acta **46**, 145–155

REIN, H., S. MARIČIĆ, G.-R. JÄNIG, S. VUK-PAVLOVIĆ, B. BENKO, O. RISTAU, and K. RUCKPAUL (1976): Biochim. Biophys. Acta **446**, 325–330

REIN, H., O. RISTAU, J. FRIEDRICH, G.-R. JÄNIG, and K. RUCKPAUL (1977): FEBS Letters **75**, 19–22

RIBEIRO, A. A., and E. A. DENNIS (1975): Biochemistry **14**, 3746–3755

RISTAU, O., H. REIN, S. GRESCHNER, G.-R. JÄNIG, and K. RUCKPAUL (1979): Acta Biol. Med. Germ. **38**, 177–185

RUCKPAUL, K., S. MARIČIĆ, G.-R. JÄNIG, B. BENKO, S. VUK-PAVLOVIĆ, and H. REIN (1976): Croat. Chem. Acta **48**, 69–86

RUCKPAUL, K. (1978): Pharmazie **33**, 310–312

RUF, H. H., and W. NASTAINCZYK (1976): Eur. J. Biochem. **66**, 139–146

SCHELER, W. (1963): Biochim. Biophys. Acta **66**, 424–433

SCHENKMAN, J. B., I. JANSSON, G. G. GIBSON, S. G. SLIGAR, and D. L. CINTI (1980) in: Biochemistry, Biophysics and Regulation of Cytochrome P-450, GUSTAFSSON, J. Å., J. CARLSTEDT-DUKE, A. MODE, and J. RAFTER, eds. Elsevier-North-Holland Biomedical Press, pp. 479–487

SLIGAR, S. G. (1976): Biochemistry **15**, 5399 to 5406

SLIGAR, S. G., and M. BRIGGS (1979): Abstracts 4th International Symposium on Microsomes and Drug Oxidation, Ann Arbor, USA, p. 76

SOLOMON, I. (1955): Phys. Rev. **99**, 559–565

STIER, A., and E. SACKMANN (1973): Biochim. Biophys. Acta **311**, 400–408

SWIFT, T. J., and R. E. CONNICK (1962): J. Chem. Phys. **37**, 307–320

TRAYLOR, T. G., and T. MINCEY (1979): Acta Biol. Med. Germ. **38**, 351–355

ULLRICH, V. (1979): Topics in Curr. Chem. **83**, 67–104

VUK-PAVLOVIĆ, S., B. BENKO, and S. MARIČIĆ (1974): Biophys. Chem. **2**, 359–368

VUK-PAVLOVIĆ, S. (1976): Biophys. Chem. **5**, 395–398

VUK-PAVLOVIĆ, S., B. BENKO, S. MARIČIĆ, G. LAHAJNAR, I. P. KURANOVA, and B. K. VAINSHTEIN (1976): Int. J. Peptide Protein Res. 8, 427–434

VUK-PAVLOVIĆ, S., V. BRAČIKA, B. BENKO, and S. MARIČIĆ (1977): Biochim. Biophys. Acta **491**, 447–456

WHITE, R. E., and M. J. COON (1980): Ann. Rev. Biochem. **49**, 315–356

WILLS, E. D. (1980) in: Biochemistry, Biophysics and Regulation of Cytochrome P-450, GUSTAFSSON, J. Å., J. CARLSTEDT-DUKE, A. MODE, and J. RAFTER, eds., Elsevier/North-Holland Biomedical Press, pp. 545 to 548

WISHNIA, A. (1960): J. Chem. Phys. **32**, 871 to 875

WUNDERLICH, F., A. RONAI, V. SPETH, J. SEELIG, and A. BLUME (1975): Biochemistry **14**, 3730–3735

Chapter 6

Cytochrome P–450 Dependent Biotransformation of Drugs and Other Xenobiotics

J. Lampe, G. Butschak and W. Scheler

Contents

1. Introduction

In the animal organism most of the foreign compounds including the majority of drugs, are metabolized. In these various enzymatic reactions the P–450 dependent monooxygenase system (MFO) plays an important role. By the biotransformation of such substances more hydrophilic metabolites are formed thus enabling a faster renal elimination. Without transformation the elimination time of lipophilic drugs would amount to a multifold of the real time either due to tubular

Abbrevations:

AHH = aryl hydrocarbon hydroxylase (monooxygenase), MFO = mixed-function oxidase, GSH = glutathione, 3-MC = 3-methylcholanthrene, P–450 = cytochrome P–450, P–448 = cytochrome P–448, PAH = polycyclic aromatic hydrocarbons, PB = phenobarbital, PBB = polybrominated biphenyls, PCB = polychlorinated biphenyls, TCDD = 2,3,7,8-tetrachlorodibenzo-p-dioxin

Table 1. Theoretical and real half-life of some drugs in the organism in dependence on their lipid solubility (REMMER, 1966; reproduced with permission from Thieme-Verlag Stuttgart)

Drug	Solubility in chloroform*	Elimination half-life	
		theoretically without metabolism	really due to the metabolism
Barbital	40	55–75 h	55–75 h
Phenobarbital	96	10–20 d	3–6 d
Pentobarbital	100	2–5 mo.	11–70 h
Glutethimide	100	2–5 mo.	7–15 h
Hexobarbital	100	2–5 mo.	5–6 h

* % extracted from water, pH 7.4, in the same volume of chloroform.

Table 2. Examples for changes of the biological activity of drugs and foreign substances by enzymatic biotransformation (REMMER, 1973; KURZ et al., 1977, supplemented)

Compounds	Metabolites
Metabolites are less effective than parent compounds	
Barbiturates	Hydroxybarbiturates
Meprobamate	Hydroxymeprobamate
Phenothiazines	Phenothiazinesulfoxides
Metabolites and parent compounds are equal effective	
Phenylbutazone	Oxyphenylbutazone
Aminopyrine	Antipyrine
Codeine	Morphine
Phenacetin	N-Acetyl-p-aminophenol
Diazepam	Oxazepam
Methylphenobarbital	Phenobarbital
Metabolites are effective, parent compounds are not	
Parathion	Paraoxon
Cyclophosphamide	Aldophosphamide
Imipramine	Desipramine
Prontosil	Sulfanilamide
Metabolites are toxic	
Benzo(a)pyrene	Arene oxide
Carbon tetrachloride	Free radical, carbene
Furosemide	Furan epoxide
Vinyl chloride	Epoxide
Halobenzene	Arene oxide
Aflatoxin	Furan epoxide
Nitrosamine	Carbonium ion

reabsorption or due to the enterohepatic circulation. This relationship is demonstrated in Table 1 for some hypnotics with different hydrophobicity.

In general the derivatives formed by biotransformation are pharmacodynamically less effective than the parent compounds (→ detoxification). On the other hand biologically active metabolites may emerge, however, from indifferent compounds (→ toxification) (REMMER, 1973; GARATTINI et al., 1975). Many metabolites are proved to have cytotoxic, teratogenic, mutagenic or carcinogenic properties (Table 2).

A qualitative change of the drug action due to biotransformation has been observed rather seldom. Zoxazolamine, for instance, and one of its metabolites, chlorozoxazone, are centrally acting muscle relaxants. Another metabolite, 6-hydroxyzoxazolamine, however, acts uricosurically (UEHLEKE, 1968), (Fig. 1).

Within the organism drugs and xenobiotics are metabolized mainly in the liver, whereas other organs such as lung, kidney, intestine, brain, placenta show a significantly lower biotransformation capacity.

Most of the enzymes involved in biotransformation are intracellular enzymes. Many of them are bound to the membranes of the endoplasmic reticulum and the nucleus. Other are localized in mitochondria, lysosomes or in the cytosol

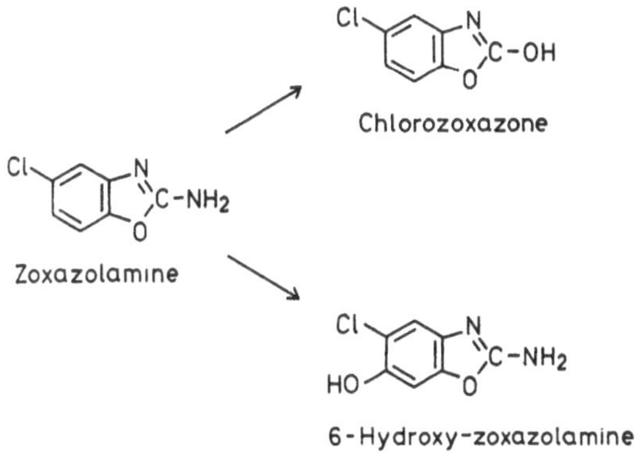

Fig. 1. Hydroxylation of zoxazolamine to metabolites of different biological activity (UEHLEKE, 1968; reproduced with permission from Verlag Chemie, Weinheim)

Table 3. General pathways of xenobiotics metabolism and localization of the enzymes in the liver (GILLETTE, 1971; supplemented according to MUKHTAR et al., 1979)

Enzymatic reaction	Localization
Phase I reactions	
Oxidations	
Hydroxylation	Microsomes, nuclear envelopes
Dealkylation	Microsomes, nuclear envelopes
Oxide formation	Microsomes, nuclear envelopes
Desulfuration	Microsomes, nuclear envelopes
Dehalogenation	Microsomes
Alcohol oxidation	Soluble, microsomes (minor)
Aldehyde oxidation	Soluble
Reductions	
Aldehyde reduction	Soluble
Azoreduction	Microsomes
Nitroreduction	Microsomes
C-Reduction	Microsomes
Hydrolyses	
De-esterification	Microsomes, soluble
De-amidation	Microsomes, soluble
Hydro-lyase (e.g. epoxide hydratase)	Microsomes, nuclear envelopes
Phase II reactions	
Glucuronide conjugation	Microsomes
Acylation	Mitochondria, soluble
Methylation	Soluble
Mercapturic acid formation	Soluble
Sulfate conjugation	Soluble

(see Table 3). Moreover, there are humoral enzyme systems which take part in the metabolism of drugs, such as blood plasma esterases catalyzing the hydrolysis of procaine, suxamethonium and other esters.

In general the enzymes metabolizing foreign substances are rather unspecific. They convert chemically different molecules in the same way. Thus a large number of foreign substances is transformed by a relatively small number of enzymes.

2. Metabolism of Xenobiotics

There are two phases in the biotransformation of foreign substances. In phase I chemical modifications of the parent compounds take place such as hydroxylation, oxidation, reduction or hydrolysis. In this phase there are either polar groups inserted into unpolar compounds or alkyl groups split off. In phase II the polar groups of the intermediate metabolites are conjugated to highly watersoluble molecules, suitable for renal excretion (Fig. 2).

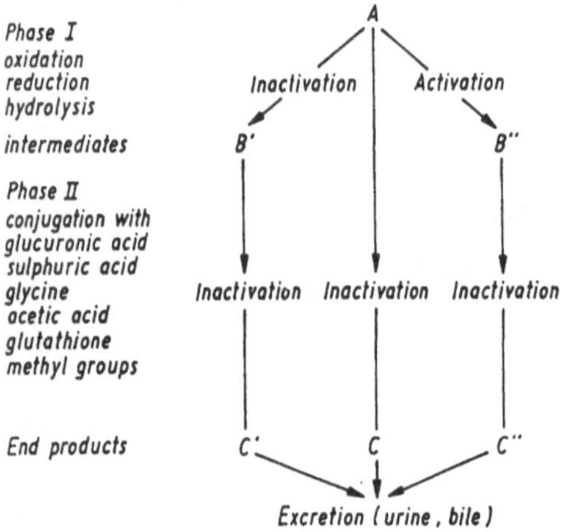

Fig. 2. Scheme of drug metabolism. The biologically active lipophilic compound A is metabolized to pharmacologically inactive and hydrophilic metabolites (C, C′ and C″) which are excreted. The biotransformation intermediates have lower (B′) or higher (B″) biological activity than the parent compound. Some metabolites are modified by conjugation in phase II (Hänninen, 1975, modified; reproduced with permission from E. Munksgaard, Copenhagen)

2.1. *Phase I*

The oxidative reactions in phase I as well as some of the reductions are catalyzed mainly by P–450 dependent MFO (Gillette et al., 1972). Table 4 represents a survey of biotransformation pathways in which the MFO system is involved.

Additionally to the P–450 dependent MFO the endoplasmic reticulum and the nucleus envelope contain a monooxygenase (amine oxidase) insensitive to CO and independent of P–450. This monooxygenase is an oxygen and NADPH dependent flavoprotein. It oxidizes secondary and tertiary amines to N-oxides (Prough and Ziegler, 1977; Patton et al., 1980).

Of particular importance with regard to hydrolytic reactions is the epoxide hydratase (EC 4.2.1.63). It transforms epoxides and arene oxides, being formed by oxidation of aliphatic and aromatic double bonds, into diols. This way highly reactive epoxides and arene oxides are inactivated (Oesch, 1979; Lu and Miwa, 1980), (cf. 5.2.). The epoxide hydratase is localized deeply inside the membrane of the endoplasmic reticulum and constitutes together with the P–450 dependent monooxygenase a stoichiometric complex (Seidegård et al., 1978).

2.2. *Phase II*

In phase II the polar intermediates are coupled to glucuronic acid, sulfuric acid, glycine, or to some other molecules. More rarely reactions are methylation, glycosylation, fatty acid conjugation, formation of urea adducts and conjugation to polypeptides (Israili et al., 1977; Jenner and Testa, 1978).

One of the most important conjugations is the coupling of glucuronic acid to foreign substances or endogenous compounds

Table 4. Major pathways of xenobiotics metabolism of phase I (MANDEL et al., 1971; PARKE, 1976; KARASHIMA et al., 1977; SIEST et al., 1978; ULLRICH et al., 1978)

Oxidation

Aromatic C-hydroxylation

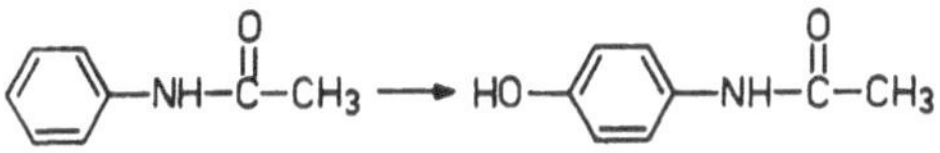

Acetanilide p-Hydroxyacetanilide

Further substrates: phenobarbital, amphetamine, benzene, biphenyl, aniline, naphthalene, 3,4-benzo(a)pyrene, 2-naphthylamine, 2-aminofluorene

Side-chain hydroxylation

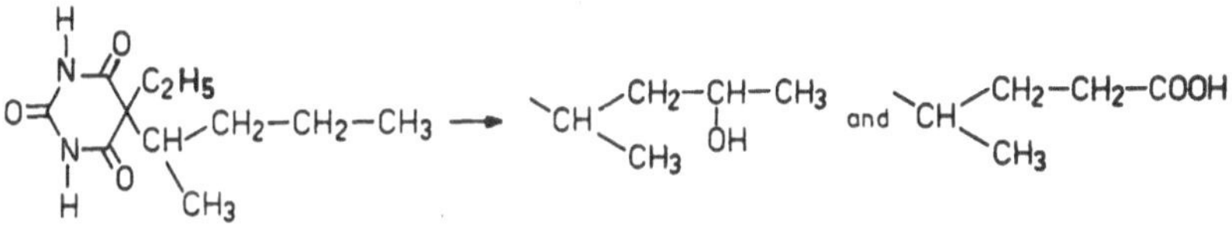

Pentobarbital Pentobarbital-(4)-ol Pentobarbital carboxylic acid

Further substrates: other barbiturates, p-nitrotoluene

N-Oxidation

$$(CH_3)_3N \longrightarrow (CH_3)_3N \rightarrow O$$

Trimethylamine Trimethylamine oxide

Further substrates: dimethylamphetamine, dimethylaniline, nicotine, promazine, morphine, guanethidine

S-Oxidation

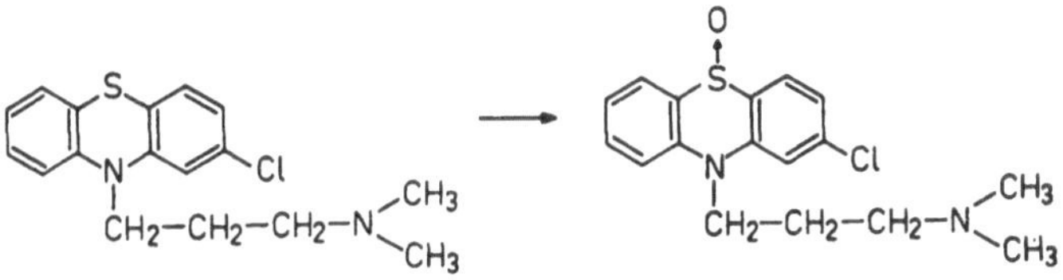

Chlorpromazine Chlorpromazine sulfoxide

Further substrates: phenothiazines

N-Dealkylation

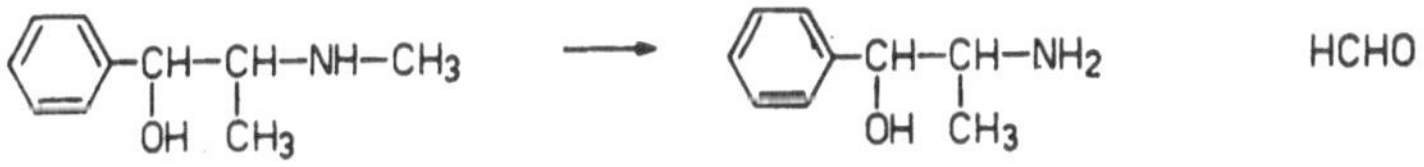

Ephedrine Norephedrine Formaldehyde

Further substrates: aminopyrine, methadone, imipramine, meperidine, methylamphetamine, methylaniline, codeine, diacetylmorphine, monomethyl-4-aminophenazone

Table 4 (continued)

O-Dealkylation

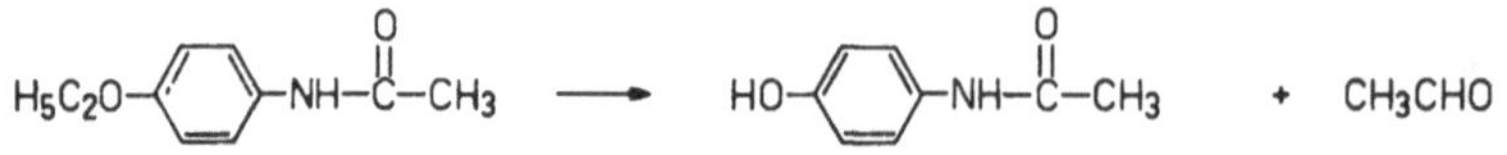

Phenacetin | p-Hydroxyacetanilide | Acetaldehyde

Further substrates: codeine, papaverine, dimethylvinphos, griseofulvin, mescaline, quinine, colchicine

S-Dealkylation

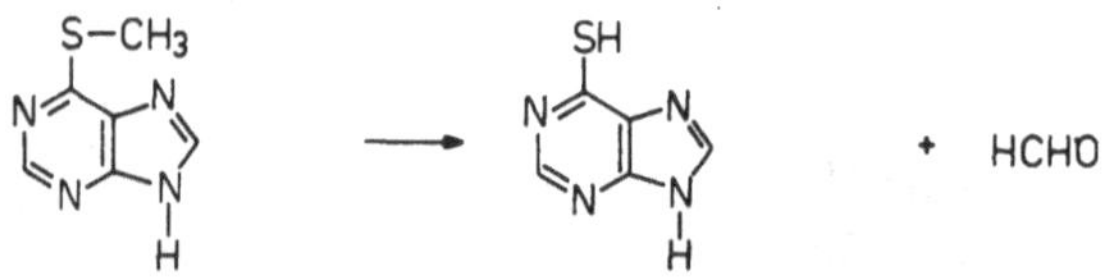

6-Methylmercaptopurine | 6-Mercaptopurine | Formaldehyde

Deamination

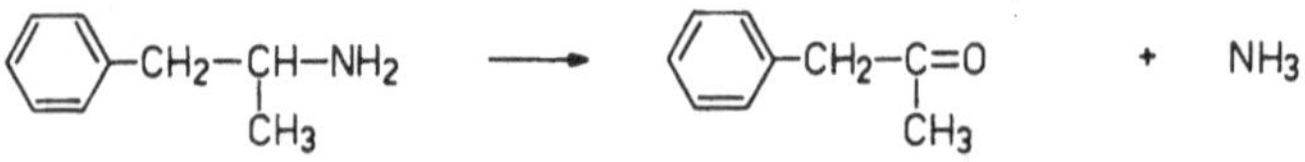

Amphetamine | Phenylacetone

Further substrates: methylamphetamine, ephedrine, mescaline

Desulfuration

Parathion | Paraoxon

Dehydrogenation

Hexachlorocyclohexane | Hexachlorocyclohexene

Denitrification

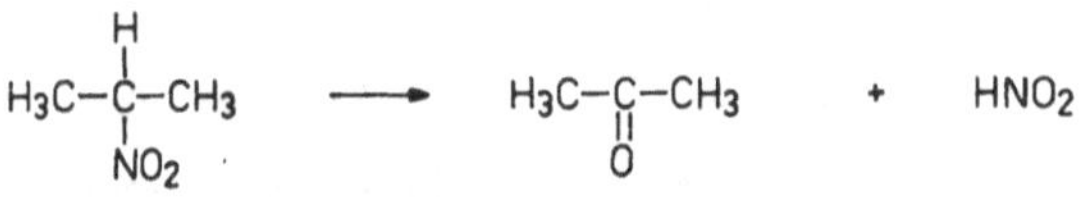

2-Nitropropane | Acetone

Table 4 (continued)

Dehalogenation

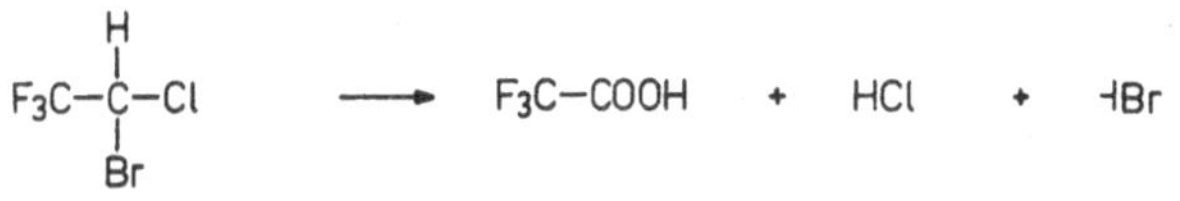

Halothane — Trifluoro-acetic acid

Reduction

Azo reduction

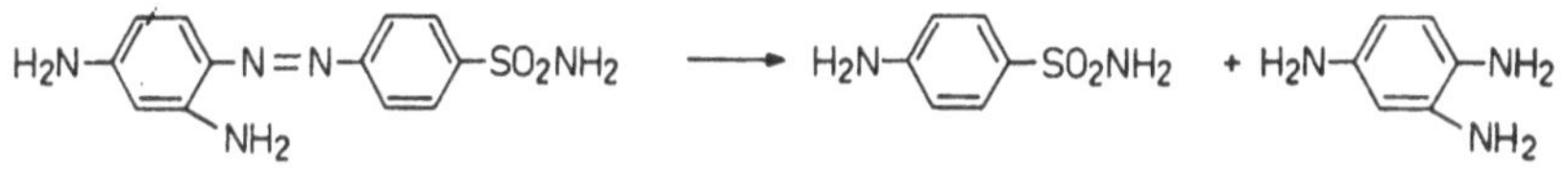

Prontosil — Sulfanil-amide — 1,2,4-Triamine benzene

Nitro reduction

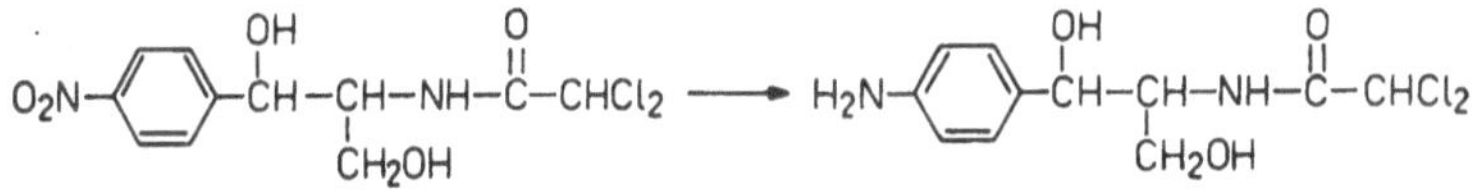

Chloramphenicol — arylamine

Carbonyl reduction

Acetophenone — 1-Phenylethanol

Hydrolysis

Ester

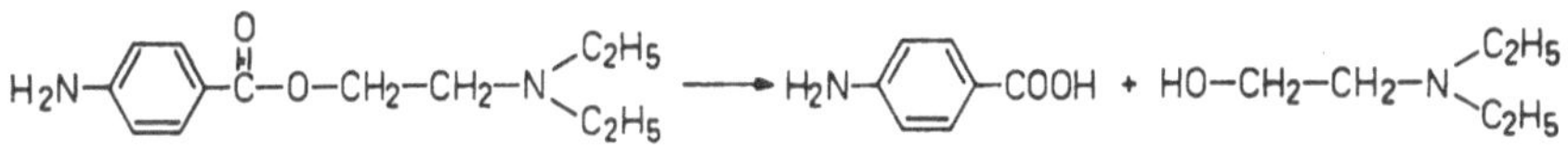

Procaine — p-Aminobencoic acid — Diethylamino-ethanol

Amide

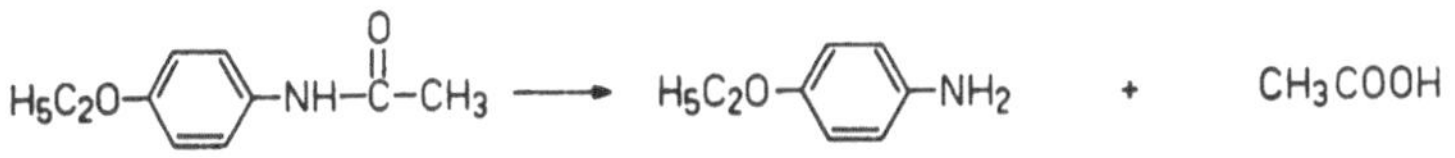

Phenacetin — Phenetidin — Acetic acid

(e.g. steroids, bilirubin). The conjugation proceeds via activated glucuronic acid and is catalyzed by UDP-glucuronosyltransferase (EC 2.4.1.17) in the endoplasmic reticulum mainly of liver cells (BOCK, 1977; DUTTON, 1978) but also in cells of kidney, lung (AITIO and VAINIO, 1976) and small intestine (JOSTING et al., 1976). The glucuronides formed are biliarly eliminated, and therefore exposed to the action of β-glucuronidases of intestinal wall and of intestinal bacterial flora. After splitting off the glucuronic acid component the parent compound respectively the intermediate metabolite can re-enter the enterohepatic circulation.

Furthermore glutathione transferases (EC 2.5.1.18) are included in the detoxification mechanisms. They represent a group of enzymes (5—7) with different, partly overlapping substrate specificity (JAKOBY and KEEN, 1977; JAKOBY, 1978). They have been observed in liver, kidney, lung, small intestine, spleen and other organs of several mammals (PINKUS et al., 1977; HILL et al., 1978). They are of importance for the GSH coupling to electrophilic compounds and thus for the detoxification of electrophilic aliphatic epoxides and arene oxides. The S-substituted GSH-adducts are transformed into S-substituted cysteine-adducts by γ-glutamyltransferase (EC 2.3.2.2) and cysteinyl-glycine dipeptidase (EC 3.4.13.6). After acetylation by N-acetyltransferases mercapturic acids emerge, which are eliminated renally.

Highly reactive electrophilic metabolites are able to bind covalently to GSH-transferase. In this case the enzyme gets inactivated. As an exemple covalent binding of 1-chloro-2,4-dinitrobenzene to the transferase has been observed. Furthermore, GSH transferases act as storage proteins for a number of hydrophobic substances, such as bilirubin, hemin, estrogens, chloramphenicol, penicillin and radiographic contrast media (JAKOBY and KEEN, 1977; JAKOBY, 1978). Due to these properties the GSH-transferases are also called ligandins.

In some cases the binding of GSH to xenobiotics leads to more toxic products. Thus, for instance, the weak mutagenic 1,2-dichloroethane is activated to a strong mutagen by GSH-transferases A and C and GSH. This compound is probably formed by a nucleophilic displacement of a chlorine atom by the -SH group of GSH. Its structure is likely related to that of mustard gas. The direct mutagenic effect of 1,2-dibromoethane is increased by conjugation with GSH, too (RANNUG et al., 1978), (see Fig. 3).

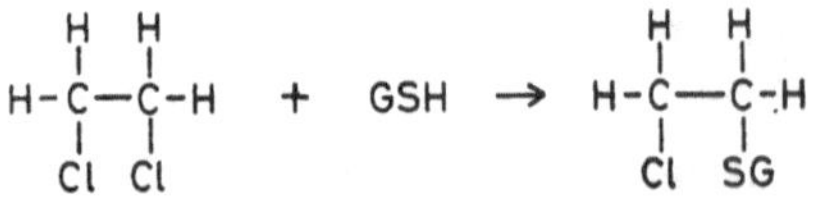

Fig. 3. Activation of 1,2-dichloroethane by glutathione to a reactive metabolite (MULDER, 1979; reproduced with permission from Elsevier/North-Holland Biomedical Press, Amsterdam)

Ester formation e.g. sulfatation represents another mechanism of activation of xenobiotics (see 5.6.2.).

3. Organ Specificity of the MFO System

Most MFO experiments are performed with liver MFO. Although liver is the main organ for biotransformation of xenobiotics approximately one third of the biotransformation activity belongs to extra-hepatic organs such as lung, kidney, skin and intestine (ACKERMANN, 1978). Furthermore, P–450 or P–448 and their corresponding enzymatic activities, have been found in trachea (COHEN et al., 1979), fore-stomach (TRIOLO et al., 1977), pancreas (SCARPELLI et al., 1980), ductus pancreaticus (HARRIS et al., 1977), colon (AUTRUP et al., 1978, 1979; STROBEL et al., 1980), brain (MARIETTA et al., 1979; HOLTZMAN and DESAUTEL, 1980), aorta (BOND et al., 1980), bone marrow

(ANDREWS et al., 1976), monocytes and lymphocytes (VAUGH et al., 1978), thrombocytes (CINTI and FEINSTEIN, 1976), mammary gland epithel (CHUANG et al., 1977), gonads (MATTISON and THORGEIRSSON, 1978) and placenta (JUCHAU, 1980).

P-450 of adrenal cortex, ovarium and testis preferably transforms steroid hormones and their precursors (GUNSALUS et al., 1975; HONECK, 1978; LU and WEST, 1978).

The P-450 content respectively P-450 dependent enzymatic activities of the various organs differ considerably. The same applies to the P-450 inducibility. Therefore results obtained with liver preparations must not be representative for other organs.

In the following some specificities of the P-450 dependent enzyme system in lung, intestine and kidney are described.

3.1. *Lung*

The P-450 dependent enzyme activities of the lung are not equally distributed between all cell types but mainly localized in type II alveolar cells and Clara cells of bronchial epithelium (BRANDENBURGER BROWN, 1974; TEEL, 1979). In human lung the content of P-450 was determined to 0.007—0.012 nmoles · mg^{-1} microsomal protein (PROUGH et al., 1977). In comparison human liver contains 0.197 ± 0.02 nmoles · mg^{-1} microsomal protein (ACKERMANN, 1972). In rabbits the amounts are 0.40 (lung) and 1.52 (liver) nmoles · mg^{-1} microsomal protein, respectively (BUENING and FRANKLIN, 1976). Although in lung of rats and rabbits the content of P-450 is lower than in liver, there is no tight correlation to the respective enzyme activities (Table 5) (BEND et al., 1972; BUENING and FRANKLIN, 1976; LITTERST et al., 1977; HOOK et al., 1979).

The xenobiotics metabolizing enzymes in lung are probably of importance in protecting lung tissues from airborne toxic chemicals. Due to the dense vascular net the lung is exposed also to foreign substances solved in the peripheral blood and takes part therefore in the total body clearance of various drugs (ROTH and WIERSMA, 1979).

As described for liver enzymes lung enzymes can also activate several xenobiotics, enabling them to react covalently with proteins and nucleic acids. This may be important for the carcinogenicity of some foreign substances, e.g. benzo(a)-pyrene (see 5.2.). The apparently high concentration of P-450 in Clara cells in contrast to its diffuse distribution in the liver may contribute to the increased susceptibility of the lung towards certain toxic chemicals which have little or no toxic effects on the liver (SERABJIT-SINGH et al., 1980). With respect to such toxic effects of chemicals there is no correlation to the absolute enzyme activity but to the relation between activating reactions, e.g. the formation of arene oxides by MFO, and detoxificating reactions, e.g. by epoxide hydratase, glutathione S-transferases and other enzymes. The activity of deactivating enzymes is much lower in the lung than

Table 5. N-Demethylation of several amines by microsomal enzymes of rabbit lung and liver (GRAM, 1973; roproduced with permission from Marcel Dekker, New York)

Substrate	nmoles HCHO · mg^{-1} protein · min^{-1}		
	Lung	Liver	Lung/Liver
Aminopyrine	$2.7^* \pm 0.4$	$6.6^* \pm 0.5$	0.40
Ethylmorphine	2.2 ± 0.6	3.5 ± 0.3	0.63
d-Benzphetamine	$8.8^* \pm 0.2$	$4.7^* \pm 0.9$	1.87
N-Methyl-p-chloroaniline	$7.2^* \pm 0.2$	$4.8^* \pm 0.4$	1.50

* $p < 0.05$ ($n = 4$)

Table 6. Kinetics parameters of NADPH-mediated covalent binding of 4-ipomeanol in incubation mixtures containing rat liver or lung microsomes (BOYD et al., 1978; reproduced with permission from Williams and Wilkins Comp., Baltimore)

Micro-somes	K_m	V_{max}	
	mM	nmoles bound $\times$ mg^{-1} microsomal protein · min^{-1}	nmoles bound $\times$ nmol^{-1} cytochrome P–450 · min^{-1}
Liver	0.42	0.45	0.48
Lung	0.029	0.51	4.85

in the liver. The ratio of the activities liver/lung is species specific (HOOK and BEND, 1976; DEPIERRE and ERNSTER, 1978; SMITH et al., 1978).

Another specificity of lung MFO in comparison to liver MFO concerns 4-ipomeanol transformation. This toxic furan compound is covalently bound to microsomal proteins of Clara cells after activation by P–450 and causes cell necrosis and pulmonary edema. In vitro studies on 4-ipomeanol showed a considerably stronger covalent binding to pulmonary than to hepatic microsomes (Table 6).

Analogous observations for in vivo experiments have been published (BOYD et al., 1978; BOYD and BURKA, 1978). Antibodies against NADH-cytochrome *c* reductase have little influence on the metabolic activation of 4-ipomeanol by liver microsomes, but they strongly inhibit the NADH and NADPH dependent activation and covalent binding to lung microsomes. The activation probably takes place by a 2-electron transfer, in which the transfer of the second electron via cytochrome b_5 is the rate limiting step in the lung microsomes (SASAME and GILLETTE, 1978).

In C57BL/6/N mice, but not in rats and Syrian golden hamsters, the organ specificity of 4-ipomeanol is age dependent. At comparable i.p. doses the amount of covalently bound metabolites in old mice is 15–20 times higher in the kidney and 2–3 times lower in the lung than in young animals. The bronchiolar necrosis is less severe at comparable doses in old mice. 4-Ipomeanol causes renal cortical necrosis only in old mice (JONES et al., 1979).

The genetic regulation of P–450 activity in human lung and its influence on the formation of pulmonary carcinoma is dealt with in 4.2.

3.2. *Intestine*

Most of the xenobiotics enter the organism via the gastrointestinal tract. The metabolism of these substances in the intestine is therefore of pharmacological and toxicological interest.

In general the enzyme activities in microsomes of the small intestine are considerably lower than in the liver. Because of the large surface of intestinal epithelia and the long passage time of xenobiotics the intestinal enzymes are remarkably involved in the metabolism of foreign substances (CHHABRA and TREDGER, 1978).

The concentration of P–450, the enzyme activities and their inducibility show partly considerable species differences. So, for instance, the hydroxylase activity towards benzo(a)pyrene decreases in the following order: guinea pig, rabbit, rat, mouse, cat (HIETANEN and VAINIO, 1973). It is further dependent on age and nutrition.

In the enterocytes of the small intestinal villi a P–450 content of 25–100 pmoles $\times$ mg^{-1} microsomal protein was found in man. The concentration decreases from duodenum to the end of ileum. The 7-ethoxycoumarin deethylase activity shows a similar dependence on the localization (HOENSCH et al., 1979). In the rat the benzo(a)pyrene hydroxylase activity decreases continuously from the small intestine to the cecum (WATTENBERG, 1972) (see Fig. 4).

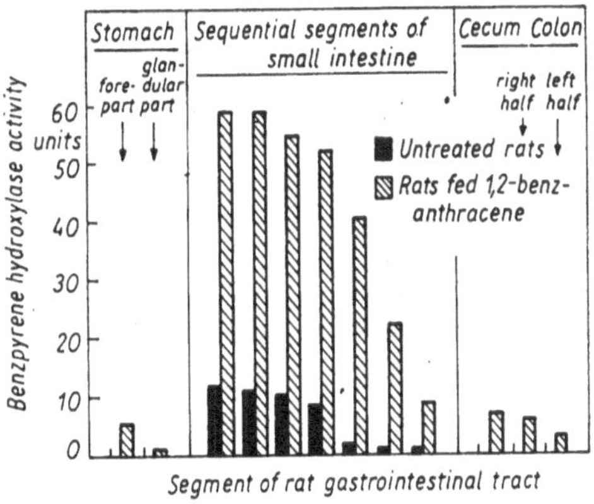

Fig. 4. Distribution of benzo(a)pyrene hydroxylase activity along the gastrointestinal tract of female untreated rats and rats given 20 mg 1,2-benz(a)anthracene by mouth 36 hours prior sacrifice. The small intestine has been divided into seven segments and differences in activity in sequential portions of the structure are shown. One unit is equivalent to the fluorescence of 1 pg of 8-hydroxy-benzo-(a)pyrene formed · mg^{-1} wet weight of tissue $\times$ min^{-1} (WATTENBERG et al., 1962; reproduced with permission from Waverly Press, Baltimore)

The concentration of P–450 in microsomal fractions from the intestine of rabbits amounts to 0.34 ± 0.03 nmoles $\times$ mg^{-1} microsomal protein. PB does not induce, benzo(a)pyrene increases the amount by 50 percent. Activity and substrate specificity of P–450 dependent enzymes differ partly in comparison to other organs. Contrary to liver and lung microsomes, for instance, no other amphetamine derivative than N-hydroxyamphetamine is metabolized by microsomes of the small intestine (BUENING and FRANKLIN, 1976). Small intestine microsomes of rabbit, guinea pig, rat, mouse and hamster contain biphenyl hydroxylase and benzo(a)pyrene hydroxylase. Ethylmorphine N-demethylase and aniline hydroxylase, however, were only proved in rabbits and guinea pigs (CHHABRA et al., 1974).

The nutritional state and inducers differently influence enzyme activities. Thus 3-MC induces benzo(a)pyrene hydroxylase 30-fold, 7-ethoxyresorufine and 7-ethoxycoumarine deethylase 20-fold, but biphenyl 2 -and 4-hydroxylases only 3-fold in the small intestine of rats. PB and pregnenolon-16α-carbonitrile do not induce or diminish several enzyme activities. Fasting of rats for 2 days causes a significant decline of the benzo(a)pyrene hydroxylase activity whereas deethylase activity could not be found at all (STOHS et al., 1976a).

In rats cholesterol-free diet caused a decrease of benzo(a)pyrene hydroxylase activity from 4.07 to 0.086 pmoles · mg^{-1} microsomal protein · min^{-1}, 2% cholesterol in the normal diet activated the enzyme approximately 70-fold in comparison to the cholesterol-free diet. The activity of UDP-glucuronosyltransferase is less influenced by changes of the cholesterol content in the diet than that of benzo(a)-pyrene hydroxylase (HIETANEN and LAITINEN, 1978). In fasted rats or in rats fed with a highly purified diet consisting of casein, starchflour, corn oil, minerals and vitamins, the benzo(a)pyrene hydroxylase of intestinal mucosa disappears but increases again if the animals are fed normally. By feeding certain cruciferous vegetables, e.g. cabbage and Brussels sprouts, the enzyme activity is raised. It is supposed that most if not all of the enzyme activity is the result of exogenous inducers in the food (see Fig. 4) (WATTENBERG, 1971; ZAMPAGLIONE and MANNERING, 1973).

The benzo(a)pyrene hydroxylase of intestinal mucosa was studied particularly thoroughly. Isolated microsomes as well as cell and tissue cultures of small intestine and colon were used. In isolated mucosa cells of small intestine of rats the enzyme is induced 30-fold by 3-MC. As a main metabolite of benzo(a)pyrene the 4,5-oxide is formed, and in comparison to liver microsomes only few dihydrodiols emerge. Only after a longer incubation time the 4,5-oxide is metabolized, too, and the dihydrodiols increase, due to the relatively low activity of the epoxide hydratase in small intestinal mucosa (STOHS et al., 1977). Similar results were obtained with isolated microsomes. Substrate specificity, inhibition and induction properties of the small intestinal xenobiotic-metabolizing enzymes differ from that of the liver (STOHS et al., 1976b).

The P–450 system is probably important for the large bowel carcinogenesis (STROBEL et al., 1980). Cultured human colon metabolizes benzo(a)pyrene by pathways similar to those found in human bronchus. The metabolites react covalently with cellular DNA. As main product an adduct between the 10-position of (±)-(7R,8S)-dihydroxy(9R,10R)epoxy-7,8,9,10-tetrahydrobenzo(a)pyrene and the 2-amino group of guanine could be identified. To a minor extent complex formation was proved between (±)-(7R,8S)-dihydroxy-(9S,10S)epoxy-7,8,9,10-tetrahydrobenzo(a)pyrene and the 2-amino group of guanine. The binding of the metabolites to DNA and proteins generally decreases from colon ascendens to the rectum (AUTRUP et al., 1978).

Intestinal cells can perform conjugation reactions in the same manner as liver cells. Substrate specificity and activity of the enzymes involved in phase II are partly different in liver and intestine (SHIRKEY et al., 1979). The cultured human colon, for instance, forms sulfate esters and GSH-conjugates with benzo(a)-pyrene metabolites but only few glucuronides (AUTRUP, 1979).

Of special interest for the intestinal metabolism of xenobiotics are also the enzyme activities of the intestinal flora. This becomes obvious in the enterohepatic circulation, when the biliarly excreted conjugates are split off by glucuronidases into glucuronic acid and the more lipophilic rest molecule, which is reabsorbed by intestinal mucosa. Enzymatic splitting and reabsorption have been ascertained by studies on germ free animals. It was found that the induction of carcinomas in colon, small intestine and breast by 3,2′-dimethyl-4-aminobiphenyl is much higher for conventionally bred rats than for germ free animals. It is supposed that the carcinogen is N-hydroxylated and coupled to glucuronic acid in the liver after its absorption. After biliar excretion the β-glucuronidases of intestinal flora and intestinal wall hydrolyze the adduct and the parent compound is then absorbed again. In germ free animals only the intestinal wall enzyme exists (REDDY and WATANABE, 1978).

Another example of the effectiveness of the intestinal flora is the metabolism of the antiparasitic agent 4-isothiocyano-4′-nitrophenylamine by intestinal bacteria. In the Ames test the compound is not mutagenic for Salmonella typhimurium neither without nor after microsomal activation. After passage of the intestinal tract, however, mutagenic metabolites are excreted in the urine. Erythromycin which decreases the intestinal flora inhibits the formation of mutagenic metabolites (BATZINGER et al., 1978).

3.3. Kidney

P–450 and P–450 depending enzyme activities are mainly localized in the cortex of the kidney. The monooxygenases in the renal medulla of rabbits are only measurable after induction by 3-MC except the lauric hydroxylase (ZENSER et al., 1978). The MFO activities in kidneys of rats and mice are considerably lower as compared to those in the liver (FREUDENTHAL et al., 1976; LITTERST et al., 1977), but phase II reactions in isolated kidney cells reach up to 50 percent of the activity of liver cells (FRY et al., 1978).

Microsomal kidney cortex MFO differs from that in the liver by the high specificity towards fatty acids (ELLIN et al., 1972; ELLIN and ORRENIUS, 1975). Besides

Table 7. Comparison of cytochrome content and monooxygenase activities in liver and kidney cortex microsomes from man and rat (JAKOBSSON and CINTI, 1973; reproduced with permission from Williams and Wilkins Comp., Baltimore)

Content or activity	Human (adult)		Male rat (adult)	
	Liver	Kidney	Liver	Kidney
Cytochrome b_5 (nmoles · mg^{-1} protein)	0.20–0.30	0.127	0.45–0.55	0.11–0.18
Cytochrome P–450 (nmoles · mg^{-1} protein)	0.25–0.30	0.030	0.6–0.8	0.12–0.17
Aminopyrine N-demethylation (nmoles HCHO · mg^{-1} protein · min^{-1})	3.2	0.35	6.0–10.0	0.1–0.3
Laurat ω- and $(\omega - 1)$-hydroxylation (nmoles product · mg^{-1} protein · min^{-1})	n.d.	0.33	3.0	3.0

n.d. not determined

the different substrate specificities of P–450 in the liver and in the kidney the pathways for the metabolism of organic compounds are probably organ specific, too (ARMBRECHT et al., 1979). Participation of an electron transfer protein similar to the nonheme iron proteins could not be excluded definitely in the microsomal hydroxylation of fatty acids (ICHIHARA et al., 1971, 1973).

The monooxygenase activities of the rat kidney do not show sex-dependent differences as observed in rat liver (LITTERST et al., 1977). The inducibility of kidney MFO differs also from that of the liver enzymes. After 3-MC treatment of rabbits the content of P–450 in the kidney cortex increases from 0.07 ± 0.003 to 0.41 ± 0.005, in the liver, however, only from 1.32 ± 0.04 to 2.13 ± 0.09 nmoles·mg^{-1} microsomal protein. The fatty acid hydroxylation is not inducible in both organs (ZENSER et al., 1978).

In human kidney as well as in human embryonal renal epithelial cells monooxygenase activities have been proved. In Table 7 P–450 and MFO activities in liver and kidney are compared. C3-hydroxylation and N1-demethylation of diazepam by human kidney microsomes are slower than by liver microsomes (ACKERMANN et al., 1976). The benzo(a)pyrene hydroxylase in kidney epithel cells from human embryos is induced by benz(a)anthracene (YAMASAKI et al., 1977).

With respect to biotransformation of drugs in man the low monooxygenase activities of kidney are quantitatively not important but they are of toxicological interest. So, for instance, the occurrence of interstitial nephritis as well as of carcinomas in the renal pelvis after extended phenacetin abuse originates from the N-hydroxylating activity of the kidney (RAAFLAUB and DUBACH, 1972; CARRO-CIAMPI, 1978).

4. Induction of Cytochrome P-450

4.1. *Introductory Remarks*

The synthesis of P–450 can be induced by a large number of chemically different substances, drugs, hormones, insecticides, carcinogens and others. For experimental investigations preferably PB, some PAH (especially 3-MC or benzo(a)pyrene), various flavones (e.g. β-naphthoflavone), PCBs and TCDD are used as inducers. PB and 3-MC type inducers cause different alterations in the liver; some of them are listed in Table 8. In general each kind of inducer provokes a specific pattern of P–450 forms. In several species (e.g. liver of rabbit and rat) P–450 (LM_2, P–450_b) can be enriched by PB and P_1–450 (P–448, LM_4, P–450_c) by PAH (COON and VATSIS, 1978; BOTELHO et al., 1979).

Table 8. Different alterations in liver after induction with phenobarbital and 3-methylcholanthrene (KURZ et al., 1977; reproduced with permission from Bibliographisches Institut, Mannheim)

Alteration	Phenobarbital	3-Methylcholanthrene
Liver weight	increased	less raised
Bile flow	raised	unchanged
Microsomal proteins	increased	less multiplied
Cytochrome P-450	increased	increased
NADPH-cytochrome c reductase	multiplied	not multiplied
NADPH-cytochrome P-450 reductase	multiplied	not multiplied
Metabolism of other drugs	in many cases accelerated	only in few cases accelerated
Steroid metabolism	accelerated	not accelerated
Maximum effect	in 2—3 days	12—24 hours

Because of different substrate specificities of the various forms of P-450 the induction leads to modifications in the metabolism of xenobiotics. This may be one cause for activating or inhibiting effects of certain inducers on the formation of malignant tumors besides the different induction of other enzymes involved in the metabolism of chemical carcinogens (MAKIURA et al., 1974; KIMURA et al., 1976; NISHIZUMI, 1976; PITOT, 1979). Various P-450 forms are induced by PCBs, whereby certain PCB isomers induce exclusively like 3-MC and others like PB. Moreover, PCB isomers were found to stimulate the synthesis of P-450 forms which normally can be induced either by PB or by 3-MC (STONARD and GREIG, 1976; KOHLI et al., 1980; PARKINSON et al., 1980).

Many of the inducers show species, strain and organ specificities and even individual specific effects have been observed. In most cases the pattern of P-450 isozymes changes in the course of ontogenesis (BRIDGES and FRY, 1979; KLINGER et al., 1979). The P-450 forms of different species and those of different ontogenetic stages generally differ in their substrate specificity and other parameters. Species and age differences of the inducing potency of a definite inducer are therefore reflected also in differences of the xenobiotics metabolism (KLINGER et al., 1979; THORGEIRSSON et al., 1979).

4.2. *Mechanism of Cytochrome P-450 Induction*

The molecular mechanism of the induction of P-450 is nearly unknown as yet. In the case of barbiturates as well as of PAH it was shown that their inducing effect depends on RNA and protein synthesis (NEBERT and GELBOIN, 1970; NEBERT and GIELEN, 1971). The induction of P-448 and of the respective AHH activity was completely inhibited at simultaneous administration of PAH and inhibitors of the RNA synthesis (e.g. actinomycin D), whereas a later addition of the inhibitor did not influence the induction. AHH induction needs only the synthesis of the heterogeneous RNA but not that of ribosomal 4—5 S RNA. Furthermore, AHH induction depends on the synthesis of poly-(A) sequences (WHITLOCK et al., 1974). Inhibitors of protein synthesis (e.g. cycloheximide) prevent the de-novo-synthesis of P-448, even if they are administered after the inducer. mRNA which is necessary for the synthesis of apocytochrome, is also formed after injection of cycloheximide (NEBERT and GELBOIN, 1970; NEBERT and GIELEN, 1971; WHITLOCK and GELBOIN, 1974). De-novo-synthesis of apocytochrome P–450 after PB or PAH administration was demonstrated gel-electrophoretically and by usage of specific antibodies (RAJAMANICKAM et al., 1975; DUBOIS and WATERMAN, 1979). Investigations in cell-free systems with RNA from animals pretreated with PB showed that the apoenzyme synthesis is correlated with an increase of the mRNA coding the synthesis of P–450. Apocytochrome P–450 emerging under these conditions has a molecular mass of about 50,000 corresponding to the microsomal enzyme (DUBOIS and WATERMAN, 1979). KUMAR and PADMANABAN (1980) using anti-P–450 and anti-P–448 antibodies, have shown that PB and 3-MC enhance the amounts of translatable mRNA for P–450 and P–448, respectively. The molecular mass of the translation products formed in vivo or in a cell free system were 47,000 for P–450 and 53,000 for P–448.

If inducers and cycloheximide are simultaneously added to inducible cell cultures and the inhibitor is removed after some hours, a significant increase in AHH activity can be observed, which is insensitive to actinomycin D. As this increase surpasses the increase after exclusive administration of the inducer, the inhibition of the protein synthesis is supposed to block the formation of a labile protein which suppresses the apocytochrome P–450 synthesis (WHITLOCK and GELBOIN, 1973, 1974). Various investigations indicate the participation of cyclic nucleotides in the induction process. The large increase in AHH activity induced by dibutyryl cAMP in certain cell cultures is sensitive to actinomycin D. Furthermore, it was shown that the induction of the MFO system in the liver is associated with the phosphorylation of an acidic nuclear protein (BLANKENSHIP and BRESNICK, 1974; YAMASAKI et al., 1975; BYUS et al., 1976).

The present knowledge indicates that transcription processes must take place in an early phase of the induction. The synthesized mRNA can be accumulated independent of the protein synthesis. After formation of the corresponding mRNA types the apoenzyme synthesis proceeds independent of the RNA synthesis and is inhibited only by inhibitors of the protein synthesis. However, it has still to be clarified whether the induction is initiated by synthesis of mRNAs specific for the various P–450 forms or by formation of mRNAs which, for instance, regulate the synthesis of the coenzyme.

The holoenzyme of P–450 is considerably later available than the apoenzyme. Inhibition of the heme synthesis diminishes the apoenzyme synthesis in vivo and in vitro. By addition of hemin this effect can be reverted (RAJAMANICKAM et al., 1975; BHAT et al., 1971; BHAT and PADMANABAN, 1978). The decreased induction of P–450 by inhibitors of the RNA synthesis (e.g. actinomycin D) may be caused by an inhibition of the heme synthesis. If sufficient amounts of heme are present, actinomycin D has no influence on the PB-induced apocytochrome P–450 synthesis (RAJAMANICKAM et al., 1975). The induction of P–450 by DDT is influenced neither by actinomycin D nor by amanitin. It is probably regulated by initiation and elongation processes on the translational level (CONAWAY, 1977). Studies on the genetic regulation revealed that P–450 synthesis is regulated by a very complicated system. An understanding of the basic mechanism is only possible if the role of the various mRNAs, their translational products and the sequence of their synthesis has been elucidated.

4.3. Genetic Regulation of Cytochrome P–450 Induction

Studies on the genetic regulation of the induction process in various mice strains show that the inducibility of P–450 can be inherited either autosomally dominantly or additively (NEBERT, 1979a, 1979b; WOOD and TAYLOR, 1979). At present the genetic analysis of the regulation of AHH induction has made great progress due to a large number of cross-breeding experiments with mice strains responsive and nonresponsive to specific AHH inducers (PAH). The gene coding the responsiveness was called Ah-locus. The present results indicate that several alleles, which are localized in more than one site, take part in the regulation of AHH induction. Induction of several further enzymes (UDP-glucuronosyltransferase, ornithine decarboxylase, and of a specific oxidoreductase localized in the cytosol) is coupled with AHH induction. The product of the Ah-locus is a cytosolic PAH-receptor which after binding of the inducer forms the receptor-inducer complex. This complex activates the structural genes coding the synthesis of various enzymes (see Fig. 5). Higher doses of TCDD initiate AHH induction also in mice strains nonresponsive to PAH. But the TCDD dosis for nonresponsive strains is 15 times higher than for responsive animals. Probably a defect receptor molecule is formed in nonresponsive strains, which has a lower affinity to inducers.

The AHH induction in other tissues is regulated similarly to that in the liver. Therefore extrahepatic tissues (e.g. pulmonary tissue or lymphocytes) can be used for genetic investigations of AHH induction as well. But one has to consider that the proportions of the P–450 subspecies can vary in different tissues. The relative contents of P_1–450 (P–448) and P–450 were found, for instance, to be 10:1 in skin, 50:1 in lung, and 1:1 in liver. Considering also the differences of organ specific contents of other enzymes, e.g. epoxide hydratase, UDP-glucuronosyltransferase, and GSH transferases very specific metabolite patterns are to be expected for each organ thus contributing to the organ specificity of chemical carcinogens.

With respect to the formation of fibrosarcomas after s.c. injection of benzo(a)pyrene there is a good correlation

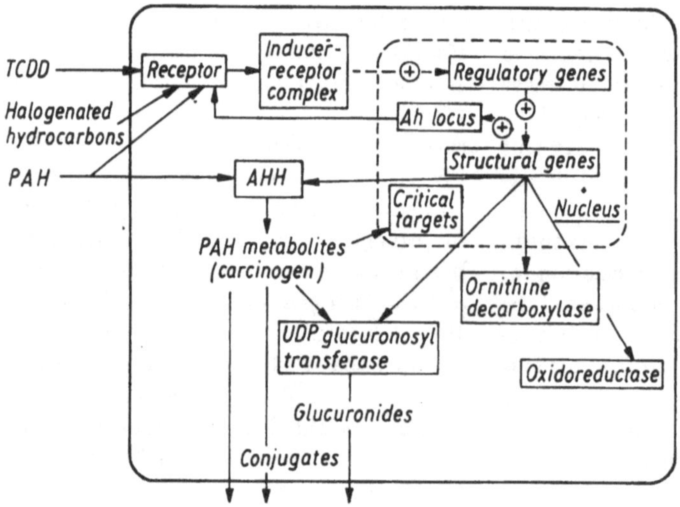

Fig. 5. A hypothetical model of the mechanism of the AHH induction and the metabolism of PAH (NEBERT, 1979b, modified; reproduced with permission from Pergamon Press, Oxford)

between AHH induction (increase of P-448 content) in the liver and tumor incidence comparing responsive and nonresponsive mice strains. Also in cross-breeding experiments tumor incidence and AHH activity are correlated.

After intratracheal 3-MC application considerably more bronchogenic carcinomas are formed in responsive mice strains in comparison to nonresponsive ones being in good correlation to the AHH activity of the pulmonary tissue. The same correlation was observed in the various cross-breeding experiments. TCDD-raised AHH activity is paralleled by the incidence of dermal fibrosarcomas in nonresponsive mice strains. The number of epidermal carcinomas which appear after local application of benzo(a)pyrene is correlated with the inducibility of AHH in various mice strains. In in vitro investigations binding of benzo(a)pyrene metabolites to DNA exclusively depends on the microsomes used but not on the origin of DNA (NEBERT et al., 1979; LEGRAVEREND et al., 1980).

The genetic regulation of enzyme activity and inducibility of AHH is probably important for human carcinogenesis, too. From studies on mitogen-stimulated human lymphocytes it became evident that the in vitro inducibility of this enzyme correlates with the elimination rate of antipyrine, aminopyrine, phenylbutazone, theophylline, phenacetin, PB and some other compounds from the blood plasma in vivo (KELLERMANN et al., 1976, 1978; KELLERMANN and LUYTEN-KELLERMANN, 1978). Correlations also exist between the hydroxylation of benzo(a)pyrene and several drugs by the human liver (KAPITULNIK et al., 1977). The half-life of a drug, e.g. antipyrine, in plasma or saliva might therefore monitor the activity of AHH. In comparison to controls the half-life of antipyrine in patients with bronchogenic carcinomas drops from 14.1 ± 2.3 to 10.8 ± 3.6 hours in saliva (KELLERMANN et al., 1978) and from 14.7 to 8.7 hours in serum (AMBRE et al., 1977). The differences are attributed to increased AHH activity. In the case, that the raised enzyme activity is not caused by the disease but genetically determined an increased metabolism of carcinogens could be associated with a higher tumor susceptibility of those patients (KELLERMANN et al., 1978).

KELLERMANN et al. (1973) found three groups in the white population of the U.S.A. with low, intermediate and high inducibility of AHH in lymphocytes. Patients with a pulmonary carcinoma predominately belonged to the two latter groups (Table 9), i.e. the susceptibility to bronchogenic carcinoma is associated with higher levels of inducible AHH activity. EMERY et al. (1978) found also in patients with pulmonary carcinomas a higher AHH inducibility of lymphocytes than in healthy control persons. Due to the high variability of the AHH activity in lymphocyte cultures these results

Table 9. Comparison of healthy control group, tumor control group and lung-cancer group for aryl hydrocarbon hydroxylase (AHH) inducibility (KELLERMANN et al., 1973; reproduced with permission from New England Journal of Medicine, Boston)

Group		AHH inducibility (%)			Gene frequencies*	
		AA[1]	AB[2]	BB[3]	AHH^a	AHH^b
Healthy control	($n = 85$)	44.7	45.9	9.4	0.676	0.324
Tumor control	($n = 46$)	43.5	45.6	10.9	0.663	0.337
Lung cancer	($n = 50$)	4.0	66.0	30.0	0.370	0.630

[1] low, [2] intermediate, [3] high, * superscripts refer to the 2 alleles

could not be reproduced in other laboratories. PRASAD et al. (1979) detected a higher inducibility of this enzyme in lymphocytes only in 4 out of 15 patients with pulmonary carcinomas. In non-cancer patients the AHH activity and inducibility in cultured macrophages correlate with the values of the lymphocytes. However, in the corresponding cells of lung cancer patients this correlation does not exist (McLEMORE et al., 1977, 1978b). The enzyme activity in lung tissue also correlates with the inducibility in the lymphocytes only in control patients, but not in patients with pulmonary carcinoma (McLEMORE et al., 1978a).

The inheritable variability of the AHH activity has been ascertained for man, too (NEBERT and ATLAS, 1978). Its use as biochemical marker for the susceptibility towards carcinogenic PAH, however, needs further investigations.

In vitro studies using somatic cell hybridization techniques proved the AHH expression in man to be coupled with the expression of 22 other enzymes. The genes of some of these enzymes (e.g. malate dehydrogenase and isocitrate dehydrogenase) are localized on chromosome 2. Therefore, the structural genes coding the AHH synthesis should also be localized there (BROWN et al., 1976). Perhaps this coupling can also be applied for the determination of the sensitivity towards carcinogenic effects of PAH.

4.4. *Clinical-pharmacological Aspects of Enzyme Induction*

The capability of xenobiotics to induce metabolizing enzymatic activities implies clinical consequences concerning pharmacological effects of drugs, more frequently an acceleration of drug metabolism than an inhibition of biotransformation. The induction is of special importance with respect to drug combinations and long lasting application. By the altered biotransformation rate quantitative differences in the profile of metabolites can even cause changes of the pharmacological action pattern of the drug combination.

The enzyme induction is dose dependent. From most drugs high doses must be applied in order to produce a clinically relevant enzyme induction. To quantitate the relations between inducing and toxic dose the induction index has been introduced, $I = \frac{\text{inducing dose}}{\text{toxic dose}}$. Only very few therapeutically significant drugs have a quotient $I < 1$. Therefore enzyme induction is of rather poor therapeutical importance (REMMER et al., 1979).

Besides drugs a variety of other xenobiotics such as insecticides, herbicides, industrial chemicals and air pollutants as well as ingredients of food and luxuries (such as caffeine, alcohol, tobacco smoke and food colors) can stimulate the biotransformation of drugs. Even the composition of food, especially its protein content, influences the rate of drug metabolism (ALVARES et al., 1976; ALVARES, 1978; BLUMBERG, 1978).

As clinical tests of enzyme induction a number of techniques have been developed (OHNHAUS and PARK, 1979; SOTANIEMI et al., 1980) (see Table 11):

1. Determination of the rate of drug metabolism (e.g. antipyrine, phenacetin). The elimination kinetics indicate the activity of the P–450 dependent enzyme system.
2. Determination of the P–450 content in biopsy material from liver.
3. The concentration of endogenous metabolites as monitor for the efficiency of the MFO system (e.g. 6β-hydroxycortisol, D-glucaric acid).
4. Determination of the activity of hepatic enzymes in serum as indicator for enzyme induction (e.g. γ-glutamyltransferase).

4.4.1. *Classes of Inducing Agents*

4.4.1.1. Drugs

In Table 10 some drugs are listed which accelerate the metabolism of other drugs by induction of P–450 in man.

The antiepileptics and rifampicin data are compiled separately because their interaction with other drugs is of peculiar clinical relevance.

Antiepileptics

The enzyme inducing properties of antiepileptics are of special importance because they are applied chronically. The enzyme inducing activity of diphenylhydantoin (DPH) is demonstrated e.g. by the following data: decrease of half-life of antipyrine from 8.7 ± 1.9 to 4.6 ± 1.2 hours, increase of antipyrine clearance from 43.9 ± 17.7 to 92.3 ± 19.5 ml · min^{-1} (PIRTTIAHO et al., 1978).

Epileptics are often treated successfully with several antiepileptics. Due to the enzyme inducing activity of DPH the elimination rate of other antiepileptics changes: The half-life of carbamazepine after single oral dose amounts to 35.6 ± 15.3 hours, its clearance to 30.4 ± 19.8 ml · min^{-1}. In a combined therapy with DPH, ethosuximide, PB, primidone, mesuximide or valproic acid the carbamazepine half-life decreases to 8.16 ± 3.32 hours and the clearance increases to 114 ± 26.7 ml · min^{-1}. The enhanced metabolism of carbamazepine is caused both by autoinduction and induction by other antiepileptics (EICHELBAUM et al., 1979).

Reversely, the DPH metabolism can be induced by other drugs. CUCINELL et al. (1965) showed that after administration of 45 to 120 mg PB the serum concentration of DPH (in chronic therapy with 200 mg DPH p.d.) was considerably reduced. For this interaction between DPH and PB great individual differences

Table 10. Selection of some compounds inducing the microsomal metabolism of drugs in man (KAHL, 1971; CONNEY and BURNS, 1972; BOLT et al., 1975; HUNTER and CHASSEAUD, 1976; BACK et al., 1979; EICHELBAUM et al., 1979; HEUBEL and NETTER, 1979; PARKE, 1979)

Inducer	Enhanced metabolism
Phenobarbital and other barbiturates	Coumarin anticoagulants, diphenylhydantoin, quinine, desmethylimipramine, chlorpromazine, salicylic acid, phenylbutazone, antipyrine, aminopyrine, griseofulvin, cortisol oestradiol, testosterone
Glutethimide	Glutethimide, warfarin, antipyrine, aminopyrine
Phenylbutazone	Digitoxin, aminopyrine, cortisol
Antipyrine	Warfarin, cortisol
Meprobamate	Meprobamate
Diphenylhydantoin	Digitoxin, antipyrine, coumarin anticoagulants, cortisol, dexamethasone, thyroxin
Carbamazepine	Diphenylhydantoin, carbamazepine, warfarin
Haloperidol	Phenindione
Diazepam	Diazepam
Griseofulvin	Coumarin anticoagulants
Spironolactone	Antipyrine, cortisol
Prednisone	Cyclophosphamide
Rifampicin	Rifampicin, tolbutamide, antipyrine, hexobarbital, methadone, oral anticoagulants, digitoxin, cortisol, ethinyloestradiol, oestradiol, norethisterone

could be established by GARRETTSON and DAYTON (1970) and MORSELLI et al. (1971).

As epilepsy needs a long lasting therapeutic regimen for a great part of epileptics treatment of other diseases with drugs becomes often necessary. The inducing properties of the antiepileptics and hypnotics can therefore modify the elimination kinetics of those drugs. Analogous findings are available e.g. for coumarin (CUCINELL et al., 1965; LEVY et al., 1970) and digitoxin (SOLOMON et al., 1971; SOLOMON and ABRAMS, 1972).

Rifampicin

The enzyme inducing properties of rifampicin have also been proved to be clinically important (Table 11). Its inducing effect on the metabolism of oral contraceptives has been examined particularly thoroughly. As mentioned the microsomal monooxygenases of the liver hydroxylate also endogenous steroid hormones as well as exogenously applied ones (CONNEY, 1967). Induction of the P–450 system accelerates the metabolism of the steroids. In the case of endogenous hormones this effect is compensated to a certain extent by increased synthesis. For exogenously applied steroids, e.g. during a contraceptive therapy, effective levels may not be reached under usual doses if their metabolism is accelerated. Women treated with rifampicin and contraceptives have often intermenstrual or breakthrough bleeding and spotting (REIMERS and JEŽEK, 1971), and in single cases they even became pregnant (NOCKE-FINCK et al., 1973).

In vitro studies with liver microsomes of patients who were treated with 600 mg rifampicin per day for 6 to 10 days affirmed the enzyme induction. In comparison to controls the hydroxylation increases fourfold in the positions 2.4 (ring A) and 6.7 (ring B) of 17α-ethinylestradiol and estradiol (increase in the hydroxylation of 17α-ethinylestradiol from 237 $\pm$ 104 to 1108 $\pm$ 487 pmoles $\cdot$ mg^{-1} microsomal protein $\cdot$ min^{-1}). The liver P–450 content rises at the same ratio. Accelerated metabolism with concomitant reduced anticonceptive effectiveness is paralleled by an increase of irreversible binding of the metabolites to microsomal proteins (BOLT et al., 1974, 1975).

Furthermore, it could be evidenced

Table 11. Influence of rifampicin on the pharmacokinetics of norethisterone and antipyrine and urinary excretion of 6β-hydroxycortisol and plasma γ-glutamyltranspeptidase activity in 8 women (BACK et al., 1979; reproduced with permission from Springer-Verlag, Heidelberg)

	During rifampicin	After rifampicin	Significance p
Norethisterone			
A.U.C.* (ng $\cdot$ ml^{-1} $\cdot$ h^{-1})	21.9 $\pm$ 5.9	37.8 $\pm$ 13.1	$<$ 0.01
plasma half-life (h)	3.2 $\pm$ 1.0	6.2 $\pm$ 1.7	$<$ 0.0025
Antipyrine			
plasma half-life (h)	7.3 $\pm$ 2.1	9.8 $\pm$ 1.7	$<$ 0.05
clearance (ml $\cdot$ h^{-1} $\cdot$ kg^{-1})	72.4 $\pm$ 24.3	42.6 $\pm$ 3.9	$<$ 0.02
6β-Hydroxycortisol excretion (μg $\cdot$ day^{-1})	1177 $\pm$ 313	341 $\pm$ 65	$<$ 0.001
γ-Glutamyltranspeptidase activity (I.U.)	29.6 $\pm$ 20.0	18.0 $\pm$ 6.6	$<$ 0.05

*A.U.C. area under the curve

that rifampicin also accelerates via enzyme induction the metabolism of digitoxin (PETERS et al., 1974), warfarin (O'REILLY, 1975), DPH (WHITING et al., 1976) and tolbutamide (ZILLY et al., 1975).

4.4.1.2. Nutrition, Alcohol and Cigarette Smoke

Nutritional constituents, alcohol and cigarette smoke are also capable to induce drug metabolizing enzymes. Recently such interrelationships have been thoroughly studied.

As influencing factors of the nutrition the composition of the food, the kind of preparation, natural constituents or exogenous contaminants have been described.

High protein low carbohydrate diet shortens the half-life of antipyrine from 17.5 to 9.2 hours (ALVARES et al., 1976). In comparison to a fatty-rich diet a protein-rich nutrition accelerates also the metabolism of antipyrine (ANDERSON et al., 1979).

The influence of the preparation mode has been ascertained especially for charcoal-broiled beef (Table 12). The enzyme inducing property originates from PAHs. LIJINSKY and SHUBIK (1964) found 8 μg benzo(a)pyrene per kg charcoal-broiled meat besides other PAHs. The plasma concentration 2 hours after application of 900 mg phenacetin is decreased from a control value of 925 to 163 ng · ml^{-1} after eating charcoal-broiled beef (PANTUCK et al., 1976).

Cabbage, especially Brussels sprouts, is known to have enzyme inducing properties attributed to the content of indole compounds. Under cabbage diet antipyrine metabolism and phenacetin pharmacokinetics are accelerated (PANTUCK et al., 1979).

A shortening of the half-life of antipyrine under chronic alcohol consumption was described by VESELL et al.

Table 12. Effect of charcoal-broiled beef diet on the mean plasma half-life, volume of distribution and metabolic clearance rate. Control diet was given for 7 days, charcoal-broiled diet for 5 days. Eight subjects received 1.8 mg/kg antipyrine and 5.0 mg/kg theophylline, respectively at the end of the test period (ALVARES et al., 1979; reproduced with permission from Marcel Dekker, New York)

Measurement	Control diet 1	Charcoal-broiled diet	Control diet 2
Plasma half-life (h)			
Antipyrine	13.7 ± 1.3	10.7 ± 1.2	13.3 ± 1.1
Theophylline	6.0 ± 0.5	4.7 ± 0.4	6.4 ± 0.7
Volume of distribution (liters)			
Antipyrine	51.7 ± 2.1	54.1 ± 2.3	51.8 ± 1.5
Theophylline	34.7 ± 1.5	34.5 ± 1.5	36.1 ± 1.7
Metabolic clearance rate (ml · min^{-1})			
Antipyrine	45.8 ± 4.2	63.2 ± 6.4	47.5 ± 4.5
Theophylline	69.8 ± 6.3	90.0 ± 9.3	69.1 ± 6.6

(1971). However, there exist controverse opinions concerning the relation between the alcohol doses and the extent of enzyme induction (PARKE, 1979).

The enzyme inducing effect of cigarette smoke has been examined extensively in animals as well as in human smokers and it is ascribed both to PAH and nicotine (BECKETT and TRIGGS, 1967). Table 13 summarizes some drugs whose metabolism is accelerated by smoking.

Detailed studies have been performed to analyze the biological consequences of cigarette smoke on enzyme activities in placentas of human smokers. Benzo(a)pyrene and zoxazolamine hydroxylases in placentas of smokers have a significantly higher activity than in those of nonsmoking women (39- and 13.4-fold, respectively; cf. Table 14, KAPITULNIK et al., 1976). 3-Methyl-4-monomethylaminoazobenzene demethylase is also induced in the placenta of smokers, whereas no activity was observed in non-smokers (WELCH et al., 1969). Benzo(a)pyrene

Table 13. Influence of smoking on in vivo biotransformation in man (JUSKO, 1979; reproduced with permission from Marcel Dekker, New York)

Compound	Major biotransformation pathway
Increased metabolic rate in smokers	
Nicotine	Hydroxylation α to N of cyclic amine
Phenacetin	O-Dealkylation
Antipyrine	Aliphatic hydroxylation
Theophylline	N-Demethylation, purine oxidation
Imipramine	N-Demethylation
Pentazocine	Allylic hydroxylation
Not affected by smoking	
Diazepam	N-Demethylation
Meperidine	N-Demethylation
Phenytoin	Aromatic hydroxylation
Nortriptyline	N-Demethylation
Warfarin	Aromatic hydroxylation
Ethanol	Alcohol dehydrogenation

monooxygenase of fetal liver is not induced by cigarette smoking pregnant women (JUCHAU, 1971). As the oxidative metabolism in the placenta is important for the hormone level and therefore also for the prenatal development, the inducing effect of smoking could provoke damages in the fetal development, which

Table 14. Effect of cigarette smoking on benzo(a)pyrene hydroxylase and zoxazolamine hydroxylase activity in human placenta. Smokers smoked 7 to 25 cigarettes daily during pregnancy (KAPITULNIK et al., 1976; reproduced with permission from Mosby Comp., St. Louis)

Proband	Benzo(a)pyrene hydroxylase (nmoles phenols formed·g^{-1}·h^{-1})	Zoxazolamine hydroxylase (nmoles $^3[H]_2O$ formed·g^{-1}·h^{-1})
Non-smokers ($n = 11$)	0.71 ± 0.08	0.43 ± 0.05
Smokers ($n = 13$)	27.91 ± 8.17	5.75 ± 1.50

are independent of the toxic effects of nicotine and other smoke constituents (HUNTER and CHASSEAUD, 1976).

4.4.1.3. Environmental Xenobiotics

Besides drugs, food constituents and luxuries a multitude of chemical compounds can act as potential enzyme inducers, for example PCB, PBB, PAH and DDT. The enzyme inducing activity of PAH has been discussed already in connection with charcoal-broiled beef and cigarette smoke.

PCBs and PBBs are compounds which are produced in large scale by the chemical industry. In workers who are occupied in aroclor 1016 (a mixture of di-, tri- and tetrachloro isomers of biphenyl) producing factories and other PCBs show significant alterations in the metabolizing rate of antipyrine (ALVARES et al., 1977; KRAMPL and KONTSEKOVÁ, 1978). The halogen derivatives were also ascertained in the food chain: in slaughter animals, in sea fish and in vegetarian food (IARC Monographs, 1978).

In workers occupied with the production of DDT enzyme induction has been observed as well (Table 15). In general, however, the environmental contamination by this pesticide does not reach the concentration necessary for P-450 induction in man (REMMER et al., 1979).

Estimating the importance of potential enzyme inducers two exposition groups have to be differentiated: particularly exposed persons (e.g. employees in certain occupational groups) and the whole population, respectively.

Table 15. Half-life of antipyrine in workers involved in the production of DDT and lindane (KOLMODIN et al., 1969)

	Exposed group	Control group
Mean ± SE (h)	7.7 ± 2.6	13.1 ± 7.5
Range (h)	2.7 – 11.7	5.2 – 35.0

4.4.2. *Practical Consequences of Enzyme Induction*

Besides the cited single examples it shall be tried to sum up possible consequences of enzyme induction under practically important points of view.

4.4.2.1. Influence on Biotransformation of Exogenous Substances

Many enzyme inducers are able to accelerate their own metabolism (autoinduction). This phenomenon has been described for barbiturates, glutethimide, meprobamate, nitroglycerol, phenylbutazone and tolbutamide (RITSCHEL, 1978). It is clinically responsible for the development to tolerance (pharmacokinetic tolerance).

In some cases biotransformation of foreign substances does not lead to their biological inactivation. Biologically nearly ineffective substances can be transformed to compounds acting as drugs or poisons. Cyclophosphamide is an important example, the cancerostatic effect of which is intensified by long-term prednisone treatment due to the enzyme induction (PARKE, 1979). Also the activation of premutagens and precarcinogens can be enhanced by enzyme induction (see 5. and Table 2).

4.4.2.2. Influence on Endogenous Metabolic Processes

P-450 dependent monooxygenases are not only important for the biotransformation of xenobiotics but also for the synthesis and metabolism of cholesterol and steroid hormones as well as for a number of other endogenous compounds. KRUSE (1968) and HUNTER et al. (1971) observed an increased vitamin D metabolism in consequence of the enzyme induction by anticonvulsants. The patients exhibited a vitamin deficiency syndrome, which in children deteriorated to a complete rachitis (KRUSE, 1968).

In new-born babies of epileptic mothers who received antiepileptics during pregnancy a vitamin K deficiency with raised bleeding tendency was observed. This deficiency, too, is connected with enzyme induction due to the therapy with anticonvulsants (DAVIES, 1970; MOUNTAIN et al., 1970).

Recently besides the already mentioned influence on the steroid hormone level a hypertriglyceridemia in consequence of enzyme induction has been observed repeatedly (GOLDBERG, 1980). MARTIN et al. (1976) related the increased serum triglyceride concentration under oral contraceptives to their enzyme inducing properties.

The microsomal enzyme induction is of further importance with respect to long-term treatment. LABADARIOS et al. (1978) observed in patients treated with anticonvulsants, phenothiazines or tricyclic antidepressants up to 5 years a decrease of the enzyme induction simultaneously to the occurrence of folic acid deficiency. The authors suggest that normal nutrition is not sufficient to meet the folic acid demand for the de novo enzyme synthesis under long-term therapy with inducers. This conclusion was supported by animal experiments. The decrease of the enzyme induction in long-term therapy with constant dosage involves the danger of overdosing.

Table 16. Concentration of total serum bilirubin in infants whose mothers were treated with 30–120 mg phenobarbital/day for ≥ 2 weeks in comparison to controls (MAURER et al., 1968; reproduced with permission from Lancet, London)

Day	Serum bilirubin (mg · 100 ml^{-1})	
	control ($n = 16$)	phenobarbital ($n = 12$)
1	3.8 ± 0.4	2.1 ± 0.4
2	5.0 ± 0.6	2.5 ± 0.4
3	5.7 ± 0.7	2.2 ± 0.4
4	5.2 ± 0.8	1.8 ± 0.4

The attempt to utilize the enzymes induction as a therapeutical principle for the acceleration of bilirubin conjugation in the treatment of hyperbilirubinemia in new-borns has not found broad application because of side effects. Treatment of pregnant women with PB several days before delivery considerably diminishes the serum bilirubin concentration of new-borns (Table 16). Such a therapy, however, provokes side effects, especially respiratory depression and aspiration. Furthermore, alterations in endogenous metabolic processes (steroid hormones) cannot be excluded completely (GLADTKE and HEIMANN, 1978).

4.4.2.3. Influence on Biochemical Diagnostic Parameters

Enzyme induction influences several clinical-chemical parameters, which may result in an impairing of their diagnostic reliability. On the other hand these changed parameters can be used for the detection of enzyme induction (GOLDBERG, 1980).

1. The reliability of the elimination kinetics of drugs (e.g. antipyrine) as parameter of the liver function is limited during enzyme induction.
2. Serum enzyme activities, indicative for hepatic diseases (e.g. γ-glutamyltransferase, alkaline phosphatase) are increased in consequence of enzyme induction. This can lead to errors, when serum enzyme activities are used as diagnostic parameters.
3. Endogenous metabolites (D-glucaric acid) and the activity of hepatic enzymes in serum (e.g. γ-glutamyltransferase) can indicate the degree of enzyme induction. They may serve as controls in the treatment with inducing drugs (e.g. antiepileptics).

5. Role of Cytochrome P-450 in the Metabolism of Some Substances of Economic, Pharmacologic and Toxicologic Significance

5.1. *Introductory Remarks*

As already mentioned MFO participates in the metabolism of numerous lipophilic substances. The water-soluble metabolites (epoxides, hydroxy-compounds) are excreted to a certain degree mainly as conjugates after binding to GSH or glucuronic acid. The epoxides or diol epoxides formed by the MFO-system, however, are strongly electrophilic and can covalently bind therefore to the nucleophilic sites of nucleic acids and proteins. Due to these reactions they become cytotoxically, mutagenically or carcinogenically effective. N-hydroxylations, catalyzed by the MFO-system as well as by P–450 independent amine oxidases, can also generate biologically active metabolites. In several cases some further enzymes localized in the endoplasmic reticulum, in the nucleic envelope or in the cytosol, respectively, are involved in the biotransformation. The biological effect of a substance or of its active metabolites finally depends on the specific reactions with the respective target molecules. DNA alterations may be reversed by repair processes.

For a comprehensive review of drug metabolism see PFEIFER (1975, 1977, 1979).

5.2. *Polycylic Aromatic Hydrocarbons*

Among the PAH there are a number of substances, which are well known as carcinogens, such as benzo(a)pyrene and 7,12-dimethylbenz(a)anthracene. PAH become biologically effective only after enzymatic activation. Different enzymes, participating in their metabolism, are

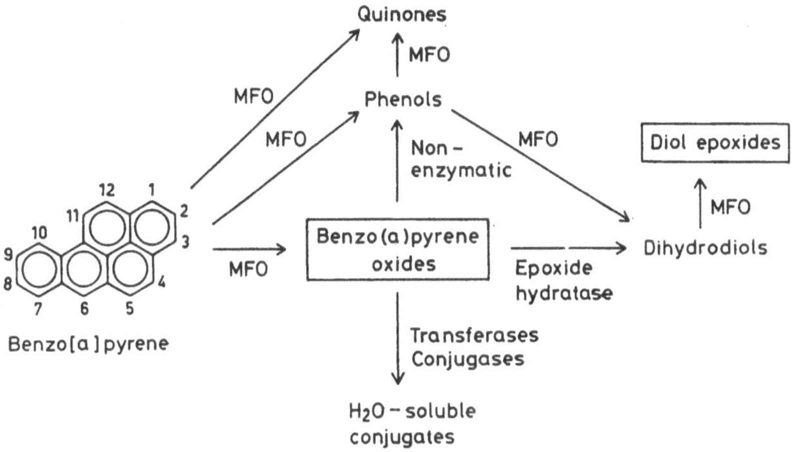

Fig. 6. Metabolism of benzo(a)pyrene (GELBOIN et al., 1976; reproduced with permission from International Agency for Research on Cancer, Lyon)

responsible for the formation of organ-specific metabolite patterns (FOX et al., 1975; GELBOIN et al., 1976). At first oxidation is catalyzed by certain P–450 subspecies (P–448 or P_1–450). Thereby various arene oxides are formed which are converted in further steps into hydroxylated derivatives. Because of the relatively high substrate specificity the PAH-oxidizing enzyme system is called aryl hydrocarbon hydroxylase (AHH). In the case of benzo(a)pyrene some of the arene oxides are strongly electrophilic and can therefore react with nucleic acids and proteins. Arene oxides are characterized by strong, direct, mutagenic effects and by a high activity concerning the malignant transformation of cell cultures. Their carcinogenic effect on mouse skin is lower than that of the parent substance (for reviews see SIMS and GROVER, 1974; GELBOIN et al., 1976; LEVIN et al., 1976; LAMPE and BUTSCHAK, 1978). The arene oxides can be metabolized to dihydrodiols by epoxide hydratase or can be transformed into water-soluble conjugates by various transferases. By a second P–448 catalyzed oxidation step diol epoxides can emerge from the dihydrodiols, some of which are strongly mutagenic and carcinogenic (MALAVEILLE et al., 1975; GELBOIN et al., 1976; SIMS, 1976; SEIDEGÅRD et al., 1977). Analysis of the DNA-adducts using radioactive precarcinogenic PAH revealed that certain stereoisomers of the diol epoxides can become ultimate mutagens or carcinogens (SIMS et al., 1974). The 7,8-diol-9,10-epoxides of benzo(a)pyrene are of particular importance. There are a number of indications that (+)-7β,8α-dihydroxy-9,10-epoxy-7,8,9,10-tetrahydrobenzo(a)pyrene acts as ultimate carcinogen in mouse skin (SLAGA et al., 1979). It is interesting that also in vitro only certain isomers of the diol epoxides are bound selectively by double-strand DNA, whereby the binding to DNA correlates with the mutagenic and carcinogenic effect (MEEHAN and STRAUB, 1979). Diol epoxides of other PAH are also known as ultimately acting metabolites (LEVIN et al., 1978). The metabolism of benzo(a)pyrene is represented schematically in Fig. 6.

Besides the group of diol epoxides hydroxymethyl derivatives and free radicals of the PAH probably play also a role as ultimate carcinogens. The MFO system is involved in their formation as well (CAVALIERI et al., 1978).

5.3. Halogenated Compounds

5.3.1. Halogenated Aliphatic Hydrocarbons

5.3.1.1. Chloroform

In the past chloroform ($CHCl_3$) was employed as anesthetic. Recently it is predominately used as solvent and for the synthesis of chlorodifluoromethane (IARC Monographs, 1979d). Occasionally it can be found as pollutant in drinking water (HOGAN et al., 1979). Chloroform is hepato- and nephrotoxic and has carcinogenic properties. Tumors in liver and kidney of mice and rats have been observed after chloroform administration (REITZ et al., 1978; IARC Monographs, 1979d). Epidemiologic investigations have evidenced a correlation between chloroform concentration in drinking water and cancer mortality in man (rectum-intestinum and bladder) (HOGAN et al., 1979). Mutagenicity tests of chloroform in presence of liver microsomes proved to be negatively for *Salmonella typhimurium* TA 1535 and 1538 (UEHLEKE et al., 1977). In mice teratogenic effect was detected (MURRAY et al., 1979).

Chloroform is oxygen- and NADPH-dependently metabolized by P–450 in vitro (SIPES et al., 1977). The K_s value for binding to rat liver microsomes under aerobic conditions was 3.0 mM (COX et al., 1976). ^{14}C-chloroform is bound irreversibly to proteins and lipids of microsomes from rats and mice whereas the ribosomal RNA is not labelled (SIPES et al., 1977; UEHLEKE et al., 1977). CO and SKF 525 A decrease the amount of covalently bound ^{14}C in proteins in vitro, pretreatment of the animals with PB favours the binding (SIPES et al., 1977; CRESTEIL et al., 1979). Chloroform is also activated to electrophilic metabolites by nuclear preparations. The metabolites are bound covalently to nuclear proteins but not to nucleic acids (DIAZ GOMEZ and CASTRO, 1980b).

Phosgen has been proved as reactive metabolite of chloroform. It is formed by the P–450 dependent oxidation of the C—H-linkage, generating trichloromethanol as an intermediate, which spontaneously disintegrates to HCl and phosgen. Phosgen reacts further with nucleophilic groups of biomacromolecules or with cysteine to form 4-carboxy-thiazolidine-2-one. A reaction with histidine is possible, too (MANSUY et al., 1977; POHL et al., 1977; CRESTEIL et al., 1979) (Fig. 7). GSH inhibits the binding to nucleophilic molecules (SIPES et al., 1977).

Chloroform inhibits the microsomal p-nitroanisole demethylase in vitro. This inhibition is mainly caused by covalent binding of chloroform metabolites to microsomal proteins and to a minor extent by oxidation of lipids. The binding of metabolites to microsomes of male rats is more pronounced than that to

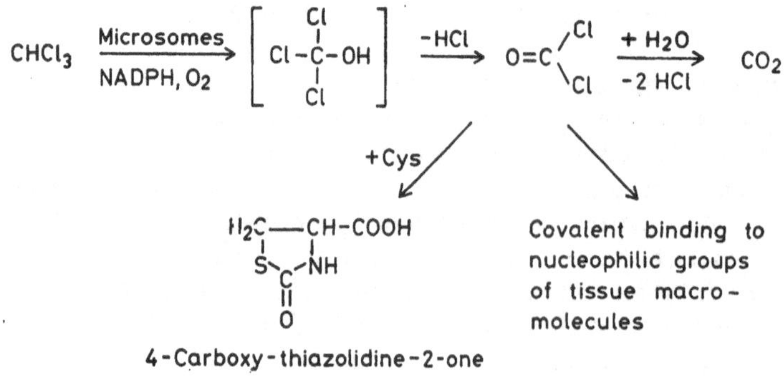

Fig. 7. Metabolism of chloroform by microsomes (MANSUY et al., 1977; reproduced with permission from Academic Press, New York)

Table 17. Irreversible binding of [^{14}C]chloroform metabolites to rat and human liver microsomes (CRESTEIL et al., 1979; reproduced with permission from Elsevier/North-Holland Biomedical Press, Amsterdam)

Species and treatment	Apparent K_m (mM)	Binding rate at 1 mM $CHCl_3$ (nmoles · mg^{-1} microsomal protein · 5 min^{-1})
Untreated rats	0.7–0.8	1.1 ± 0.3
Phenobarbital-treated rats	1.4 ± 0.1	♂ 3.2 ± 1.0 ♀ 1.9 ± 0.2
Humans	0.5–0.7	0.2–0.7

females (see Table 17). PB-pretreatment of rats results in increases of K_m, K_s and V_{max}. The mechanism of hepatotoxicity in man probably is similar, as the reactive metabolites are bound to human liver microsomes, too (CRESTEIL et al., 1979).

The splitting of the C—H-linkage is the rate-limiting step for both the formation of phosgen and hepatotoxicity (POHL and KRISHNA, 1978). In the formation of this metabolite a deuterium isotope effect could be revealed. In rats the effect of $CHCl_3/CDCl_3$ was determined in vivo to be 2.10 ± 0.06 (POHL et al., 1979). This result corresponds to the values obtained in vitro and also explains the lower toxicity of $CDCl_3$ in comparison to $CHCl_3$ (POHL and KRISHNA, 1978; McCARTY et al., 1979).

The metabolism of chloroform was repeatedly investigated in vivo. Pretreatment of rats with PB results in increased lipid peroxidation and centrolobular liver necrosis (SCHOLLER, 1970; BROWN jr., 1972). This process is paralleled by a decrease of GSH, probably by binding to reactive chloroform metabolites (DOCKS and KRISHNA, 1976).

The nephrotoxicity of chloroform is different in certain mice strains and genetically determined. It was shown that the amount of metabolites bound to lipids and proteins of microsomes, mitochondria and cytosol biomacromolecules of DBA/2J male mice is greater than for C57BL/6J males. These findings are in good agreement with the higher nephrotoxic and letal effect of chloroform in DBA/2J mice (CLEMENS et al., 1979). Inducers of P-450 influence the chloroform metabolism differently. So PBB raises its hepato- and nephrotoxicity in mice, evidenced by a decrease of the LD_{50} in the first 24 hours from 1.28 to 0.62 ml chloroform/kg after pretreatment with 100 ppm PBB (KLUWE and HOOK, 1978). 3-MC, TCDD and PCB diminish the nephrotoxicity whereas PB is assumed to be without effect (KLUWE et al., 1978). In contrary to this finding the covalent binding of chloroform metabolites in the liver is raised by PB treatment, but diminished in the kidney. After enzyme induction possibly the major part of chloroform is metabolized in the liver and only to a minor extent in the kidney (ILETT et al., 1973). However, the nephrotoxic metabolites of chloroform are formed in the kidney itself and are not tansferred from the liver to the kidney by blood circulation (KLUWE et al., 1978).

In vivo studies in man have proved CO_2 as a metabolite of chloroform. The metabolism is supposed to proceed according to the reaction

$$2CHCl_3 + 2H_2O + O_2 \rightarrow 2CO_2 + 6HCl$$

The cleavage of the C—Cl bond, however, and the intermediate occurrence of methylene dichloride as initial reaction product could be not ascertained (FRY et al., 1972). As another metabolite of the halomethanes bromoform, iodoform and chloroform CO was proved in rats. The velocity of CO formation decreases according to the sequence iodoform > bromoform > dibromochloromethane > dichlorobromomethane > chloroform. Analogously as in the case of $CHCl_3$ also $CDBr_3$ is more slowly metabolized than $CHBr_3$ (ANDERS et al., 1978).

5.3.1.2. Carbon Tetrachloride

Carbon tetrachloride is strongly hepatotoxic and induces tumors such as hepatocellular carcinomas and mammary adenocarcinomas in mice and rats. Carbon tetrachloride is foetotoxic (IARC Monographs, 1979c), but no mutagenic effects could be shown in bacteria (UEHLEKE et al., 1977).

The toxic effect is based on metabolic activation to reactive intermediates by P–450 and on lipid peroxidation (RECKNAGEL, 1967). Due to the P–450 dependence the toxicity of carbon tetrachloride is influenced by many factors such as nutrition and enzyme induction (JUDAH et al., 1970; LITTERST et al., 1973) (Table 18). In the lung of mice carbon tetrachloride causes striking morphologic changes in the Clara cells and occasionally cellular necrosis (BOYD et al., 1980).

Carbon tetrachloride is bound to microsomes showing a type 1 difference spectrum. With rabbit liver microsomes K_s amounts to 0.5 mM. After reduction with dithionite the affinity increases more than one order of magnitude ($K_s = 20$ to 40 μM) (UEHLEKE et al., 1973).

After administration of ^{14}C-carbon tetrachloride to rats or during incubation with liver microsomes the radioactivity is irreversibly bound to proteins and lipids, whereas ribosomal RNA is not labelled (REYNOLDS, 1967; SIPES et al., 1977; UEHLEKE et al., 1977). Nuclear rat liver preparations are able to activate carbon tetrachloride anaerobically to reactive metabolites which bind to lipids and proteins (DIAZ GOMEZ and CASTRO, 1980a). There exists a good correlation between the intensity of carbon tetrachloride activation and P–450 content in various tissues of rat (VILLARRUEL et al., 1977). The covalent binding of the carbon tetrachloride metabolites to microsomal proteins is decreased by dithiocarb and SKF 525A (SIEGERS et al., 1978).

Carbon tetrachloride administration to rats results in a decrease both of the concentration of liver P–450 and the respective enzyme activities. Pretreatment of the animals with PB increases this effect as well as the lipid peroxidation (REYNOLDS et al., 1972; VAINIO et al., 1976). In vitro P–450 is destructed also anaerobically independent of lipid peroxidation (YAMAZOE et al., 1979; DE GROOT and HAAS, 1980).

P–450 catalyzes the biotransformation of carbon tetrachloride both oxidatively and reductively and furthermore generates free radicals or carbenes. The toxic effect is probably caused by the trichloromethyl free radical $\cdot CCl_3$ (SLATER, 1966; ULLRICH, 1977) which has been demonstrated by means of the spin-trapping tech-

Table 18. The influence of nutrition and enzyme induction on microsomal hydroxylation and toxicity of carbon tetrachloride in rats (JUDAH et al., 1970; reproduced with permission from American Journal of Medicine, New York)

Diet	Treatment	Aminophenazone demethylase (μmoles · g^{-1} liver · h^{-1})	LD_{50} (ml CCl_4/kg body weight)
Protein-free	—	0.1	14.7
Protein-free	DDT	0.6	4.3
Standard-diet	—	0.7	6.4
Standard-diet	Phenobarbital	2.6	0.5

nique in in vivo experiments (Lai et al., 1979; Poyer et al., 1980) as well as in liver microsomes (Poyer et al., 1978, 1980). According to other investigations the ESR-signal is attributed to a carbon-centered dienyl lipid radical (Kalyanaraman et al., 1979).

Anaerobically the $\cdot CCl_3$ radical may undergo several reactions, e.g. formation of chloroform (Reiner et al., 1972), binding to proteins and lipids (Sipes et al., 1977) or reductive dechlorination to carbon monoxide, presumably via the carbene CCl_2 (Wolf et al., 1977). CO, O_2, metyrapone and SKF 525A inhibit the formation of chloroform and the covalent binding to proteins (Reiner et al., 1972; Sipes et al., 1977). Under physiological conditions the chloroform formation is probably very low. Aerobically the free radical presumably is transformed via the instable intermediate Cl_3COH to phosgen. CO_2 formation could result from hydrolytic dechlorination of phosgen (Shah et al., 1979).

Moreover, the free radical arising from the metabolism of carbon tetrachloride attacks the double bonds of unsaturated fatty acids, thus initiating their oxidative destruction (Recknagel and Ghoshal, 1966; Klaassen and Plaa, 1969). The extent of lipid peroxidation can be determined in vivo by the expiration of ethane and pentane (Lindstrom and Anders, 1978; Sagai and Tappel, 1979), and in vitro by determination of the diene conjugates formed (Recknagel and Ghoshal, 1966). The mechanism of lipid peroxidation has been elucidated (Danni et al., 1977). However, the covalent binding of the metabolites of carbon tetrachloride to proteins and lipids may be more important for the toxic effect of carbon tetrachloride than the lipid peroxidation (Villarruel et al., 1977; De Toranzo et al., 1978).

The much higher toxicity of carbon tetrachloride compared with chloroform or carbon tetrafluoride is due to the lower dissociation energy which favours the cleavage of the C—Cl-bond. Among the halogenated methane derivatives only carbon bromide trifluoride has a higher toxicity than carbon tetrachloride (Table 19) (Koch et al., 1974). Free radical scavengers such as promethazine, propyl gallate and (+)-catechin inhibit the lipid peroxidation (Poli et al., 1978). Vitamin E has a protecting effect, too (Sagai and Tappel, 1978).

5.3.1.3. Halothane

Halothane (2-bromo-2-chloro-1,1,1-trifluoroethane) is used as a narcotic since 1956. A transient increase of the serum xanthine oxidase activity — a highly sensitive marker for acute liver damage —

Table 19. Relation of bond dissociation energies of haloalkanes to their effectiveness as pro-oxidants on rat liver microsomal lipid peroxidation (Koch et al., 1974; reproduced with permission from Pergamon Press, Oxford)

Reaction	Bond dissociation energy (kcal/mole)	Relative molar activity*	Pro-oxidant activity relative to CCl_4**
$CHCl_3 \rightarrow \cdot CCl_3 + \cdot H$	90	7	10
$CCl_4 \rightarrow \cdot CCl_3 + \cdot Cl$	68	100	100
$CBrCl_3 \rightarrow \cdot CCl_3 + \cdot Br$	49	3,000	20,000

* Relative molar activity in stimulating production of malonaldehyde in vitro.

** Pro-oxidant activity was determined in vitro with a liver microsome-supernatant system supplemented with a NADPH generating system and glutathione.

was found in a relatively large number of patients (GILER et al., 1977). Fortunately only in one of approximately 30,000 cases a deleterious massive hepatic necrosis is observed after anesthesia (BUNKER et al., 1969).

Liver damage after halothane application has also been described in laboratory animals. Repeated exposition to halothane leads to liver necrosis in guinea pigs (HUGHES and LANG, 1972). In rats it occurred only after pretreatment with PB (STENGER and JOHNSON, 1972; REYNOLDS and MOSLEN, 1974; RAO, 1977; JEE et al., 1980) or with aroclor 1254 (REYNOLDS et al., 1976). The liver necrosis is accompanied by a decrease of the liver P–450 content.

After inhalation of ^{14}C-halothane by rats the main portion of radioactivity is bound irreversibly to liver proteins and a small portion to lipids. DNA (RAO, 1977) and RNA (UEHLEKE et al., 1977) are not labelled.

Mutagenicity tests of halothane on *Salmonella typhimurium* and *Escherichia coli* were negative, also in the presence of rat or rabbit liver microsomes (IARC Monographs, 1976; UEHLEKE et al., 1977). Other authors, however, found some metabolites of halothane genotoxic and mutagenic in *Bacillus subtilis* and *Salmonella typhimurium*, respectively (GARRO and PHILLIPS, 1978; SACHDEV et al., 1980). Carcinogenic activity of halothane could not be demonstrated till now (IARC Monographs, 1976).

Halothane is metabolized by P–450. There exists a good correlation between the P–450 content and the formation of the metabolite trifluoroacetic acid. CO and SKF 525A inhibit the biotransformation (KARASHIMA et al., 1977). P–450 in microsomes shows a type 1 difference spectrum in the presence of halothane. Under aerobic conditions the K_s value of halothane binding to liver microsomes of rabbits pretreated with PB amounts to 0.3 mM, and significantly decreases after reduction with dithionite to 3–6 μM. During the anaerobic incubation with NADPH-reduced microsomes metabolites of ^{14}C-halothane are covalently bound to proteins. Metyrapone, CO and O_2 interfere with the halothane binding whereas SKF 525A does not (UEHLEKE et al., 1973).

Halothane can be metabolized both oxidatively and reductively. In the first case trifluoroacetic acid is formed via an unstable hydroxylated intermediate and subsequent hydrolysis (MANSUY et al., 1974):

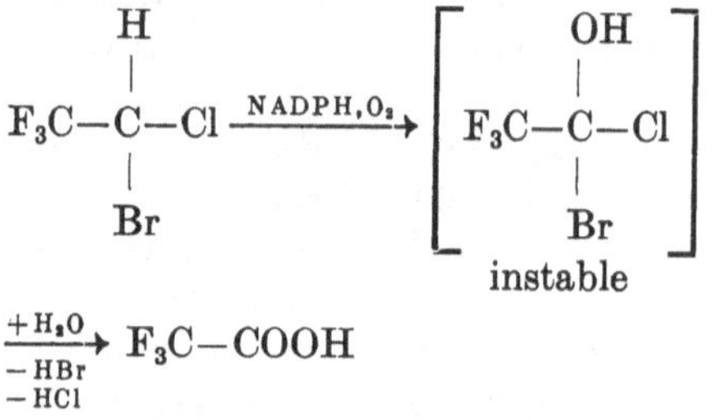

VAN DYKE and WOOD (1973) propose a metabolic pathway in which chlorine and bromine are split off and the formed trifluoroacetaldehyde is oxidized to trifluoroacetic acid:

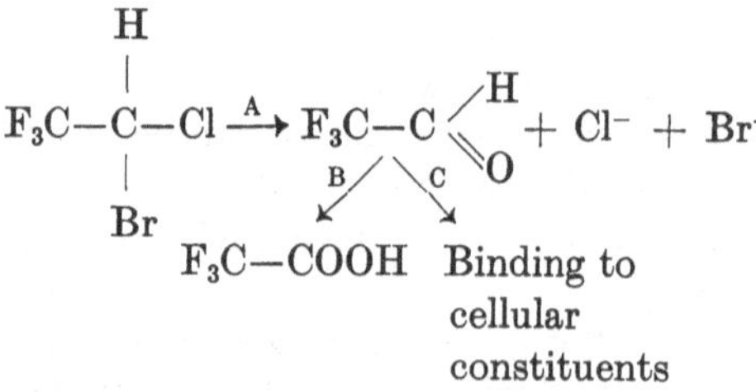

The reaction pathways A and B are enzymatic ones and require O_2, while C probably is non-enzymatic. Trifluoroacetaldehyde is highly reactive and presumably is bound to cellular macromolecules. Under decrease of the blood supply of the liver the binding of labelled halothane metabolites was increased. Pathway A possibly reacts faster than B. However, trifluoroethanol and trifluoroacetaldehyde have not been proved directly as metabolites up to now. In man after administration of ^{14}C-trifluoroethanol more than 80 per cent of the radioactivity were found as trifluoroacetate in the urine indicating that trifluoroethanol should be an intermediate of the halothane metabolism (Cascorbi and Blake, 1971).

Dithionite-reduced liver microsomes convert halothane into a trifluoromethylcarbene complex which is supposed to be formed via a two electron reduction step of halothane:

$$F_3C{-}\underset{\displaystyle Br}{\overset{\displaystyle H}{\underset{|}{\overset{|}{C}}}}{-}Cl \xrightarrow[-Cl^-,\,-Br^-]{+2e} F_3C{-}\bar{C}H$$

This carbene binds to microsomal proteins and also destroys P–450 (Mansuy et al., 1974).

Furthermore halothane increases the microsomal lipid peroxidation in PB-induced rats indicating the formation of free radicals (Brown Jr., 1972; Reynolds and Moslen, 1975) as already assumed by Stier (1968). The formation of conjugated dienes is correlated with the binding of halothane metabolites to microsomal phospholipids under nitrogen atmosphere (Wood et al., 1976).

It is assumed that the metabolites formed by reductive biotransformation are potentially more hepatotoxic than those formed via the oxidative pathway. The amount of covalently bound metabolites is increased under hypoxic conditions. In hypoxic rats (7% O_2) covalent binding of ^{14}C-halothane metabolites to microsomal lipids increases threefold in comparison to 40% O_2. The lipid/protein binding ratio in control animals revealed an average value of 0.76, in hypoxic animals of 3.24. At the same time the plasma fluorine level increases. The fluorine release from halothane is catalyzed by the reductive metabolism (Widger et al., 1976).

Jee et al. (1980) observed that halothane in PB-induced male rats is hepatotoxic only under hypoxic conditions (14% O_2 in the respiratory air). The degree of morphological and functional alterations correlates with the concentrations both of hepatic P–450 and of inspired halothane. As hepatotoxicity is dose-dependent, a hypersensitivity reaction as a cause of the toxicity is unlikely. Halothane reacts toxically already after a short exposition. A longer exposition intensifies its toxicity not significantly because halothane possibly inhibits its own biotransformation. SKF 525A and metyrapone reduce hepatotoxicity if administered prior to halothane. Cysteine, cystamine and N-acetylcysteine are protectives when administered 4 hours after ending anesthesia.

Cohen et al. (1975) found three major metabolites of halothane in human urine: trifluoroacetic acid, N-trifluoroacetyl-2-aminoethanol and N-acetyl-S-(2-bromo-2-chloro-1,2-difluoroethyl)-L-cysteine. The occurrence of cysteine and ethanolamine conjugates implies the presence of reactive intermediates. The cysteine conjugate is supposed to result from the reductively formed bromochloro-difluoroethylene with GSH, while the ethanolamine complex is produced by the reaction of the intermediately occurring trifluoroacetyl radical with phosphatidylethanolamine of the membrane lipids. The reaction pathway to trifluoroacetic acid is oxidative (Fig. 8). The renal elimination is probably caused by degraded protein macromolecules containg covalently bound halothane metabolites and hepatic lipids.

The active factors provoking halothane hepatitis in man are still unknown. There is no direct correlation between the extent of covalent binding and hepatic necrosis. Furthermore it has been discussed that the metabolites formed under reductive or oxygen-deficient conditions might induce hypersensitivity accelerating hepatic cellular destruction (Brown and Sipes, 1977; Vaughan et al., 1978). Moreover,

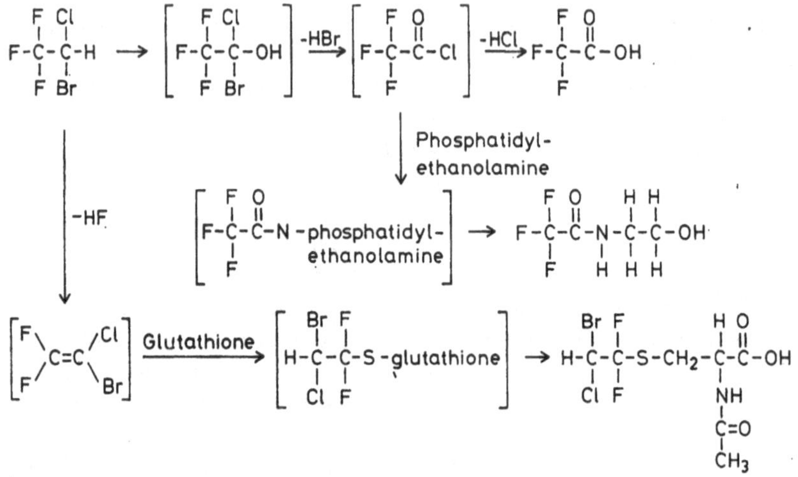

Fig. 8. Proposed metabolic routes for the formation of the three major urinary metabolites of halothane (Cohen et al., 1975; reproduced with permission from Lippincott/Harper and Row, Philadelphia)

as biotransformation of halothane is genetically controlled therefore individual differences of halothane toxicity could be caused by distinct enzyme patterns in man (Cascorbi et al., 1971).

Halothane liver damages in man occur only in few cases. They do not justify its general refusal as anesthetic. However, it is recommended to use other anesthetics for patients who, after halothane anesthesia, reacted with icterus or functional disorders of the liver. Furthermore, it is necessary during narcosis to take care for sufficient perfusion and oxygenation of the liver (Widger et al., 1976; Fischer and Scholler, 1977), as halothane may diminish the blood supply of the liver (Gattiker et al., 1966; Benumof et al., 1976). The reactive metabolites are supposed to be formed especially under hypoxic conditions. The formation of trifluoromethylcarbene, for instance, increases strongly if the oxygen concentration of the cells drops below 50 μM (Nastainczyk and Ullrich, 1978).

As the coincidence of enzyme induction and hypoxia yields a higher risk, a premedication with inducing drugs should be avoided if halothane anesthesia is planed. It is still unknown which inducers especially increase the reductive halothane metabolism (Cousins, 1980). Possibly the introduction of deuterated halothane into anesthesia reduces the side effects because $CF_3CClBrD$ is metabolized more slowly than $CF_3CClBrH$ (McCarty et al., 1979). Sipes et al. (1980), however, found that both the reductive metabolism and the hepatotoxicity of deuterated halothane are not significantly reduced in rats.

5.3.2. Vinylchloride and Other Monomers of Synthetic Polymers

From some monomeric substances used as starting material for synthetic polymers toxic metabolites may emerge from the catalytic action of P–450. In this connection vinylchloride (VC) is investigated best. For most species including man it is strongly cytotoxic, mutagenic and carcinogenic as well as embryotoxic and teratogenic (IARC Monographs, 1979a). In a first step VC is oxidized to chloroethylene oxide, which can be bound to nucleophilic macromolecules of the cell or transformed into chloroacetaldehyde. Chloroethylene oxide and chloroacetaldehyde can react with DNA

(GREEN and HATHWAY, 1978) and they act as strong mutagens (BARTSCH et al., 1976). By oxidation to monochloroacetic acid chloroacetaldehyde is inactivated (see Fig. 9). The intermediate occurrence of chloroethylene oxide was proved by in vitro investigations (BARTSCH et al., 1976; IARC Monographs, 1979a). The importance of P–450 for the metabolism of VC was experimentally demonstrated, e.g. the VC reaction with liver microsomes gives a type 1 difference spectrum and stimulates the CO-inhibited NADPH consumption. Investigations on microsomes from rats pretreated with PB or 3-MC showed that VC is mainly oxidized by P–450 while P–448 plays only a minor role. The binding of VC to P–450 is completely inhibited by SKF 525A, not, however, the increase in the CO-sensitive NADPH consumption caused by addition of VC (IVANETICH et al., 1977). The metabolism of VC in vivo is reduced by inhibitors of the MFO, e.g. 3-bromo-phenyl-4(5)-imidazole or 6-nitro-1,2,3-benzothiazole. Enzyme inducers such as PB intensify the hepatotoxic or mutagenic effect of VC. The binding of ^{14}C-VC to proteins and nucleic acids in vitro requires

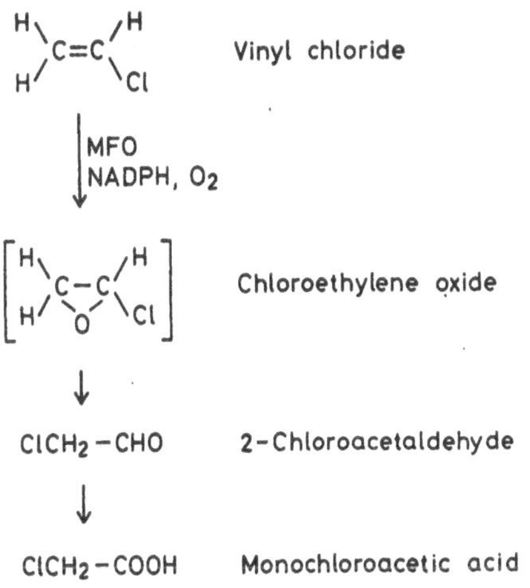

Fig. 9. Metabolism of vinyl chloride (BARTSCH et al., 1976; reproduced with permission from International Agency for Research on Cancer, Lyon)

microsomal enzymes, NADPH and O_2 (IARC Monographs, 1979a).

Vinylidene chloride, which is related to VC, is probably also oxidized by P–450 to an epoxide which can be bound to biologically important nucleophils or be transformed into the corresponding acylchloride. Vinylidene chloride is hepatotoxic, in various bacterial test systems mutagenic and carcinogenic in rats and mice (IARC Monographs, 1979a).

Chlorobutadiene, which is used for the production of synthetic rubber, acts directly mutagenic in bacterial systems. The mutagenic effect is increased by adding the 9,000 g supernatant from mouse liver; pretreatment of the animals with PB intensifies this effect. From these findings a P–450 dependent epoxide formation is supposed (BARTSCH et al., 1976). Possible carcinogenic effects of 2-chlorobutadiene need further investigations.

Chloroprene, a substance required for the production of plastics, is mutagenic in various test systems. Carcinogenic effects have not been proved as yet. The mutagenic effect in *Salmonella typhimurium* is increased by adding liver preparations of mouse and man. The occurring metabolite reacts as alkylating agent (IARC Monographs, 1979a).

Styrol is mutagenic for *Salmonella typhimurium* TA 1535 and TA 100 in the presence of the 9,000 g liver supernatant. Furthermore, chromosomal aberrations have been determined in rat and man. No carcinogenic effect was observed till now. The intermediately formed styrol oxide proved to be the toxic substance. The metabolism of styrol is stimulated by pretreating the animals with PB and inhibited by SKF 525A (LINNAINMAA et al., 1978; IARC Monographs, 1979a).

5.3.3. Phosphoric Acid Esters of Halogenated Alcohols (Flame Retardants)

Phosphoric acid esters of halogenated alcohols reduce the inflammability of textiles and plastics. Therefore they are used for the impregnation of children's clothings, especially of polyester or polyester acetate made sleeping suits (Blum and Ames, 1977). Commonly used are tris(1,3-dichloro-2-propyl)-phosphate (Fyrol FR2) and tris(2,3-dibromopropyl)phosphate (Tris-BP). These flame retardants and also the halogenated propanoles are mutagens. Structure-action investigations evidenced that phosphates with adjacent β and γ halogen atoms in the alkyl moiety, e.g. Tris-BP, are particularly mutagenic and that bromoalkyl derivatives induce more revertants in the Ames test than chloroalkyl derivatives (Carr and Rosenkranz, 1978; Nakamura et al., 1979).

Generally, these compounds become mutagenic only after metabolic activation, or its mutagenicity is increased thereby. Fyrol FR2 and the enzymatic hydrolysis product 1,3-dichloro-2-propanol are activated by liver homogenate from PB- or PCB-induced mice. The metabolites are mutagenic in *Salmonella typhimurium*. The metabolite 1,3-dichloro-2-propanone is a potent directly acting mutagen and probably a carcinogen, too (Gold et al., 1978).

After activation Tris-BP is mutagenic for *Salmonella typhimurium* TA 1535 and TA 100 but not for TA 1538, i.e. it induces base-pair substitution mutations. The metabolite 2,3-dibromopropanol detected in rats is mutagenic without metabolic activation, but mutagenicity is increased by liver homogenate. Highly concentrated Tris-BP is mutagenic even without activation (Prival et al., 1977). It is metabolized by P–450. As reactive metabolite probably a 2-keto-derivative is formed which can react with nucleophilic DNA groups thus leading to mutagenic events (Søderlund et al., 1979).

Tris-BP is also mutagenic for *Drosophila melanogaster* (Berkowitz, 1978) and the Chinese hamster cell line V79 (Furukawa et al., 1978). It induces II to III translocations and sister chromatid exchanges, respectively. In growing human KB cells Tris-BP causes repairable single-strand breaks of the DNA (Gutter and Rosenkranz, 1977). Moreover, cytotoxic effects such as growth inhibition of H-4-II-E cells from Reuber rat hepatoma H-35 have been observed. In isolated hepatocytes of rats Tris-BP inhibits the protein and DNA synthesis (Søderlund and Dybing, 1979).

Tris-BP is carcinogenic in rats and mice (Abelson, 1977). After its feeding renal tubular cell adenomas and carcinomas are observed which emerged from the proximate convoluted tubular epithelium (Reznik et al., 1979). In female ICR/Ha Swiss mice benignant and malignant tumors of skin, forestomach, oral cavity, lung, liver and kidney are induced after cutaneous application (van Duuren et al., 1978). Mouse skin tumors were also obtained in an initiator-promoter experiment using Tris-BP and 12-O-tetradecanoylphorbol-13-acetate, respectively (Haroz, 1979). For man carcinogenicity has not been described as yet (IARC Monographs, 1979e).

Tris-BP is absorbed by human skin. Children wearing pyjamas treated with Tris-BP excreted up to 29 ng/ml of the mutagenic metabolite 2,3-dibromopropanol in urine (Blum et al., 1978).

Because of its mutagenic and carcinogenic effect the U.S. Consumer Product Safety Commission banned the usage of Tris-BP for children's clothing in 1977. As substitutes tetrabromophthalic anhydride and tetrabromophthalic acid can be used. These compounds are not mutagenic for *Salmonella typhimurium* TA 1535, TA 100 and TA 1537 neither without nor after microsomal activation (MacGregor and Friedman, 1977).

For another flame retardant, tetrakis-(hydroxymethyl)phosphonium chloride (THPC), no carcinogenic activity has been observed up to now (van Duuren et al., 1978). However, Loewengart and van Duuren (1977) proved the substance to be an active tumor promoter, using 7,12-dimethylbenz(a)anthracene as initiator. Furthermore, EHRLICH et al. (1980) demonstrated mutagenicity and cell transformation for THPC and the corresponding sulfate in mammalian cell cultures.

5.3.4. *Halogenated Cyclic Hydrocarbons*

5.3.4.1. Halobenzenes

Halogenated benzenes are relatively inert chemicals, which are used e.g. as fungicides, plasticizer and flame retardants and are released as air pollutants during the production of halogenated hydrocarbons. Hepatotoxicity has been proved for most halobenzenes. Additionally hexachlorobenzene (HCB) has also carcinogenic properties. It induces liver tumors in mice and hamsters (CABRAL et al., 1977, 1979).

The toxicity of halobenzenes is mainly based on their active metabolites formed by P-450. After administration of bromo-, chloro-, iodo-, fluoro-, o-dichloro- and p-dichlorobenzene their corresponding metabolites are covalently bound to liver proteins provoking a more or less marked liver necrosis. The major amount of the metabolites is found in the necrotic centrolobular hepatocytes. There exists a correlation between the degree of severity of the necrosis and the extent of metabolites binding (BRODIE et al., 1971; REID and KRISHNA, 1973).

The importance of MFO for the activation of the halogenated benzenes becomes obvious after treating the animals with inducers and inhibitors of the MFO system. If rats are pretreated with PB, more covalently bound metabolites are observed as a result of the raised halobenzenes metabolism. There exist differences in toxicity between the various derivatives due to differences in biotransformation rates (BRODIE et al., 1971; REID and KRISHNA, 1973). Table 20 shows the influence of enzyme induction on the formation of liver necrosis by halobenzenes.

Another inducer, 3-MC, protects rats

Table 20. Effects of phenobarbital pretreatment on centrolobular necrosis in rats (BRODIE et al., 1971; reproduced with permission from the authors)

Compound	Dose per rat	Severity of centrolobular necrosis	
		Control rats	Phenobarbital-treated
Bromobenzene	0.03 ml	Glycogen loss to minimal necrosis	Extensive necrosis
Iodobenzene	0.02 ml	Normal to minimal necrosis	Extensive to massive necrosis
Chlorobenzene	0.04 ml	Glycogen loss to minimal necrosis	Extensive to massive necrosis
Fluorobenzene	0.20 ml	No specific lesions	Extensive to massive necrosis
1,2-Dichloro-benzene	0.03 ml	Glycogen loss to minimal necrosis	Glycogen loss to massive necrosis
1,3-Dichloro-benzene	0.03 ml	Normal to minimal necrosis (rarely)	Glycogen loss to extensive necrosis
1,4-Dichloro-benzene	0.1 g	Little or no effect	Little or no effect
Benzene	0.25 ml	No specific lesions	No specific lesions

to a certain degree from liver necrosis. Despite the increase of bromobenzene metabolism the covalent binding of its metabolites to liver macromolecules is diminished (Reid et al., 1971a). 3-MC induces transformation of bromobenzene into the 2,3-epoxide derivative which is converted to 2-bromophenol or is hydrated to bromophenyldihydrodiol and bromocatechol by epoxide hydratase. These metabolites are excreted renally. PB on the other hand induces the metabolism in direction to the more toxic 3,4-epoxide (Zampaglione et al., 1973; Lau and Zannoni, 1979).

Metabolism and hepatotoxicity of substituted bromobenzenes depend on the type of the substituent in opposition. Bromobenzenes with electron-withdrawing substituents are very potent liver toxicants, whereas substituents with electron-releasing properties are less toxic (Toranzo et al., 1977).

Inhibitors of P–450 such as SKF 525A, piperonylbutoxide and dithiocarb diminish the activation of bromobenzenes and thus decrease the extent both of the covalently bound metabolites and the liver necrosis (Reid et al., 1971a, b; Reid and Krishna, 1973; Younes et al., 1979).

In vitro halobenzenes are metabolized by isolated hepatocytes or liver microsomes (Sipes et al., 1971; Thor et al., 1978a, b). By pretreatment of the animals with PB or 3-MC the in vitro turnover of bromobenzene is accelerated (Table 21).

The pathways of bromobenzene metabolism are illustrated by Fig. 10. It is oxidized to the 3,4-epoxide derivative by P–450 which can be transformed non-enzymatically into p-bromophenol or is metabolized by GSH-transferase — a PB-inducible enzyme (Marniemi and Parki, 1975; Down and Chasseaud, 1979) — to a GSH conjugate as well as by epoxide hydratase to a diol. These intermediates are further converted either enzymatically or non-enzymatically to 3,4-dihydro-3-hydroxy-4-S-acetylcysteinyl bromobenzene and to 3,4-dihydroxy-bromobenzene, respectively (Brodie et al., 1971; Gillette, 1976). The mercapturic acids or their unstable precursors are excreted renally (Knight and Young, 1958) or biliary. Pretreatment of the animals with PB or SKF 525A increases or decreases the biliar excretion (Sipes et al., 1974). In Table 22 the relative amounts of renal excreted mercapturic acids and catechols formed from the GSH conjugates or dihydrodiols are summarized. The excretion of p-bromophenylmercapturic acid is raised by pretreatment with PB (Reid et al., 1971b).

Bromobenzene and other halobenzenes cause a rapid decrease of GSH in the liver cell. The cytotoxic activity of the

Table 21. K_m and V_{max} values for the metabolism of bromobenzene by the 9,000 g supernatant fraction of rat liver (Zampaglione et al., 1973)

Induction	K_m (mM)	V_{max} (nmoles · g^{-1} liver · min^{-1})
—	2.04 ± 0.20	45.5 ± 5
Phenobarbital	0.90 ± 0.15	400 ± 20
3-Methyl-cholanthrene	1.43 ± 0.20	120 ± 10

Table 22. Renal elimination of the catechol and mercapturic acid derivatives of various halogenated benzenes in rabbits (Azous et al., 1953; reproduced with permission from Biochemical Society, London)

Compound	Percentage of dose excreted	
	as catechol derivatives	as mercapturic acid
Benzene	4.1	1.3
Fluorobenzene	21	3
Chlorobenzene	37	28
Bromobenzene	30	22
Iodobenzene	21	23
o-Dichlorobenzene	8	8
Phenol	0.5	0

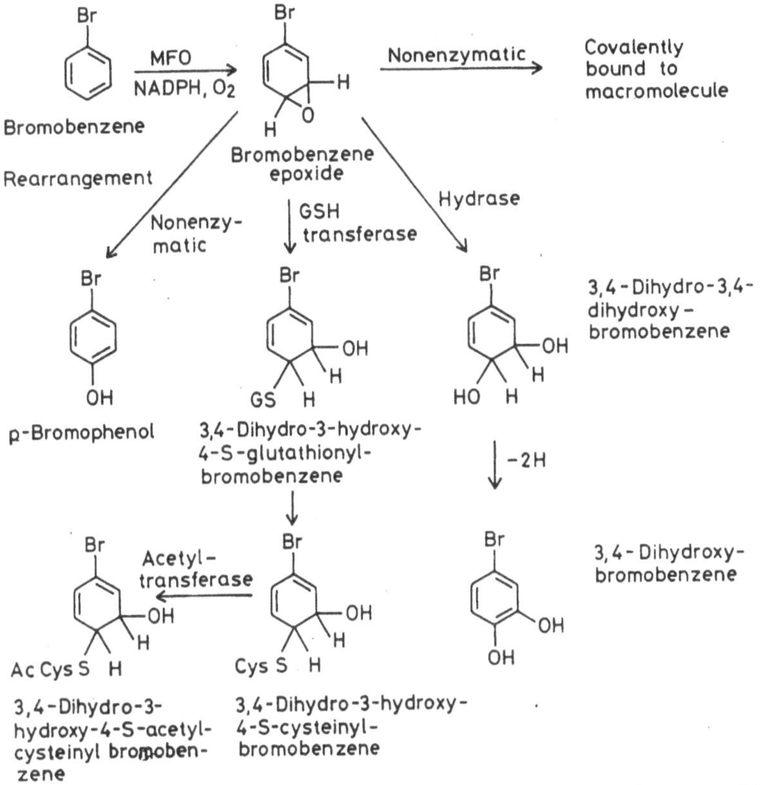

Fig. 10. Pathways of bromobenzene metabolism (GILLETTE, 1976; reprinted from Federation Proc. with permission)

halobenzenes depends therefore on the ratio between the concentration of reactive metabolites on the one hand and the GSH level on the other hand. The electrophilic bromobenzene-3,4-epoxide reacts easily with GSH, thus diminishing the extent of liver necrosis. After administration of small non-toxic doses of bromobenzene (0.05 mmol/kg rat) about 70 per cent of the occurring epoxides are conjugated to GSH (ZAMPAGLIONE et al., 1973). If the GSH content is low, however, the epoxides react to a higher extent with cellular proteins (molecular weight 20,000 and smaller than 13,000) (SUN and DENT, 1979), provoking liver necrosis. Pretreatment of the animals with diethylmaleate deprives the liver cells of GSH thus intensifying liver necrosis by halobenzenes (REID et al., 1971b).

In cultures of hepatocytes a diminuation of the GSH concentration to about 40 per cent of the initial values is caused by bromobenzene. The cells, however, retain their viability. In an amino acid deficient medium bromobenzene provokes a decrease of the GSH content to less than 5 per cent of control values and the cytotoxic effect increases correspondingly (THOR et al., 1978b).

In addition to the decrease of the cellular GSH concentration, to the alkylation of coenzymes and to the inhibition of cytosolic enzymes bromobenzene also causes marked structural and functional alterations of the plasma membrane, which is supposed to be the primary target of the bromobenzene-induced cytotoxicity (THOR and ORRENIUS, 1980).

Cysteine and cysteamine inhibit the formation of protein-bound metabolites of bromobenzene in microsomes. Methionine does not affect the metabolism of bromobenzene in microsomes, but inhibits its toxicity for hepatocytes, probably by an enhanced synthesis of GSH (THOR et al., 1978a). Addition of cysteine and N-acetyl-cysteine to the medium also protects hepatocytes (THOR et al., 1979). In rats cysteamine and cysteine diminish the extent of liver necrosis caused by bromobenzene, even if they are administered up to 12 hours after uptake indicating a possibility for therapeutical treatment of halobenzene intoxication (DE FERREYRA et al., 1979).

5.3.4.2. Further Halogenated Cyclic Hydrocarbons

Besides the halobenzenes numerous halogenated cyclic hydrocarbons play an important role in industry and agriculture, others occur as intermediate products or contaminants. Wellknown representatives of this family of substances are DDT, dieldrin, mirex, kepone (chlorodecone) as well as halogenated dibenzofuranes, naphthalenes, biphenyls, terphenyls and dibenzodioxines. The PCBs and PBBs have been examined most thoroughly. From the halogenated dibenzodioxines tetrachlorodibenzodioxine (TCDD) is well-known for its high toxicity. Many halogenated cyclic hydrocarbons show toxic effects, e.g. lesions of liver, nervous system and skin, as well as mutagenic and carcinogenic effects (McCONNELL and MOORE, 1979). SCHOENY et al. (1979), however, were not able to find mutagenicity in *Salmonella typhimurium* neither for aroclor 1254 — a commercial PCB mixture — nor for the insecticides mirex and kepone. The substances mentioned did not increase dominant lethal mutations in rats (GREEN et al., 1975; KHERA et al., 1976; SIMON et al., 1978). Liver tumors have been observed in rodents after administration of DDT, dieldrin, mirex, kepone, PCB including aroclor 1254, PBBs, TCDD and hexachlorodibenzodioxine. TCDD induces tumors also in lung, heart and nasal sphere (palate and nasal turbinates). A large part of the halogenated cyclic hydrocarbons have not yet been examined with respect to carcinogenic effects (IARC Monographs, 1978, 1979b).

Most of the representatives of this substance class exhibit toxic effects only after biotransformation. There are some exceptions such as TCDD. The mechanism of the toxic action of these substances is still unknown. Their intensive induction effect on the MFO system may play a role. The halogenated cyclic hydrocarbons probably are oxidized by the P–450 dependent MFO system into epoxides or arene oxides — as already shown for halobenzenes. The epoxides can be bound covalently to proteins and nucleic acids or rearranged to hydroxy-compounds, which then can be hydroxylated at further sites. Other reactive derivatives are formed by epoxide hydratase, which catalyzes the conversion of epoxides into dihydrodiols. The emerging metabolites are conjugated with glucuronic acid or GSH and subsequently excreted. In most cases cytotoxic, mutagenic and carcinogenic effects probably are caused by binding of the epoxides to cellular macromolecules. The occurrence of methylsulphone derivatives of PCBs indicates the binding of the arene oxides to methionyl residues of proteins which are then split off (YOSHIMURA and YOSHIHARA, 1976; ALLEN and NORBACK, 1977; IARC Monographs, 1978, 1979b). The metabolic pathways for 2,2′,5,5′-tetrachlorobiphenyl are shown in Fig. 11.

The hydroxylation can also proceed in connection with a dehalogenation. Degree, position and type of halogenation influence the metabolism of the halogenated biphenyls. Generally, higher substituted biphenyls are more slowly metabolized than those with a smaller number of substituents (MATTHEWS and KATO, 1979).

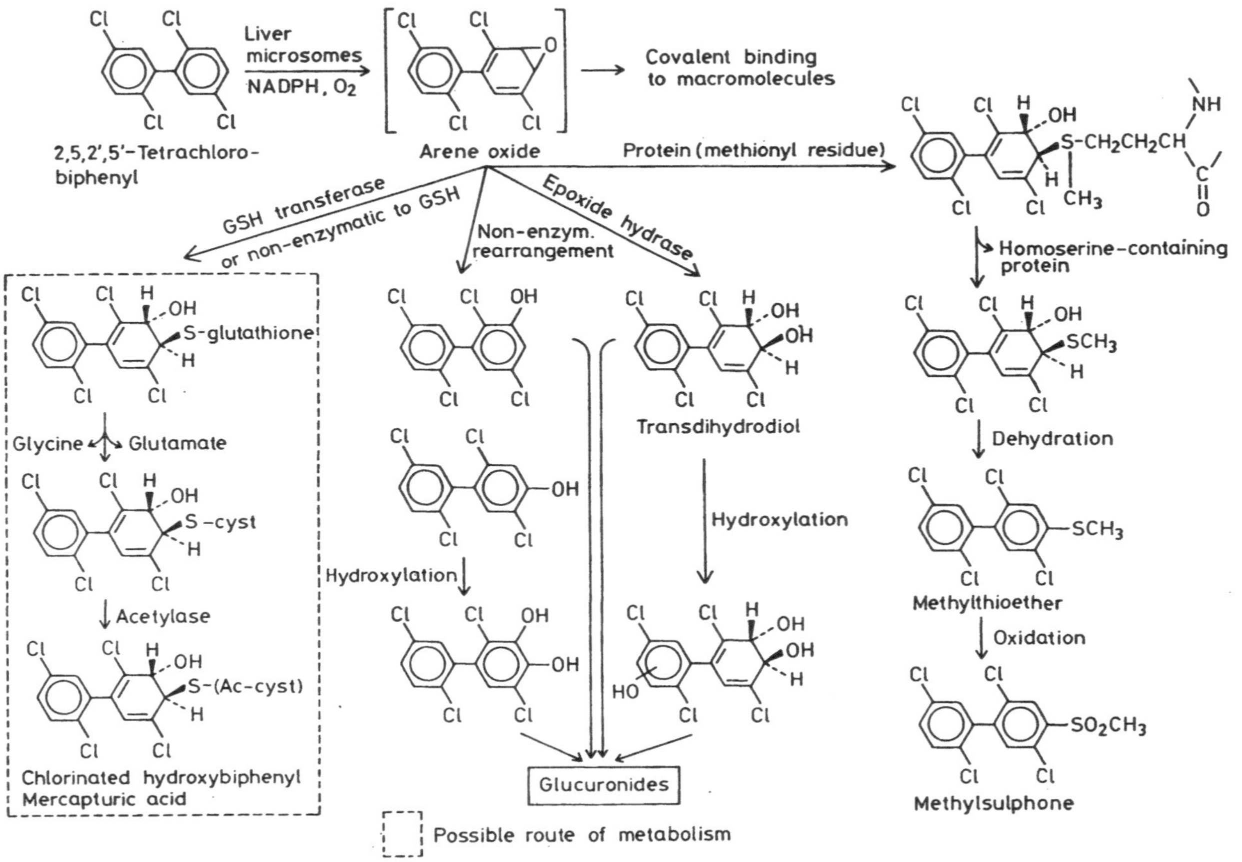

Fig. 11. Metabolic pathways for 2,2′,5,5′-tetrachlorobiphenyl (IARC Monographs, 1978; reproduced with permission from International Agency for Research on Cancer, Lyon)

5.4. *Furosemide*

Furosemide (2-furylmethylamino-4-chloro-5-sulfamoylbenzoic acid) is a diuretic with a relatively wide therapeutic range. In higher doses it is also used for the forced diuresis in the therapy of poisoning by drugs. However, high doses have side effects because of oxidative activation of the diuretic. After resorption furosemide is bound to plasma proteins and is eliminated either biliary as a glucuronide or renally in unconjugated form (MITCHELL et al., 1976; SPITZNAGLE et al., 1977). The biliary excretion is doese-dependent and saturable. Under saturation conditions of elimination furosemide is oxidatively metabolized (SPITZNAGLE et al., 1977).

Furosemide and other 2-substituted furanes are metabolized in liver and kidney of rat and mouse but also by isolated microsomes (MITCHELL et al., 1974a; MCMURTRY and MITCHELL, 1977) involving P–450 but not P–448 (WIRTH et al., 1975). Biologically active intermediates are able to bind covalently to proteins of liver and kidney cells thus inducing necrosis. The binding to mouse liver microsomes occurs with an apparent K_m of 0.11 mM and V_{max} of 0.21—0.24 nmoles · mg^{-1} protein · min^{-1} (WIRTH et al., 1975). Human liver microsomes bind 0.168 nmoles · mg^{-1} protein $\times$ min^{-1} (DYBING, 1977).

Investigations using [^{35}S]-, [α-^{2}H]-, [α,α'-^{2}H]- and [^{14}C]-furosemide have evidenced an arene oxide intermediate as reactive metabolite which is bound covalently via the furane ring to protein (WIRTH et al., 1975; MITCHELL et al., 1976). Fig. 12 represents the presumed mechanism of metabolic activation of furosemide.

The inhibition of P–450 synthesis by $CoCl_2$ and the inhibition of P–450 enzymatic activity by piperonylbutoxide and 2-naphthylisothiocyanate protect mice against liver necrosis caused by furosemide. Antibodies against NADPH-cytochrome c reductase diminish the covalent binding of the metabolites to microsomal proteins (MITCHELL et al., 1974a). The animals pretreatment with PB enhances the metabolism of furosemide, while covalent binding and liver necrosis increase only to a small degree. Probably inactivation and elimination are accele-

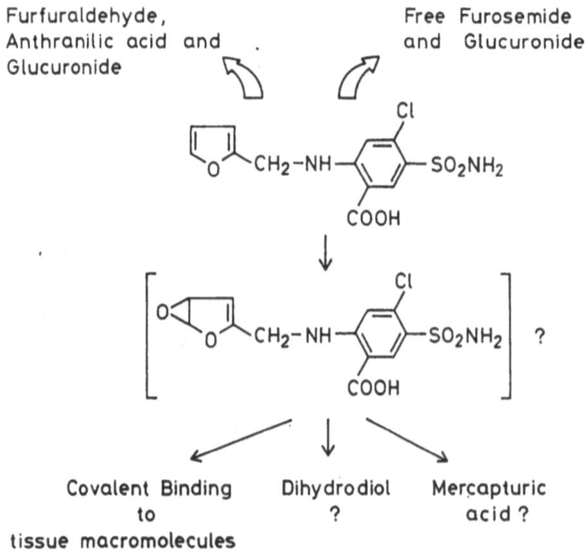

Fig. 12. Proposed pathways for the metabolism of furosemide (MITCHELL et al., 1976; reproduced with permission from Williams and Wilkins Comp., Baltimore)

rated simultaneously so that the concentration of toxic metabolites remains nearly constant (MITCHELL et al., 1976). With respect to furosemide metabolism 3-MC is ineffective in mice (WIRTH et al., 1975), in rats, however, severe furosemide liver necrosis is provoked by 3-MC pretreatment (MITCHELL et al., 1976). The inhibitor of epoxide hydratase — 1,2-epoxy-3,3,3-trichloropropane -- increases the covalent binding of furosemide metabolites in vitro (WIRTH et al., 1975), whereas GSH, cysteine and cysteamine diminish it (MITCHELL et al., 1976).

Hepato- and nephrotoxicity of furosemide is observed only after surmounting threshold doses. This threshold depends on the elimination capacity of liver and kidney as well as on the pretreatment with enzyme inducers or inhibitors. The therapeutic doses in man amount to 40—600 mg per day. Within this range no liver damages have to be feared. Only in the case of renal diseases with delayed renal elimination of furosemide (it is excreted by tubular secretion) leading to higher concentrations of free furosemide liver damages cannot be excluded. These consequences should also be taken into consideration when up to 2,000 mg furosemide are administered in the case of forced diuresis (DYBING, 1977; DYBING and MITCHELL, 1977).

5.5. *Aflatoxins*

Among the mycotoxins aflatoxins B_1 and G_1 are known to be strong liver carcinogens. Being themselves biologically ineffective the aflatoxins are enzymatically activated by P–450 causing oxidation of the 2,3-double bond of the terminal furane ring (see Fig. 13). The emerging 2,3-epoxides are unstable and assumed react immediately with cellular nucleophilic molecules. Therefore they could neither be directly proved nor synthesized as yet. In aqueous environment they are converted into the corresponding dihydrodiols. Indications for the occurrence of 2,3-epoxides are obtained

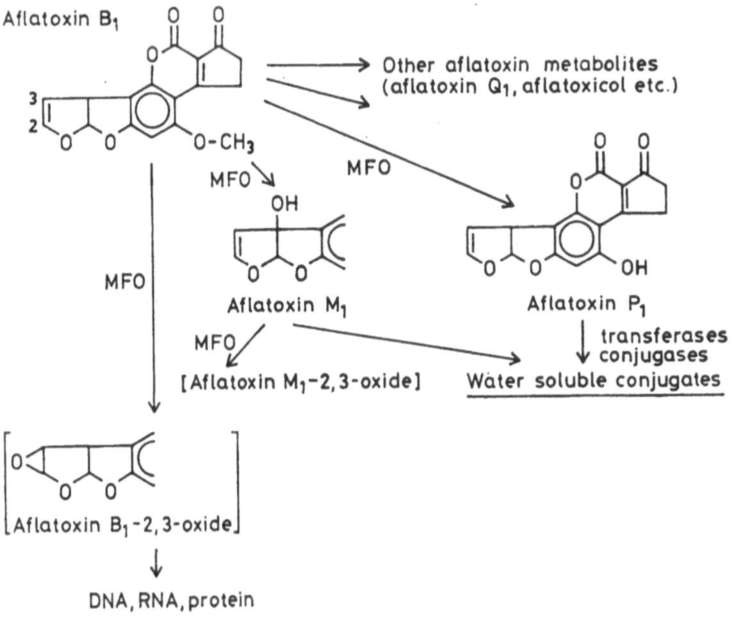

Fig. 13. Metabolism of aflatoxin B_1 (GARNER, 1978, modified; reproduced with permission from Karger, Basel)

from the following investigations: 1. The incubation of aflatoxin B_1 with liver microsomes and NADPH leads to the formation of 2,3-dihydro-2,3-dihydroxy-aflatoxin B_1. Addition of DNA prevents the formation of the dihydroxydiols. 2. After in vivo application of radioactive aflatoxin B_1 both the dihydrodiol derivative and 2,3-dihydro-2-(guan-7-yl)-3-hydroxyaflatoxin B_1 were found in the DNA hydrolysates. Incubation of aflatoxin B_1 with microsomes, NADPH and DNA gives the same hydrolysis products (Lin et al., 1978). Microsomes from PB-pretreated rats and hamsters produce a 2- to 4-fold increase in the amount of DNA bound metabolites. In the absence of DNA a corresponding increase of the formed dihydrodiol in the incubation medium was observed. 3-MC, however, induces a different pathway of metabolism which leads to less active metabolites, e.g. a hydroxylation product called aflatoxin M_1. The induction of the P–448 dependent aflatoxin B_1 hydroxylase is paralleled by the induction of AHH in PAH- inducible mice strains. That means that this process is regulated via the Ah-locus while the P–450 dependent activation of aflatoxin B_1 is independent of the Ah-locus (Gurtoo et al., 1978a).

Besides these metabolites several other metabolic products with less mutagenic or carcinogenic effectivity may be formed (Gurtoo et al., 1978b; Neal and Colley, 1978). Species differences of the aflatoxin B_1 metabolism were shown by Ueno et al. (1980). After i.p. administration of [^{14}C]aflatoxin B_1 to rats, mice and guinea pigs the binding to nucleic acids in the liver was more extensive than to proteins. The amount of bound metabolites and the specific radioactivity of RNA, DNA and proteins was different in each of these species.

5.6. Amines

5.6.1. Phenacetin and Acetaminophen (Paracetamol)

Acetaminophen is the active metabolite of phenacetin and widely used as analgetic and antipyretic drug. The therapeutic dose, ranging between 600 and 1,800 mg, is not hepatotoxic. At acute overdosages, e.g. in suicide attempts, liver damages up to necrosis have been observed. The acute toxic dose is assumed to be about 15 g acetaminophen (Mitchell et al., 1974b). Liver damages are indicated if the plasma half-life of acetaminophen exceeds 4 hours (Prescott et al., 1971). Besides acute liver damages also diseases of the kidney such as chronic interstitial nephritis and tumors are observed after chronic abuse. Correlations between the total amount of phenacetin taken by the patients and the incidence of carcinomas of the urinary tract have been verified (Carro-Ciampi, 1978; Mihatsch et al., 1980). In the case of interstitial nephritis the participation of drug induced allergic reactions has been discussed, too (Raaflaub and Dubach, 1972).

Phenacetin administered in vivo increased the number of chromosomal aberrations in rat lymphocytes (Granberg-Öhman et al., 1980). The phenacetin metabolite N-hydroxyphenacetin is mutagenic in *Salmonella typhimurium* TA 100 after microsomal activation. Another metabolite — N-acetoxyphenacetin — acts also mutagenically in TA 100 and TA 98 after activation (Shudo et al., 1978).

Acetaminophen is preferentially metabolized in the liver. The major part is bound to sulfuric and glucuronic acid and renally eliminated. About 3.8% of acetaminophen administered in therapeutical doses is metabolized oxidatively (Mitchell et al., 1974b, c.f. Fig. 14). An electrophilic metabolite produced by P–450 reacts with GSH and is further

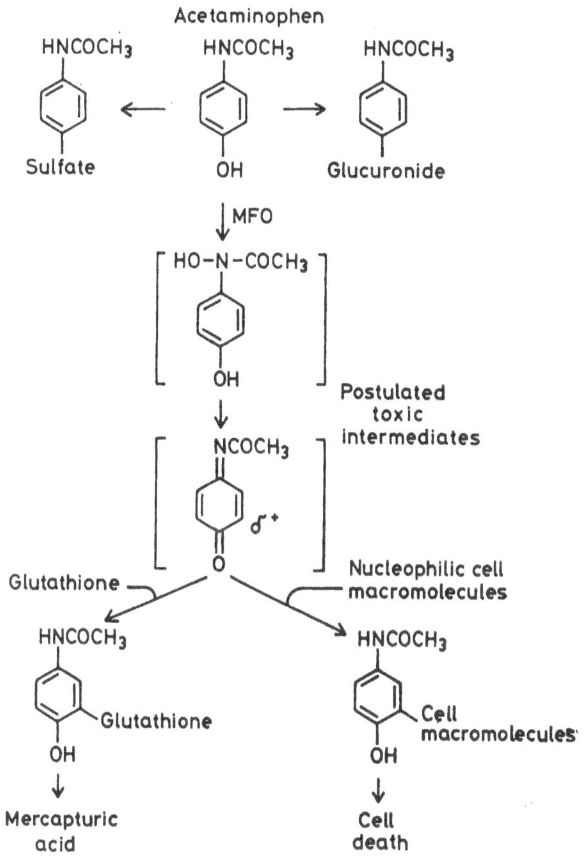

Fig. 14. Pathways of acetaminophen metabolism (MITCHELL et al., 1975; reproduced with permission from Springer-Verlag, Heidelberg)

transformed to a derivative of mercapturic acid, which is excreted renally. If the metabolite concentration exceeds the GSH content in the liver cell, the metabolite reacts with nucleophilic groups of proteins causing cell destruction (GILLETTE, 1976). The structure of the active intermediates is not exactly known as yet. N-acetyl-p-quinoneimine is regarded to be one of the active metabolites (MINER and KISSINGER, 1979). Recent investigations indicate that the covalently bound metabolite is not derived from N-hydroxyacetaminophen (HINSON et al., 1979a; NELSON et al., 1980), although the oxidation of the amide nitrogen of acetaminophen is essential for hepatotoxicity (NELSON et al., 1978a).

Phenacetin is dealkylated by P-450, too. The deuterization in α-methylene position diminishes dealkylation to acetaminophen, lowers the GSH decrease in the liver cell, reduces the amount of covalent binding and the incidence of liver necrosis. The formation of methemoglobin is favoured indicating the methemoglobin formation being caused by the phenetidine structure. The hepatotoxicity, however, is due to acetaminophen (NELSON et al., 1978b). Minor pathways of phenacetin metabolism are epoxidation and conversions of the glucuronidated and sulfated derivatives of N-hydroxy-phenacetin (HINSON et al., 1977, 1979b).

The interrelationship between the acetaminophen metabolism, GSH-concentration, covalent linkage to macromolecules and liver necrosis in mice and rats have been thoroughly examined. There exist correlations between the amount of metabolites bound to microsomal, mitochondrial and soluble proteins and the severity of liver necrosis. The covalent binding of the metabolites to cellular proteins as preceded by a diminution of the cellular GSH to about 20%. Covalent protein binding and GSH decrease are inversely correlated (Fig. 15). Induction by PB or release of GSH by diethylmaleate increase both the covalent binding and liver necrosis. Inhibition of P-450 by piperonylbutoxide as well as reduction of the P-450 synthesis by $CoCl_2$ cause a decrease of the acetaminophen toxicity (JOLLOW et al., 1973; MITCHELL et al., 1973a, b; POTTER et al., 1973).

Besides the liver also the kidney is target organ for acetaminophen. In mice and rats GSH depletion, covalent binding of the metabolites and necrosis of the kidney are observed. Contrary to hepatotoxicity pretreatment of the animals with 3-MC hardly raises nephrotoxicity (McMURTRY et al., 1978; MUDGE et al., 1978).

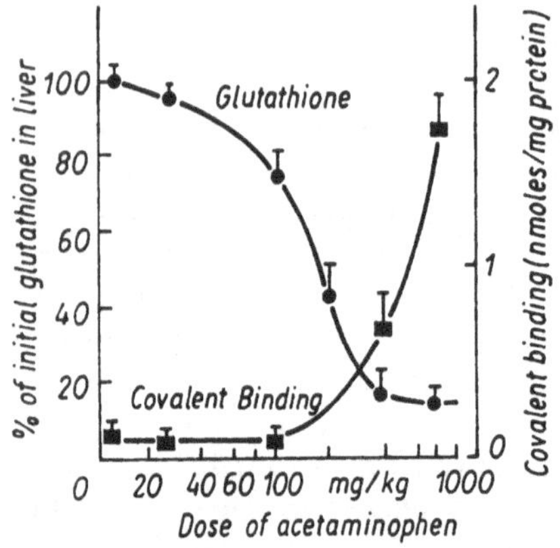

Fig. 15. Hepatic glutathione level and covalent binding to hepatic proteins after increasing doses of acetaminophen. Glutathione concentration (•) and covalent binding (■) of ^{3}H-acetaminophen were determined 120 min after administration. Points are means ± S.E.M. of at least three mice (MITCHELL et al., 1973b; reproduced with permission from Williams and Wilkins Comp., Baltimore)

Derived from the results of animal experiments the following therapeutical possibilities may be taken into consideration in the case of acetaminophen intoxications:

1. Inhibition of the activating monooxygenase system
2. Stimulation of the conjugation reactions
3. Inhibition of the covalent binding of active metabolites to cellular proteins
4. Application of sulfhydryl compounds
5. Enhancement of the GSH concentration in liver cells.

In mice 30 per cent increase of acetaminophen LD_{50} was proved by postmedication of metyrapone — an inhibitor of P-450. Simultaneously the liver GSH content decreased only to about 65% compared with 95% under administration of 375 mg acetaminophen/kg without metyrapone (FOSSA et al., 1979). Metyrapone also causes an increase of the survival rate of mice treated with 400 mg acetaminophen/kg from 38 to 78–86% (GOLDSTEIN and NELSON, 1979). Dithiocarb lowers acetaminophen toxicity and diminishes GSH depletion. It also favours the renal excretion of acetaminophen conjugates probably caused by inhibition of the MFO system thus decreasing the formation of reactive metabolites (SIEGERS et al., 1976). Application of cysteamine diminished liver damages after acetaminophen intoxication in man (PRESCOTT et al., 1974).

5.6.2. Aromatic Amines and Azo-Compounds

Within this group there are a large number of mutagenic and carcinogenic substances. Some of them, e.g. 2-naphthylamine, have been proved to be carcinogenic also for man. The biotransformation of aromatic amines and azo-compounds leads both to C- and N-hydroxylated derivatives. While the P-450 dependent C-hydroxylation in this group of substan-

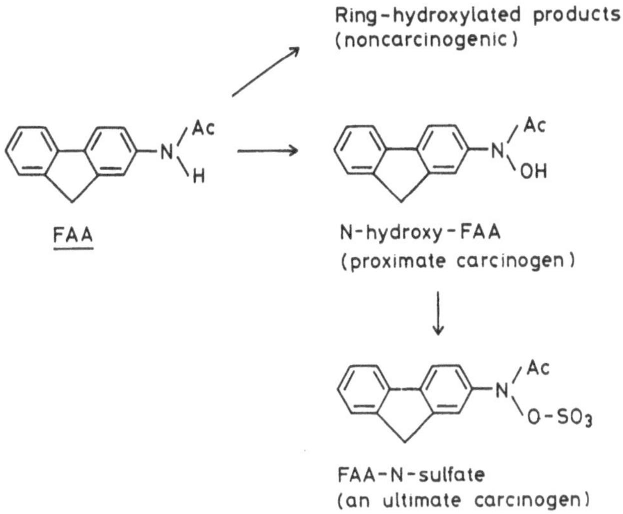

Fig. 16. Metabolism of 2-acetylaminofluorene (FAA) (CONNEY and LEVIN, 1974; reproduced with permission from International Agency for Research on Cancer, Lyon)

ces exclusively generates inactive products the N-hydroxylation may produce proximate carcinogens and mutagens. The ultimately effective metabolites in most cases emerge from esterification with sulfate and acyl groups catalyzed by cytosolic enzymes (e.g. sulfotransferases and transacetylases) (MILLER and MILLER, 1976). The biotransformation of aromatic amines is exemplified by 2-acetylaminofluorene (AAF) in Fig. 16. From the N-hydroxy-AAF the strongly electrophilic N-hydroxy-AAF-sulfate is formed by a sulfotransferase. This metabolite is assumed to be an ultimate mutagen and carcinogen. Further ultimately effective metabolites probably are N-acetoxy-2-aminofluorene emerging from cooperative catalysis of deacetylase and transacetylase as well as N-acetoxy-AAF and 2-nitrosofluorene formed by peroxidase (CONNEY and LEVIN, 1974; MILLER and MILLER, 1976).

N-hydroxylation can proceed in different ways, whereby several P-450 subspecies may be involved. Furthermore, the P-450 independent microsomal amine oxidase takes part in the N-hydroxylation of numerous secondary and tertiary amines (see 2.1.). The distinguished species specificity (CONNEY and LEVIN, 1974) of the AAF-hydroxylation points to the involvement of different enzymes in the various hydroxylation reactions. So for instance, in rat N- as well as C-hydroxylations are induced by 3-MC while this inducer in hamster excludingly stimulates N-hydroxylations (LOTLIKAR et al., 1978a). Possibly there are different P-450 forms induced in the various species. The results of numerous investigations allow the conclusion that N-hydroxylation of AAF is exclusively catalyzed by the P-450 dependent monooxygenase system. Both C- and N-hydroxylations of this substance are inhibited by CO (THORGEIRSSON et al., 1973). Both hydroxylation processes are catalyzed by a monooxygenase system reconstituted from purified components, in which P-450 (P-448) from 3-MC-pretreated rats is multifold more active than from untreated animals (LOTLIKAR and ZALESKI, 1975). An immuno-γ-globulin directed against NADPH-cytochrome P-450 reductase was proved to be a strong inhibitor of the N-hydroxylation of AAF. The substrate specificity of the individual N-hydroxylating enzymes (P-450, P-448 and amine oxidase) directs their participation in the metabolism of different substrates. So for instance,

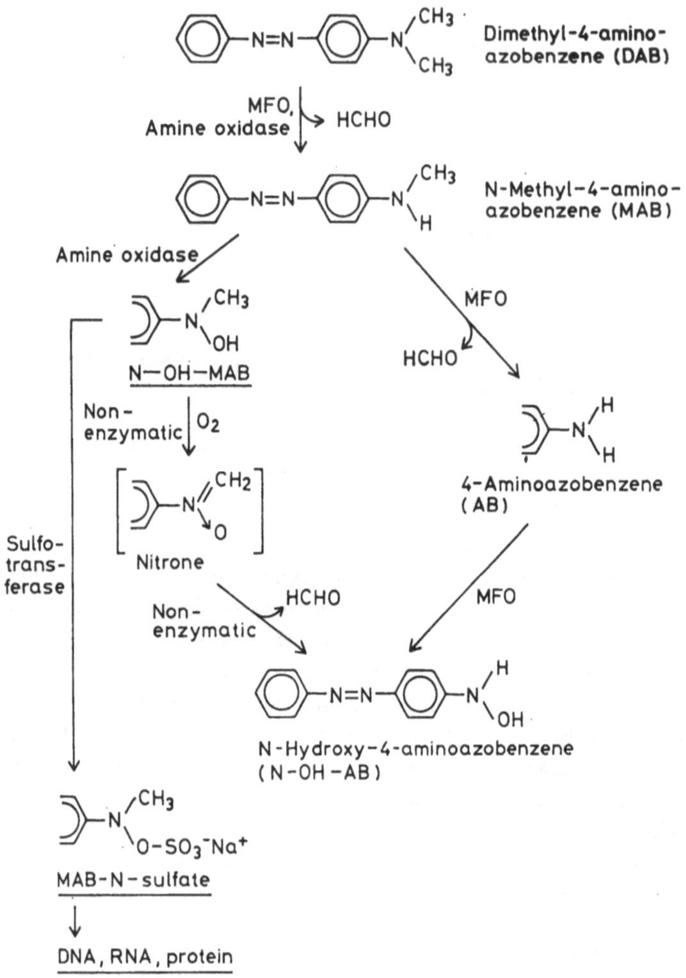

Fig. 17. Metabolism of dimethyl-4-aminoazobenzene (MILLER and MILLER, 1976, supplemented; reproduced with permission from International Agency for Research on Cancer, Lyon)

N-acetylamines are not converted by purified microsomal amine oxidase whereas 2-naphthylamine is oxidized to 2-naphthylhydroxylamine (THORGEIRSSON et al., 1973).

For monomethylaminoazobenzene (MAB) two different N-hydroxylation pathways have been proved (MILLER and MILLER, 1976) (see Fig. 17). The P–450 dependent reaction leads to the biologically inactive aminoazobenzene, which is further hydroxylated to inactive N-hydroxy-aminoazobenzene by the MFO system. The formation of proximate and ultimate carcinogens and mutagens is P–450 independent. This could be proved by the use of specific inhibitors of the MFO system. Moreover, this pathway of N-hydroxylation is not favoured by cumene hydroperoxide in the absence of NADPH. MAB is oxidized – probably by the microsomal amine oxidase – to N-hydroxy-MAB, which then rearranges non-enzymatically to formaldehyde and N-hydroxy-aminoazobenzene. From the highly reactive N-hydroxy-MAB the corresponding sulfuric acid ester can be formed by a sulfotransferase, which in vitro reacts with various nucleosides (e.g. guanosine) or nucleic acids. The addition of sodium sulfate to the diet of rats fed with 3'-methyl-dimethyl-aminoazobenzene causes an increase of the incidence of liver tumors. These investigations suggest that N-hydroxy-MAB-sulfate or similar sulfates act as ultimate carcinogens (BLUNEK and CROWTHER, 1975; MILLER and MILLER, 1976).

5.6.3. Substituted Aryldiamines (Hair Dyes)

Substituted aryldiamines are constituents of cosmetics. Some of them are used as hair dyes. A number of these compounds act as mutagens in the *Salmonella typhimurium* system. So after biotransformation of 2,4-diaminoanisole (2,4-DAA), 4-nitro-o-phenylenediamine (4-NOPD), 2-nitro-p-phenylenediamine (2-NPPD), 2,5-diaminoanisole, 2-amino-5-nitrophenol, m-phenylenediamine, o-phenylenediamine, 2-amino-4-nitrophenol and 2,5-toluenediamine by the S9-fraction from rat liver frameshift mutants were induced in *Salmonella typhimurium* TA 1538 (AMES et al., 1975). After application of 2-NPPD and 4-NOPD rats excrete urinary metabolites mutagenic for *Salmonella typhimurium* (AMMENHEUSER and WARREN, 1979). They also induce chromosomal aberrations and cell transformations in various cell cultures (SEARLE et al., 1975; BENEDICT, 1976; KIRKLAND and VENITT, 1976). 2,4-toluenediamine and 4-NOPD are mutagenic in *Drosophila* (FAHMY and FAHMY, 1977).

BURNETT et al. (1979), however, did not find substances in urines from women collected after hair dying which are mutagenic for *Salmonella typhimurium* after microsomal activation.

Some aryldiamines are observed to be carcinogenic and teratogenic. 2,4-Toluenediamine for instance caused hepatocellular carcinomas and mammary tumors in rats after oral application (ITO et al., 1969; BURNETT et al., 1975; CARDY, 1979). This compound therefore was prohibited to be used as a hair dye. In female mice multiple eosinophilic hepatocellular adenomas are induced by 2-NPPD after oral application (REZNIK and WARD, 1979), while 2,4-DAA-sulfate leads to the formation of follicular cell

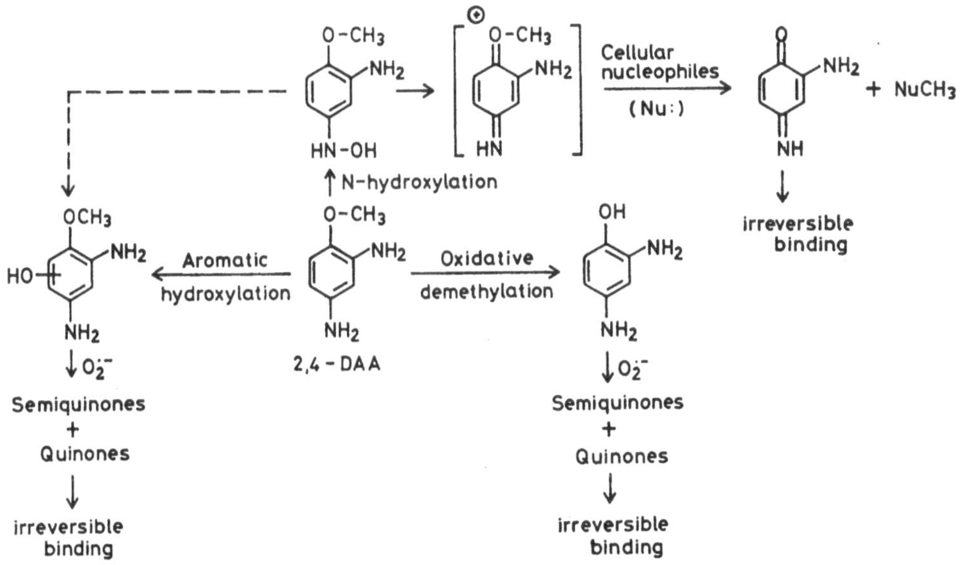

Fig. 18. Tentative scheme for microsomal metabolism of 2,4-diaminoanisole (DYBING et al., 1979; reproduced with permission from Pergamon Press, Oxford)

carcinomas of the thyroid glands in male and female rats (WARD et al., 1979). For 2,5-toluenediamine a teratogenic effect was proved in mice (INOUYE and MURAKAMI, 1977). After cutaneous application of a number of hair dyes no toxic and carcinogenic effects were observed (BURNETT et al., 1975, 1980).

The P-450 dependent monooxygenase system considerably participates in the biotransformation of aryldiamines, e.g. 2,4-DAA. Numerous investigations with inducers support this fact. The mutation frequency in the *Salmonella typhimurium* system after activation of 2,4-DAA by the S9-fraction of liver and kidney of the PAH-responsive mouse strains (e.g. C57BL/6J/BOM) increases especially after pretreatment with β-naphthoflavone and 3-MC (DYBING and THORGEIRSSON, 1977; AUNE and DYBING, 1979). AUNE et al. (1980) could demonstrate that both microsomal and nucleic fractions from rat liver after β-naphthoflavone pretreatment of the animals can convert 2,4-DAA into mutagenic metabolites. Nuclei of untreated rats or of PB-pretreated ones, however, showed no effect. Thus the activation of 2,4-DAA seems to be catalyzed by P-448. Inhibitors of the MFO system inhibit the metabolic activation of the arylamines.

Besides the microsomal MFO system and that of the nucleic membrane also cytosolic enzymes probably participate in the activation of these substances (DYBING and THORGEIRSSON, 1977). The biotransformation pathways of aryldiamines are represented schematically by the 2,4-DAA metabolism in Fig. 18. The toxicological importance of the various metabolites, however, is almost unknown as yet. For the mutagenic effect probably N-hydroxylation plays an important role (DYBING et al., 1979).

5.7. Nitrosamines and Nitrosamides

Numerous N-nitroso compounds are cytotoxic, mutagenic and carcinogenic. While nitrosamides (e.g. N-methyl-1-nitrosourea) are directly effective, most nitrosamines have to be activated enzymatically whereby one of the two side chains is split off (for review see MAGEE et al., 1975; LAMPE and BUTSCHAK, 1978). The mechanism of this

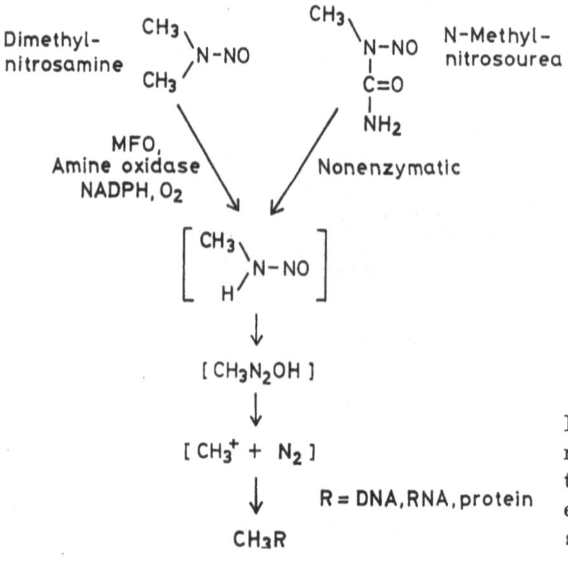

Fig. 19. Biotransformation of dimethylnitrosamine and N-methyl-1-nitrosourea to the alkylating carbonium ion (MAGEE et al., 1975; reproduced with permission from Springer-Verlag, Heidelberg)

process has not been completely elucidated as yet. It is assumed, that at first a P–450 dependent oxidation at the α-C-atom of one of the two side chains takes place, which is followed by the cleavage of the C—N-bond and by formation of alkyldiazohydroxide and aldehyde. Finally from the alkyldiazohydroxide intermediate a carbonium ion emerges which can react with cellular macromolecules (Park and Archer, 1978). The enzymatic and non-enzymatic conversion of dimethylnitrosamine and 1-methyl-1-nitrosourea is shown in Fig. 19. Diethylnitrosamine probably is analogously converted. Recent studies on dimethylnitrosamine indicate that P–450 as well as P–448 may participate in the enzymatic activation whereby their specific contribution depends on the substrate concentration (Lotlikar et al., 1978b) and the nature of substrates (Bartsch et al., 1975).

For nitrosamines with longer side chains (e.g. propyl or butyl) also a P–450 dependent β-oxidation — besides α-hydroxylation — has been observed (Park and Archer, 1978).

The dealkylation of dialkylnitrosamines can also be catalyzed by a P–450 independent amine oxidase (Lake et al., 1976a, b; Prough and Ziegler, 1977; Godoy et al., 1978; Lotlikar et al., 1978b).

6. Conclusions

P–450 plays a main role in the biotransformation of many xenobiotics including drugs, mutagens and carcinogens as well as endogenous substances such as steroids and prostaglandins. The P–450 dependent enzyme system catalyzes the conversion of lipophilic compounds to more hydrophilic metabolites thus enabling renal elimination, often after conjugation of the hydrophilic metabolites to glucuronic or sulfuric acid or GSH and other compounds. In many cases toxic metabolites such as epoxides, radicals or N-hydroxylated compounds are generated. Many of these intermediates are mutagenic and/or carcinogenic. In some cases also pharmacodynamically active compounds are formed from inactive precursors.

The species- and organspecificity of P–450 dependent enzyme activities leads both to species- and organspecific metabolite patterns and biological effects, respectively. Therefore no direct conclusions can be drawn from animal experiments with regard to the drug metabolism in man and peculiarities of biological effects in human beings.

The P–450 dependent enzyme activities are influenced by several factors such as genetic disposition, age, sex, nutrition, and hormonal state. The variability of the MFO system is mainly caused by differences in the inducibility of individual P–450 forms by xenobiotics and endogenous compounds, such as hormones. The inducibility depends on specific genetic loci, for instance on the Ah-locus in the case of P_1–450 (P–448).

Sometimes the induction of P–450 plays an important role in pharmacotherapy. Clinically relevant alterations of the metabolism of drugs and of endogenous compounds (e.g. steroid hormones) may occur: 1. Under administration of inducing drugs which are applied in high doses. 2. In chronic therapy with inducing drugs. 3. During abuse of drugs, alcohol and tobacco.

The studies of the P–450 dependent reactions have contributed to the understanding of biological peculiarities of toxic substances and their species- and organspecificity. Especially the mechanism of chemical carcinogenesis could be made more transparent. Because P–450 participates in both detoxification and activation to toxic products it is ambiguous to utilize either induction or inhibition of this enzyme system for preventing mutagenic and carcinogenic activities of xenobiotics. Consequences of manipulations of the MFO system are not predictable as yet.

References

ABELSON, P. H. (1977): Science **197**, 113

ACKERMANN, E. (1972): Biochem. Pharmacol. **21**, 2169–2180

ACKERMANN, E. (1978): Pharmazie **33**, 400–406

ACKERMANN, E., K. RICHTER, and S. SAGE (1976): Biochem. Pharmacol. **25**, 1557–1560

AITIO, A., and H. VAINIO (1976): Acta Pharmacol. Toxicol. **39**, 555–561

ALLEN, J. R., and D. H. NORBACK (1977) in: Origins of Human Cancer, Book A, Incidence of Cancer in Humans (H. H. HIATT, J. D. WATSON, and J. A. WINSTEN, eds.), Cold Spring Harbor, N.Y., Cold Spring Harbor Laboratory, pp. 173–186

ALVARES, A. P. (1978): Clin. Pharmacokinet. **3**, 462–477

ALVARES, A. P., K. E. ANDERSON, A. H. CONNEY, and A. KAPPAS (1976): Proc. Natl. Acad. Sci. U.S. **73**, 2501–2504

ALVARES, A. P., A. FISCHBEIN, K. E. ANDERSON, and A. KAPPAS (1977): Clin. Pharmacol. Ther. **22**, 140–146

ALVARES, A. P., E. J. PANTUCK, K. E. ANDERSON, A. KAPPAS, and A. H. CONNEY (1979): Drug Metab. Rev. **9**, 185–205

AMBRE, J., D. GRAEFF, F. BURES, D. HAUPT, and K. DEASON (1977): J. Med. **8**, 57–70

AMES, B. N., H. O. KAMMEN, and E. YAMASAKI (1975): Proc. Natl. Acad. Sci. U.S. **72**, 2423–2427

AMMENHEUSER, M. M., and M. E. WARREN (1979): Mutat. Res. **66**, 241–245

ANDERS, M. W., J. L. STEVENS, R. W. SPRAGUE, Z. SHAATH, and A. E. AHMED (1978): Drug Metab. Dispos. **6**, 556–560

ANDERSON, K. E., A. H. CONNEY, and A. KAPPAS (1979): Clin. Pharmacol. Ther. **26**, 493–501

ANDREWS, L. S., B. R. SONAWANE, and S. J. YAFFE (1976): Res. Commun. Chem. Pathol. Pharmacol. **15**, 319–330

ARMBRECHT, H. J., L. S. BIRNBAUM, T. V. ZENSER, M. B. MATTAMMAL, and B. B. DAVIS (1979): Arch. Biochem. Biophys. **197**, 277–284

AUNE, T., and E. DYBING (1979): Biochem. Pharmacol. **28**, 2791–2797

AUNE, T., E. DYBING, and S. D. NELSON (1980): Chem.-Biol. Interact. **31**, 35–49

AUTRUP, H. (1979): Biochem. Pharmacol. **28**, 1727–1730

AUTRUP, H., C. C. HARRIS, B. F. TRUMP, and A. M. JEFFREY (1978): Cancer Res. **38**, 3689–3696

AZOUZ, W. M., D. V. PARKE, and R.T. WILLIAMS (1953): Biochem. J. **55**, 146–151

BACK, D. J., A. M. BRECKENRIDGE, F. CRAWFORD, M. MACIVER, M. L'E. ORME, B. K. PARK, P. H. ROWE, and E. SMITH (1979): J. Clin. Pharmacol. **15**, 193–197

BARTSCH, H., C. MALAVEILLE, and R. MONTESANO (1975): Cancer Res. **35**, 644–651

BARTSCH, H., C. MALAVEILLE, and R. MONTESANO (1976): IARC Sci. Publ. **12**, 467–486

BATZINGER, R. P., E. BUEDING, B. S. REDDY, and J. H. WEISBURGER (1978): Cancer Res. **38**, 608–612

BECKETT, A. H., and E. J. TRIGGS (1967): Nature **216**, 587

BEND, J. R., G. E. R. HOOK, R. E. EASTERLING, T. E. GRAM, and J. R. FOUTS (1972): J. Pharmacol. Exp. Ther. **183**, 206–217

BENEDICT, W. F. (1976): Nature **260**, 368–369

BENUMOF, J. L., J. J. BOOKSTEIN, L. J. SAIDMAN, and R. HARRIS (1976): Anesthesiology **45**, 545–551

BERKOWITZ, S. (1978): Mutat. Res. **57**, 385–387

BHAT, K. S., and G. PADMANABAN (1978): FEBS Lett. **89**, 337–340

BHAT, K. S., M. K. SARDANA, and G. PADMANABAN (1977): Biochem. J. **164**, 295–303

BLANKENSHIP, J., and E. BRESNICK (1974): Biochim. Biophys. Acta **340**, 218–227

BLUM, A., and B. N. AMES (1977): Science **195**, 17–23

BLUM, A., M. D. GOLD, B. N. AMES, C. KENYON, F. R. JONES, E. A. HETT, R. C. DOUGHERTY, E. C. HORNING, I. DZIDIC, D. I. CARROLL, R. N. STILLWELL, and J.-P. THENOT (1978): Science **201**, 1020–1023

BLUMBERG, W. E. (1978): Quart. Rev. Biophys. **11**, 481–542

BLUNEK, J. M., and C. E. CROWTHER (1975): Eur. J. Cancer **11**, 23–32

Bock, K. W. (1977): Arch. Toxicol. **39**, 77–85
Bolt, H. M., H. Kappus, and M. Bolt (1974): Lancet I, 1280–1281
Bolt, H. M., H. Kappus, and M. Bolt (1975): Eur. J. Clin. Pharmacol. **8**, 301–307
Bond, J. A., H.-Y. L. Yang, M. W. Majesky, E. P. Benditt, and M. R. Juchau (1980): Toxicol. Appl. Pharmacol. **52**, 323–335
Botelho, L. H., D. E. Ryan, and W. Levin (1979): J. Biol. Chem. **254**, 5635–5640
Boyd, M. R., and L. T. Burka (1978): J. Pharmacol. Exp. Ther. **207**, 687–697
Boyd, M. R., L. T. Burka, B. J. Wilson, and H. A. Sasame (1978): J. Pharmacol. Exp. Ther. **207**, 677–686
Boyd, M. R., C. N. Statham, and N. S. Longo (1980): J. Pharmacol. Exp. Ther. **212**, 109–114
Brandenburger Brown, E. A. (1974): Drug Metab. Rev. **3**, 33–87
Bridges, J. W., and J. R. Fry (1979) in: The Induction of Drug Metabolism (R. W. Estabrook, and E. Lindenlaub, eds.), pp. 453–469, Schattauer, Stuttgart, New York
Brodie, B. B., W. D. Reid, A. K. Cho, G. Sipes, G. Krishna, and J. R. Gillette (1971): Proc. Natl. Acad. Sci. U.S. **68**, 160–164
Brown Jr., B. R., (1972): Anesthesiology **36**, 458–465
Brown, B. R., and I. G. Sipes (1977): Biochem. Pharmacol. **26**, 2091–2094
Brown, S., F. J. Wiebel, H. V. Gelboin, and J. D. Minna (1976): Proc. Natl. Acad. Sci. U.S. **73**, 4628–4632
Buening, M. K., and M. R. Franklin (1976): Drug Metab. Dispos. **4**, 556–561
Bunker, J. P., W. H. Forrest Jr., F. Mosteller, and L. D. Vandam (eds.) (1969): The National Halothane Study. A Study of the Possible Association Between Halothane Anesthesia and Postoperative Hepatic Necrosis, National Institutes of Health, National Institute of General Medical Sciences, Bethesda, Md.
Burnett, C. M., C. M. Fuchs, and J. F. Corbett (1979): Drug Chem. Toxicol. **2**, 283–293
Burnett, C., M. M. Jacobs, A. Seppala, and P. Shubik (1980): J. Toxicol. Environm. Hlth **6**, 247–257
Burnett, C., B. Lanman, R. Giovacchini, G. Wolcott, R. Scala, and M. Keplinger (1975): Food Cosmet. Toxicol. **13**, 353–357
Byus, C. V., M. Costa, I. G. Sipes, B. B. Brodie, and D. H. Russell (1976): Proc. Natl. Acad. Sci. U.S. **73**, 1241–1245
Cabral, J. R. P., T. Mollner, F. Raitano, and P. Shubik (1979): Int. J. Cancer **23**, 47–51
Cabral, J. R. P., P. Shubik, T. Mollner, and F. Raitano (1977): Nature **269**, 510–511
Cardy, R. H. (1979): J. Natl. Cancer Inst. **62**, 1107–1116
Carr, H. S., and H. S. Rosenkranz (1978): Mutat. Res. **57**, 381–384
Carro-Ciampi, G. (1978): Toxicology **10**, 311–339
Cascorbi, H. F., and D. A. Blake (1971): Anesthesiology **35**, 493–495
Cascorbi, H. F., E. S. Vesell, D. A. Blake, and M. Helrich (1971): Ann. N.Y. Acad. Sci. **179**, 244–248
Cavalieri, E., R. Roth, E. Rogan, C. Grandjean, and J. Althoff (1978) in: Carcinogenesis, Vol. **3**, Polynuclear Aromatic Hydrocarbons, pp. 273–284, (P. W. Jones and R. I. Freudenthal, eds.), Raven Press, New York
Chhabra, R. S., R. J. Pohl, and J. R. Fouts (1974): Drug Metab. Dispos. **2**, 443–447
Chhabra, R. S., and J. M. Tredger (1978) in: Nutrition and Drug Interrelations, pp. 253–277 (J. N. Hathcock and J. Coon, eds.), Academic Press, New York
Chuang, A. H. L., E. F. Howard, and E. Bresnick (1977): Chem.-Biol. Interact. **17**, 9–16
Cinti, D. L., and M. B. Feinstein (1976): Biochem. Biophys. Res. Commun. **73**, 171–179
Clemens, T. L., R. N. Hill, L. P. Bullock, W. D. Johnson, L. G. Sultatos, and E. S. Vesell (1979): Toxicol. Appl. Pharmacol. **48**, 117–130
Cohen, E. N., J. R. Trudell, H. N. Edmunds, and E. Watson (1975): Anesthesiology **43**, 392–401
Cohen, G. M., A. C. Marchok, P. Nettesheim, V. E. Steele, F. Nelson, S. Huang, and J. K. Selkirk (1979): Cancer Res. **39**, 1980–1984
Conaway, C. C., B. V. Madhukar, and F. Matsumura (1977): Environm. Res. **14**, 305–321

CONNEY, A. H. (1967): Pharmacol. Rev. **19**, 317—366
CONNEY, A. H., and J. J. BURNS (1972): Science **178**, 576—586
CONNEY, A. H., and W. LEVIN (1974): IARC Sci. Publ. **10**, 3—22
COON, M. J., and K. P. VATSIS (1978) in: Polycyclic Hydrocarbons and Cancer (H. V. GELBOIN and P. O. Ts'o, eds.), Vol. **1**, pp. 335—360, Academic Press, New York
COUSINS, M. J. (1980): Drugs **19**, 1—6
COX, P. J., L. J. KING, and D. V. PARKE (1976): Xenobiotica **6**, 363—375
CRESTEIL, TH., PH. BEAUNE, J. P. LEROUX, M. LANGE, and D. MANSUY (1979): Chem.-Biol. Interact. **24**, 153—165
CUCINELL, S. A., A. H. CONNEY, M. SANSUR, and J. J. BURNS (1965): Clin. Pharmacol. Ther. **6**, 420—429
DANNI, O., B. C. SAWYER, and T. F. SLATER (1977): Biochem. Soc. Trans. **5**, 1029—1032
DAVIES, P. P. (1970): Lancet I, 413
DE FERREYRA, E. C., O. M. DE FENOS, A. S. BERNACCHI, C. R. DE CASTRO, and J. A. CASTRO (1979): Toxicol. Appl. Pharmacol. **48**, 221—228
DE GROOT, H., and W. HAAS (1980): FEBS Lett. **115**, 253—256
DE PIERRE, J. W., and L. ERNSTER (1978): Biochim. Biophys. Acta **473**, 149—186
DE TORANZO, E. G. D., M. I. DIAZ GÓMEZ, and J. A. CASTRO (1978): Res. Commun. Chem. Pathol. Pharmacol. **19**, 347—352
DIAZ GOMEZ, M. I., and J. A. CASTRO (1980a): Res. Commun. Chem. Pathol. Pharmacol. **27**, 191—194
DIAZ GOMEZ, M. I., and J. A. CASTRO (1980b): Cancer Lett. **9**, 213—218
DOCKS, E. L., and G. KRISHNA (1976): Exp. Molec. Pathol. **24**, 13—22
DOWN, W. H., and L. F. CHASSEAUD (1979): Biochem. Pharmacol. **28**, 3525—3528
DUBOIS, R. N., and M. R. WATERMAN (1979): Biochem. Biophys. Res. Commun. **90**, 150—157
DUTTON, G. J. (1978): Ann. Rev. Pharmacol. Toxicol. **18**, 17—35
DYBING, E. (1977): Acta Pharmacol. Toxicol. **41**, 89—93
DYBING, E., and J. R. MITCHELL (1977): Acta Pharmacol. Toxicol. **41**, Suppl. 2, 263—272
DYBING, E., and S. S. THORGEIRSSON (1977): Biochem. Pharmacol. **26**, 729—734
DYBING, E., T. AUNE, and S. D. NELSON (1979): Biochem. Pharmacol. **28**, 43—50
EHRLICH, K., A. HULETT, and T. TURNHAM (1980): J. Toxicol. Environm. Hlth **6**, 259—271
EICHELBAUM, M., K. W. KÖTHE, F. HOFFMANN, and G. E. VON UNRUH (1979): Clin. Pharmacol. Ther. **26**, 366—371
ELLIN, Å., S. V. JAKOBSSON, J. B. SCHENKMAN, and S. ORRENIUS (1972): Arch. Biochem. Biophys. **150**, 64—71
ELLIN, Å., and S. ORRENIUS (1975): Molec. Cell. Biochem. **8**, 69—79
EMERY, A. E. H., R. ANAND, N. DANFORD, W. DUNCAN, and L. PATON (1978): Lancet I, 470—472
FAHMY, M. J., and O. G. FAHMY (1977): Mutat. Res. **56**, 31—38
FISCHER, W., and K. L. SCHOLLER (1977): Med. Klin. **72**, 1959—1965
FOSSA, A. A., W. M. HADLEY, and J. L. BORN (1979): Toxicol. Lett. **4**, 379—384
FOX, C. H., J. K. SELKIRK, F. M. PRICE, R. G. CROY, K. K. STANFORD, and M. COTTLER-FOX (1975): Cancer Res. **35**, 3551—3557
FREUDENTHAL, R. I., A. P. LEBER, D. C. EMMERLING, G. A. KERCHNER, and A. A. OVEJERA (1976): Res. Commun. Chem. Pathol. Pharmacol. **15**, 267—278
FRY, B. J., T. TAYLOR, and D. E. HATHWAY (1972): Arch. Int. Pharmacodyn. Thér. **196**, 98—111
FRY, J. R., P. WIEBKIN, J. KAO, C. A. JONES, J. GWYNN, and J. W. BRIDGES (1978): Xenobiotica **8**, 113—120
FURUKAWA, M., S. R. SIRIANNI, J. C. TAN, and C. C. HUANG (1978): J. Natl. Cancer Inst. **60**, 1179—1181
GARATTINI, S., F. MARCUCCI, and E. MUSSINI (1975) in: Handbuch der experimentellen Pharmakologie, Vol. **28**, Part 3, Concepts in Biochemical Pharmacology (J. R. GILLETTE, J. R. MITCHELL, and P. S. RANDALL, eds.), pp. 113—129, Springer, Berlin, Heidelberg, New York
GARNER, R. C. (1978): Wld Rev. Nutr. Diet. **29**, 178—186
GARRETTSON, L. K., and P. L. DAYTON (1970): Clin. Pharmacol. Ther. **11**, 674—679
GARRO, A. J., and R. A. PHILLIPS (1978): Mutat. Res. **54**, 17—22
GATTIKER, R., A. D. SESSLER, R. O. LUNDBORG, and H. J. C. SWAN (1966): Anaesthesist **15**, 151—154

Gelboin, H. V., T. Okuda, J. Selkirk, N. Nemoto, S. K. Yang, P. J. Wiebel, J. P. Whitlock, H. J. Rapp, and R. C. Bast (1976): IARC Sci. Publ. **12**, 225–247

Giler, Sh., Y. Eshel, J. Pinkhas, E. Ventura, E. Levy, I. Urca, O. Sperling, and A. De Vries (1977): Experientia **33**, 1356–1358

Gillette, J. R. (1971): Ann. N.Y. Acad. Sci. **179**, 43–66

Gillette, J. R. (1976): Fed. Proc. **35**, 1142–1147

Gillette, J. R., D. C. Davis, and H. A. Sasame (1972): Ann. Rev. Pharmacol. **12**, 57–84

Gladtke, E., and G. Heimann (1978) in: Methoden der Klinischen Pharmakologie (H.-P. Kuemmerle, ed.), pp. 713–727, Urban & Schwarzenberg, München, Wien, Baltimore

Godoy, H. M., M. I. Diaz Gomez, and J. A. Castro (1978): J. Natl. Cancer Inst. **61**, 1285–1289

Gold, M. D., A. Blum, and B. N. Ames (1978): Science **200**, 785–787

Goldberg, D. M. (1980): Clin. Chem. **26**, 691–699

Goldstein, M., and E. B. Nelson (1979): Res. Commun. Chem. Pathol. Pharmacol. **23**, 203–206

Gram, T. E. (1973): Drug Metab. Rev. **2**, 1–32

Granberg-Öhman, I., S. Johansson, and A. Hjerpe (1980): Mutat. Res. **79**, 13–18

Green, S., F. M. Sauro, and L. Friedman (1975): Food Cosmet. Toxicol. **13**, 507–510

Green, T., and D. E. Hathway (1978): Chem.-Biol. Interact. **22**, 211–224

Gunsalus, I. C., T. C. Pederson, and S. G. Sligar (1975): Ann. Rev. Biochem. **44**, 377–407

Gurtoo, H. L., R. P. Dahms, P. Kanter, and J. B. Vaught (1978a): J. Biol. Chem. **253**, 3952–3961

Gurtoo, H. L., R. P. Dahms, and B. Paigen (1978b): Biochem. Biophys. Res. Commun. **81**, 965–972

Gutter, B., and H. S. Rosenkranz (1977): Mutat. Res. **56**, 89–90

Hänninen, O. (1975): Acta Pharmacol. Toxicol. **36**, Suppl. 2, 3–20

Haroz, R. K. (1979): Toxicol. Appl. Pharmacol. **48**, Part 2, A 151

Harris, C. C., H. Autrup, G. Stoner, S. K. Yang, J. C. Leutz, H. V. Gelboin, J. K. Selkirk, R. J. Connor, L. A. Barrett, R. T. Jones, E. McDowell, and B. F. Trump (1977): Cancer Res. **37**, 3349–3355

Heubel, F., and K. J. Netter (1979): Biochem. Pharmacol. **28**, 3373–3378

Hietanen, E., and M. Laitinen (1978): Biochem. Pharmacol. **27**, 1095–1097

Hietanen, E. and H. Vainio (1973): Acta Pharmacol. Toxicol. **33**, 57–64

Hill, D. L., T.-W. Shih, T. P. Johnston, and R. F. Struck (1978): Cancer Res. **38**, 2438–2442

Hinson, J. A., S. D. Nelson, and J. R. Gillette (1979b): Molec. Pharmacol. **15**, 419–427

Hinson, J. A., S. D. Nelson, and J. R. Mitchell (1977): Molec. Pharmacol. **13**, 625–633

Hinson, J. A., L. R. Pohl, and J. R. Gillette (1979a): Life Sci. **24**, 2133–2138

Hoensch, H. P., R. Hutt, and F. Hartmann (1979): Z. Gastroenterol. **17**, 600

Hogan, M. D., Pi-Y. Chi, D. G. Hoel, and T. J. Mitchell (1979): J. Environm. Pathol. Toxicol. **2**, 873–887

Holtzman, D., and M. Desautel (1980): J. Neurochem. **34**, 1535–1537

Honeck, H. (1978): Pharmazie **33**, 317–321

Hook, G. E. R., and J. R. Bend (1976): Life Sci. **18**, 279–290

Hook, G. E. R., J. R. Bend, D. Hoel, J. R. Fouts, and T. E. Gram (1979): J. Pharmacol. Exp. Ther. **182**, 474–490

Hughes, H. C., Jr. and C. M. Lang (1972): Anesthesiology **36**, 466–471

Hunter, J., and L. F. Chasseaud (1976): Clinical Aspects of Microsomal Enzyme Induction, in: Progress in Drug Metabolism, Vol. 1 (J. W. Bridges and L. F. Chasseaud, eds.), pp. 129–191, Wiley and Sons, London

Hunter, J., J. D. Maxwell, D. A. Stewart, V. Parsons, and R. Williams (1971): Brit. Med. J. **4**, 202–204

Ichihara, K., E. Kusunose, and M. Kusunose (1971): Biochim. Biophys. Acta **239**, 178–189

Ichihara, K., E. Kusunose, and M. Kusunose (1973): Eur. J. Biochem. **38**, 463–472

Ilett, K. F., W. D. Reid, I. G. Sipes, and G. Krishna (1973): Exp. Molec. Pathol. **19**, 215–229

Inouye, M., and U. Murakami (1977): Food Cosmet. Toxicol. **15**, 447–451

ISRAILI, Z. H., P. G. DAYTON, and J. R. KIECHEL (1977): Drug Metab. Dispos. **5**, 411–414

ITO, N., Y. HIASA, Y. KONISHI, and M. MARUGAMI (1969): Cancer Res. **29**, 1137–1145

IVANETICH, K. M., I. ARONSON, and I. D. KATZ (1977): Biochem. Biophys. Res. Commun. **74**, 1411–1418

JAKOBSSON, S. V., and D. L. CINTI (1973): J. Pharmacol. Exp. Ther. **185**, 226–234

JAKOBY, W. B. (1978): Adv. Enzymol. **46**, 383–414

JAKOBY, W. B., and J. H. KEEN (1977): Trends Biochem. Sci. **2**, 229–231

JEE, R. C., I. G. SIPES, A. J. GANDOLFI, and B. R. BROWN Jr. (1980): Toxicol. Appl. Pharmacol. **52**, 267–277

JENNER, P., and B. TESTA (1978): Xenobiotica **8**, 1–26

JOLLOW, D. J., J. R. MITCHELL, W. Z. POTTER, D. C. DAVIS, J. R. GILLETTE, and B. B. BRODIE (1973): J. Pharmacol. Exp. Ther. **187**, 195–202

JONES, R., C. ALLEGRA, and M. BOYD (1979): Toxicol. Appl. Pharmacol. **48**, Part 2, A 129

JOSTING, D., D. WINNE, and K. W. BOCK (1976): Biochem. Pharmacol. **25**, 613–616

JUCHAU, M. R. (1971): Toxicol. Appl. Pharmacol. **18**, 665–675

JUCHAU, M. R. (1980): Pharmacol. Ther. **8**, 501–524

JUDAH, J. D., A. E. M. MCLEAN, and E. K. MCLEAN (1970): Amer. J. Med. **49**, 609–616

JUSKO, W. J. (1979): Drug Metab. Rev. **9**, 221–236

KAHL, G. F. (1971): Klin. Wschr. **49**, 384–396

KALYANARAMAN, B., R. P. MASON, E. PEREZ-REYES, C. F. CHIGNELL, C. R. WOLF, and R. M. PHILPOT (1979): Biochem. Biophys. Res. Commun. **89**, 1065–1072

KAPITULNIK, J., W. LEVIN, P. J. POPPERS, J. E. TOMASZEWSKI, D. M. JERINA, and A. H. CONNEY (1976): Clin. Pharmacol. Ther. **20**, 557–564

KAPITULNIK, J., P. J. POPPERS, and A. H. CONNEY (1977): Clin. Pharmacol. Ther. **21**, 166–176

KARASHIMA, D., Y. HIROKATA, A. SHIGEMATSU, and T. FURUKAWA (1977): J. Pharmacol. Exp. Ther. **203**, 409–416

KELLERMANN, G., and M. LUYTEN-KELLERMANN (1978): Pharmacology **17**, 191–200

KELLERMANN, G., M. LUYTEN-KELLERMANN, M. G. HORNING, and M. STAFFORD (1976): Clin. Pharmacol. Ther. **20**, 72–80

KELLERMANN, G., M. LUYTEN-KELLERMANN, J. R. JETT, H. L. MOSES, and R. S. FONTANA (1978): Human Genet. **45**, Suppl. 1, 161–168

KELLERMANN, G., C. R. SHAW, and M. LUYTEN-KELLERMANN (1973): New Engl. J. Med. **289**, 934–937

KHERA, K. S., D. C. VILLENEUVE, G. TERRY, L. PANOPIO, L. NASH, and G. TRIVETT (1976): Food Cosmet. Toxicol. **14**, 25–29

KIMURA, N. T., T. KANEMATSU, and T. BABA (1976): Z. Krebsforsch. **87**, 257–266

KIRKLAND, D. J., and S. VENITT (1976): Mutat. Res. **40**, 47–56

KLAASSEN, C. D., and G. L. PLAA (1969): Biochem. Pharmacol. **18**, 2019–2027

KLINGER, W., D. MUELLER, and U. KLEEBERG (1979) in: The Induction of Drug Metabolism (R. W. ESTABROOK and E. LINDENLAUB, eds.), pp. 517–544, Schattauer, Stuttgart, New York

KLUWE, W. M., and J. B. HOOK (1978): Toxicol. Appl. Pharmacol. **45**, 861–869

KLUWE, W. M., K. M. MCCORMACK, and J. B. HOOK (1978): J. Pharmacol. Exp. Ther. **207**, 566–573

KNIGHT, R. H., and L. YOUNG (1958): Biochem. J. **70**, 111–119

KOCH, R. R., E. A. GLENDE Jr., and R. O. RECKNAGEL (1974): Biochem. Pharmacol. **23**, 2907–2915

KOHLI, K. K., R. M. PHILPOT, P. W. ALBRO, and J. D. MCKINNEY (1980): Life Sci. **26**, 945–952

KOLMODIN, B., D. L. AZARNOFF, and F. SJÖQVIST (1969): Clin. Pharmacol. Ther. **10**, 638–642

KRAMPL, V., and M. KONTSEKOVÁ (1978): Bull. Environm. Cont. Toxicol. **20**, 191–198

KRUSE, R. (1968): Mschr. Kinderheilk. **116**, 378–381

KUMAR, A., and G. PADMANABAN (1980): J. Biol. Chem. **255**, 522–525

KURZ, H., H.-G. NEUMANN, W. FORTH, D. HENSCHLER, and W. RUMMEL (1977) in: Allgemeine und spezielle Pharmakologie und Toxikologie, 2nd ed. (W. FORTH, D. HENSCHLER, and W. RUMMEL, eds.), pp. 1–72, Bibliographisches Institut Mannheim, Wien, Zürich, B. I.-Wissenschaftsverlag

LABADARIOS, D., J. W. T. DICKERSON, and D. V. PARKE (1978): Brit. J. Clin. Pharmacol. 5, 167–173

LAI, E. K., P. B. McCAY, T. NOGUCHI, and K.-L. FONG (1979): Biochem. Pharmacol. 28, 2231–2235

LAKE, B. G., J. C. PHILLIPS, C. E. HEADING, and S. D. GANGOLLI (1976a): Toxicology 5, 297–309

LAKE, B. G., J. C. PHILLIPS, S. D. GANGOLLI, and A. G. LLOYD (1976b): Biochem. Soc. Trans. 4, 684–685

LAMPE, J., and G. BUTSCHAK (1978): Pharmazie 33, 407–411

LAU, S. S., and V. G. ZANNONI (1979): Toxicol. Appl. Pharmacol. 50, 309–318

LEGRAVEREND, C., B. MANSOUR, D. W. NEBERT, and J. M. HOLLAND (1980): Pharmacology 20, 242–255

LEVIN, W., D. R. THAKKER, A. W. WOOD, R. L. CHANG, R. E. LEHR, D. M. JERINA, and A. H. CONNEY (1978): Cancer. Res. 38, 1705–1710

LEVIN, W., A. W. WOOD, H. YAGI, P. M. DANSETTE, D. M. JERINA, and A. H. CONNEY (1976): Proc. Natl. Acad. Sci. U.S. 73, 243–247

LEVY, G., R. A. O'REILLY, P. M. AGGELER, and G. M. KEECH (1970): Clin. Pharmacol. Ther. 11, 372–377

LIJINSKY, W., and P. SHUBIK (1964): Science 145, 53–55

LIN, J. K., K. A. KENNAN, E. C. MILLER, and J. A. MILLER (1978): Cancer. Res. 38, 2424–2428

LINDSTROM, T. D., and M. W. ANDERS (1978): Biochem. Pharmacol. 27, 563–567

LINNAINMAA, K., T. MERETOJA, M. SORSA, and H. VAINIO (1978): Mutat. Res. 58, 277–286

LITTERST, C. L., T. M. FARBER, and E. J. VAN LOON (1973): Toxicol. Appl. Pharmacol. 25, 354–362

LITTERST, C. L., E. G. MIMNAUGH, and T. E. GRAM (1977): Biochem. Pharmacol. 26, 749–755

LOEWENGART, G., and B. L. VAN DUUREN (1977): J. Toxicol. Environm. Hlth 2, 539–546

LOTLIKAR, P. D., Y. S. HONG, and W. J. BALDY (1978a): Toxicol. Lett. 2, 135–139

LOTLIKAR, P. D., Y. S. HONG, and W. J. BALDY (1978b): Cancer Lett. 4, 355–361

LOTLIKAR, P. D., and K. ZALESKI (1975): Biochem. J. 150, 561–564

LU, A. Y. H., and G. T. MIWA (1980): Ann. Rev. Pharmacol. Toxicol. 20, 513–531

LU, A. Y. H., and S. B. WEST (1978): Pharmacol. Ther. A 2, 337–358

MACGREGOR, J. T., and M. FRIEDMAN (1977): Mutat. Res. 56, 81–84

MAGEE, P. N., A. E. PEGG, and P. F. SWANN (1975) in: Handbuch der allgemeinen Pathologie, Geschwülste II, Virale und chemische Carcinogenese (E. GRUNDMANN, ed.), pp. 329–419, Springer, Berlin

MAKIURA, S., H. AOE, S. SUGIHARA, K. HIRAO, M. ARAI, and N. ITO (1974): J. Natl. Cancer Inst. 53, 1253–1257

MALAVEILLE, C., H. BARTSCH, P. L. GROVER, and P. SIMS (1975): Biochem. Biophys. Res. Commun. 66, 693–700

MANDEL, H. G. (1971) in: Fundamentals of Drug Metabolism and Drug Disposition (B. N. LADU, H. G. MANDEL, and E. L. WAY, eds.), pp. 149–181, Williams and Wilkins, Baltimore

MANSUY, D., P. BEAUNE, T. CRESTEIL, M. LANGE, and J.-P. LEROUX (1977): Biochem. Biophys. Res. Commun. 79, 513–517

MANSUY, D., W. NASTAINCZYK, and V. ULLRICH (1974): Naunyn-Schmiedebergs Arch. Pharmacol. 285, 315–324

MARIETTA, M. P., E. S. VESELL, R. D. HARTMAN, J. WEISZ, and B. DVORCHIK (1979): J. Pharmacol. Exp. Ther. 208, 271–279

MARNIEMI, J., and M. G. PARKKI (1975): Biochem. Pharmacol. 24, 1569–1572

MARTIN, J. V., P. J. MARTIN, and D. M. GOLDBERG (1976): Lancet I, 1107–1108

MATTHEWS, H. B., and S. KATO (1979): Ann. N.Y. Acad. Sci. 320, 131–137

MATTISON, D. R., and S. S. THORGEIRSSON (1978): Cancer Res. 38, 1368–1373

MAURER, H. M., J. A. WOLF, M. FINSTER, P. J. POPPERS, E. PANTUCK, R. KUNTZMAN, and A. H. CONNEY (1968): Lancet II, 122–124

McCARTY, L. P., R. S. MALEK, and E. R. LARSEN (1979): Anesthesiology 51, 106–110

McCONNELL, E. E., and J. A. MOORE (1979): Ann. N.Y. Acad. Sci. 320, 138–150

McLEMORE, T. L., R. R. MARTIN, D. L. BUSBEE, R. C. RICHIE, R. R. SPRINGER, K. L. TOPPELL, and E. T. CANTRELL (1977): Cancer. Res. 37, 1175–1181

McLemore, T. L., R. R. Martin, L. R. Pickard, R. R. Springer, N. P. Wray, K. L. Toppell, K. L. Mattox, G. A. Guinn, E. T. Cantrell, and D. L. Busbee (1978a): Cancer **41**, 2292–2300

McLemore, T. L., R. R. Martin, N. P. Wray, E. T. Cantrell, and D. L. Busbee (1978b): Cancer Res. **38**, 3805–3811

McMurtry, R. J., and J. R. Mitchell (1977): Toxicol. Appl. Pharmacol. **42**, 285–300

McMurtry, R. J., W. R. Snodgrass, and J. R. Mitchell (1978): Toxicol. Appl. Pharmacol. **46**, 87–100

Meehan, T., and K. Straub (1979): Nature **277**, 410–412

Mihatsch, M. J., T. Manz, C. Knüsli, H. O. Hofer, M. Rist, R. Guetg, G. Rutishauser, and H. U. Zollinger (1980): Schweiz. Med. Wschr. **110**, 255–264

Miller, J. A., and E. C. Miller (1976): IARC Sci. Publ. **12**, 153–176

Miner, D. J., and P. T. Kissinger (1979): Biochem. Pharmacol. **28**, 3285–3290

Mitchell, J. R., D. J. Jollow, W. Z. Potter, D. C. Davis, J. R. Gillette, and B. B. Brodie (1973a): J. Pharmacol. Exp. Ther. **187**, 185–194

Mitchell, J. R., D. J. Jollow, W. Z. Potter, J. R. Gillette, and B. B. Brodie (1973b): J. Pharmacol. Exp. Ther. **187**, 211–217

Mitchell, J. R., W. L. Nelson, W. Z. Potter, H. A. Sasame, and D. J. Jollow (1976): J. Pharmacol. Exp. Ther. **199**, 41–52

Mitchell, J. R., W. Z. Potter, J. A. Hinson, and D. J. Jollow (1974a): Nature **251**, 508–511

Mitchell, J. R., W. Z. Potter, J. A. Hinson, W. R. Snodgrass, J. A. Timbrell, and J. R. Gillette (1975) in: Handbuch der experimentellen Pharmakologie, Vol. **28**, Part 3, Concepts in Biochemical Pharmacology (J. R. Gillette, J. R. Mitchell, and P. S. Randall, eds.), pp. 383–419, Springer, Berlin–Heidelberg–New York

Mitchell, J. R., S. S. Thorgeirsson, W. Z. Potter, D. J. Jollow, and H. Keiser (1974b): Clin. Pharmacol. Ther. **16**, 676–684

Morselli, P. L., M. Rizzo, and S. Garattini (1971): Ann. N.Y. Acad. Sci. **179**, 88–107

Mountain, K. R., J. Hirsh, and A. S. Gallus (1970): Lancet I, 265–268

Mudge, G. H., M. W. Gemborys, and G. G. Duggin (1978): J. Pharmacol. Exp. Ther. **206**, 218–226

Mukhtar, H., T. H. Elmamlouk, and J. R. Bend (1979): Arch. Biochem. Biophys. **192**, 10–21

Mulder, G. J. (1979): Trends Biochem. Sci. **4**, 86–90

Murray, F. J., B. A. Schwetz, J. G. McBride, and R. E. Staples (1979): Toxicol. Appl. Pharmacol. **50**, 515–522

Nakamura, A., N. Tateno, S. Kojima, M.-A. Kaniwa, and T. Kawamura (1979): Mutat. Res. **66**, 373–380

Nastainczyk, W., and V. Ullrich (1978): Biochem. Pharmacol. **27**, 387–392

Neal, G. E., and P. J. Colley (1978): Biochem. J. **174**, 839–851

Nebert, D. W. (1979a) in: The Induction of Drug Metabolism (R. W. Estabrook and E. Lindenlaub, eds.), pp. 419–452, Schattauer, Stuttgart, New York

Nebert, D. W. (1979b): Pharmacol. Ther. **6**, 395–417

Nebert, D. W., and S. A. Atlas (1978): Human Genet. Suppl. **1**, 149–160

Nebert, D. W., and H. V. Gelboin (1970): J. Biol. Chem. **245**, 160–168

Nebert, D. W., and J. E. Gielen (1971): J. Biol. Chem. **246**, 5199–5206

Nebert, D. W., R. C. Levitt, and O. Pelkonen (1979) in: Carcinogens: Identification and Mechanisms of Action (A. C. Griffin and C. R. Shaw, eds.), pp. 157–185, Raven Press, New York

Nelson, S. D., A. J. Forte, and D. C. Dahlin (1980): Biochem. Pharmacol. **29**, 1617–1620

Nelson, S. D., A. J. Forte, and R. J. McMurtry (1978a): Res. Commun. Chem. Pathol. Pharmacol. **22**, 61–71

Nelson, S. D., W. A. Garland, J. R. Mitchell, Y. Vaishnav, C. N. Statham, and A. R. Buckpitt (1978b): Drug Metab. Dispos. **6**, 363–367

Nishizumi, M. (1976): Cancer Lett. **2**, 11

Nocke-Finck, L., H. Breuer, and D. Reimers (1973): Dtsch. Med. Wschr. **98**, 1521–1523

Oesch, F. (1979) in: Progress in Drug Metabolism (J. W. Bridges and L. F. Chasseaud, eds.), Vol. **3**, pp. 253–301, Wiley and Sons, Chichester, New York

Ohnhaus, E. E., and B. K. Park (1979): Eur. J. Clin. Pharmacol. **15**, 139–145

O'Reilly, R. A. (1975): Ann. Int. Med. **83**, 506–508

PANTUCK, E. J., K.-C. HSIAO, A. H. CONNEY, W. A. GARLAND, A. KAPPAS, K. E. ANDERSON, and A. P. ALVARES (1976): Science **194**, 1055–1057

PANTUCK, E. J., C. B. PANTUCK, W. A. GARLAND, B. H. MIN, L. W. WATTENBERG, K. E. ANDERSON, A. KAPPAS, and A. H. CONNEY (1979): Clin. Pharmacol. Ther. **25**, 88–95

PARK, K. K., and M. C. ARCHER (1978): Chem.-Biol. Interact. **22**, 83–90

PARKE, D. V. (1976): Chem. Ind. 380–388

PARKE, D. V. (1979) in: Drug Toxicity (J. W. GORROD, ed.), pp. 133–150, London

PARKINSON, A., R. COCKERLINE, and S. SAFE (1980): Chem.-Biol. Interact. **29**, 277–289

PATTON, S. E., G. M. ROSEN, E. J. RAUCKMAN, D. G. GRAHAM, B. SMALL, and D. M. ZIEGLER (1980): Molec. Pharmacol. **18**, 151–156

PETERS, U., T.-U. HAUSAMEN, and F. GROSSE-BROCKHOFF (1974): Dtsch. Med. Wschr. **99**, 2381–2386

PETERSON, R. G., and B. H. RUMACK (1977): J. Amer. Med. Ass. **237**, 2406–2407

PFEIFER, S. (1975, 1977, 1979): Biotransformation von Arzneimitteln, Vol. 1, 2 and 3, VEB Verlag Volk und Gesundheit, Berlin

PINKUS, L. M., J. N. KETLEY, and W. B. JAKOBY (1977): Biochem. Pharmacol. **26**, 2359–2363

PIRTTIAHO, H. I., E. A. SOTANIEMI, J. T. AHOKAS, and U. PITKÄNEN (1978): Brit. J. Clin. Pharmacol. **6**, 273–278

PITOT, H. C. (1979) in: The Induction of Drug Metabolism (R. W. ESTABROOK and E. LINDENLAUB, eds.), pp. 471–497, Schattauer, Stuttgart, New York

POHL, L. R., B. BHOOSHAN, N. F. WHITTACKER, and G. KRISHNA (1977): Biochem. Biophys. Res. Commun. **79**, 684–691

POHL, L. R., J. W. GEORGE, J. L. MARTIN, and G. KRISHNA (1979): Biochem. Pharmacol. **28**, 561–563

POHL, L. R., and G. KRISHNA (1978): Life Sci. **23**, 1067–1072

POLI, G., M. POLI CHIONO, T. F. SLATER, M. U. DIANZANI, and E. GRAVELA (1978): Biochem. Soc. Trans. **6**, 589–591

POTTER, W. Z., D. C. DAVIS, J. R. MITCHELL, D. J. JOLLOW, J. R. GILLETTE, and B. B. BRODIE (1973): J. Pharmacol. Exp. Ther. **187**, 203–210

POYER, J. L., P. B. MCCAY, E. K. LAI, E. G. JANZEN, and E. R. DAVIS (1980): Biochem. Biophys. Res. Commun. **94**, 1154–1160

POYER, J. L., R. A. FLOYD, P. B. MCCAY, E. G. JANZEN, and E. R. DAVIS (1978): Biochim. Biophys. Acta **539**, 402–409

PRASAD, N., R. PRASAD, J. E. HARRELL, J. THORNBY, J. H. LIEM, P. T. HUDGINS, and G. A. GUINN (1979): Int. J. Cancer **24**, 415–420

PRESCOTT, L. F., R. W. NEWTON, C. P. SWAINSON, N. WRIGHT, A. R. W. FORREST, and H. MATTHEW (1974): Lancet I, 588–592

PRESCOTT, L. F., N. WRIGHT, P. ROSCOL, and S. S. BROWN (1971): Lancet I, 519–522

PRIVAL, M. J., E. C. MCCOY, B. GUTTER, and H. S. ROSENKRANZ (1977): Science **195**, 76–78

PROUGH, R. A., Z. SIPAL, and S. W. JAKOBSSON (1977): Life Sci. **21**, 1629–1636

PROUGH, R. A., and D. M. ZIEGLER (1977): Arch. Biochem. Biophys. **180**, 363–373

RAAFLAUB, J., and U. C. DUBACH (1972): Klin. Wschr. **55**, 489–497

RAJAMANICKAM, C., M. R. S. RAO, and G. PADMANABAN (1975): J. Biol. Chem. **250**, 2305–2310

RANNUG, U., A. SUNDVALL, and C. RAMEL (1978): Chem.-Biol. Interact. **20**, 1–16

RAO, G. S. (1977): J. Med. Chem. **20**, 262–265

RECKNAGEL, R. O. (1967): Pharmacol. Rev. **19**, 145–208

RECKNAGEL, R. O., and A. K. GHOSHAL (1966): Lab. Invest. **15**, 132–148

REDDY, B. S., and K. WATANABE (1978): J. Natl. Cancer Inst. **61**, 1269–1271

REID, W. D., B. CHRISTIE, M. EICHELBAUM, and G. KRISHNA (1971a): Exp. Molec. Pathol. **15**, 363–372

REID, W. D., B. CHRISTIE, G. KRISHNA, J. R. MITCHELL, J. MOSKOWITZ, and B. B. BRODIE (1971b): Pharmacology **6**, 41–55

REID, W. D., and G. KRISHNA (1973): Exp. Molec. Pathol. **18**, 80–99

REIMERS, D., and A. JEŽEK (1971): Praxis Pneumol. **25**, 255–262

REINER, O., S. ATHANASSOPOULOS, K. H. HELLMER, R. E. MURRAY, and H. UEHLEKE (1972): Arch. Toxikol. **29**, 219–233

REITZ, R. H., P. J. GEHRING, and C. N. PARK (1978): Food Cosmet. Toxicol. **16**, 511–514

REMMER, H. (1966): Dtsch. Med. Wsch. **91**, 289–296

REMMER, H. (1973): Arzneimittel-Forsch. **23**, 1150–1154
REMMER, H., R. FLEISCHMANN, and W. KUNZ (1979) in: The Induction of Drug Metabolism (R. W. ESTABROOK and E. LINDENLAUB, eds.), pp. 555–581, Schattauer, Stuttgart
REYNOLDS, E. S. (1967): J. Pharmacol. Exp. Ther. **155**, 117–126
REYNOLDS, E. S., and M. T. MOSLEN (1974): Biochem. Pharmacol. **23**, 189–195
REYNOLDS, E. S., and M. T. MOSLEN (1975): Biochem. Pharmacol. **24**, 2075–2081
REYNOLDS, E. S., M. T. MOSLEN, and S. SZABO (1976): Fed. Proc. **35**, 376
REYNOLDS, E. S., H. J. REE, and M. T. MOSLEN (1972): Lab. Invest. **26**, 290–299
REZNIK, G., and J. M. WARD (1979): Food Cosmet. Toxicol. **17**, 493–500
REZNIK, G., J. M. WARD, J. F. HARDISTY, and A. RUSSFIELD (1979): J. Natl. Cancer Inst. **63**, 205–212
RITSCHEL, W. A. (1978) in: Methoden der klinischen Pharmakologie (H.-P. KUEMMERLE, ed.), pp. 133–166, Urban & Schwarzenberg, München
ROTH, R. A., and D. A. WIERSMA (1979): Clin. Pharmacokinet. **4**, 355–367
SACHDEV, K., E. N. COHEN, and V. F. SIMMON (1980): Anesthesiology **53**, 31–39
SAGAI, M., and A. L. TAPPEL (1978): Toxicol. Lett. **2**, 149–155
SAGAI, M., and A. L. TAPPEL (1979): Toxicol. Appl. Pharmacol. **49**, 283–291
SASAME, H. A., and J. R. GILLETTE (1978): Biochem. Biophys. Res. Commun. **84**, 389–395
SCARPELLI, D. G., M. S. RAO, V. SUBBARAO, M. BEVERSLUIS, D. P. GURKA, and P. F. HOLLENBERG (1980): Cancer. Res. **40**, 67–74
SCHOENY, R. S., C. C. SMITH, and J. C. LOPER (1979): Mutat. Res. **68**, 125–132
SCHOLLER, K. L. (1970): Brit. J. Anaesth. **42**, 603–605
SEARLE, C. E., D. G. HARNDEN, S. VENITT, and O. H. B. GYDE (1975): Nature **255**, 506–507
SEIDEGÅRD, J., J. W. DE PIERRE, M. S. MORON, K. A. M. JOHANNESEN, and L. ERNSTER (1977): Cancer Res. **37**, 1075–1087
SEIDEGÅRD, J., M. S. MORON, L. C. ERIKSSON, and J. W. DEPIERRE (1978): Biochim. Biophys. Acta **543**, 29–40
SERABJIT-SINGH, C. J., C. R. WOLF, R. M. PHILPOT, and C. G. PLOPPER (1980): Science **207**, 1469–1470
SHAH, H., S. P. HARTMAN, and S. WEINHOUSE (1979): Cancer Res. **39**, 3942–3947
SHIRKEY, R. J., J. KAO, J. R. FRY, and J. W. BRIDGES (1979): Biochem. Pharmacol. **28**, 1461–1466
SHUDO, K., T. OHTA, Y. ORIHARA, T. OKAMOTO, M. NAGAO, Y. TAKAHASHI, and T. SUGIMURA (1978): Mutat. Res. **58**, 367–370
SIEGERS, C.-P., J. G. FILSER, and H. M. BOLT (1978): Toxicol. Appl. Pharmacol. **46**, 709–716
SIEGERS, C.-P., O. STRUBELT, A. SCHÜTT, and M. VÖLPEL (1976): Naunyn-Schmiedebergs Arch. Pharmacol. **293**, Suppl. R66
SIEȘT, G., A. M. BATT, and J. M. ZIEGLER (1978): Lyon Pharm. **29**, 347–364
SIMON, G. S., B. R. KIPPS, R. G. TARDIFF, and J. F. BORZELLECA (1978): Toxicol. Appl. Pharmacol. **45**, 330–331
SIMS, P. (1976): IARC Sci. Publ. **12**, 211–224
SIMS, P., and P. L. GROVER (1974): Adv. Cancer Res. **20**, 165–274
SIMS, P., P. L. GROVER, A. SWAISLAND, K. PAL, and H. HEWER (1974): Nature **252** 326–328
SIPES, I. G., A. J. GANDOLFI, L. R. POHL, G. KRISHNA, and B. R. BROWN Jr. (1980): J. Pharmacol. Exp. Ther. **214**, 716–720
SIPES, I. G., P. L. GIGON, and G. KRISHNA (1974): Biochem. Pharmacol. **23**, 451–455
SIPES, G., G. KRISHNA, and B. B. BRODIE (1971): Fed. Proc. **30**, 439
SIPES, I. G., G. KRISHNA, and J. R. GILLETTE (1977): Life Sci. **20**, 1541–1548
SLAGA, T. J., W. J. BRACKEN, G. GLEASON, W. LEVIN, H. YAGI, D. M. JERINA, and A. H. CONNEY (1979): Cancer Res. **39**, 67–71
SLATER, T. F. (1966): Nature **209**, 36–40
SMITH, B. R., J. H. MAGUIRE, L. M. BALL, and J. R. BEND (1978): Fed. Proc. **37**, 2480–2484
SØDERLUND, E. J., and E. DYBING (1979): Acta Pharmacol. Toxicol. **44**, 284–288
SØDERLUND, E. J., S. D. NELSON, and E. DYBING (1979): Acta Pharmacol. Toxicol. **45**, 112–121
SOLOMON, H. M., and W. B. ABRAMS (1972): Amer. Heart J. **83**, 277–280
SOLOMON, H. M., S. REICH, N. SPIRT, and W. B. ABRAMS (1971): Ann. N.Y. Acad. Sci. **179**, 362–369

SOTANIEMI, E. A., R. O. PELKONEN, and M. PUUKKA (1980): Eur. J. Clin. Pharmacol. **17**, 267–274

SPITZNAGLE, L. A., P. J. WIRTH, S. W. BOOBIS, S. S. THORGEIRSSON, and W. L. NELSON (1977): Toxicol. Appl. Pharmacol. **39**, 283–294

STENGER, R. J., and E. A. JOHNSON (1972): Proc. Soc. Exp. Biol. Med. **140**, 1319–1324

STIER, A. (1968): Anesthesiology **29**, 388–390

STOHS, S. J., R. C. GRAFSTRÖM, M. D. BURKE, P. W. MOLDÉUS, and S. G. ORRENIUS (1976 b) Arch. Biochem. Biophys. **177**, 105–116

STOHS, S. J., R. C. GRAFSTRÖM, M. D. BURKE, and S. ORRENIUS (1976 a): Drug Metab. Dispos. **4**, 517–521

STOHS, S. J., R. C. GRAFSTRÖM, M. D. BURKE, and S. ORRENIUS (1977): Arch. Biochem. Biophys. **179**, 71–80

STONARD, M. D., and J. B. GREIG (1976): Chem.-Biol. Interact. **15**, 365–379

STROBEL, H. W., W.-F. FANG, and R. J. OSHINSKY (1980): Cancer **45**, 1060–1065

SUN, J. D., and J. G. DENT (1979): Toxicol. Appl. Pharmacol. **48**, Part 2, A129

TEEL, R. W. (1979): Cancer Lett. **7**, 349–355

THOR, H., P. MOLDÉUS, R. HERMANSON, J. HÖGBERG, D. J. REED, and S. ORRENIUS (1978 a): Arch. Biochem. Biophys. **188**, 122–129

THOR, H., P. MOLDÉUS, A. KRISTOFERSON, J. HÖGBERG, D. J. REED, and S. ORRENIUS (1978 b): Arch. Biochem. Biophys. **188**, 114–121

THOR, H., P. MOLDÉUS, and S. ORRENIUS (1979): Arch. Biochem. Biophys. **192**, 405–413

THOR, H., and S. ORRENIUS (1980): Arch. Toxicol. **44**, 31–43

THORGEIRSSON, S. S., S. A. ATLAS, A. R. BOOBIS, and J. S. FELTON (1979): Biochem. Pharmacol. **28**, 217–226

THORGEIRSSON, S. S., D. J. JOLLOW, H. A. SASAME, I. GREEN, and J. R. MITCHELL (1973): Molec. Pharmacol. **9**, 398–404

TORANZO, E. G., T. GILLESSE, M. MENDENHALL, G. J. TRAIGER, P. G. RILEY, R. P. HANZLIK, and R. A. WILEY (1977): Toxicol. Appl. Pharmacol. **40**, 415–425

TRIOLO, A. J., G. E. APONTE, and D. L. HERR (1977): Cancer Res. **37**, 3018–3021

UEHLEKE, H. (1968): Mitt. Dtsch. Pharm. Gesell. **38**, 1–15

UEHLEKE, H., K. H. HELLMER, and S. TABARELLI-POPLAWSKI (1973): Naunyn-Schmiedebergs Arch. Pharmacol. **279**, 39–52

UEHLEKE, H., T. WERNER, H. GREIM, and M. KRÄMER (1977): Xenobiotica **7**, 393–400

UENO, I., L. FRIEDMAN, and C. L. STONE (1980): Toxicol. Appl. Pharmacol. **52**, 177–180

ULLRICH, V. (1977): Arzneimittel-Forsch. **27**, 1821–1827

ULLRICH, V., G. HERMANN, and P. WEBER (1978): Biochem. Pharmacol. **27**, 2301–2304

VAINIO, H., M. G. PARKKI, and J. MARNIEMI (1976): Xenobiotica **6**, 599–604

VAN DUUREN, B. L., G. LOEWENGART, I. SEIDMAN, A. C. SMITH and S. MELCHIONNE (1978): Cancer Res. **38**, 3236–3240

VAN DYKE, R. A., and C. L. WOOD (1973): Anesthesiology **38**, 328–332

VAUGH, J. B., H. L. GURTOO, B. PAIGEN, J. MINOWADA, and P. SARTORI (1978): Cancer Lett. **5**, 261–268

VAUGHAN, R. W., I. G. SIPES, and B. R. BROWN Jr. (1978): Life Sci. **23**, 2447–2462

VESELL, E. S., J. G. PAGE, and G. T. PASSANANTI (1971): Clin. Pharmacol. Ther. **12**, 192–201

VILLARRUEL, M. del C., E. G. D. DE TORANZO, and J. A. CASTRO (1977): Toxicol. Appl. Pharmacol. **41**, 337–344

WARD, J. M., S. F. STINSON, J. F. HARDISTY, B. Y. COCKRELL, and D. W. HAYDEN (1979): J. Natl. Cancer Inst. **62**, 1067–1073

WATTENBERG, L. W. (1971): Cancer **28**, 99–102

WATTENBERG, L. W. (1972): Toxicol. Appl. Pharmacol. **23**, 741–748

WATTENBERG, L. W., J. L. LEONG, and P. J. STRAND (1962): Cancer Res. **22**, 1120–1125

WELCH, R. M., Y. E. HARRISON, B. W. GOMMI, P. J. POPPERS, M. FINSTER, and A. H. CONNEY (1969): Clin. Pharmacol. Ther. **10**, 100–109

WHITING, B., M. ULLAH, and P. A. MEREDITH (1976): Brit. Med. J. 471

WHITLOCK Jr., J. P., and H. V. GELBOIN (1973): J. Biol. Chem. **248**, 6114–6121

WHITLOCK Jr., J. P., and H. V. GELBOIN (1974): J. Biol. Chem. **249**, 2616–2623

WHITLOCK Jr., J. P., H. MILLER, and H. V. GELBOIN (1974): J. Cell Biol. **63**, 136–145

WIDGER, L. A., A. J. GANDOLFI, and R. A. VAN DYKE (1976): Anesthesiology **44**, 197–201

WIRTH, P. J., C. J. BETTIS, and W. L. NELSON (1975): Molec. Pharmacol. **12**, 759–768

WOLF, C. R., D. MANSUY, W. NASTAINCZYK, G. DEUTSCHMANN, and V. ULLRICH (1977): Molec. Pharmacol. **13**, 698–705

WOOD, C. L., A. J. GANDOLFI, and R. A. VAN DYKE (1976): Drug. Metab. Dispos. **4**, 305–313

WOOD, A. W., and B. A. TAYLOR (1979): J. Biol. Chem. **254**, 5647–5651

YAMASAKI, H., E. HUBERMAN, and L. SACHS (1975): J. Biol. Chem. **250**, 7766–7770

YAMASAKI, H., E. HUBERMAN, and L. SACHS (1977): Int. J. Cancer **19**, 378–382

YAMAZOE, Y., M. SUGIURA, T. KAMATAKI, and R. KATO (1979): Japan J. Pharmacol. **29**, 715–721

YOSHIMURA, H., and S. YOSHIHARA (1976) in: PCB Poisoning and Pollution (K. HIGUCHI, ed.), pp. 41–67, Kodansha Ltd., Tokyo

YOUNES, M., C.-P. SIEGERS, and J. G. FILSER (1979): Arch. Toxicol. **42**, 289–293

ZAMPAGLIONE, N., D. J. JOLLOW, J. R. MITCHELL, B. STRIPP, M. HAMRICK, and J. R. GILLETTE (1973): J. Pharmacol. Exp. Ther. **187**, 218–227

ZAMPAGLIONE, N. G., and G. J. MANNERING (1973): J. Pharmacol. Exp. Ther. **185**, 676–685

ZENSER, T. V., M. B. MATTAMMAL, and B. B. DAVIS (1978): J. Pharmacol. Exp. Ther. **207**, 719–725

ZILLY, W., D. D. BREIMER, and E. RICHTER (1975): Eur. J. Clin. Pharmacol. **9**, 219–227

General considerations on volatile anaesthetics (1976) in: IARC Monographs on the Evaluation of Carcinogenic Risk of Chemicals to Man, Vol. 11, pp. 285–293, International Agency for Research on Cancer, Lyon

Polychlorinated Biphenyls and Polybrominated Biphenyls (1978) in: IARC Monographs on the Evaluation of the Carcinogenic Risk of Chemicals to Humans, Vol. 18, International Agency for Research on Cancer, Lyon

Some Monomers, Plastics and Synthetic Elastomers, and Acrolein (1979a) in: IARC Monographs on the Evaluation of the Carcinogenic Risk of Chemicals to Humans, Vol. 19, International Agency for Research on Cancer, Lyon

Some Halogenated Hydrocarbons (1979b) in: IARC Monographs on the Evaluation of the Carcinogenic Risk of Chemicals to Humans, Vol. 20, International Agency for Research on Cancer, Lyon

Carbon tetrachloride (1979c) in: Some Halogenated Hydrocarbons, IARC Monographs on the Evaluation of the Carcinogenic Risk of Chemicals to Humans, Vol. 20, pp. 371–399, International Agency for Research on Cancer, Lyon

Chloroform (1979d) in: Some Halogenated Hydrocarbons, IARC Monographs on the Evaluation of the Carcinogenic Risk of Chemicals to Humans, Vol. 20, pp. 401–427, International Agency for Research on Cancer, Lyon

Tris(2,3-dibromopropyl) phosphate (1979e) in: Some Halogenated Hydrocarbons, IARC Monographs on the Evaluation of the Carcinogenic Risk of Chemicals to Humans, Vol. 20, pp. 575–588, International Agency for Research on Cancer, Lyon

Chapter 7

Cytochrome P-450 of Microorganisms

H.-G. MÜLLER, W.-H. SCHUNCK,
P. RIEGE and H. HONECK

Contents

1. Introduction

The general biological importance of monooxygenases consists in their capacity to attack unactivated C—H-bonds, an ability, which is lacking other enzymes. Therefore, it cannot surprise, that they are integrated in many biosynthetic and degradative pathways both in procaryotes and eucaryotes. A number of special metabolic capacities are joint to these enzymes. The persistent interest of biochemists and microbiologists for these subgroup of oxidoreductases can be established by the following facts.

Monooxygenases hold key positions in the metabolism of aerobic microorganisms which play an essential role in the biological carbon cycle. These microorganisms are able to degradate and to some extent utilize for growth inert compounds, which originate continuously as final products of animal and plant metabolism and of microbial fermentation processes, first of all methane, further aliphatic, alicyclic, aromatic, and polycyclic hydrocarbons and heterocyclics.

In this century the ecological importance of this capacity of several aerobic microorganisms increased yet, as additional amounts of the mentioned and other recalcitrant organic compounds were world-wide spread and their number and quantities are growing permanently by the output of the developing industry. So far these xenobiotics can be degraded by biological systems monooxygenatic reactions are often involved in this process, frequently they initiate it.

Furthermore, monooxygenases are integrated in the biosynthesis of amino acids and secondary metabolites in microorganisms.

Since the detection of monooxygenases by MASON (1955) extensive studies in numerous laboratories demonstrated that P–450 is a very essential type of monooxygenases both in relation to the importance of the catalyzed reactions and to the extensive occurrence in animals, plants, fungi, and bacteria. Most investigators concentrated their attention to the mammalian P–450, especially of the liver. Therefore, only a very small number of the detected microbial P–450 species were studied more thoroughly.

With regard to the limited scope of the review it is not possible to discuss all results about the P–450 forms of the different microorganisms in detail. The main view was directed to the molecular properties and to the reaction mechanism. Other aspects were taken into consideration so far it seemed to be necessary for a comprehensive understanding. Reviews from other points of view given recently by WISEMAN (1977), ROSAZZA and SMITH (1979), SMITH and DAVIS (1980), and WISEMAN (1980).

2. Historical Overview

Already in 1964 LINDENMAYER and SMITH demonstrated P–450 in cell extracts from *Saccharomyces cerevisiae* and they recognized the similarity of this yeast hemoprotein with liver microsomal P–450. However, the first detection was not followed immediately by further investigations. Only ISHIDATE et al. (1969a, b) showed in more detail the influence of oxygen on the P–450 content in this yeast. But, the biological function remained unclear and hasn't been understood completely up to now, in spite of essential advances in the 1970s, mainly by intense studies in the laboratories of YOSHIDA and WISEMAN (see 4.2).

In the meantime, 2 soluble P–450 species were detected by APPLEBY (1967) in the N_2-fixing bacteroid *Rhizobium japonicum* and by GUNSALUS' group in a camphor-utilizing strain of the soil bacterium *Pseudomonas putida* (KATAGIRI et al. 1968). Due to the easy availability of the P–450 of *P. putida*, its discovery was the starting-point of extensive investigations about the molecular properties and the reaction mechanism, influencing decisively the whole research on P–450 and delivering an important part of our present general knowledge about the P–450 containing enzyme systems (see 3.1).

The discoveries that P–450 is formed in alkane-assimilating cells both of a Corynebacterium (CARDINI and JURTSHUK, 1968) and of a *Candida tropicalis* strain (LEBEAULT et al. 1971) were further important steps, stimulating investigations in different laboratories. For several yeast strains, which are able to grow on n-alkanes as the only carbon source (Tab. 1), it was suggested that this growth substrate acts as the inducer of P–450. The alkane-induced P–450 was regarded as responsible for catalyzing the first step of alkane degradation (CARDINI and JURTSHUK, 1968, 1970, LEBEAULT et al. 1971).

A similar involvement in the first step of growth substrate utilization seems to be the physiological function of P–450 in many of the procaryotic and eucaryotic microorganisms listed in Table 1. Accordingly, P–450 containing enzyme systems with very different substrate specificity were found.

3. Cytochrome P-450 of Bacteria

All P–450 proteins, which are known hitherto, are constituents of multiprotein complexes. Several well investigated P–450

Table 1. Microbial cytochrome P–450 systems

Microorganism	Physiological function or catalyzed reaction	References
Acinetobacter spec.	induced by n-alkanes	MÜLLER et al. 1978 ASPERGER et al. 1981
Bacillus megaterium	15β-hydroxylation of 3-oxo-Δ^4steroids (a, c)	BERG et al. 1975, 1978
Bacillus megaterium	(ω-2)-fatty acid hydroxylation (b)	MATSON et al. 1977
Corynebacterium sp.	n-alkane hydroxylation (b)	CARDINI and JURTSHUK 1970
Corynebacterium sp.	involvement in cholesterol-degradation (c)	YAMAMOTO et al. 1974
Nocardia NH1	N-demethylation of phenolic ethers (c)	CARTWRIGHT et al. 1971
Pseudomonas putida	Camphor-5-exo-hydroxylation (a)	KATAGIRI et al. 1968,
Rhizobium japonicum	unknown	APPLEBY 1967 (and personal communication 1979)
Streptomyces erythreus	involvement in biosynthesis of erythromycine-antibiotics: hydroxylation of 6-Deoxyerythronolide	VYGANTAS and CORCORAN 1974
Brettanomyces anomalus	unknown	KÄRENLAMPI et al. 1980b
Candida tropicalis LM7	hydroxylation of n-alkanes, fatty acids and drugs (b)	LEBEAULT et al. 1971 DUPPEL et al. 1973
Candida tropicalis 101	hydroxylation of n-alkanes and fatty acids (b)	GALLO et al. 1971 BERTRAND et al. 1979
Candida guilliermondii	not determined, but induced by n-alkanes	TITTELBACH et al. 1976 MAUERSBERGER et al. 1980
Candida pulcherima	by n-alkanes	TAKAGI et al. 1980a
Candida maltosa	by n-alkanes	TAKAGI et al. 1980c
Debaryomyces hansenii	unknown	KÄRENLAMPI et al. 1980b
Hansenula anomala	unknown	KÄRENLAMPI et al. 1980b
Kluyveromyces fragilis	unknown	KÄRENLAMPI et al. 1980b
Lodderomyces elongisporus	terminal hydroxylation of n-alkanes (a, b)	SCHUNCK et al. 1978 RIEGE et al. 1981
Pichia fermentans	unknown	KÄRENLAMPI et al. 1980b
Saccharomyces cerevisiae	—	LINDENMAYER and SMITH 1964
	14α-demethylation of lanosterol (a)	AOYAMA and YOSHIDA 1978a, b
	hydroxylation of benzo(a)pyren (c)	WOODS and WISEMAN 1980
	hydroxylation of biphenyl (c)	WISEMAN et al. 1975
	activation of promutagens (c)	CALLEN and PHILPOT 1977
	metabolism of halogenated hydrocarbons (c)	CALLEN et al. 1980
Saccharomyces carlsbergensis	unknown	CARTLEDGE et al. 1972
Saccharomyces bayanus	unknown	KÄRENLAMPI et al. 1980b
Saccharomyces chevalieri	unknown	KÄRENLAMPI et al. 1980b

Microorganism	Physiological function or catalyzed reaction	References
Saccharomyces italicus	unknown	Kärenlampi et al. 1980b
Saccharomyces uvarum AH15	unknown	Kärenlampi et al. 1980b
Saccharomycopsis lipolytica	hydroxylation of n-alkanes (b)	Delaisse and Nyns 1974
Schizosaccharomyces pombe	unknown	Poole et al. 1974
Schizosaccharomyces japonicus	unknown	Kärenlampi et al. 1980b
Torulopsis spec.	ω- or ω-1 hydroxylation of fatty acids, terminal hydroxylation of n-alkanes (b)	Heinz et al. 1970
Torulopsis candida	not determined, but induced by n-alkanes	Ilchenko et al. 1980
Torulopsis dattila	unknown	Kärenlampi et al. 1980b
Torulopsis glabrata	unknown	Kärenlampi et al. 1980b
Acremonium sp.	oxidation of ethane, propane, butane	Davies et al. 1976
Claviceps purpurea	involvement in alkaloid biosynthesis	Ambike et al. 1970
Cunninghamella bainieri	hydroxylation of arylhydrocarbons (b)	Ferris et al. 1976
Cunninghamella elegans	hydroxylation of naphthalene (b)	Cerniglia and Gibson 1978a
Penicillium patulum	involvement in patulin biosynthesis: hydroxylation of the methyl group of m-cresol	Murphy et al. 1974
Rhizopus nigricans	11α-hydroxylation of progesterone (c)	Breskvar and Hudnik-Plevnik 1977
Pseudomonas spec.	hydroxylation of linalool	Gunsalus 1982
Aspergillus ochraceus	11α-hydroxylation of progesterone	Jayanthi et al. 1982

footnotes: hydroxylase activities were determined with (a) highly purified reconstituted system
(b) microsomes or partially purified system
(c) cells or cellfree system

containing monooxygenase systems of bacteria turned out to consist of 3 proteins, a flavoprotein, containing FAD or FMN as the prosthetic group, an iron-sulfur protein, both for the e^--transfer, and P–450 as terminal oxidase. In this respect, they are more similar to the mitochondrial than to the microsomal P–450 systems of mammalian cells. For P–450 containing systems of Nocardia (see 3.4) and Corynebacterium (3.5) the lacking of iron-sulfur proteins was suggested. However, the knowledge about these and the most other P–450 systems of bacteria, except the camphor-hydroxylating P–450 of *Pseudomonas putida*, is rather poor. Therefore, it is unclear up to now, whether the existence of an iron-sulfur protein in bacterial P–450 systems is a general rule or not.

3.1. The Camphor-Hydroxylating Enzyme System of Pseudomonas Putida

As already mentioned the camphor-hydroxylating enzyme system of *P. putida* is the best known P–450-containing monooxygenase system of microorganisms investigated so far and has been repeatedly discussed in excellent reviews (GUNSALUS et al. 1974, 1975, ULLRICH and DUPPEL 1975, ISHIMURA 1978, GUNSALUS and SLIGAR 1978). Therefore, in this article only the most important results are surveyed.

Several *Pseudomonas putida* strains were isolated, which have the capability to grow on camphor as the only carbon source. This unconventional substrate causes in the bacteria the induction of a stereoselective camphor-hydroxylating monooxygenase system with P–450_{CAM} as the terminal oxidase. The enzyme is soluble and catalyzes an early step of the camphor catabolism. In position C_5 the methylene group is hydroxylated and the exo-alcohol is produced (HEDEGAARD and GUNSALUS 1965):

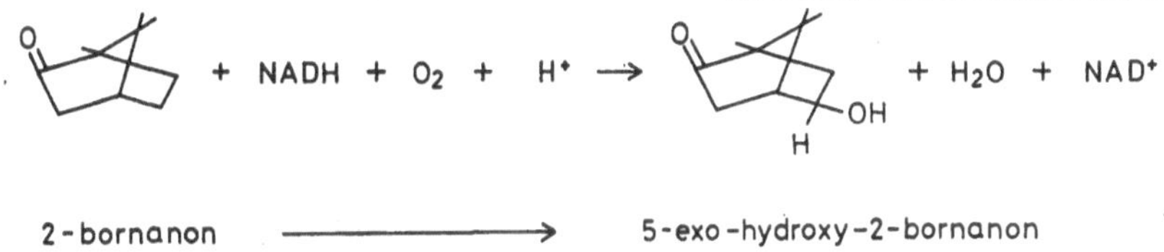

3.1.1. Molecular Properties of the Individual Constituents

The camphor-hydroxylating enzyme system of *P. putida* consists of three proteins: the P–450_{CAM} as terminal oxidase, the putidaredoxin as the immediate electron transmitter and the NADH-putidaredoxin reductase as acceptor of the electrons provided by NADH (KATAGIRI et al. 1968). All 3 proteins are readily soluble and were obtained in a homogeneous state. A comprehensive description of the purification procedures of the three proteins were published recently by GUNSALUS and WAGNER (1978). The recombination of all 3 constituents results in the highly active camphor hydroxylase with a turnover of 1200 min^{-1} (GUNSALUS et al. 1974). This value is orders of magnitudes greater than the hydroxylation rates of mammalian P–450, which are ranging between 10 and 0.1 min^{-1} for most of the substrates, except benzphetamine with a turnover number of 63 min^{-1} in the reconstituted LM_2 system (COON and VATSIS 1978). P–450_{CAM} and all other P–450 species investigated in detail up to now are characterized by common physical properties, which destinguish the whole group and all other hemoproteins. In Fig. 1 the most important optical spectra of the P–450_{CAM} are summarized and the coefficients of maxima are scheduled in Table 2.

The most striking and commonly known feature of P–450 is the strong red shift of the Soret band of the CO-complex to the 450 nm range. The CO-complex of P–450_{CAM} exhibits its absorption maximum at 447 or 446 nm in dependence on the presence or absence of the substrate. By addition of camphor to the substrate-free ferric P–450_{CAM} a blue shift of the Soret band from 417 to 391 nm occurs and new bands appear near 500 and 650 nm. Simultaneously, the α- and the β-bands (567 and 538 nm, respectively) disappear. These spectral changes reflect a variation of the spin state and represent a further conspicuous characteristic of P–450. In the EPR spectrum this spin state change is detectable only at deep temperature. The oxidized substrate-free P–450_{CAM} exhibits an EPR signal with g values 2.45, 2.26, and 1.91, which is typical for the low spin state of the ferric heme iron. Upon addition of the substrate a high spin signal with g values 8, 4, and 1.8 appears. However, only 60% can be detected in the high-spin state, the rest

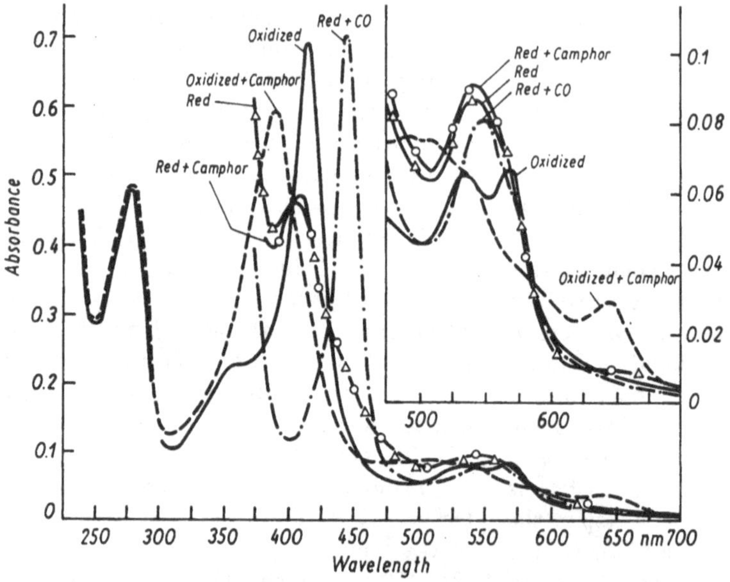

Fig. 1. Absorption spectra of P–450$_{CAM}$ of *Ps. putida* (GUNSALUS et al. 1974) (c = 300 μg per ml in 50 mM potassium phosphate, pH 7.0, 100 μM D-camphor, lightpath 1 cm, 5 cm in the inset); reproduced with permission from Academie Press, Inc.

remains in the low-spin state (GUNSALUS et al. 1974). The highly negative redox potential and its change in the course of the reaction cycle is a third characteristic and distinguishing feature of each P–450 species. The determination of the redox potential in the substrate-free low-spin state of P–450$_{CAM}$ yields an E_0 of −300 mV. Upon camphor addition a drastic change occurs and a value of −170 mV is measured (GUNSALUS and SLIGAR 1978). A careful analysis of the equilibria of the spin states, of the oxidation-reduction state and of the substrate binding suggests that the substrate binding equilibrium and the oxidation-reduction equilibrium is modulated by the spin state. These relationships are virtually the basis of the very effective coupling of electron transfer and substrate hydroxylation (for details see GUNSALUS and SLIGAR 1978).

P–450$_{CAM}$ purified from camphor-induced cells of *P. putida*, strain Pp G 786, is a single polypeptide chain and has a molecular weight in the range of 44000 to 46000, depending on the method of determination. It contains 1 mole protoporphyrin IX per mole P–450 as prosthetic group (YU et al. 1974a). The heme can be removed by dropping a solution of P–450$_{CAM}$ into an acetone/HCl mixture and the apoprotein thus obtained can be reconstituted with added heme in greater than 66% yield (YU and GUNSALUS 1974b). P–450$_{CAM}$ is very stable in the pH range between 6.0 and 8.0, especially in the presence of substrate. The stabilizing effect of camphor is also true for the heat-stability. At 47°C, camphor-free P–450$_{CAM}$ rapidly lost activity in neutral phosphate buffer, while the substrate complex was stable for more than 20 minutes (YU et al. 1974).

On treatment of P–450 with different compounds as acetone, sodium deoxycholate, guanidine hydrochloride, urea, or mercurial compounds catalytically inactive derivatives arise, showing a Soret maximum at 420 nm and an increased absorption in the α- and β-band region in the reduced state CO-complex. All these inactive forms are termed cytochrome P–420.

Table 2. Spectral properties of cytochrome P-450$_{CAM}$ of *Pseudomonas putida* (1–4)

Form	Wavelength (nm)	E_{mM} (mM^{-1} cm^{-1})	Form	Wavelength (nm)	E_{mM} (mM^{-1} cm^{-1})
ferric	280	68.3	ferric,	280	63.3
	360	36.7	substrate bound	391	102
	417	115		510	13.0
	535	11.6		540	11.2
	569	11.9		646	5.4
ferrous	408	76.7	ferrous,	408	86.5
	540	15.1	substrate bound	542	16.0
ferrous,	447	121	ferrous,	446	120
CO bound	550	14.3	CO and substrate bound	550	14.0
			(ferrous + CO + substrate)-	446	88.0
			(ferrous + substrate)	446–490	92.8

1 Values from GUNSALUS and SLIGAR, 1978; reproduced with permission from John Wiley and Sons, Inc.

2 Determined in 50 mM potassium phosphate, pH 7.0 at 25°C

3 Reduction with $Na_2S_2O_3$ 4 Substrate = 10^{-4} MD-(+)-camphor

However, small differences of their spectral properties show that they are caused by different modifications. Thus, P-420 cannot be considered as an uniform molecular species.

YU and GUNSALUS (1974a) described an effective method for the reconversion of P-420 to catalytically active P-450$_{CAM}$ by addition of cysteine or other sulfhydryl compounds. Hitherto this is the only convincing example of a successful reactivation of a P-420 to its P-450 form.

The iron-sulfur-protein putidaredoxin is a single polypeptide chain with a molecular weight of 12.500 consisting of 106 amino acids. The primary structure has been published (TANAKA et al. 1974). 4 cysteine residues connect the two inorganic labile sulfur and iron atoms of the prosthetic group and the polypeptide chain like other ferredoxins. Following the nomenclature such proteins like the putidaredoxin are signed as $Fe_2S_2-CyS_4$-proteins. The absorption spectra of the oxidized and the reduced putidaredoxin (fig. 2) are similar to those of other $Fe_2S_2-CyS_4$-proteins. The most important result concerning the role of putidaredoxin was its two fold function as electron transmitter and as effector of the actual substrate hydroxylation. In the latter, the putidaredoxin can be replaced even by low-molecular compounds (see 3.1.2).

The complete reaction cycle of the monooxygenase system requires two electrons. However, the transfer proceeds in two one-electron steps. In the oxidized state both iron atoms of the electron transfer protein can not be distinguished. They are in the high-spin ferric state and coupled to a diamagnetic ground state, $S = 0$. In contrast the reduced putidaredoxin shows an EPR signal with g values 1.94 and 2.01. This paramagnetic activity originates from the two spins, $S = \frac{5}{2}$ and $S = 2$, of the two iron atoms, which are coupled antiferromagnetically to a total spin of $S = \frac{1}{2}$. GUNSALUS and coworkers (WILSON et al. 1973, GUNSALUS et al. 1974) succeeded in preparing derivatives of the native putidaredoxin, which

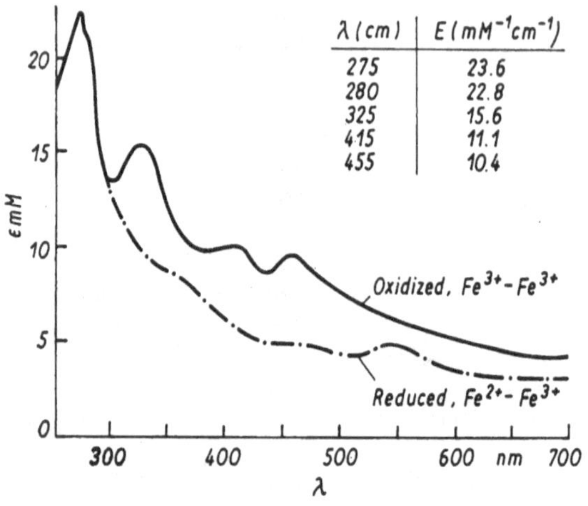

Fig. 2. Absorption spectra of putidaredoxin of *P. putida* (GUNSALUS et al. 1974, coefficients from GUNSALUS and SLIGAR 1978); reproduced with permission from Academic Press, Inc.

were applied to the elucidation of the mode of action and of the relationships with the monooxygenase. For different physical investigations the original iron and sulfur atoms of the active center have been replaced by isotopes of the same elements or the sulfur even by selenium. The selenium-derivative, for instance, exhibits a red-shifted absorption spectrum, which has been assigned to a metal-ligand charge transfer complex. By using this modified putidaredoxin in the reconstituted enzyme system the hydroxylating activity is diminished, whereas the NADH-oxidation rate is comparable with that shown by the native iron-sulfur-protein. The up to now most interesting derivative was obtained by gentle exoproteolytic digestion with carboxypeptidase A (GUNSALUS et al. 1974). This protease is known for being unable to split peptide bonds of basic amino acids. The C-terminus of the putidaredoxin has the sequence ...Arg. Gln.Trp.COOH. The carboxypeptidatic proteolysis results in the formation of a Des-Trp-putidaredoxin, which can be monitored by the decrease of the tryptophan fluorescence of the protein. Simultaneously, the hydroxylase activity of a reconstituted system is lost and the K_M value changes from 4.1×10^{-6} M to 9.1×10^{-5} M of the modified protein. All these results evidence the great importance of the putidaredoxin C-terminus for the P–450_{CAM}-putidaredoxin interaction (GUNSALUS et al. 1974). Recently, SLIGAR et al. (1980a, b) have attributed an important role in the process of oxygen activation to the terminal carboxyl group.

The NADH-putidaredoxin reductase is also a single polypeptide protein (molecular weight 45000) containing 1 FAD per molecule as prosthetic group. The most essential molecular parameters are published (GUNSALUS and SLIGAR 1978), however, the knowledge about this enzyme is small as compared with the two other components of the camphor-hydroxylating enzyme system. The reductase is strongly NADH-specific. Determinations of the redox potential revealed a value of −320 mV (GUNSALUS and SLIGAR 1978). For the actual substrate hydroxylation the reductase is not directly necessary. The addition of a sufficient amount of reduced putidaredoxin to the P–450_{CAM}-substrate complex results in the product formation.

In the course of the reductase reaction the electron transfer has to be transmitted from the 2-electron step, as it is performed by the NADH-oxidation, to the 1-electron step of the putidaredoxin reduction. Up to now reliable results giving an insight into this process were not available. Neither the stoichiometry of the reductase-putidaredoxin comlpex nor any intermediate state of the prosthetic group characterizing the catalytic cycle are known (GUNSALUS and SLIGAR 1978).

3.1.2. The Reaction Cycle

Fig. 3 demonstrates the reaction cycle of the camphor-hydroxylating enzyme system as it has been elucidated by GUNSALUS and his group. It can be assumed at the present knowledge that all the P–450 catalyzed reactions follow the same basic scheme.

After splitting off the product from the enzyme at the end of a complete reaction cycle the $P\text{–}450_{CAM}$ is in the ferric state (m^0). In this form the $P\text{–}450_{CAM}$ is also obtained by the isolation procedure. The first step, the substrate binding, is a very rapid one:

$$m^0 + S \rightarrow m^{0s}$$

The on-rate was found to be $k_1 = 3.1 \times 10^{-5}\ M^{-1}\ sec^{-1}$ and the off-rate $k_{-1} = 19\ sec^{-1}$ resulting in an equilibrium

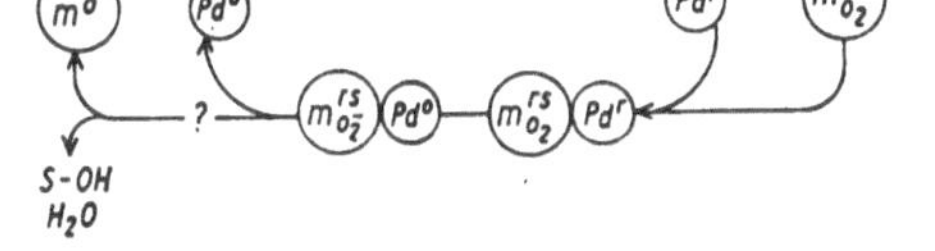

Fig. 3. Reaction cycle as described for $P\text{–}450_{CAM}$ of *Ps. putida* (GUNSALUS & SLIGAR 1978) (m^0 = oxidized monooxygenase; m^{0s} = oxidized monoocygenase, substrate bound; m^{rs} = monooxygenase, substrate bound; Pd = putidaredoxin; S = substrate); reproduced with permission from John Wiley and Sons, Inc.

constant $K_D = \frac{k_{-1}}{k_1} = 0.61 \times 10^{-6}$ M. From the spectrophotometric titration of the substrate binding a K_D für *D*-camphor of about $3 \cdot 10^{-6}$ M has been determined (YU et al. 1974). As already mentioned concomitant events of this step are the low-spin high-spin transition and a change of the redox potential from -300 mV to -170 mV (GUNSALUS and SLIGAR 1978). The substrate binding reaction seems to differ in some relation between the $P\text{–}450_{CAM}$ and the $P\text{–}450_{LM2}$ of the rabbit liver. An increasing K^+-concentration stabilizes the camphor binding (PETERSON 1971), but doesn't influence the substrate binding of the $P\text{–}450_{LM2}$ (WHITE and COON 1980). Obviously, $P\text{–}450_{CAM}$ takes up a proton from the medium in the course of this reaction (SLIGAR and GUNSALUS 1979) whereas $P\text{–}450_{LM2}$ doesn't (WHITE and COON 1980). Corresponding studies on other microbial P–450 forms are lacking.

The transfer of the first electron follows as the next step:

$$m^{0s} + Pd^r \rightarrow m^{rs} + Pd^0$$

A more detailed analysis of this reaction offered the possibility to solve it into two sub-reactions

$$m^{0s} + Pd^r \rightarrow m^{0s} \cdot Pd^r$$

$$m^{0s} \cdot Pd^r \rightarrow m^{rs} \cdot Pd^0 \rightarrow m^{rs} + Pd^0$$

The on-rate and the off-rate of the complex formation have been determined as 3×10^8 and $150\ M^{-1}\ sec^{-1}$ respectively (PEDERSON and GUNSALUS 1977). Hydrophobic interactions are responsible for this complex formation. A low ionic strength favors an electrostatic repulsion, because both proteins exhibit an acidic isoelectric point. The actual electron transfer, $m^{0s} \cdot Pd^r \rightarrow m^{rs} \cdot Pd^0$, proceeds more slowly. GUNSALUS and SLIGAR (1978) found a rate of $18\ sec^{-1}$. Simultaneously, the separation of the dienzyme complex is initiated. The reduction of the enzyme-substrate complex results in its further stabilization. Besides the redox potential modification in the course of the substrate binding and the kinetic control of the forward electron-transfer rate this increased stability of the enzyme-substrate complex in the reduced state is an additional guarantee for the coupling of e^--transfer and hydroxylation.

In relation to the electron transfer, there exists a great difference between this bacterial P–450 system and the microsomal P–450 forms of eucaryotes. This has to be emphasized. In the former an electron transfer from the iron-sulfur cluster of the putidaredoxin to the heme prosthetic group is realized, the latter utilize the direct transfer from the reductase to the terminal oxidase.

The first reduction step is followed by the formation of the dioxygen-ferrous-intermediate:

$$m^{rs} + O_2 \rightarrow m^{rs}_{O_2}$$

The on-rate was found to be 1.7×10^6 $M^{-1}\,sec^{-1}$, the off-rate $1\,sec^{-1}$ for a dissociation constant of $0.6\,\mu M$ (Gunsalus and Sligar 1978). This intermediate can be involved in different reactions. On the one hand, a slow dissociation proceeds. Moreover, in the absence of the second reduction equivalent the oxygenated $P\text{–}450_{CAM}$ can decay on the autoxidative way releasing O_2^- with the very slow rate of $0.01\,sec^{-1}$ (Gunsalus and Sligar 1978). Nucleophilic substances such as cyanide, azide, thiocyanates and cyanate accelerate this process (Sligar et al. 1976).

Apparently, in the liver microsomal system the formation of the superoxide anion is a more favoured reaction. Even in the presence of substrate about half of the oxygen is transformed to peroxide, which should result mainly from dismutation of superoxide (White and Coon 1980).

At the $P\text{–}450_{CAM}$, however, the hydroxylation of the substrate is the most effective reaction, which proceeds with the ferrousdioxygen complex in the presence of the putidaredoxin.

$$m^{rs}_{O_2} + Pd^r \rightarrow m^0 + Pd^0 + H_2O$$
$$+ S{-}OH$$

The rate of this final step has been determined with $17-20\,sec^{-1}$ by Gunsalus et al. (1974) and should be limiting for the reaction. In the reconstituted hydroxylase system the steady-state turnover was found to be $1200\,min^{-1}$.

In this step, which is initiated by the transfer of the second electron, the decisive events of the whole monooxygenase reaction take place, the oxygen splitting and the insertion of one atom oxygen into the organic substrate molecule replacing an unactivated C—H bond by a C—OH group. Unfortunately, no precise informations are available up to now, neither from the $P\text{–}450_{CAM}$ nor from other P–450 systems. Ideas and suggestions published to this subject at P–450 in general are discussed in more detail by White and Coon (1980) and in another chapter of this monography.

Obligatory intermediate for the completion of the $P\text{–}450_{CAM}$ reaction cycle is again a di-enzyme complex between the reduced putidaredoxin which has to deliver the second electron and the $P\text{–}450_{CAM}$ existing now in the camphor- and oxygen-bound ferrous form. In this complex putidaredoxin exerts a twofold function. It is both an electron donor and an effector of the mixed function oxidation. Different rubredoxins, cyt b_5, their apoproteins and even dithiols as dithiothreitol and lipoic acid can also act as an effector (Lipscomb et al. 1976). Concerning the first phase of oxygen splitting, Sligar et al. (1980a, b) have submitted an interesting proposal. The C-terminal carboxylate ion of the putidaredoxin acylates the $\underset{\displaystyle S}{\overset{}{Fe^{2+}}}{-}O{-}O^-$ complex, resulting in a protein-bound peroxy-acid, $\underset{S}{\underset{|}{Fe^{2+}}}-O-O-\underset{O}{\underset{\|}{C}}-R$, as the first intermediate of the oxygen activation. This proposal is supported by experiments with lipoic acid as effector. Using $^{18}O_2$ both in the 5-OH camphor and in the lipoic acid half of heavy oxygen was incorporated and the produced water was found to be unlabeled. This peroxyacid intermediate suggested by SLIGAR et al. (1980a, b) is in agreement with different proposed mechanisms (see WHITE and COON 1980) of monooxygenase catalysis. However, this result obtained with a model effector is not yet conformed with putidaredoxin.

Like $P\text{–}450_{LM}$ the $P\text{–}450_{CAM}$ can also catalyze peroxide- and peracid-driven hydroxylation reactions. This process occurs in the absence of O_2 and NADH and cannot be inhibited by CO. In relation to the $P\text{–}450_{CAM}$ this reaction was discussed in detail by GUNSALUS and SLIGAR (1978).

3.2. The Steroid-Hydroxylating Enzyme System of Bacillus Megaterium

In 1975 GUSTAFSSON and co-workers (BERG et al. 1975) demonstrated, that the 15-β-hydroxylation of 3-oxo-Δ^4-steroids in Bacillus megaterium (ATCC 13368) is catalyzed by a P–450 containing enzyme system. Further investigations showed similarities both to the camphor hydroxylating enzyme system from *Pseudomonas putida* and to the steroid hydroxylating enzyme system of adrenal mitochondria (BERG et al. 1976, 1979a, b). Like those, the P–450 dependent steroid 15-β-hydroxylase system from *Bacillus megaterium* has been resolved into three components a) a NADPH-dependent FMN-containing flavoprotein (megaredoxin reductase), b) an iron-sulfur protein (megaredoxin) and c) $P\text{–}450_{meg}$. Recombination experiments revealed the pathway of e^--transfer to P–450: NADPH → reductase → iron sulfur protein → $P\text{–}450_{meg}$ (BERG et al. 1976). In accordance, all P–450 contai-

Table 3. Hydroxylation of various steroids by cell-free extracts of *Bacillus megaterium*[1])

Substrate	Hydroxylated product	Specific enzyme activity nmol/mg protein/ 30 min
4-Androstene-3,17-dione	15β-Hydroxyandrostenedione	0.7
Testosterone	15β-Hydroxytestosterone	0.4
Progesterone	6β-Hydroxyprogesterone	0.4
	15β-Hydroxyprogesterone	1.6
20α-Dihydroprogesterone	15β-Hydroxy-20α-dihydroprogesterone	1.4
17α-Hydroxyprogesterone	15β, 17α-Dihydroxyprogesterone	1.5
Deoxycorticosterone	15β-Hydroxydeoxycorticosterone	2.0
Corticosterone	15β-Hydroxycorticosterone	0.06

Following steroids were not metabolized:
Estrone, Estradiol, Estriol, Dehydroepiandrosterone, Prenenolone; 5α-Dihydrotestosterone; 5α-Androstane-3,17β-diol; 5β-Pregnane-3,20α-diol; 5α-Androstane-3,17β-diol 3,17-disulfate

[1]) BERG et al. 1976; reproduced with permission from the American Society of Biological Chemists, Inc.

ning steroid hydroxylases described so far, have a NADPH-dependent flavoprotein as the first component of the e^--transfer. The individual components of the *Bacillus megaterium* system can be exchanged by the corresponding proteins of the adrenal mitochondrial 11-β-hydroxylase system (BERG et al. 1976), although the prosthetic groups of the reductases differ, for this mammalian reductase contains FAD and the bacterial enzyme FMN.

The 15-β-hydroxylase system has a high substrate specificity. Only 3-oxo-Δ^4-steroids are hydroxylated, mainly in the 15-β-position (Tab. 3, BERG et al. 1976). Replacing the e^--transfer components by oxidizing agents as $NaJO_4$ or $NaClO_2$ the stereospecificity of the reaction is changed. Both the 15β- and the 15α-hydroxylation products appear (BERG et al. 1975, 1979a). The inhibition experiments characterize the P–450_{meg} as a typical member of this group of hemoproteins, for the reaction is strongly inhibited by CO, SKF 525-A and metyrapone. On the other hand, imidazole is a weak inhibitor and CN^- does not inhibit at all.

The P–450_{meg} purified to homogeneity (17 nmol P–450/mg protein) has an isoelectric point of 4.9. Analytical centrifugation in sucrose density gradient delivered a homogeneous sedimentation peak with a sedimentation coefficient of 3.3 (BERG et al. 1976, 1979a).

Using the results of the amino acid analysis and of the comparative gel electrophoresis a molecular weight of 52×10^3 daltons was calculated (BERG et al. 1979a, b). The heme content was determined to be 0.94 nmol heme/nmol P–450 (BERG et al. 1976). The spectral characteristics of P–450_{meg} (see Table 4) and the EPR-signals of $g = 2.41$, 2.24 and 1.91 are typical for a low-spin P–450. The fluorescence spectrum showed a tryptophan residue in a relatively apolar environment (BERG et al. 1979a).

The FMN-containing megaredoxin reductase has a molecular weight of $55 \times 10^3 - 60 \times 10^3$ daltons. It is able to reduce DCIP (950 units/mg protein) but not cytochrome c (BERG et al. 1976, 1979a).

Table 4. Spectral properties of cytochrome P–450_{meg} of *Bacillus megaterium*[1])

Form	Wavelength (nm)
ferric	417
	534
	571
ferrous	412
	544
ferrous	450
+ CO	550

[1]) BERG et al. 1979b; reproduced with permission from the American Society of Biological Chemists, Inc.

By means of amino acid analysis for the megaredoxin a molecular weight of 13458 daltons has been determined (BERG 1980). The ESR spectrum shows the typical values for an iron-sulfur protein −1.9, 1.93 and 2.06 (BERG et al. 1976). All these results provide evidence for the similarity of the camphor-hydroxylating (see 3.1) and the 15-β-steroid-hydroxylating enzyme system.

3.3. The Fatty Acid-Hydroxylating Enzyme System of Bacillus Megaterium

In a series of papers FULCO and co-workers (MIURA and FULCO 1974, HARE and FULCO 1975, HO and FULCO 1976, MATSON et al. 1977) reported on the NADPH- and O_2-dependent hydroxylation of fatty acids by a P–450 containing enzyme system in another Bacillus megaterium strain (ATCC 14581). They found mainly $(\omega - 2)$-, $(\omega - 1)$-, and $(\omega - 3)$ hydroxylation products, but in no case the terminal hydroxy fatty acid was detectable (MIURA and FULCO 1974). Since the reaction was inhibited by CO, the involvement of P–450 was assumed. Furthermore, it was concluded that the

($\omega - 2$) hydroxylase system differs from the 15β-hydroxylase system (see 3.2), because the former is unable to hydroxylate progesterone (MATSON et al. 1977). A more detailed characterization of the P-450 is lacking, for the P-450 could be only partially purified to 0.2 nmol/mg protein. The involvement of an iron-sulfur protein in the fatty acid hydroxylation was concluded from the inhibition by low concentration of p-hydroxymercuribenzoate and the reversion by ferredoxin, cysteine or glutathion (HARE and FULCO 1975).

Another reaction catalyzed by this partially purified enzyme fraction could be the epoxidation of unsaturated fatty acids (BUCHANAN and FULCO 1978). After incubation of palmitoleic acid with the impure enzyme preparation the investigators identified 9,10-epoxipalmitate and 9,10-dihydroxypalmitate as reaction products. However, there is no evidence hitherto that P-450 is actually involved in the epoxidation reaction.

3.4. The O-dealkylating Enzyme System of Nocardia Sp.

CARTWRIGHT et al. (1971) showed that in *Nocardia* NH1 an O-demethylase system is induced by growth on 4-methoxy-3-hydroxybenzoate (iso-vanillate). This bacterial monooxygenase system specifically dealkylates para-alkylphenyl ethers and has been resolved into two fractions: one fraction contained P-450, the other a NADH-dependent flavin-containing reductase (BROADBENT and CARTWRIGHT 1974).

The $P\text{-}450_{npd}$ could be purified to homogeneity (BROADBENT and CARTWRIGHT, 1974). A molecular weight of about 45×10^3 daltons was estimated by amino acid analysis (424 residues/molecule) and SDS gel electrophoresis. The isoelectric point is at 4.94. The spectral characteristics so far obtained are like those of $P\text{-}450_{CAM}$. In the oxidized state the Soret peak of the absolute

Table 5. Spectral properties of cytochrome $P\text{-}450_{npd}$ of *Nocardia* NH1[1-3])

Form	Wavelength (nm)	Form	Wavelength
ferric	417	ferric,	394
	535	substrate bound	540
	567		650
ferrous	412	ferrous,	408
	535	substrate bound	540
		ferrous,	448
		CO bound	545

[1]) BROADBENT and CARTWRIGHT 1974; reproduced with permission from Faculty Press Cambridge
[2]) estimated at 0.5 mg P-450/ml in Tris/HCl, pH 7.4 [3]) substrate = 10^{-6} M iso-vanillate

spectrum has its maximum at 417 nm. The CO-complex exhibits the Soret absorption maximum at 448 nm (for details see Table 5). Iso-vanillate and 4-methoxybenzoate induce a type-I difference spectrum, while both 2- and 3-methoxybenzoate and camphor do not influence the spectrum. The substrate-induced low-spin high-spin transition, which was observed in the optical spectra at ambient temperature could not be detected in ESR measurements at 1.5 °K. From ESR measurements, however, it was concluded that the thiol group of a cysteine residue may act as the 6th ligand of the heme group (BROADBENT and CARTWRIGHT, 1974).

The NADH-dependent reductase could be purified only partially (CARTWRIGHT and BROADBENT 1974). Therefore, the question cannot be decided definitely, whether an iron-sulfur protein is necessary for the e^--transfer or not.

3.5. Alkane- and Cholesterol-Hydroxylating Enzyme Systems of Corynebacterium

CARDINI and JURTSHUK (1968, 1970) reported on a P-450 which was induced in *Corynebacterium* by growth on n-octane. The hydroxylase system, which oxidizes n-octane in the presence of NADH and O_2 to 1-octanol could be separated by $(NH_4)_2SO_4$ precipitation into 2 fractions (CARDINI and JURTSHUK (1970), a P–450 containing and a second one, oxidizing NADH. P–450 was slightly purified (0.25 nmol/mg protein). It was reduced enzymatically by NADH in the presence of the second fraction. YAMAMOTO et al. (1974) reported the P–450 formation in a *Corynebacterium* strain after growth on cholesterol. Using a 40% pure preparation the molecular weight was estimated with 50000 daltons. The absolute spectrum of the oxidized form exhibited peaks at 393 nm, 510 nm and 645 nm, suggesting a high-spin P–450. An iron-sulfur protein seems to be missing.

Table 6. Spectral properties of *Rhizobium japonicum*[1]) bacteroid cyt P–450 Rh

Form	P–450 Rh_a nm	E_{mM}	P–450 Rh_b nm	E_{mM}	P–450 Rh_c nm	E_{mM}
ferric	280	144	279	122	280	82
	358	36.3	396	86	417	99
	416	114	525	11.4	522	11.6
	532	11.6	560	8.7	566	10.8
	564	12.0			645	2.8
ferrous	413	91	412	88	408	86
	543	15.9	542	16.1	543	14.9
ferrous,	449	112	449	128	447	104
CO bound	549	15.7	549	14.6	551	13.4
ferrous, CO bound, difference	450	84	449	102	448	79

[1]) APPLEBY 1978a; with permission from Academic Press

3.6. *P–450 of Nitrogen-Fixing Bacteroids*

Although P–450Rh as first soluble bacterial P–450 has been discovered in the nitrogen fixing form (bacteroids) of *Rhizobium japonicum* from soybean root nodules by APPLEBY already in 1967 (APPLEBY 1967, 1969), the biological function remains unclear up to now. P–450RH shows with typical substrates, ligands and inhibitors of other P–450 species the characteristic difference spectra (APPLEBY and DANIEL 1973; KRETOVICH et al. 1974), but till now a monooxygenatic enzyme activity has not been observed (APPLEBY by letter, 1979). Likewise a specific electron transfer system hitherto was not described. Since P–450Rh in aerobic grown *Rhizobia* which are unable to fix nitrogen, is contained only in a very small amount (DANIEL and APPLEBY, 1972) it was assumed, that P–450Rh is involved in the nitrogen fixation process. But hitherto the assumed relation to the nitrogen fixation has not been proved definitely. Studies with N-phenylimidazol (APPLEBY et al. 1975) and other specific inhibitors of P–450 (KRETOVICH et al. 1974) suggest, that bacteroid P–450 is directly involved (as terminal oxidase, peroxidase or as oxygen carrier) in the bacteroid respiration (APPLEBY 1978b).

Anion-exchange chromatography (APPLEBY and DANIEL, 1973; APPLEBY 1978a) resulted into at least 3 distinct P–450Rh forms (a, b, c), the spectral properties of which are summarized in Tab. 6. The same heterogeneity of P–450 was observed in *Rhizobium lupini* bacteroids (INOZENTSEVA et al., 1978).

The optical (APPLEBY and DANIEL 1973, APPLEBY 1978a) and ESR-spectra (PEISACH et al. 1972) show, that P–450Rha is a low-spin hemoprotein, whereas P–450Rhb contains a large portion of highspin form and P–450Rhc is a mixed-spin hemoprotein. Molecular weights of 45000 were estimated by gel filtration for all 3 subspecies (APPLEBY, 1978a). Immunochemically the P–450Rh (a, b, c) show a close relationship to P–450_{CAM} and P–450_{LM2} (DUS et al. 1976). The amino acid composition of P–450Rhc, which was further partially purified by affinity chromatography on phenobarbitalsepharose also shows a close correlation to the composition of P–450_{CAM} and P–450_{LM2} (DUS et al. 1976).

4. P-450 of Eucaryotic Microorganisms

P–450 containing monooxygenase systems have been found in several classes of eucaryotic microorganisms and they are involved in very different metabolic pathways (Tab. 1). Most of the eucaryotic P–450 species are inducible enzymes. Frequently, the growth substrate itself acts as the inducer like in the alkane-assimilating yeasts. In contrast to that the formation of P–450 in *Saccharomyces cerevisiae* obviously depends on other factors and is not directly caused by the action of a specific extracellular inducer.

The available data suggest that there exist similarities between the fungal and the liver microsomal P–450 system with respect to the intracellular localization and the structure of the enzyme system. Thus, the P–450 systems of alkane assimilating yeasts and of *Saccharomyces cerevisiae* are bound to a microsomal membrane and consist of a NADPH-P–450 reductase with FAD and FMN as prosthetic groups and of P–450 as the terminal oxidase.

The P–450 systems of *Cunninghamella* species seems to exhibit a similar broad substrate specificity (see 4.3) as it is known for the liver microsomal system. However, in connection with their involvement in specific metabolic pathways the substrate specificity of most of the other microbial P–450 systems seems to be much more limited.

4.1. P–450 of Alkane-Assimilating Yeasts

4.1.1. Induction and Physiological Function

The induction of P–450 in yeasts using n-alkanes as the only source of carbon and energy has been described for a number of species (see Table 1). As illustrated in Fig. 4 a strong correlation seems to exist between the growth course on n-alkanes and the P–450 formation: the lag-phase is characterized by a large increase of the cellular P–450 content, in the exponential phase the specific P–450 content reaches a constant level; after consumption of the alkane substrate the P–450 content declines in the stationary phase. Besides long-chain n-alkanes also alkenes, alkanols and fatty acids act in a different degree as inducers of the P–450 in different yeast strains (Gilewicz et al. 1979, Ilchenko et al. 1980, Takagi et al. 1980a). With *Lodderomyces elongisporus* it has been demonstrated that the induction is significantly dependent on the chain length of the aliphatic hydrocarbons. Tetradecane and 1-tetradecene are the most effective inducers. Generally, the branching of the alkyl

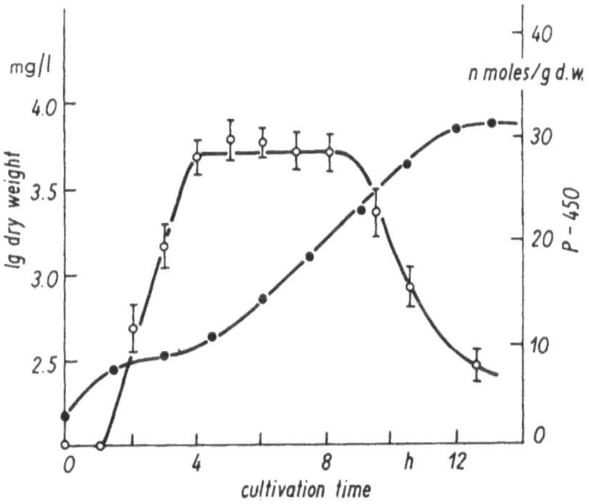

Fig. 4. P–450 concentration in alkane-utilizing *C. guilliermondii* in dependence on the growth phase (Mauersberger and Matyashova 1980); reproduced with permission from the Isdatelstvo Nauka, Moscow

chain as well as the terminal introduction of electrophilic atoms decrease the induction effect (MAUERSBERGER et al. 1981).

The P-450 induction requires the de-novo-synthesis of protein and of lipid, as it was shown by inhibition with cycloheximide (TAKAGI et al. 1980a, b, MAUERSBERGER and MATYASHOVA 1980) and cerulenin, respectively (TAKAGI et al. 1980a, b). Glucose, but not glycerol or galactose acts as a catabolite repressor (TAKAGI et al. 1980a, b, MAUERSBERGER and MATYASHOVA 1980, MAUERSBERGER et al. 1981). Furthermore, the induction process is influenced by the oxygen concentration of the medium (GMÜNDER und FIECHTER 1978, GMÜNDER 1979, MAUERSBERGER et al. 1980). As shown in Fig. 5, in the range between 100 and 20% of saturation no dependence can be observed and a relatively small level of 25 to 35 n moles/g dry weight was reached. However, lower pO_2 values cause a strong

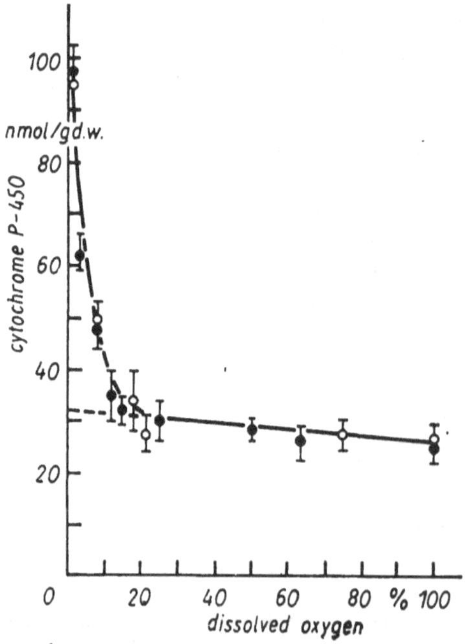

Fig. 5. P-450 concentration in alkane-utilizing *C. guilliermondii* in the exponential growth phase in dependence on the pO_2 (MAUERSBERGER et al. 1980); reproduced with permission from Springer Verlag

increase of the cellular P-450 concentration up to about 100 n moles/g dry weight (MAUERSBERGER et al. 1980).

Since the detection of cytochrome P-450 in the alkane-grown yeast cells of *Candida tropicalis* (LEBEAULT et al. 1971, GALLO et al. 1971) it was postulated that the P-450 system catalyzes the terminal hydroxylation of n-alkanes. This reaction is generally accepted now to be the first step of the terminal degradative pathway which proceeds in many of the alkane-utilizing microorganisms (KLUG and MARKOVETZ 1971; SCHUNCK et al. 1978b, FUKUI and TANAKA 1979, MIALL 1980):

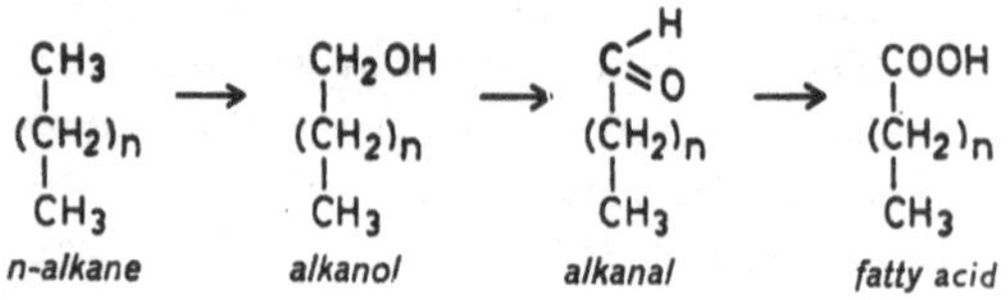

Available data, especially the reconstitution of the alkane-hydroxylating system from its purified protein constituents (RIEGE et al. 1981), suggest that this is really the physiological function of the alkane-induced P-450 system in the yeast cells. However, up to now no explanation was found for the weak in-vitro activities (about 1—5 n moles converted substrate/nmole P-450 · min) of the microsomal and reconstituted systems which are at least two orders of magnitude lower than the in-vivo activity calculated from the alkane oxidation rate and of the P-450 content of Lodderomyces elongisporus cells (unpublished results).

4.1.2. Structure and Enzymatic Properties of the Cytochrome P-450 System

In *Candida tropicalis* (GALLO et al. 1971, 1973, 1976, GILEWICZ et al. 1979), in *C. guilliermondii* (MAUERSBERGER et al. 1980), in *Lodderomyces elongisporus* (SCHUNCK et al. 1978a, MÜLLER et al. 1979), and in *Endomycopsis lipolytica* (DELAISSE and NYNS 1974) the P-450

Table 7. Characteristic constituents of the microsomal fraction from alkane-grown yeasts

Species	Cyt P–450 (nmol/mg protein)	Cyt b_5 (nmol/mg protein)	NADPH-cyt c(P–450) reductase (µmol/mg × min)	Hydroxylase-activity (nmol/mg × min)
Lodderomyces elongisporus[1])	0.2–0.3	0.2–0.3	0.7–0.9	0.1–0.3*)
Candida tropicalis[2])	0.12	0.2	0.7	5**)

*) determined with hexadecane as substrate **) determined with laurate as substrate

[1]) MÜLLER et al. 1978 [2]) Gilewicz et al. 1979; reproduced with permission from the National Research Council of Canada

containing alkane hydroxylating enzyme system is bound to a membrane fraction, which sediments at 100000 xg. This membrane fraction contains the same constituents which are characteristic for the electron transfer system of the endoplasmic reticulum of mammalian liver: P–450, NADPH-P–450 reductase, cytochrome b_5 and NADH-dependent cytochrome b_5 reductase (see Table 7).

At first time the alkane hydroxylating enzyme system from *C. tropicalis* was resolved by DUPPEL et al. (1973) into three components: P–450, NADPH-P–450 reductase and a thermostable lipid fraction. These authors were able to reconstitute the poorly purified components to the active enzyme system. The yeast reductase and the lipid could be replaced by the corresponding fractions from rat liver microsomes (DUPPEL et al. 1973). BERTRAND et al. (1979a) were also able to reconstitute the alkane hydroxylase activity by mixing partially purified P–$450_{C.t.}$ and NADPH-cyt c reductase. In *L. elongisporus* the involvement of P–450 in alkane hydroxylation has been proved combining the electrophoretically homogeneous hemoprotein with the highly purified NADPH-cyt P–450 reductase (Table 8). The enhancement of the enzymatic activity by nonionic detergents like Präwozell W-ON 100 or Triton x-100 suggest a lipid requirement of the enzyme system.

Besides oxygen the alkane-hydroxylating enzyme system requires a pyridine nucleotide as cosubstrate (LEBEAULT et al. 1971, GALLO et al. 1971, 1973, SCHUNCK et al. 1978a). Using microsomes or a poorly purified solubilized fraction NADPH is much more effective than NADH. However, a strong synergistic action of both pyridine nucleotides was

Table 8. The reconstitution of the alkane monooxygenase system of *Lodderomyces elongisporus*

Component omitted from the complete system	Hydroxylase activity nmoles product / nmole P–450 · min
	5.64
Cyt. P–450	0
reductase	0
Präwozell W-ON 100	0.80
NADPH	0

The complete system contained in a total volume of 1.0 ml: 0.18 n moles cyt. P–450, 0.67 units (µ moles reduced cyt. c/min) NADPH-cyt. P–450 reductase, 160 n moles of the nonionic detergent Präwozell W-ON 100, 250 nmoles NADPH, 100 nmoles 1-^{14}C hexadecane (300000 DPM), 200 µ moles potassium phosphate buffer pH 7.25 and 20% glycerol. After a preincubation for 10 min at 30 °C the reaction was started with NADPH. The incubation was stopped after 10 minutes, and extracted for product determination as described by SCHUNCK et al., 1978.

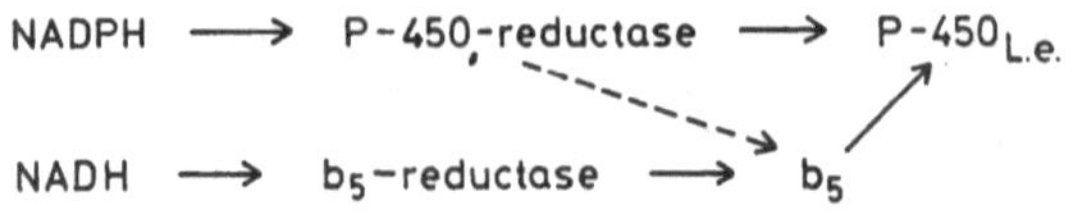

Fig. 6. Possible e^--transfer pathways to $P–450_{L.e.}$

observed (DUPPEL et al. 1973, MÜLLER et al. 1978, 1979). The mechanism of this effect is not clarified up to now. First kinetic studies of the yeast microsomal P–450 (RIEGE and SMETTAN, in preparation) suggest, that two pathways of e^--transfer exist, a direct one via the NADPH-cyt P–450 reductase and a second one, involving the cyt b_5-system (Fig. 6). Microsomal $P–450_{L.e.}$ is reducible aerobically both by NADPH and by NADH to the same extent as by dithionite. The rate constant with NADPH is about 2.5 fold higher than with NADH. In contrast to the substrate hydroxylation rate the reduction is not accelerated using a mixture of both reduced pyridine nucleotides (Table 9). Therefore, the synergistic action is presumably not connected with the first reduction step of P–450 in the reaction cycle.

Table 9. Comparison of the rates of both the first reduction reaction and the overall substrate conversion for microsomal P–450 from *Lodderomyces elongisporus*[1,2]

Electron donor	Reduction rate s^{-1}	Hexadecane conversion nmoles product / nmole P–450 · h
$Na_2S_2O_4$	0.33[3]	—
NADPH, 1 mM	1.66[3]	7.24 ± 0.47
NADH, 1 mM	0.64[3]	1.16 ± 0.17
NADPH, 1 mM + NADH, 1 mM	1.08[4] 0.35	17.9 ± 0.3

[1]) RIEGE and SMETTAN in preparation

[2]) The reduction of cytochrome P–450 by NADPH, NADH or dithionite was studied in a DURRUM stopped flow spectrophotometer by following the formation of the CO complex of ferrous cytochrome P–450 at 450 nm. Microsomal protein was suspended in 100 mM potassium phosphate buffer, pH 7.25, and mixed in the stopped flow apparatus with the equal volume of the same buffer containing the reducing agent and CO.

[3]) monophasic

[4]) biphasic

The substrate specificity of the alkane-induced P–450 species from yeasts has not been sufficiently clarified as yet. With reconstituted systems only the hydroxylation of laurate (BERTRAND et al. 1979a), hexadecane (RIEGE et al. 1981) and other aliphatic hydrocarbons (unpublished) have been detected. Microsomes from *C. tropicalis* hydroxylate alkanes and fatty acids in the ω-position, however, the latter are converted at higher rates (LEBEAULT et al. 1971). The terminal oxidation of long-chain n-alkanes was proved also with microsomes from *L. elongisporus*, where the highest oxidation rates were found with hexadecane and octadecane. Besides the P–450 system the membrane preparations contain high activities of alcohol- and aldehyd dehydrogenases. Consequently, fatty acids were the first detectable oxidation products of n-alkanes. The postulated primary alcohol was proved only by using a P–450 system reconstituted from its highly purified constituents (unpublished).

Furthermore, DUPPEL et al. (1973) were able to detect the N-demethylation of aminopyrine, benzphetamine and ethylmorphine by the partially purified preparation of *C. tropicalis* containing 0.4 nmoles P–450/mg protein, p-nitroanisole, the only o-methyl compound tested by the investigators, was inactive, DUPPEL et al. (1973) concluded that the $P–450_{C.t.}$ "hydroxylates a variety of fatty acids, hydrocarbons and drugs". In relation to the latter group of compounds we cannot

Table 10. Substrate induced spectral changes of purified P–450 of *Lodderomyces elongisporus*[1])

Substrate added	Peak (nm)	Trough (nm)	Spectral change type
hexadecane	386	418	I
aminopyrine	419	387	reverse I
benzphetamine	420	386	reverse I
imidazole	432	410	II
aniline	430	410	II
estradiol	420	385	reverse I
deoxycorticosterone	421	386	reverse I
SKF 525-A	430	408	II

[1]) Müller et al. 1979

agree without reservations. The observed conversion rates of the three drugs were low, ranging between 1 and 0.1 nmoles/nmol P–450. The actual involvement of P–450 in this reactions has not been proved. In fact, the N-demethylation of benzphetamine by microsomes of *L. elongisporus* is not inhibited by CO (Müller et al. 1979) and no conversion of benzphetamine, aminopyrine and ethylmorphine was detected with the reconstituted P–450 system of *L. elongisporus* (unpublished results). Moreover, it was found that the $P\text{–}450_{L.e.}$ differs in its substrate binding behaviour from the liver microsomal P–450. Thus, among the substances investigated only hexadecane induced a type-I spectral change (Table 10) which is produced generally only by substrates of the mixed function oxidation. Imidazole, aminopyrine, benzphetamine, deoxycorticosterone, estradiol, aniline and SKF 525 A induce type-II or modified type II (reverse type-I) spectral changes. Therefore, further thorough investigations are indisponsable for resolving the question of substrate specificity. We favour the opinion that the alkane-induced P–450 species of yeasts are highly specific for long-chain alkanes and alkyl residues.

4.1.3. Purification and Molecular Properties of the Individual Constituents

Attempts to purify the individual proteins of the alkane-hydroxylating enzyme system were little successful for a long time. Duppel et al. (1973) obtained a poorly purified P–450 fraction (0.4 nmoles/mg protein) from the cellfree supernatant of *C. tropicalis*. Efforts in the laboratory of Azoulay (Bertrand et al. 1979b) resulted in a 13-fold purification of $P\text{–}450_{C.t.}$ from microsomal membranes, delivering a specific P–450 content of 1.2 nmoles/mg protein.

Recently, in an effective procedure the $P\text{–}450_{L.e.}$ was obtained in a homogeneous state, as shown by means of SDS polyacrylamide gel electrophoresis (Riege et al. 1981): Starting from a P–450 content of 0.05–0.07 nmoles/mg protein in the yeast cells a microsomal fraction with 0.2–0.3 nmoles/mg protein was obtained by a Ca^{2+}-mediated low speed sedimentation. In this state, $P\text{–}450_{L.e.}$ can be stored at −20 °C for some weeks without considerable loss of its specific content. After cholate solubilization, chromatography on ω-NH_2-octyl Sepharose 4B, and on caclcium phosphate gel P–450 preparations with a specific content of 12–17 nmoles/mg protein were obtained. No P–420 could be detected. The yield was about 20% related to the microsomal fraction. The $P\text{–}450_{L.e.}$ separated from most of the detergents tends to aggregate. By SDS-PAGE an apparent molecular weight of about 53000 was estimated, demanding a specific content of 18,8 nmoles/mg protein for the pure protein. The difference between the experimentally detected and the theoretical value is presumably due to the presence of a significant part of apoprotein in the electrophoretically homogeneous protein.

The absorption spectrum of the alkaline pyridine ferrohemochrome shows maxima at 555, 523 and 417 nm, suggesting that

Table 11. Spectral properties of P–450 of alkane-grown *Lodderomyces elongisporus*[1,2]

Form	Wavelength (nm)
ferric	568
	536
	417
ferrous	Plateau at 570–530
	414
ferrous, CO-complex	Plateau at 555
	447
ferrous, CO-difference	447

[1]) RIEGE et al. 1981; reproduced with permission from Academic Press, Inc.

[2]) Determined in 200 mM potassium phosphate buffer, pH 7.25, containing 20% glycerol, 0.4% Tween 20, 0.5% Na-cholate and 0.5 mM DTE

protoporphyrin IX as in other P–450 forms is the prosthetic group. A stoichiometry of 1 mole heme per mole P–450 has been determined (RIEGE et al. 1981).

The spectral properties (Table 11) characterize the P–$450_{L.e.}$ as a typical member of this group of hemoproteins. The Soret peak of the ferric form at 417 nm indicates the low spin state. This is in agreement with P–450_{LM2} (HAUGEN and COON 1976) of the rabbit liver. The Soret maximum of the reduced CO complex is localized at a relatively short wavelength (at 447 nm; RIEGE et al. 1981). Within the cell P–$450_{L.e.}$ is mainly in the ferrous form.

Recently, MANSUY et al. (1980) investigated spectral characteristics of P–$450_{C.t.}$ in the microsomal form under the influence of different compounds. Both ferric and ferrous state ligands of P–450 produce similar spectral changes as with liver microsomal P–450. At ambient temperature a spin-state equilibrium exists with about 30% high-spin. The investigators failed in transforming the P–$450_{C.t.}$ to the high-spin state by adding the substrate or other hydrophobic compounds. The addition of cumene hydroperoxide leads to a maximum at 444 nm in the difference spectrum, which can be considered, in agreement with the results of RAHIMTULA et al. (1974) at the liver microsomal P–450, as sign of a reactive ferrylcomplex of the P–450. No information was presented, whether this complex has enzymatic activity (MANSUY et al. 1980).

The purified P–$450_{L.e.}$ can be reduced enzymatically by the addition of highly purified homologous NADPH-cyt c (P–450) reductase and NADPH. To get nearly 100% of reduction as judged by the formation of the CO complex a further addition of the substrate, e.g. hexadecane, is necessary (RIEGE et al. 1981). Unlike P–450_{LM}, which is rapidly autoxidizable, both the P–$450_{C.t.}$ and P–$450_{L.e.}$ are stable in the reduced state under aerobic conditions (DUPPEL et al. 1973; RIEGE and SMETTAN, in preparation). It has not been clarified, wether this pecularity has any importance for the reaction mechanisms of these P–450 forms.

The purification of the NADPH-cytochrome c (P–450) reductase of *L. elongisporus* was achieved in an effective 2-step chromatography in presence of Präwozell W-ON 100 and cholate on DEAE-Sephacel and hydroxylapatite after solubilization with deoxycholate. The procedure yields spectrally pure reductase with a specific activity of 55 μ moles reduced cyt c/mg protein x min, which is homogeneous in SDS-PAGE (HONECK et al. 1982).

BERTRAND et al. (1980) were successful in the purification of NADPH-cyt c (P–450) reductase of *C. tropicalis*, using cholate solubilization and column chromatography on Phenyl-Sepharose CL-4B, DEAE-cellulose, hydroxylapatite and Sephadex G 150. The purified reductase showed an activity of 56 μ moles cyt c/mg protein × min. In contrast to the reductase solubilized by gentle proteolysis (BERTRAND et al. 1979b), the detergent solubilized enzyme preparation was able to transfer electrons to P–$450_{C.t.}$.

4.2. Cytochrome P–450 of Saccharomyces Cerevisiae

4.2.1. Induction and Physiological Function

Obviously, the formation of P–450 in *Saccharomyces cerevisiae* depends on completely different factors than in alkane assimilating yeasts. The former is facultative anaerobic and can modify its metabolism in dependence on both the oxygen supply and the carbon source. A careful study on the relation between the oxygen concentration in the growth medium and the cellular P–450 content of *S. cerevisiae* was carried out by RODGERS and STEWART (1973) showing the highest P–450 concentration under semianaerobic conditions. KÄRENLAMPI et al. (1981) were able to show that strongly fermentable sugars as glucose, fructose, mannose in high concentration guarantee the accumulation of P–450. Moreover, the utilization of galactose, mannose, or maltose, sugars leading to a simultaneous fermentation and respiration, resulted in a P–450 formation. Obviously, the occurrence of P–450 in *S. cerevisiae* and related yeasts (KÄRENLAMPI et al. 1980a, b) seems to be linked with the physiological state of fermentation. A detailed discussion of the different aspects of the P–450 induction in *S. cerevisiae* was recently given by TAGAWA (1978).

The biological function of P–450 from *S. cerevisiae* has remained unclear for a long time. ALEXANDER et al. (1974) assumed involvement of P–450 in late stages of ergosterol biosynthesis. This hypothesis has been confirmed recently by OHBA et al. (1978), showing that the oxygen- and NADPH-dependent conversion of lanosterol to 4.4-dimethylzymosterol is inhibited by P–450 antibodies. Some time later, the oxidative removal of the 14α-methyl group of lanosterol by a reconstituted system of P–450 and P–450-reductase was demonstrated (AOYAMA and YOSHIDA, 1978a). Fig. 8 shows the way proposed by the authors for removal of the 14α-methyl group. The subsequent demethylations of the 4,4-dimethylzymosterol to zymosterol in the yeast microsomes are catalyzed by a CN^--sensitive enzyme system (OHBA et al. 1978):

$$\text{lanosterol} \xrightarrow{\text{P-450}} \text{4,4-dimethylzymosterol} \xrightarrow{CN^-\text{-sensitive}} \text{4-methylzymosterol} \xrightarrow{CN^-\text{-sensitive}} \text{zymosterol.}$$

Nevertheless, in our opinion, the biological importance of the P–450 induction

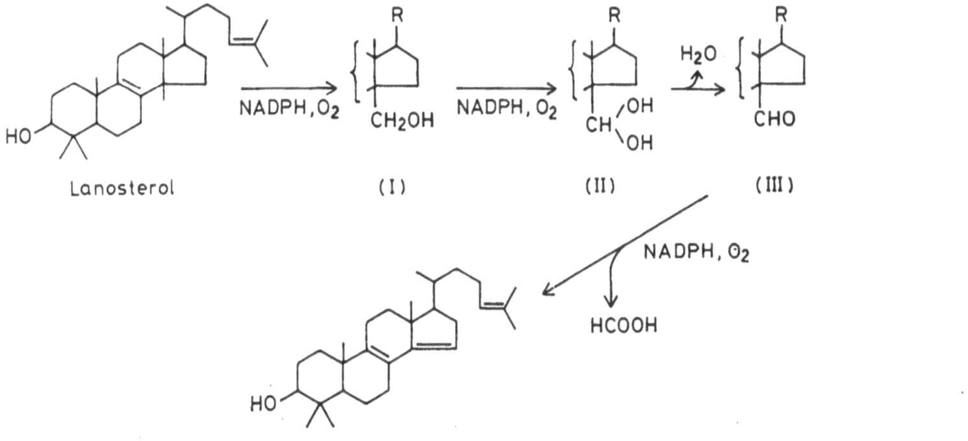

Fig. 7. Postulated pathway for the 14α-demethylation of lanosterol (AOYAMA and YOSHIDA 1978a)

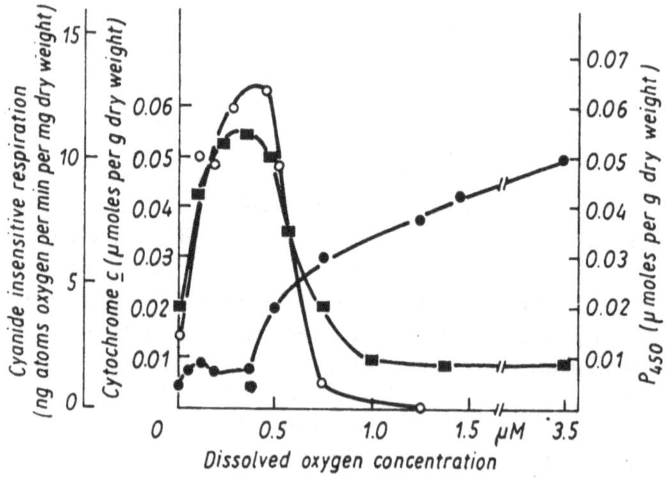

Fig. 8. Cellular content of cytochrome c (•), P–450 (○), and cyanid insensitive respiration of *S. cerevisiae* cells (■) as a function of oxygen concentration in the growth medium (Rogers and Stewart 1973); reproduced with permission from the American Society for Mikrobiology

in *S. cerevisiae* is not yet completely understood up to now, for the necessity of the enhanced ergosterol biosynthesis actually remains unclear.

4.2.2. Structure of the Cytochrome P–450 Containing Enzyme System and Enzymatic Properties

After the detection of the P–450 in cell extracts of *Saccharomyces* cerevisiae at first Ishidate et al. (1969) reported that this hemoprotein is bound to a particulate fraction of yeast cells. Later on Yoshida and co-workers isolated and purified the components of the microsomal e^--transfer system (Yoshida et al. 1974a) of anaerobically cultured yeast: cytochrome b_5 (Yoshida et al. 1974b), P–450 (Yoshida et al. 1977), NADH-cytochrome b_5-reductase (Kubota et al. 1977) and NADPH-P–450 reductase (Aoyama et al. 1978b).

About the enzymatic properties of this hemoprotein not so many facts are available because no hydroxylation reaction could be attached for a long time. Besides the 14α-demethylation of lanosterol (Aoyama and Yoshida, 1978b, see 4.2.1, Fig. 7) some further reactions were detected recently which are typical for the P–450 of liver microsomes.

Callen et al. (1977, 1978, 1980) showed that glucose-grown yeast cells with a high P–450 content (38.5 n mol/g dry weight) have a strongly enhanced mutation rate when exposed to promutagens or carcinogens. This mutation yield can be further increased by addition of cumene hydroperoxide. In control experiments with galactose as carbon source, in which the cells contain very little P–$450_{S.c.}$, the mutation rate was not increased. These findings suggest that the enhanced genetic activity is related to the P–450-dependent mixed-function oxidation. At the described activation of the promutagens and carcinogens the following reactions occur: dealkylation of dimethylnitrosamine, N-hydroxylation of β-naphthylamine and ethyl-carbamate, epoxidation of aflatoxin B_1, and conversion of halogenated hydrocarbons. Furthermore, Wiseman et al. (1975) and Woods and Wiseman (1980) reported on the conversion of biphenyl and of benzo(a)pyrene by cellfree extracts of *S. cerevisiae*. Biphenyl is hydroxylated in the 4-position. 3-hydroxybenzo(a)pyrene, 9-hydroxybenzo(a)pyrene, and, especially, 7,8 dihydrodiol-benzo(a)pyrene were identified as products of the benzo(a)pyrene. The K_m-value of benzo(a)pyrene hydroxylation was decreased during the growth of the yeast in the presence of benzo(a)pyrene, suggesting the induction of a different form of P–450, which preferably converts this aromatic hydrocarbon.

However, not all strains of *S. cerevisiae* seems to possess the capacity to trans-from xenobiotics in the same extent, for other investigators (Kärenlampi, personal communication, Kuhl 1980, Wiseman 1980) were not able to confirm these results. For that reason, in our opinion, further studies are necessary for clearing up the substrate specificity of P–$450_{S.c.}$, preferably using a reconstituted system of purified constituents.

4.2.3. Purification of the Individual Constituents and Molecular Properties

The P–$450_{S.c.}$ was purified to a specific content of 12–15 n moles/mg protein, using cholate solubilization, ammonium sulfate precipitation, and column chromatography on w–NH_2-hexyl Sepharose 4B, hydroxylapatite and CM-Sephadex (Yoshida et al. 1977). The molecular weight was estimated with 51000 daltons by SDS gel electrophoresis. Based on that value a specific content of 19.5 n moles/mg protein can be calculated for the pure protein. No evidence was found for different P–450 forms (Yoshida et al. 1977). Protoporphyrin IX was determined as prosthetic group (Yoshida and Kumaoka 1975).

Table 12 presents the spectral parameters of P–$450_{S.c.}$. The purified P–450 is present in the Fe^{3+} form. Like the alkane induced P–$450_{L.e.}$ (Riege et al. 1981) and P–$450_{C.t.}$ (Mansuy et al. 1980) it shows a typical low-spin state absorption spectrum. However, using the method of Jefcoate (1969), Yoshida and Kumaoka were able to show that

Table 12. Spectral properties of P–450 of *Saccharomyces cerevisiae*[1])

	Absorption peak (nm)	Millimolar absorbancy ($mM^{-1}\,cm^{-1}$)
ferric	575	11
	540	11
	418	113
	360	35
ferrous	550	13
	412	71
ferrous, CO complex	555	14
	447	117
ferrous, CO-difference	448–500	92

[1]) Yoshida et al. (1977); reproduced with permission from Academic Press, Inc.

P–$450_{S.c.}$ in the microsomal fraction is mainly present in the high-spin form. The low-spin absorption spectrum of the purified hemoprotein is explained by the use of Triton X-100 during the preparation acting as type II compound (Yoshida 1978). The low-spin state of the partially purified P–$450_{S.c.}$ can be partly reverted to the high-spin state by increasing the ionic strength. Addition of Triton X-100 or alcohols of different chain lengths reverts this spectral changes (Yoshida and Kumaoka 1975).

Substrate induced spectral changes can also be observed with microsomal P–$450_{S.c.}$. However, only lanosterol (Aoyama and Yoshida 1978a) and benzo(a)pyrene (Woods and Wiseman 1980) induce a type-I difference spectrum. Both substances are hydroxylated by P–$450_{S.c.}$ (see 4.2.2). All the other substances studied so far induce type-II or reverse type-I (modified type-II) spectral changes (Yoshida and Kumaoka 1975, Yoshida 1978). Typical type-I substrates of P–450_{LM}, such as alkanes and fatty acids, induce no spectral change (Yoshida 1978). Barbiturates, also a group of type-I substances of P–450_{LM}, induce a reverse type-I binding spectrum, which the authors referred to the presence of the high-spin form of P–450 in yeast microsomes (YOSHIDA and KUMAOKA 1975).

The CO complex of $P\text{–}450_{S.c.}$ exhibits a Soret band at 447—448 nm. However, the similarity of yeast P–450 with P–448 from liver microsomes based on these findings is not confirmed by other spectral properties (YOSHIDA and KUMAOKA 1975). The ratio between the 455 nm Soret peak and the 430 nm Soret peak of the ethylisocyanide complex of the reduced hemoprotein is lower than with P–448, however, it is larger than with the phenobarbital induced P–450 of hepatic microsomes.

The relative intensities of these peaks are dependent on the pH of the medium and indicate that reduced $P\text{–}450_{S.c.}$ exists in two interconvertible forms (YOSHIDA and KUMAOKA 1975, YOSHIDA 1978).

Like the liver microsomal P–450 also the yeast $P\text{–}450_{S.c.}$ can be inactivated to P–420 by several organic solvents, high concentrations of detergents, high concentrations of chaotropic anions and mercurials (YOSHIDA 1978, YOSHIDA and KUMAOKA 1975, YOSHIDA et al. 1977).

A cytochrome c reductase was purified to homogeneity and finally characterized as NADPH P–450 reductase (AOYAMA et al. 1978). The enzyme has many properties in common with that from liver microsomes. The molecular weight determined from SDS gel electrophoresis is 83×10^3 dalton; FAD and FMN have been determined as prosthetic groups. The flavoprotein reduces P–450, solubilized cytochrome b_5 from yeast, and cytochrome c (specific activity = 120—140 μ mol cyt c/mg protein × min) and shows, like the NADPH-P-450 reductase from the liver microsomal fraction, a menandione (vitamin K_3)-dependent cytochrome oxidase activity (KUBOTA et al. 1977, AOYAMA et al. 1978).

As already mentioned (see 4.2.1), the addition of this reductase to the partially purified $P\text{–}450_{S.c.}$ resulted in the reconstitution of the lanosterol-demethylase activity (AOYAMA and YOSHIDA 1978a).

4.3. Cytochrome P–450 Containing Steroid and Arylhydrocarbon Hydroxylases of Other Eucaryotic Microorganisms

Little is known about the molecular properties of P–450 systems occurring in other eucaryotic microorganisms, in comparison to *Candida* and *Saccharomyces* yeasts (SCHUNCK et al. 1978b). The refered studies listed in Tab. 1 mainly dealt with the induction and detection of P–450 and the characterization of the enzymatic reaction of the P–450 systems which are mainly bound to a membrane fraction. The purification of such a P–450 species has not yet been described.

In *Rhizopus nigricans*, P–450 is bound to the membrane fraction sedimenting at 105000 xg; the specific content is 2 nmol P–450/mg protein (BRESKVAR and HUDNIK-PLEVNIK 1977). The 11α-hydroxylation of progesterone by the cellfree supernatant is strongly inhibited by CO. The reversibility of this inhibition by irradiation with light at a predominant wavelength of $\lambda = 450$ nm shows the involvement of P–450 in this hydroxylation reaction. The P–450 containing membrane fraction, however, lacks hydroxylase activity; and it has to be assumed that the hydroxylating system is distributed between membrane structures and cytosol (BRESKVAR and HUDNIK-PLEVNIK 1977).

FERRIS et al. (1976) detected P–450 and a NADPH-dependent cyt. c reductase as constituents of the arylhydrocarbon hydroxylase system from *Cunninghamella bainieri*. This enzyme system catalyzes the O-demethylation of 4-nitroanisole and anisole and the hydroxylation of anisole, aniline and naphthalin. NADPH and O_2 are essential cosubstrates of the

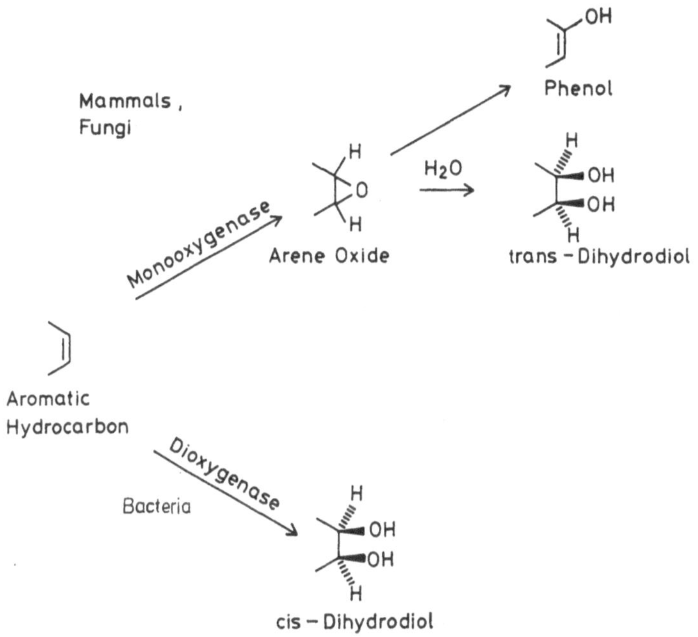

Fig. 9. Pathways of aromatic hydrocarbon oxidation utilized by mammals and fungi and by bacteria (Taking pattern from CERNIGLIA et al. 1978, modified)

enzymatic reaction which is, on the other hand, strongly inhibited by CO, metyrapon and SKF 525 A, whereas KCN acts only as a weak inhibitor. The findings of the inhibition experiments are typical for the involvement of P–450.

In the extensive studies on the metabolism of aromatic hydrocarbons by fungi, CERNIGLIA and GIBSON (1978a) detected P–450 in microsomal preparations of the naphthalene-precultured filamentous fungus *Cunninghamella elegans*. The P–450 containing membrane fraction of this microorganism oxidizes, in the presence of NADPH and O_2, naphthalene to 1,2-dihydroxy-1,2-dihydronaphthalene, 1-naphthol, and 2-naphthol. The strong inhibition of the naphthalene metabolism of fungal microsomes by CO, SKF 525-A and metyrapon suggest that P–450 is responsible for the naphthalene oxidation in *Cunninghamella elegans*.

These and further results (CERNIGLIA and GIBSON, 1978b) show the correspondence of routes of naphthalene oxidation both in mammalia and fungi, which is different from that in bacteria.

5. The Chloroperoxidase of *Caldariomyces fumago*

The studies performed by HAGER and his coworkers on the biosynthesis of Caldariomycin (2,2-dichloro-1,3-cyclopentandione) (SHAW and HAGER, 1959, 1961, BECKWITH et al. 1963) led in 1966 to the isolation and crystallization of a chloroperoxidase (chloride-hydrogen peroxide oxidoreductase; CPO EC 1.11.1.10) from the culture medium of *Caldariomyces fumago* (MORRIS and HAGER 1966). Preparation (HOLLENBERG and HAGER 1978), molecular (MORRIS and HAGER 1966), catalytic and spectral (see below) properties of this hemoprotein have been described in detail.

The purified enzyme is a monomeric glycoprotein with a molecular weight of about 42000 daltons and contains 1 iron protoporphyrin IX per molecule (MORRIS and HAGER 1966). As stressed by HAGER and his coworkers the CPO is quite similar to other protoheme peroxidases in many of its physical and catalytic properties but exhibits one striking difference in its ability to catalyze chlorination reactions at low pH. Therefore, they suggested that the CPO must differ significantly from other peroxidases in the vicinity of the active site (CHAMPION et al. 1973). Studies performed since 1973 have disclosed that the CPO shares a suprising number of unusual physicochemical properties with the cytochromes of P–450 type. From the results it may be concluded that P–450 and CPO have very similar if not identical axial ligands to their heme prosthetic group and that an "active oxygen species" can be formed in the catalytic cycle of the CPO which resembles that of P–450 enzymes in structure and reactivity.

5.1. Substrate and Reaction Specificity

The enzymatic properties of the chloroperoxidase show a broad specificity both with regard to the substrates converted and to the types of reactions being catalyzed by it (HOLLENBERG and HAGER 1978, THOMAS et al. 1970a, b). It catalyzes at least 3 different types of reactions (equ. 1—3 according to (HOLLENBERG and HAGER 1978)):

$$AH + X^- + H_2O_2 + H^+ \rightarrow AX + 2\,H_2O \qquad (1)$$

$$BH_2 + H_2O_2 \rightarrow B + 2\,H_2O \qquad (2)$$

$$2\,H_2O_2 \rightarrow O_2 + 2\,H_2O \qquad (3)$$

In the presence of H_2O_2, a halogen anion ($X^- = Cl^-$, Br^- or J^-) and of an appropriate halogen acceptor (AH, e.g. β-keto acids, aromatic esters) the enzyme catalyzes the formation of a covalent carbon-halogen bond (equ. 1). The complex formation of CPO with H_2O_2 and the halogen anions is accompanied by characteristic spectral changes (THOMAS et al. 1970a, b, HOLLENBERG and HAGER 1973).

In addition to the well studied halogenation reaction, CPO catalyzes the peroxidatic oxidation of classical peroxidase substrates ($= BH_2$, equ. 2) (THOMAS et al. 1970a, b) and exhibits catalatic activity (equ. 3) (THOMAS et al. 1970a, b).

Because of the different pH dependencies of these three enzymatic activities it was assumed that CPO may be present in two active forms, an "acidic" responsible for the halogenation reaction and a "neutral" one catalyzing the reactions 2 and 3. Transition from one form into the other occure between pH 5 and pH 3.

Recently published EPR studies by HOLLENBERG et al. (1980) led to the conclusion that the CPO exists in the pH range where it is enzymatically active in the halogenation reaction as a mixture of high and low spin ferric forms.

A pH-dependent transition to a predominantly high spin form (g values of 7.64, 4.29 and 1.78) was found when the pH was lowered to 1.5 and to a low spin form (g values near 2.61, 2.26 and 1.83) when the pH was raised to 5. At pH 3 the binding of halide ions to the CPO results in an equilibrium shift to the high spin form (HOLLENBERG et al. 1980).

5.2. Similarities to Cytochrome P–450

HOLLENBERG and HAGER (1973) reported that the reduced CO complex of the CPO reveals a similar strong red shift of the Soret band as it is characteristic for the P–450 proteins. This finding was the starting point of a thorough comparison of both hemoproteins using a variety of physical methods.

Table 13 shows the very similar parameters of CPO and P–450_{CAM} in the

Table 13. Spectral properties of chloroperoxidase in comparison to P–450$_{CAM}$[1])

Form	CPO-Soret peak wavelength (nm)	mM	P–450$_{CAM}$-Soret peak wavelength (nm)
ferric	396	91	391[2])
ferrous	406	77.5	408
CO-complex	443	123	446
+ CN^-	435	82	436
+ NO	430	80	435
+ N-phenylimidazol	424	75	420

[1]) HOLLENBERG and HAGER 1973; reproduced with permission from the American Society of Biological Chemists, Inc.
[2]) camphor-bound high-spin form

optical absorption spectra. Moreover, remarkable similarities were found by MÖSSBAUER (CHAMPION et al. 1973, 1975), RAMAN (CHAMPION et al. 1976), and magnetic circular dichroism spectroscopy (DAWSON et al. 1976). The EPR spectrum of the high spin form of the CPO shows that it exhibits like P–450 an extremely large rhombic distortion (HOLLENBERG et al. 1980). The EPR spectra of the low spin forms indicate a resemblance of the both hemoproteins in the electronic environment of the heme iron.

A detailed discussion of the similarities between CPO and P–450 and their structural basis was given by CHAMPION et al. (1975), CHIANG et al. (1975), DAWSON et al. (1976) and HOLLENBERG et al. (1980). Accordingly for both proteins an axial ligand to the heme prosthetic group has to be assumed which has the property of high electron density and large π-electron back donation to the iron (HOLLENBERG et al. 1980). Comparisons with porphyrin model complexes led to the hypothesis that a thiolate anion may function as the fifth heme iron ligand. Particulary conclusive the presence of an axial sulfur ligand was demonstrated by MCD spectra (DAWSON et al. 1976) and by the analysis of the extended X-ray absorption fine structure (EXAFS) of CPO, P–450 and porphyrin model complexes (CRAMER et al. 1978). From the EXAFS studies the iron-sulfur distance was calculated to be 2.19 Å in low spin P–450$_{LM2}$ and 2.30 Å in high spin CPO. Moreover, HOLLENBERG et al. (1980) found out that the EPR properties of CPO and the hydrosulfide complex of ferric myoglobin are very similar. Therefore the authors expect a negatively charged sulfur ligand to the heme iron of the CPO which is, however, not as electron rich as cysteinyl sulfur. It is interesting to note that this is really the same conclusion as it was drawn from quantum-chemical calculations carried out to explain the peculiar spectral properties of P–450 proteins (REIN et al., this volume). The postulated electronic properties of the sulfur ligand could be realized by an uncommon amino acid (HOLLENBERG et al. 1980) or by an involvement of cysteinyl sulfur in a hydrogen bridge (REIN et al., this volume).

In the amino acid composition, there are considerable differences between the CPO (MORRIS and HAGER 1966) and P–450$_{CAM}$ (GUNSALUS et al. 1974). CPO contains only 2 half-cysteine residues, which were found to be linked to a disulfide bridge (CHIANG et al. 1975). Thus, the strongly favoured hypothesis suggesting a thiolate as iron ligand would be ruled out. However, DAWSON et al. (1976) and ULLRICH et al. (1979) point to the possibility that the disulfide bridge may be an artifact in the denaturated state, and the impossibility to detect chemically a free SH group in the native state may be due to its shielding in the protein molecule.

On the other hand, O'KEEFFE et al. (1978) emphasize the possibility that the characteristic spectral properties of CPO and P–450 might not be produced by thiolate as iron ligand, but also by special cis-effects.

Based on the very similar structure of the active sites, as reflected by the spectroscopical data, it is suggested that there are certain common features also in the catalytic properties of CPO and P–450 proteins.

It was proposed that the "active oxygen species" in the catalytic cycles of both hemoproteins are nearly identical in structure and reactivity (MAKINO et al. 1976, REMBA et al. 1979, CORBETT et al. 1980).

It was found that CPO catalyzes like P–450 the N-oxidation of arylamines, although with a much higher activity (CORBETT et al. 1980).

On the other hand for liver microsomal P–450 peroxidatic activity and the catalysis of peroxide-dependent hydroxylation reactions were demonstrated (CH 8 f. chapter). Recently, KEDDERIS et al. 1980, reported for the first time that the CPO can catalyze the peroxide-dependent dealkylation of several N-methyl arylamines in a manner similar to that described for liver microsomal P–450. According to the authors this reaction can be written in the following equation:

$$\mathrm{ROOH} + \mathrm{R'R''{-}N{-}CH_3} \rightarrow \mathrm{R{-}OH} + \mathrm{R'R''{-}NH} + \mathrm{HCHO}$$

where ROOH is the peroxide and $R'R''-N-CH_3$ is the N-methyl arylamine substrate.

For the ethyl hydrogen peroxide-supported dealkylation of N,N-dimethylaniline by CPO a turnover number of 1476 was found. In comparison to the rates of dealkylation reactions catalyzed by microsomal P–450 this represent a very high activity.

Differences exist, moreover, with respect to the substrate specificity of the two hemoproteins.

From future investigations comparing CPO and P–450 an important insight may be expected how the different catalytic properties of enzymes which contain the same prosthetic heme group are determined by the structure of their active sites. On the other hand future studies on the mechanism of the halogenation reactions catalyzed by the CPO will be stimulated by the environmental concern of some halometabolites and the demonstration of related enzyme systems in marine organisms. It has been assumed that these enzymes are involved in the biosynthesis of the widespread occurring bromine and chlorine containing organic compounds in numerous marine organisms (THEILER et al. 1978).

6. Outlook

In this chapter we attempted to summarize the most essential current knowledge about the P–450 of microorganisms, mainly concerning the molecular and enzymatic properties and the reaction mechanism. Certainly, this topic will remain in the midpoint of P–450 research in the next years. Especially, the clarification of the intermediate states and the intermolecular interactions between the constituents of the monooxygenase system will challenge biochemists and biophysicists also in future. P–450 species of microbial origin, first of all the $P\text{–}450_{CAM}$ system, will be included in these investigations. However, the importance of other topics will grow. The elucidation of genetic, epigenetic and metabolic regulation of monooxygenase systems is only in the very beginning.

Little attention was payed hitherto to the interactions of P-450 with the enzymes following in the catabolic pathway(s). Depending on the exact question, bacteria or yeasts may be very appropriate objects for investigating these problems. Furthermore, it will be necessary to extend the number of strains, being able to hydroxylate xenobiotics or lipophilic substances of biological origin. Very certainly, many wait for their discovery yet and the advances in genetics will facilitate the access to useful strains. Moreover, the multiplicity of enzymatic hydroxylations in microorganisms is not completely known. In combination with an enlarged knowledge about the P-450 regulation novel possibilities may reveal from this for taking advantage of the selective conversion of inert compounds for the production of valuable chemicals and pharmaceuticals. A first step on this way could to be the enzymatic ethylene oxide synthesis developed by the Cetus Corporation (U.S.A.).

References

ALEXANDER, K. T. W., K. A. MITROPOULOS, and G. F. GIBBONS (1974): Biochem. Biophys. Res. Commun. **60**, 460–467

AMBIKE, S. H., R. M. BAXTER, and N. D. ZAHID (1970): Phytochemistry, **9**, 1953 to 1958

AOYAMA, Y., and Y. YOSHIDA (1978a): Biochem. Biophys. Res. Commun. **85**, 28–34

AOYAMA, Y., and Y. YOSHIDA (1978b): Biochem. Biophys. Res. Commun. **82**, 33–38

AOYAMA, Y., Y. YOSHIDA, S. KUBOTA, H. KUMAOKA, and A. FURIMICHI (1978): Arch. Biochem. Biophys. **185**, 362–369

APPLEBY, C. A. (1967): Biochim. Biophys. Acta **147**, 399–402

APPLEBY, C. A. (1969): Biochim. Biophys. Acta **172**, 71–87

APPLEBY, C. A., and R. M. DANIEL (1973): in "Oxidases and related redox system" (T. T. KING, H. S. MASON, and M. MORRISON, eds.) Univ. Park Press, Baltimore, Maryland, p. 515–528

APPLEBY, C. A., G. L. TURNER, and P. K. MACNICOL (1975): Biochim. Biophys. Acta **387**, 461–474

APPLEBY, C. A. (1978a): Meth. Enzymol. **52**, 157–166

APPLEBY, C. A. (1978b): Proc. FEBS Meet. 1977, Vol. 48, Colloqu. B 6, Oxford, New York, p. 11–20

ASPERGER, O., A. NAUMANN, and H.-P. KLEBER, (1981): FEMS Microbiol Letters **11**, 309–312

BECKWITH, J. R., R. CLARK, and L. P. HAGER (1963): J. Biol. Chem. **238**, 3086–3090

BERG, A., K. CARLSTRÖM, J.-Å. GUSTAFSSON, and M. INGELMAN-SUNDBERG, (1975): Biochem. Biophys. Res. Commun. **66**, 1414–1423

BERG, A., J.-Å. GUSTAFSSON, and M. INGELMAN-SUNDBERG, K. CARLSTRÖM (1976): J. Biol. Chem. **251**, 2831–2838

BERG, A., M. INGELMAN-SUNDBERG, and J.-Å. GUSTAFSSON (1979a): Acta biol. med. germ. **38**, 333–344

BERG, A., M. INGELMAN-SUNDBERG, and J.-Å. GUSTAFSSON (1979b): J. Biol. Chem. **254**, 5264–5272

BERG, A., (1980): in Biochemistry, Biophysics and Regulation of Cytochrome P-450 (Gustafsson, J.-Å. et al. ed.) Elsevier Amsterdam p. 121–124

BERTRAND, J. C., M. GILEWICZ, H. BAZIN, M. ZACEK, and E. AZOULAY (1979a): FEBS-Letters **105**, 143–146

BERTRAND, J. C., H. BAZIN, M. ZACEK, K. GILEWICZ, and E. AZOULAY (1979b): Eur. J. Biochem. **93**, 237–243

BERTRAND, J. C., M. GILEWICZ, H. BAZIN, and E. AZOULAY (1980): Biochem. Biophys. Res. Commun. **94**, 880–893

BRESKVAR, K., and T. HUDNIK-PLEVNIK (1977): Biochem. Biophys. Res. Commun. **74**, 1192–1197

BRESKVAR, K., and T. HUDNIK-PLEVNIK (1978): J. Steroid Biochem. **9**, 131–134

BROADBENT, D. A., and N. J. CARTWRIGHT (1974): Microbios **9**, 119–130

BUCHANAN, J. F., and A. J. FULCO (1978): Biochem. Biophys. Res. Commun. **85**, 1254 to 1260

CALLEN, D. F., and R. M. PHILPOT (1977): Mutation Research, **45**, 305–324

CALLEN, D. F., C. R. WOLF, and R. M. PHILPOT (1978): Biochem. Biophys. Res. Commun. **83**, 14–20

CALLEN, D. F., C. R. WOLF, and R. M. PHILPOT (1980): Mutation Res. **77**, 55–63

Cardini, G., and P. Jutrshuk (1968): J. biol. Chem. **243**, 6070—6072
Cardini, G., and P. Jutrshuk (1970): J. biol. Chem. **245**, 2789—2796
Cartledge, T. G., D. L. Loyd, M. Erecinska, and B. Chance (1972): Biochem. J. **130**, 739—747
Cartwright, N. J., and D. A. Broadbent (1974): Microbios, **10**, 87—96
Cartwright, N. J., K. Holdan, and D. A. Broadbent, (1971): Microbios, **3**, 113—130
Cerniglia, C. E. and D. T. Gibson (1978a): Arch. Biochem. Biophys. **186**, 121—127
Cerniglia, C. E., and D. T. Gibson (1978b): Arch. Microbiol. **117**, 135—143
Champion, P. M., E. Munck, P. Debrunner, P. F. Hollenberg, and L. P. Hager (1973): Biochemistry **12**, 426—435
Champion, P. M., R. Chiang, E. Munck, P. Debrunner, and L. P. Hager (1975): Biochemistry **14**, 4159—4166
Champion, P. M., R. D. Remba, R. Chiang, D. B. Fitchen, and L. P. Hager (1976): Biochim. Biophys. Acta **446**, 486—492
Chiang, R., T. Rand-Meir, R. Makino, and L. P. Hager (1976): J. Biol. Chem. **251**, 6340 to 6346
Chiang, R., R. Makino, W. R. Spomer, and L. P. Hager (1975): Biochemistry **14**, 4166 to 4171
Coon, M. J., and K. P. Vatsis (1978): Polycyclic Hydrocarbons and Cancer **1**, 335 to 360
Corbett, M. D., B. R. Chipko, and A. O. Batchelor (1980): Biochem. J. **187**, 893 to 903
Cramer, S. P., J. H. Dawson, K. O. Hodgson, and L. P. Hager (1978): J. Amer. Chem. Soc. **100**, 7283—7290
Daniel, R. M., and C. A. Appleby (1972): Biochim. Biophys. Acta **275**, 347—354
Dawson, J. H., J. R. Trudell, G. Barth, R. E. Linder, E. Bunnenberg, C. Djerassi, R. Chiang, and L. P. Hager (1976): J. Amer. Chem. Soc. **98**, 3709—3710
Davies, J. S., A. M. Wellmann, and J. E. Zajic (1976): Appl. Environm. Microbiol. **32**, 14—20
Delaisse, J. M., and E. J. Nyns (1974): Arch. internat. Physiol. Biochem. **82**, 179
Demanova, N. F., E. R. Davydov, and A. D. Gololobov (1980): Appl. Biochem. Microbiol. (Moscow) **16**, 5—12
Duppel, W., J.-M. Lebeault, and M. J. Coon (1973): Eur. J. Biochem. **36**, 583—592
Dus, K., R. Goewert, C. C. Weaver, D. Carey, and C. A. Appleby (1976): Biochem. Biophys. Res. Commun. **69**, 437—445
Ferris, J. P., L. H. MacDonald, M. A. Patrie, and M. A. Martin (1976): Arch. Biochem. Biophys. **175**, 443—452
Fukui, S., and A. Tanaka (1979): in Fiechter, A. (ed.) Microbiology applied to Biotechnology, Dechema-Monographien Bd. 83, Verlag Chemie Weinheim
Gallo, M., J. C. Bertrand, and E. Azoulay (1971): FEBS-Letters, **19**, 45—49
Gallo, M., J. C. Bertrand, B. Roche, and E. Azoulay (1973): Biochim. Biophys. Acta (Amsterdam) **296**, 624—638
Gallo, M., B. Roche, and E. Azoulay (1976): Biochim. Biophys. Acta (Amsterdam) **419**, 425—434
Gilewicz, M., M. Zacek, J.-C. Bertrand, and E. Azoulay (1979): Canad. J. Microbiol. **25**, 201—206
Gmünder, F., and A. Fiechter (1978): Abstr. XIIth Internat. Congr. Microbiol. Munich p. 128
Gmünder, F. (1979): Assimilation von Hexadekan durch *Candida tropicalis*, Dissertation Nr. 6427 ETH Zürich
Griffin, B. W., and J. A. Peterson (1972): Biochemistry **11**, 4740—46
Gunsalus, I. C. (1982): 4th Internat. Conf. on Cytochrome P-450 in Kuopio Abstracts
Gunsalus, I. C., J. R. Mercks, J. D. Lipscomb, P. G. Debrunner, and E. Münch (1974): in O. Hayaishi (ed.) Molecular Mechanisms of Oxygen Activation, Academic Press, New York p. 559—613
Gunsalus, I. C., T. C. Pederson, and S. G. Sligar (1975): Ann. Rev. Biochemistry **44**, 377—407
Gunsalus, I. C., and S. G. Sligar (1978): Adv. Enzymology **47**, 1—44
Gunsalus, I. C., and G. C. Wagner (1978): in Colowick, S. P., and N. O. Kaplan: Methods in Enzymology **52**, 166—188
Hager, L. P., D. R. Morris, F. S. Brown, and H. Eberwein (1966): J. Biol. Chem. **241**, 1769—1777
Hare, R. S., and A. J. Fulco (1975): Biochem. Biophys. Res. Commun. **65**, 665—672
Haugen, D. A., and M. J. Coon (1976): J. Biol. Chem. **251**, 7929—7939
Hedegarrd, J., and I. C. Gunsalus (1965): J. Biol. Chem. **240**, 4038—4043
Heinz, E., A. D. Tulloch, and J. F. T. Spen-

CER (1970): Biochim. Biophys. Acta **202**, 49–55

HO, P. P., and A. J. FULCO (1976): Biochim. Biophys. Acta **431**, 249–256

HOLLENBERG, P. E., and L. P. HAGER (1973): J. Biol. Chem. **248**, 2630–2633

HOLLENBERG, P. E., and L. P. HAGER (1978): in: Meth. Enzymol. **52**, 521–529

HOLLENBERG, P. E., L. P. HAGER, W. E. BLUMBERG, and J. PEISACH (1980): J. Biol. Chem. **255**, 4801–4807

HONECK, H., W.-H. SCHUNCK, P. RIEGE, and H.-G. MÜLLER (1981): Biochem. Biophys. Res. Commun. **106**, 1318–1324

ILCHENKO, A. D., S. MAUERSBERGER, R. N. MATYASHOVA, and A. B. LOSINOV (1980): Mikrobiologija (Moscow) **49**, 452–458

INOZENTSEVA, I. A., S. S. MELIK-SARKISYAN, and V. L. KRETOVICH (1978): Dokl. Akad. Nauk. SSSR **240**, 1468–1471

ISHIDATE, K., K. KAWAGUCHI, and K. TAGAWA (1969a): J. Biochem. (Tokyo) **65**, 375 to 384

ISHIDATE, K., K. KAWAGUCHI, and K. TAGAWA (1969b): J. Biochem. (Tokyo) **65**, 385–392

ISHIMURA, Y. (1978): in Cytochrome P-450 (SATO, R., and T. OMURA, ed., p. 209–227, Kodanska Tokyo)

JAYANTHI, C. R., P. MADYASTHA and K. M. MADAYASTHA (1982): Biochem. Biophys. Res. Commun. **106**, 1262–1269

JEFCOATE, C. D. E., J. L. GOUSLAR, and R. CABABRESE (1969): Biochemistry **8**, 3455 to 3463

KATAGIRI, M., B. N. GANGULI, and I. C. GUNSALUS (1968): J. Biol. Chem. **243**, 3543 to 3546

KÄRENLAMPI, S., E. MARIN, and O. HÄNNINEN (1980a): in GUSTAFSSON, J.-Å. et al. (ed.) Biochemistry, Biophysics and Regulation of Cytochrome P-450 Elsevier/North-Holland Biomedical Press, p. 247–250

KÄRENLAMPI, S., E. MARIN, and O. HÄNNINEN (1980b): J. Gen. Microbiol. **120**, 529 to 533

KÄRENLAMPI, O., E. MARIN, and O. HÄNNINEN (1981): Biochem. J. **194**, 407–413

KEDDERIS, G. L., P. R. KOOP, and P. F. HOLLENBERG (1980): J. Biol. Chem. **255**, 10174–10182

KLUG, M. J., and A. J. MARKOVETZ (1971): Adv. Microbiol. Physiol. **5**, 1–43

KRETOVICH, W. L., S. S. MELIK-SARKYSYAN, M. V. RAIKCHINSTEIN, and A. I. ARCHAKOV (1974): FEBS-Lett. **44**, 305–308

KUBOTA, S., Y. YOSHIDA, and M. KUMAOKA (1977): J. Biochem. (Tokyo) **81**, 187–195

KUHL, R. (1980): Zytochrom P-450 in der Hefe *Saccharomyces cerevisiae*, Diss. Akad. d. Wiss. der DDR Berlin

LEBEAULT, C. M., E. T. LODE, and M. J. COON (1971): Biochem. Biophys. Res. Commun. **42**, 413–419

LINDENMAYER, A., and L. SMITH (1964): Biochim. Biophys. Acta (Amsterdam) **93**, 445–461

LIPSCOMB, J. D., S. G. SLIGAR, M. J. NAMTVEDT, and I. C. GUNSALUS (1976): J. Biol. Chem. **251**, 1116–1124

MAKINO, R., R. CHIANG and L. P. HAGER (1976): Biochemistry **15**, 4748–4754

MANSUY, D., M. CARLIER, J.-C. BERTRAND, and E. AZOULAY (1980): Eur. J. Biochem. **109**, 103–108

MASON, H. S., W. L. FOWLKS, and E. PETERSON (1955): J. Amer. Chem. Soc. **77**, 2914 to 2917

MATSON, R. S., R. S. HARE, and A. J. FULCO (1977): Biochim. Biophys. Acta **487**, 487 to 494

MAUERSBERGER, S., R. N. MATYASHOVA, H.-G. MÜLLER, and A. B. LOSINOV (1980): Europ. J. Appl. Microbiol. Biotechnol. **9**, 285–294

MAUERSBERGER, S., and R. N. MATYASHOVA (1980): Mikrobiologija (Moscow) **49**, 571–577

MAUERSBERGER, S., W.-H. SCHUNCK, and H.-G. MÜLLER (1981): Z. Allgem. Mikrobiol. **21**, 313–321

MIALL, L. M. (1980): in HARRISON, D. E. F., I. J. HIGGINS, and R. WATKINSON (ed.): Hydrocarbons in Biotechnology, Heyden & Son, London, p. 25–34

MIURA, Y., and A. J. FULCO (1974): J. biol. Chem. **249**, 1880–1888

MIYAKE, Y., Y. NAKAMURA, N. TAKAYAMA, and K. HORIKE (1975): J. Biochem. (Tokyo) **78**, 773–781

MORRIS, D. R., and L. P. HAGER (1966): J. Biol. Chem. **241**, 1763–1768

MÜLLER, H.-G., H. HONECK, S. MAUERSBERGER, P. RIEGE, W. H. SCHUNCK, H. TERYTZE, and E. KÄRGEL (1978): Abstr. XII. Internat. Congr. Microbiol. München, p. 127

MÜLLER, H.-G., W.-H. SCHUNCK, P. RIEGE, and H. HONECK (1979): Acta biol. med. germ. **38**, 345–349

MURPHY, G., G. VOGEL, G. KRIPPAHL, and F. LYNEN (1974): Eur. J. Biochem. **49**, 433 to 438

OHBA, M., R. SATO, Y. YOSHIDA, T. NISHONO, and H. KATSUKI (1978): Biochem. Biophys. Res. Commun. **85**, 21–27
O'KEEFFE, D. H., R. E. EBEL, J. A. PETERSON, J. C. MAXWELL, and W. S. CAUGHEY (1978): Biochemistry **17**, 5845–5852 (1978)
PEISACH, J., C. A. APPLEBY, and W. E. BLUMBERG (1979): Arch. Biochem. Biophys. **150**, 725–732
PETERSON, J. A. (1971): Arch. Biochem. Biophys. **144**, 678–693
POOLE, R. K., D. LLOYD, and B. CHANCE (1974): Biochem. J. **138**, 201–210
RAHIMTULA, A. D., P. J. O'BRIEN, E. G. HRYCAY, J. A. PETERSON, and R. ESTABROOK (1974): Biochem. Biophys. Res. Commun., **60**, 695–702
REMBA, R. D., P. M. CHAMPION, D. B. FITCHEN, R. CHIANG, L. P. HAGER (1979): Biochemistry **18**, 2280–2290
RIEGE, P., W.-H. SCHUNCK, H. HONECK, and H.-G. MÜLLER (1981): Biochem. Biophys. Res. Commun. **98**, 527–534
RIEGE, P., G. SMETTAN: in prep.
ROGERS, P. J., and P. R. STEWART (1973): J. Bact. **115**, 88–97
ROSAZZA, J. P., and R. V. SMITH (1979): Adv. Appl. Microbiol. **25**, 169–208
SCHUNCK, W.-H., P. RIEGE, R. BLASIG, H. HONECK, and H.-G. MÜLLER (1978a): Acta biol. med. germ. **37**, K3–K7
SCHUNCK, W.-H., P. RIEGE, and R. KUHL (1978b): Pharmazie **33**, 412–414
SHAW, P. D., and L. P. HAGER (1959): J. Amer. Chem. Soc. **81**, 1011–1012
SHAW, P. D., and L. P. HAGER (1961): J. Biol. Chem. **236**, 1626–1630
SLIGAR, S., P. G. DEBRUNNER, J. E. LIPSCOMB, M. J. NAMTVEDT, and I. C. GUNSALUS (1974): Proc. Nat. Acad. Sci. USA **71**, 3906 to 3910
SLIGAR, S. G., M. J. NAMTVEDT, V. R. ELLIS, and I. C. GUNSALUS, (1976): Hoppe-Seylers Z. phys. Chem. **357**, 1056
SLIGAR, S., M. BESMAN, M. GELB, P. GOULD, D. HEIMBROOK, and D. PEARSON (1980a): in Biochemistry, Biophysics and Regulation of Cytochrome P-450 (GUSTAFSSON, J.-Å. et al. ed.) Elsevier Amsterdam, p. 379–382
SLIGAR, S., and I. C. GUNSALUS (1979): Biochemistry **18**, 2290–2295
SLIGAR, S., J. KENNEDY, and D. PEARSON (1980b): Proc. Nat. Acad. Sci. USA **77**, 1240 to 1244
SMITH, R. V., and P. J. DAVIS (1980): Adv. in Biochem. Engineering **14**, 61–100
TAGAWA, K. (1978): in "Cytochrome P-450" (R. SATO, and T. OMURA ed.) 202–208, Kodansha (Academic Press), Tokyo
TAKAGI, M., K. MORIYA, and K. YANO (1980a): Cell. Molec. Biol. **25**, 363–370
TAKAGI, M., K. MORIYA, and K. YANO (1980b): Cell. Molec. Biol. **25**, 371–375
TAKAGI, M., H. KAWAMURA, K. MORIYA, and K. YANO (1980c): Abstr. 5th Internat. Yeast Symp. London/Ontario, SV 7–8
TANAKA, M., HANIU, M., YASANOBU, K. T., DUS, K., and I. C. GUNSALUS (1974): J. Biol. Chem. **249**, 3689–3701
THEILER, R., J. C. COOK, L. P. HAGER, J. F. SINDA (1978): Science **202**, 1094–1096
THOMAS, J. A., D. R. MORRIS, and L. P. HAGER (1970a): J. Biol. Chem. **245**, 3129 to 3134
THOMAS, J. A., D. R. MORRIS, and L. P. HAGER (1970b): J. Biol. Chem. **245**, 3135 to 3142
TITTELBACH, M., H.-G. ROTHE, und H. WEIDE (1976): Z. allg. Mikrobiologie **16**, 155–156
TYSON, C. A., J. D., LIPSCOMB, and I. C. GUNSALUS (1972): J. Biol. Chem. **247**, 5777 to 5784
ULLRICH, V., and W. DUPPEL (1975): in P. D. BOYER (ed.): The Enzymes Vol. **12**, Part B. Academic Press, New York p. 253 to 294
ULLRICH, V., H. SAKURAI, and H. H. RUF (1979): Acta biol. med. germ. **38**, 287–297
VYGANTAS, A. M., and J. W. CORCORAN (1974): Fed. Proceedings **33**, 1233
WHITE, R. E., and M. J. COON (1980): Ann. Rev. Biochem. **49**, 315–356
WILSON, G. S., J. C. M. TSIBRIS, and I. C. GUNSALUS (1973): J. Biol. Chem. **248**, 6059 to 6061
WISEMAN, A. (1977): Topics in Enzyme and Fermentation Technology **1**, 172–184
WISEMAN, A. (1980): Trends in Biochemical Sciences **5**, 102–104
WISEMAN, A., J. A. GONDAL, and P. SIMS (1975): Biochem. Soc. Trans. **3**, 278–281
WISEMAN, A., and J. A. GONDAL (1975): Biochem. Trans. **3**, 977
WISEMAN, A., and L. F. J. WOODS (1980): Biochem. Soc. Trans. **8**, 200–201
WOODS, L. F. J., and A. WISEMAN (1980): Biochim. Biophys. Acta **613**, 52–61
YAMAMOTO, A., S. TAKEMORI, and M. KATA-

GIRI (1974): Seikagaku (Japanese), **46**, 534 cited in Yoshida (1978)
YOSHIDA, Y. (1978): in "Cytochrome P-450" (ed. R. SATO and T. OMURA), 194–202, Kodanska, Academic Press, Tokyo
YOSHIDA, Y., H. KUMAOKA, and R. SATO (1974a): J. Biochem. (Tokyo) **75**, 1201 to 1210
YOSHIDA, Y., H. KUMAOKA, and R. SATO (1974b): J. Biochem. (Tokyo) **75**, 1211–1213
YOSHIDA, Y., and H. KUMAOKA (1975): J. Biochem. (Tokyo) **78**, 785–794
YOSHIDA, Y., Y. AOYAMA, H. KUMAOKA, and S. KUBOTA (1977): Biochem. Biophys. Res. Commun. **78**, 1005–1010
YU, C.-A., I. C. GUNSALUS, M. KATAGIRI, K. SUHARA, and S. TAKEMORI (1974a): J. Biol. Chem. **249**, 94–101
YU, C.-A., and I. C. GUNSALUS (1974b): J. Biol. Chem. **249**, 102–106
YU, C.-A., and I. C. GUNSALUS (1974c): J. Biol. Chem. **249**, 107–110

Chapter 8

Aspects of Application of Cytochrome P-450 and Related Systems in Substrate Hydroxylation and Conversion Processes

P. Mohr, F. Scheller, R. Renneberg, M. Kühn, K. Pommerening, F. Schubert and W. Scheler

Contents

1. Introduction

Among the hemoproteins the cytochromes P-450 are of considerable practical interest because the various P-450 types are able to catalyze specific chemical conversion reactions of organic substrates by molecular oxygen not only in vivo but also under in vitro conditions (1). This so-called mixed function oxidation reaction needs additionally two electrons for each substrate molecule. In the living cells these electrons are provided by reduced pyridine nucleotides.

$$R-H + O_2 + NADPH + H^+ \xrightarrow{'/P-450/} R-OH + H_2O + NADP^+ \quad (1)$$

P-450 enzymes are widely spread in animals, plants and microorganisms. Most of the investigations have been made with P-450 preparations isolated e.g. from liver microsomes (P-450_{LM}), suprarenal glands (P-450_{AC}), and various yeast strains. All known P-450 enzyme systems consist of an electron transfer component and a terminal oxidase. Occurrence and properties of some P-450 systems are summarized in Table 1. In all cases the terminal oxidase is an iron protoporphyrin IX — protein complex. Differences exist only in the chemical structure and the conformation of the protein component. Table 1 shows furthermore that NADH and/or NADPH generally act as the physiological electron donor, whereas the electron transferring proteins are different in the various species. The specific hydroxylation abilities of P-450 enzymes are attractive for practical application in substrate conversion processes and analytical systems, respectively. At present this field is in a pioneering stage of development. There are several main directions of prac-

Abbreviations:
P-450 — cytochrome P-450; LM — liver microsomal; AC — adrenocortical; HRP — horseradish peroxidase; IPP = Iron (III) protoporphyrin IX

Table 1. Occurrence, composition and properties of some selected P–450 systems

Species	Electron donor	Electron transfer system		Substrate(s)	References
		flavoprotein	non heme iron protein		
Rat, rabbit (liver microsomes)	NADPH	NADPH P–450 reductase	—	drugs	Lu et al. (1972)
Bovine (suprarenal gland) (mitrochondria)	NADPH	adrenodoxin reductase	adrenodoxin	steroids in position 11β	Omura et al. (1966)
Pseudomonas putida	NADH	putidaredoxin reductase	putidaredoxin	camphor in position 5-exo	Katagiri et al. 1968)
Bacillus megaterium	NADPH	megaredoxin reductase	megaredoxin	steroids in position 15β	Berg et al. (1976)
Lodderomyces elongisporus (microsomes)[1]	NADPH	NADPH specific reductase	—	alkanes	Riege et al. (1981), Müller et al. (1979)
Candida tropicalis (microsomes)	NADPH	NADPH specific reductase	—	alkanes and fatty acids in β-position	Duppel et al. (1973)

[1]) in first publications designated as *Candida guilliermondii*

tical interest:

a) Some selective hydroxylation or side chain cleavage reactions of organic compounds (e.g. steroids) (Cooper et al., 1979; Akhrem et al., 1979) in preparative scale might be useful for the pharmaceutical industry.

b) The ability of the P–450$_{LM}$-system to catalyze the conversion of drugs and other foreign substances into water soluble compounds, which can be excreted by the kidney, opens the possibility to use these enzymes for an extracorporal metabolic detoxification of blood in the case of drug overdosage or other toxic damages (Brunner and Loesgen, 1977; Denti et al., 1976).

c) In a similar way "metabolizing reactors" can be used under in vitro conditions to analyze the metabolic pattern of a prospective drug in order to synthesize and investigate the metabolites with respect, to possible mutagenic or carcinogenic activities.

d) Development of sensors, enzyme electrodes and test systems on the basis of P–450 for the determination and control of substrate concentrations (Scheller et al., 1977a; Schubert et al., 1980) as well as for diagnostic purposes (Ames, 1976).

P–450 systems, however, are relatively instable and their preparation is very expensive, too. The costs of such devices and applications are furthermore increased by the required quantities of the cofactors NADH and NADPH, respectively. For this reason the research work with regard to a prospective utilization of the principle of P–450 catalysis now proceeds into five directions:

1. Stabilization of the enzyme system by immobilization of microsomes or of the isolated components (Table 2).

2. Regeneration of the required cofactors.

Table 2. Immobilization of P–450

Species	Matrix	Substrate	References
Rat (LM)[1]	CNBr-activated Sepharose 4B	aminopyrine, benzphetamine	BAESS et al. (1975), JÄNIG et al. (1977), PFEIL et al. (1977)
Rabbit (LM)[1]	CNBr-activated Sepharose 4B (coimmobilization with UDP-glucuronyltransferase)	paracetamol, hexan	BRUNNER and LOESGEN (1977)
Rat (LM)[2]	Cellulose and silicone hollow fibers	p-nitroanisole, hexobarbital	COHEN et al. (1977)
Rat/Rabbit (LM)[1]	polymethacrylate containing imidazole groups	benzphetamine, aniline	POMMERENING (unpublished results)
Rat (LM)[1]	Biogel P–300	aniline[3]	ERYOMIN et al. (1979)
Rat (LM)[1]	nylon	aniline[3]	Metelitsa (personal communication)
Rabbit (LM)[2]	Ca-alginate gel	aminopyrine, ethylmorphine	SCHUBERT et al. (1982)
Rabbit (LM)[2]	gelatin	aminopyrine, aniline	SCHUBERT et al. (1980)
Rat (LM)[2]	polyvinylpyrrolidone hydrogel	aniline, p-nitroanisole, phenobarbital, diazepam	DENTI et al. (1976)
Rabbit (LM)[1]	CNBr-activated Sepharose, acrylamide/acrylic acid copolymer, liquid-lipid membrane vesicles	benzphetamine, octanoate	BRUNNER et al. (1980)
Rabit (LM)[1]	albumin-modified Sepharose 4B	aniline[3]	ERYOMIN et al. (1980)
Rabbit (LM)	CNBr-activated Sepharose (coimmobilization with UDP-glucuronyltransferase)	ethylmorphine, p-nitroanisole	YOST et al. (1980)
Bovine (AC)[1]	Sepharose 4B	deoxycorticosteron	JÄNIG et al. (1977)
Bovine (AC)[1]	Sepharose 4B	cholesterol	AKHREM et al. (1977)

[1]) solubilized
[2]) microsomes
[3]) H_2O_2-dependent, catalyzed by the isolated terminal oxidase

3. Substitution of the electron donor and the electron transfer components by simpler principles (SCHELLER et al., 1977b; MOHR et al., 1978; SCHELLER et al., 1979).
4. Preparation of iron porphyrin complexes possessing P–450-like catalytic properties (MOHR et al., 1978, 1979; KÜHN and MOHR, 1981).
5. Development of suitable enzyme reactors and analytical systems as well as improvement of the reaction conditions

(e.g. solubility of substrates and products, increase of the turnover). It is the aim of the following sections to discuss these directions more in detail and to draw some conclusions concerning their future development.

2. The Present State of Application of Cytochrome P-450

2.1. *Immobilization of the Enzyme System*

For many practical applications it is useful to have an enzyme in a form which is water-insoluble. This can be attained by coupling the enzyme to supporting materials (synthetic polymers, glass beads, molecular sieves, biopolymers in their natural or chemically modified form). Various methods are used for the immobilization, such as (a) covalent or non-covalent attachment to a matrix, (b) crosslinking the enzyme molecules with bi- or multifunctional agents (e.g. glutaraldehyde), (c) entrapment in gels, hollow fibers and microcapsules.

An immobilized enzyme can be easily arranged in reactor devices and separated from soluble low molecular weight compounds (e.g. substrates, products, buffer ions). Immobilized P-450 was used by several authors in enzyme reactors.

A review of the various active preparations described in the literature is given in Table 2. The enzymatic activity of these preparations decreases within a few hours in the hydroxylation reactions. By entrapment in gelatin membranes and alginate gels (where about 65% and 100%, respectively, of the activity could be retained) the functional stability of the immobilized P-450 corresponds to that of the microsomal suspension (SCHUBERT et al., 1980, 1982). PVP-hydrogel entrapment results in a considerable storage stabilization of the microsomal enzyme system (DENTI et al., 1976). With regard to practical applications, however, further improvement of the stability of immobilized P-450 is essential.

2.2. *Cofactor Regeneration*

In the reaction cycle of the P-450 catalysis two electrons are transferred from NADH or NADPH to molecular oxygen. The dioxygen molecule is splitted off and one oxygen atom is transmitted to the substrate while the other one is reduced to H_2O. Furthermore, the cofactor is oxidized to NAD^+ or $NADP^+$ in this reaction. In order to overcome the need of cofactor excess in enzyme reactors the oxidized cofactors have to be regenerated continuously corresponding to (2).

$$\left.\begin{matrix} NAD^+ \\ NADP^+ \end{matrix}\right\} + H^+ + 2e \longrightarrow \left\{\begin{matrix} NADH \\ NADPH \end{matrix}\right. \quad (2)$$

This can be achieved by chemical, enzymatic and electrochemical methods with soluble as well as immobilized cofactors.

Chemical methods: The chemical reduction of oxidized cofactors can be carried out by various reducing agents. $Na_2S_2O_4$, for example, is able to reduce NAD^+ in a two-electron step to 1.4-NADH whereas NAD^+ is reduced by $NaBH_4$ to a mixture of isomers (KAWAI and SCRIMGEOUR, 1972). The enzyme system, however, is attacked in these reactions and successively inactivated.

Enzymatic methods: For the enzymatic regeneration of NADPH and NADH all enzyme reactions can be utilized in which NADPH (or NADH) is formed from the respective oxidized nucleotides. Such reactions have to be coupled with the P-450 catalyzed reaction, whereby further problems arise with regard to practical applications.

Several enzymes have been preferentially used in the research work in this respect. The continuous enzymatic regeneration of NADH was carried out by means of dehydrogenase catalyzed reactions (DAVIES et al., 1974; ANONYMOUS, 1974; SKINNER, 1975; KIRSTEIN et al., 1981), for example:

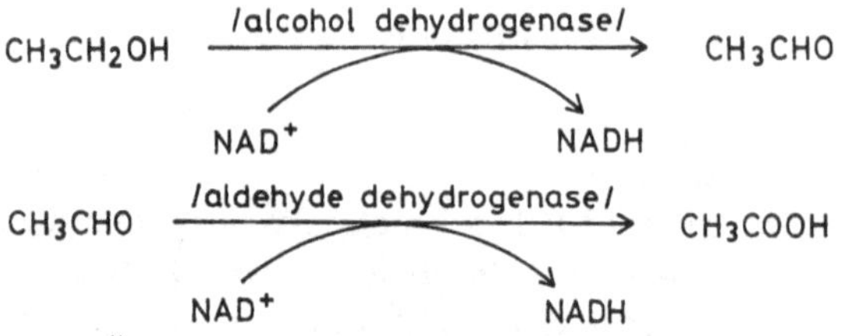

(3a)

(3b)

as well as

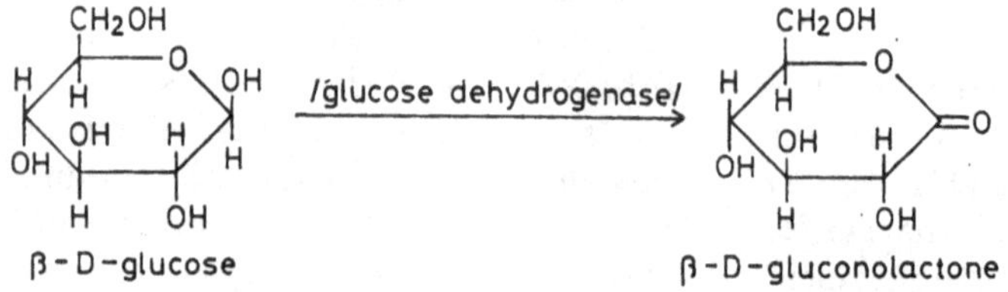

(3c)

DAVIES et al. (1974) utilized this principle under reactor conditions for the conversion of pyruvate to alanine (in laboratory scale). In these investigations the required cofactor NADH was immobilized on Sepharose and could be continuously regenerated by addition of galactose and galactose dehydrogenase.

The regeneration of NADPH can be performed by hexokinase and glucose-6-phosphate dehydrogenase catalyzed reactions in two consecutive steps:

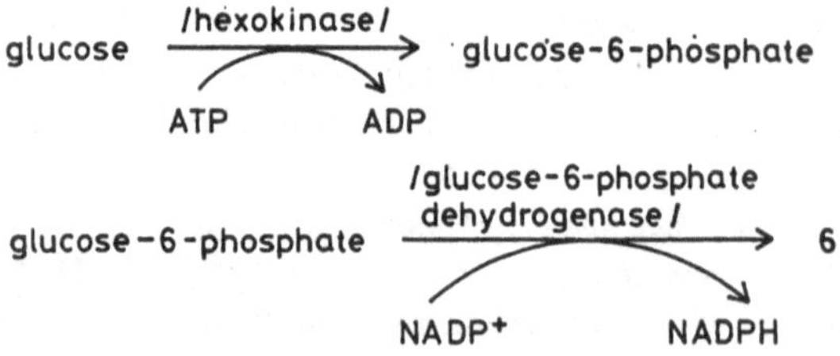

(4a)

(4b)

Another possibility for the enzymatic reduction of $NADP^+$ to NADPH is given by use of isocitric acid dehydrogenase (SOFER et al., 1974, 1975):

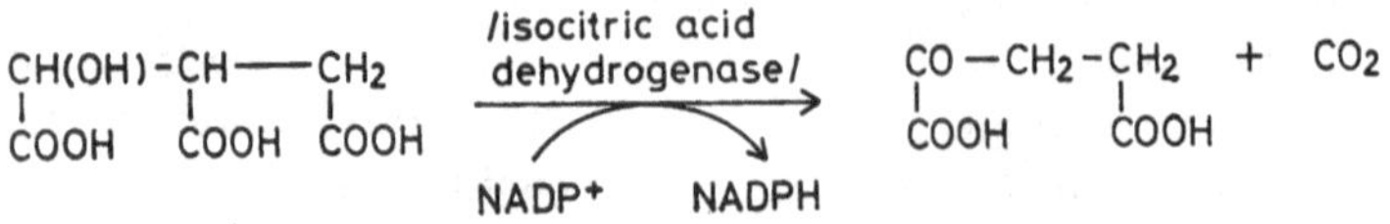

(5)

Finally, the glucose dehydrogenase catalyzed dehydrogenation of glucose can also be taken into consideration to regenerate NADPH (COHEN et al., 1977). The reaction, however, is rather slow in comparison to the reduction of NAD^+ under the same conditions.

Electrochemical regeneration of NADPH: The cathodic reduction of $NADP^+$ has the advantage that it works without any additional enzymes or enzyme systems. The application of this process, however, raises various new problems, some of which shall be mentioned.

The electrochemical reduction of soluble pyridine nucleotides was studied by KE (1956), POWNING and KRATZING (1957), CUNNINGHAM and UNDERWOOD (1966), JANIK and ELVING (1968), THEVENOT and HAMMOYA (1971), SCHMAKEL et al. (1975), THEVENOT and BUVET (1972), ELVING

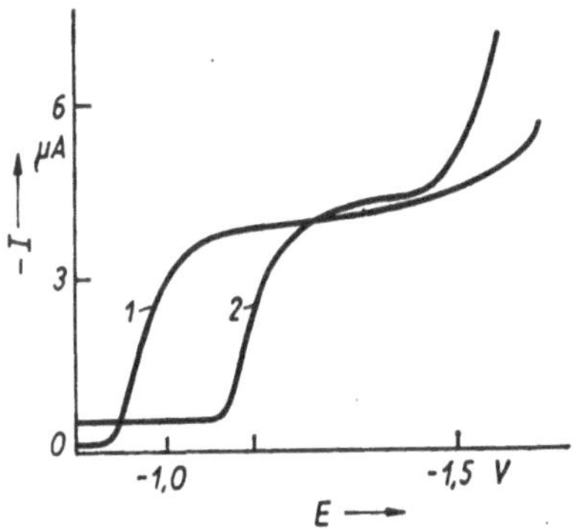

Fig. 1. Polarograms of $NADP^+$ (SCHELLER 1980) 1. $[NADP^+] = 0.5$ mM; 2. $[NADP^+] = 0.5$ mM; [P-450] = 10 μM. pH 8; 0.1 M phosphate buffer

(1976) and SCHELLER (1980). The d.c. polarogram of $NADP^+$ shows a reduction step with a pH- independent half wave potential at −970 mV (Fig. 1, curve 1). This step is shifted in cathodic direction to about −1150 mV in presence of liver microsomes (Fig. 1, curve 2). The preparative electrolysis in the limiting current range of the first reduction step (about −1300 mV) generates NADP-radicals (Fig. 2, II) which dimerize within few milliseconds as it is demonstrated schematically in Fig. 2 (II and IV) (CUNNINGHAM and UNDERWOOD, 1966; JANIK and ELVING, 1968). This dimerization is irreversible and gives probably 4.4′-isomers. The reaction products are coenzymatically inactive and cannot be reduced at the electrode — at positive potentials, however, they can be reoxidized to $NADP^+$, as it is shown in Fig. 2, I, (THEVENOT and BUVET, 1972). Further possible side reactions being responsible for the formation of "inactive" products are the reduction to an 2.2-dihydropyridine derivative (Fig. 2, III) as well as a hydrolytic cleavage of the $-CO-NH_2$ bond. Therefore, a continuous loss of the coenzymatic activity is observed in this process. All these results demonstrate that the electrochemical reduction is less specific than the enzymatic reduction.

The d.c. polarograms show furthermore that the formation of NADPH takes place only at the potential of the second step. Since at this potential the dimers are not

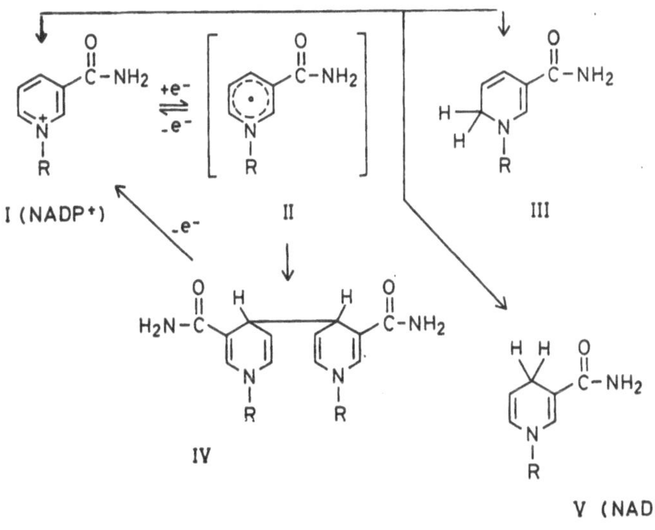

Fig. 2. Scheme of the cathodic reduction of $NADP^+$ (ELVING, 1976)

Table 3. Preparative cathodic reduction of NAD^+ and $NADP^+$, respectively

Electrode	Potential (V)	Conditions	Product of the electrolysis	References
Pd, Ni, Ag, Hg	−1.4	NAD^+	inactive	KE (1956)
Hg	−1.4	tris buffer	37% activity¹)	
Pb	−1.4	pH 7.5	56% activity¹)	
Hg	−1.7	$NADP^+$, pyrophosphate, pH 9.2, 2°C	94% activity¹)	POWNING and KRATZING (1957)
Hg	−1.8	$NADP^+$ NAD^+	20% activity¹) 54% NADH	CUNNINGHAM and UNDERWOOD (1966)
Hg	−1.8	carbonate, $(C_2H_5)_4N^+Cl^-$ pH 9	35% 1.6-NADH 11% dimers	ELVING (1976)
Hg	−1.8	$NADP^+$, $(C_2H_5)_4N^+Cl^-$, carbonate, pH 8—10, 20°C	30% reduced product²)	SCHELLER (1980)

¹) tested by enzymatic reoxidation (KE, 1956); by means of alcohol dehydrogenase (POWNING and KRATZING, 1957), by means of diaphorase and 2.6-dichlorophenol (CUNNINGHAM and UNDERWOOD, 1966)

²) tested in the P–450 catalyzed aniline hydroxylation

reduced a direct transfer of the two electrons has to be assumed. That means that the transfer of the second electron occurs more rapidly than the dimerization. The experimental data quoted in the literature are contradictory concerning the potential of the second step. The found activites amount up to 94% of the intrinsic activity as it is shown in Table 3.

With regard to an application of the P–450 system in enzyme reactors the cofactor was coupled to a polymeric matrix. The immobilized cofactor can be easily separated from other low molecular weight compounds (substrates, products) by ultrafiltration or other separation methods. Furthermore by spatial isolation of the NADH- or NADPH-molecules a dimerization of the intermediary formed products can be prevented. It is suitable to couple the cofactor molecule via the position 6 or 8 of the adenine part to the matrix. Inactivation of the cofactor occurs, however, when the OH-groups of the ribose part are involved in the coupling reactions (BERGER, 1976).

AIZAWA et al. (1976a, b) coupled NAD^+ via CMC-metho-p-toluolsulfonate on alginic acid. The adduct can be precipitated in acidic solution by a decrease in pH. This behaviour is especially advantageous for separation procedures. Immobilized NAD^+ shows in 0.4 M $(C_2H_5)_4NCl$ — solution (pH 9.5) a polarogram with two step of the same altitude. In comparison to the free NAD^+, the first step was shifted by about 150 mV in cathodic direction.

Preparative electrolysis at −1.75 V reduces 50% of NAD^+ within 24 h. The product was coenzymatically active in reducing dichlorophenol-indophenol in the presence of the enzyme diaphorase. The anodic oxidation of the electrochemically prepared NADH provided 95% NAD^+ within 3 h. It was active in the alcohol dehydrogenase catalyzed reaction (AIZAWA et al., 1976b). The authors estimated this result as evidence that the dimerization of the partially reduced cofactor can be prevented by immobilization. The problem of the formation of immobilized 1.6-NADH in this system is not mentioned.

Investigations with immobilized $NADP^+$ were performed with dialdehyde Sephadex-coupled $NADP^+$ (POMMERENING et al., 1974). The solubilized product (molecular weight 2000 to 5000) gives two reduction steps with half-wave potentials corresponding to those of free NADPH (Fig. 3). No significant changes of the electrochemical properties of $NADP^+$ are observed in consequence of the immobilization. It can be assumed, therefore, that the accessibility of the electrode surface for the pyridine part is not hindered. On the other hand coupling of NADPH to the macromolecular support sterically hinders its interaction with the NADPH-dependent cytochrome P-450 reductase. Consequently, the reduction rate of ferric P-450 by the immobilized cofactor is significantly decreased in comparison to free NADPH (Fig. 3).

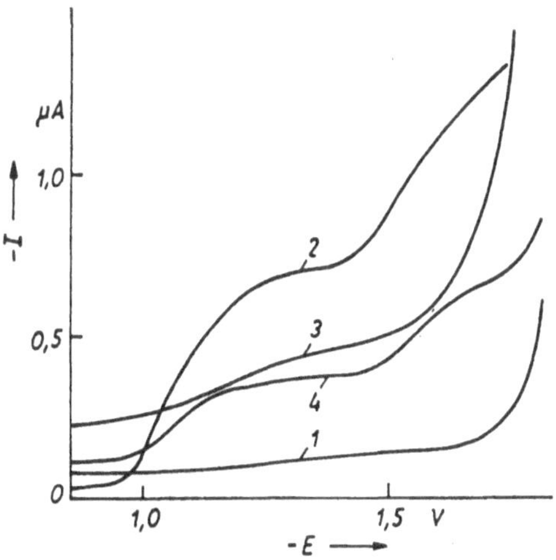

Fig. 3. Polarograms of immobilized $NADP^+$ (SCHELLER, 1980)

1. basic solution; 2. basic solution +0.3 mM immobilized $NADP^+$; 3. 6 h after electrolizing; 4. basic solution +0.1 mM $NADP^+$.
$[(C_2H_5)_4NCl] = 400$ mM; $[Na_2CO_3] = 50$ mM; $[NaHCO_3] = 50$ mM; pH 9.3

The preparative reduction of immobilized $NADP^+$ at a Hg-pool electrode at −1.8 V provided 80% NADPH within two hours.

Quantum chemical calculations carried out by PULLMAN (1964) concerning the reduction of nicotinamide suggest that the 1.4-dihydropyridine derivative is formed by a direct transfer of a hydride ion. In successive one-electron steps, however, a mixture of 1.2- and 1.4-dihydropyridine derivatives is formed. Analogously to the hydride ion transfer to nicotinamide a two-electron transfer seems to favour the formation of 1.4-NADH. In such a manner KWE (1974) succeeded in reducing NAD^+ completely to the coenzymatically active NADH by use of a two-electron transferring pteridin mediator. This mediator is able to transfer the electrons in a homogeneous redox reaction. This reaction, however, is paralleled by the direct cathodic reduction of NAD^+ forming inactive products.

The often utilized redox compound methylviologen is not suitable for NADPH regeneration because it is an one-electron mediator. On the other hand the reduction of $NADP^+$ was possible in the redox chain methylviologen$^+$/ferredoxin $NADP^+$ reductase in which methylviologen^{2+} was prior reduced at the electrode. The product obtained in this way was able to catalyze the CO_2 fixation in presence of a chlorophyll extract of chloroplasts (WELIKY et al., 1972).

Evaluating the results in the field of electrochemical cofactor regeneration it can be established:

a) The reduction of $NADP^+$ or NAD^+ to NADPH or NADH can be carried out at a Hg-electrode and takes place at about −1.8 V (Table 3). At this potential, however, the cosubstrate O_2 is also reduced at the electrode diminishing the yield of reduced cofactor. In order to overcome this problem the regeneration of the cofactor has to be carried out in absence

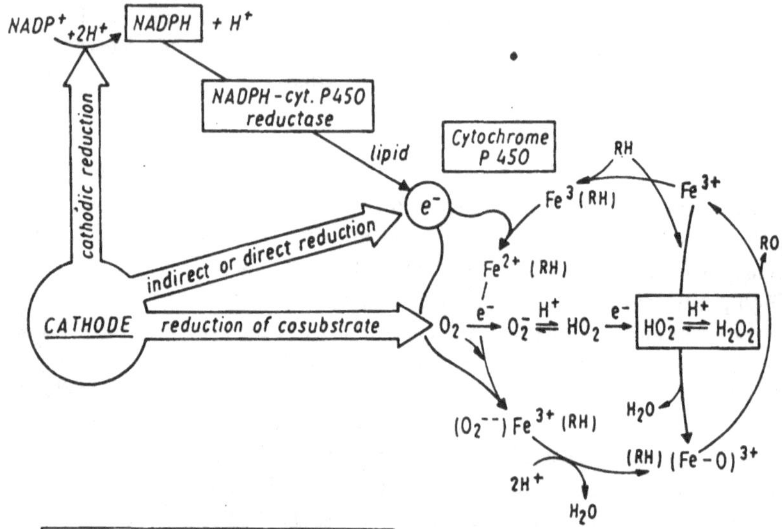

a

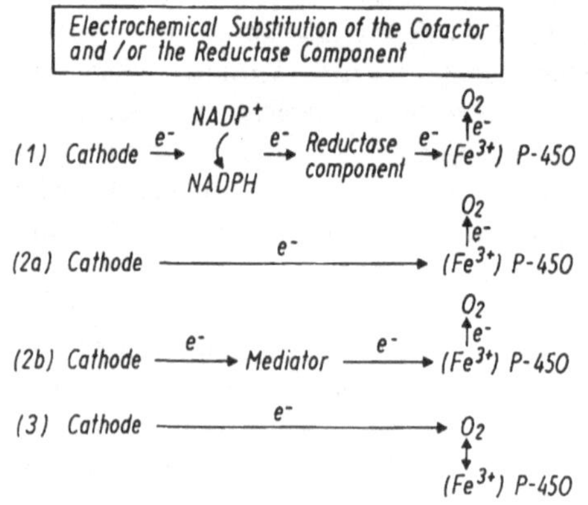

Fig. 4. (a) Variants of coupling the P–450 system with the cathode
(b) Possibilities to introduce electrons into the P–450/substrate system by means of a cathode (1) reduction of the cofactor. (2a) direct, (2b) indirect reduction of iron via methylviologen; (3) reduction of the cosubstrate oxygen.

of oxygen. Therefore, this process has to be separated from P–450 catalyzed substrate conversion.

b) The second problem consists in the occurrence of side reactions as it has been discussed above in detail. Further research work, therefore, is necessary to optimize the cofactor regeneration with regard to practical application in enzyme reactors. Direct coupling of $NADP^+$ or NAD^+ to the electrode surface (Yao and Musha, 1979) and the use of chemically modified electrodes should lead to further progress (Scheller et al., 1979).

2.3. Substitution of the Electron Donor and Transfer Components

2.3.1. Direct and Indirect Reduction of the Terminal Oxidase

It is also possible to transfer electrons to the enzyme substrate complex by direct or indirect reduction of the heme iron of the terminal oxidase using electrochemical or other methods (Fig. 4). The heme iron of P–450 in its ferric form is not reducible cathodically in intact liver microsomes. The access of the electrode is probably hindered by the entrapment of the terminal oxidase in the membrane of the microsomes as it is demonstrated schematically in Fig. 5.

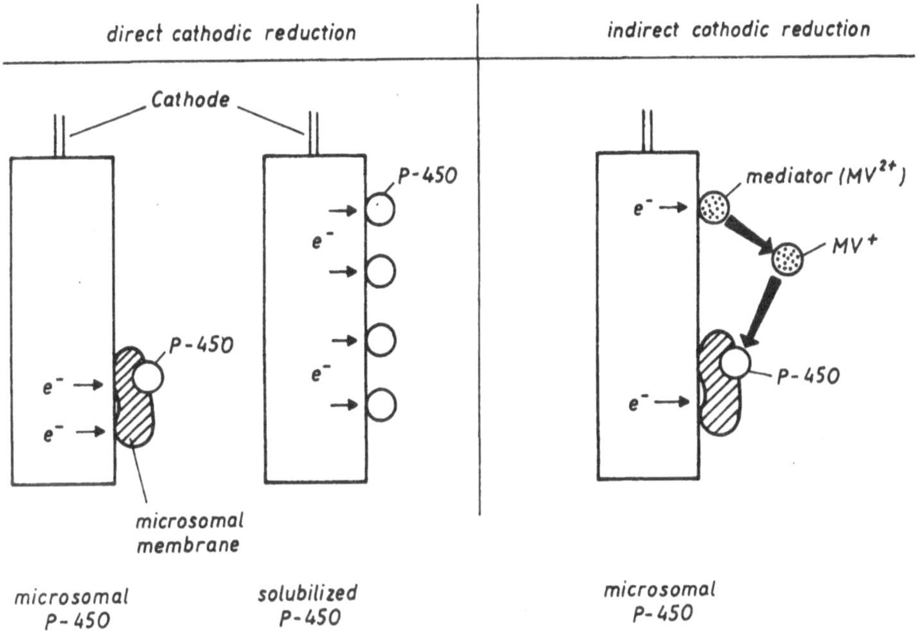

Fig. 5. Schematic representation of the direct and indirect cathodic reduction of microsomal and solubilized P–450

In order to accelerate the electron transfer from the electrode to the enzyme in intact microsomes methylviologen was employed as a mediator (Fig. 5). It can be reduced at the mercury electrode at a potential of −800 mV vs. S.C.E. to a blue coloured radical which quantitatively reduces the ferric heme iron of P–450 (cf. Fig. 4).

In its solubilized form, however, the reduction of the ferric heme iron of P–450 can be performed by means of electrolysis at a mercury pool electrode without inactivation which is evidenced by optical absorption spectroscopy (Scheller et al., 1977b). In comparison with the reduction by $Na_2S_2O_4$ the extent of reduction is up to 10% in this case. The use of a carbon or Ni cathode causes denaturation and generates P–420 with ferrous heme iron (Kusnetsov et al., 1979). 4.4-Dipyridyl accelerates the electron transfer from the electrode to P–450 molecules (Archakov et al. 1981).

At present all these procedures are difficult to handle due to the extreme oxygen sensitivity of both P–450 and the reduced mediator, respectively. Therefore, further research work is necessary to ascertain the technical applicability of this reduction method.

2.3.2. Electrochemical Reduction of the Cosubstrate O_2

The preceding sections (2.2. and 2.3.1.) dealt with the insertion of electrons into the P–450/substrate complex (a) by reduction of the pyridine nucleotides as well as (b) by direct or indirect reduction of the terminal oxidase. Finally, electrons can also be inserted (c) by direct cathodic reduction of the molecular oxygen as it is demonstrated in Fig. 4.

Molecular oxygen is reduced at a mercury electrode in a potential region between −0.2 V and −0.8 V to the superoxide anion O_2^-. This reduction product is immediately implied in further reactions leading to the anion HO_2^- Or to H_2O_2 (6) (CHEVALET et al., 1972).

$$O_2 \xrightarrow{e^-} O_2^- \xrightarrow{H^+} HO_2^{\bullet} \xrightarrow{e^-} HO_2^- \xrightarrow{H^+} H_2O_2 \quad (6)$$

The stability of the intermediary formed O_2^- increases with increasing pH. The half-life, for example, amounts at pH 10 to about 0.2 sec. and at pH 12 to about 2 sec. The O_2^- species, therefore, is relatively stable.

The transfer of the second electron during the oxygen reduction is not prevented by the protein component (BACH and MILLER, 1970) or by the microsomal fraction (SCHELLER et al., 1976) in contrast to such surfactants like quinoline or triphenylphosphine oxide which can form compact films (CHEVALET et al., 1972; DIVISEK and KASTENIG, 1975).

In the presence of microsomes the d.c. polarographic limiting current reaches the same value as in pure buffer solution, indicating equal numbers of transferred electrons in both cases. In the presence of microsomes, however, the step of oxygen reduction is shifted in the cathodic direction being caused by its interaction with the enzyme (Fig. 6).

2.3.3. Peroxide Supported Catalysis of Cytochrome P-450

It has been shown by several authors that P–450 does not act only as a monooxygenase activating molecular oxygen but also catalyzes hydroxylations and other conversion reactions of organic substrates supported by various hydroperoxides (KADLUBAR et al., 1973; HRYCAY and O'BRIEN, 1972; 1974; HRYCAY et al. 1975; GUSTAFSSON et al., 1976; RAHIMTULA and O'BRIEN, 1974, 1975; AKHREM et al., 1977), H_2O_2 (SCHELLER et al., 1976; NORDBLOM et al., 1976) as well as $NaIO_4$, $NaClO_2$ (HRYCAY et al., 1975, 1976; GUSTAFSSON and BERGMAN, 1976) C_6H_5IO (LICHTENBERGER et al., 1976) etc. as cosubstrates. These observations have led to the hypothesis that the pathway of the P–450 catalysis involves a common series of reactive intermediates when NADPH and O_2 or hydroperoxides serve as the initiator of the reactions.

For a practical application the peroxide supported reactions is of interest because:

- expensive cofactors and cofactor regenerating systems are replaced by peroxides,
- in many cases a higher rate of product formation is reached in comparison to the NADPH supported reaction.

Till now it is generally accepted that NADPH and hydroperoxide supported reactions have analogous species of activated oxygen and a part of the reaction sequence in common. In most cases, therefore, same substrates should be con-

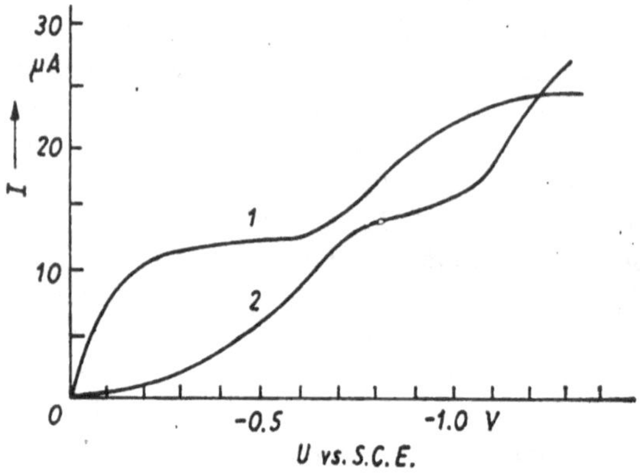

Fig. 6. Oxygen reduction at a dropping mercury electrode.
1. 0.1 M phosphate buffer; pH = 7.4; O_2-saturated 2. background solution +4 μM P–450

verted to same products in both reaction types. In the last time, however, several lines of evidence have cast doubts on the reliability of this hypothesis (despite a number of common characteristics) (WHITE and COON, 1980). The expected similar product distribution could not be found; different spectrally active intermediates of P–450 were identified in dependence on the used oxidant. An uniform compound I intermediate was not ascertainable. In both reaction types not only the oxenoide mechanism but also a yet unknown "1-electron-oxidation" mechanism seems to take part. An illustration of this hypothesis is presented in Figure 7b. It could be shown that the nature of the used oxidant has a considerable influence on the distribution of the oxenoide or 1-electron (radicalic) oxidation products formed in P–450 catalyzed reactions. In substrate conversion processes supported by organic hydroperoxides the 1-electron (radicalic) mechanism seems to dominate.

Contrary to organic hydroperoxides the hydrogen peroxide supported reaction can be considered as a real model for the "oxenoide" NADPH-dependent reaction of P–450 because identical products and a similar product distribution have been found in demethylation reactions as well as in the aniline, camphor and benzo(α)pyrene hydroxylation, demonstrated in Fig. 7, (RENNEBERG et al.,, 1981a). It was not possible, however, to hydroxylate aliphatic and cycloaliphatic compounds by H_2O_2 in presence of P–450. This may be caused by low binding affinity of P–450 towards H_2O_2.

2.4. *Enzyme Reactors*

At present there are only few reports concerning the application of P–450 in enzyme reactors. The mitochondrial steroid hydroxylating P–450 system of adrenal cortex was used to cleave the side chain of cholesterol (COOPER et al., 1973; AKHREM et al., 1979). AKHREM et al. (1979) immobilized adrenodoxin to CNBr-activated Sepharose 4B and reconstituted

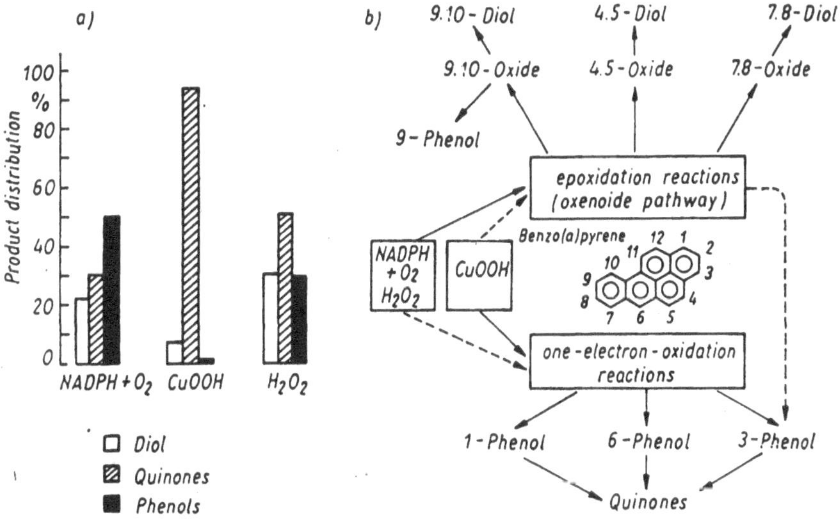

Fig. 7. (a) Distribution of the main organic soluble products generated during the oxidation of benzo(a)pyrene (RENNEBERG et al., 1981a) CuOOH = cumene hydroperoxide
(b) Postulated reaction scheme for the oxidation of benzo(α)pyrene by either NADPH and oxygen, hydrogen peroxide, or cumene hydroperoxide.

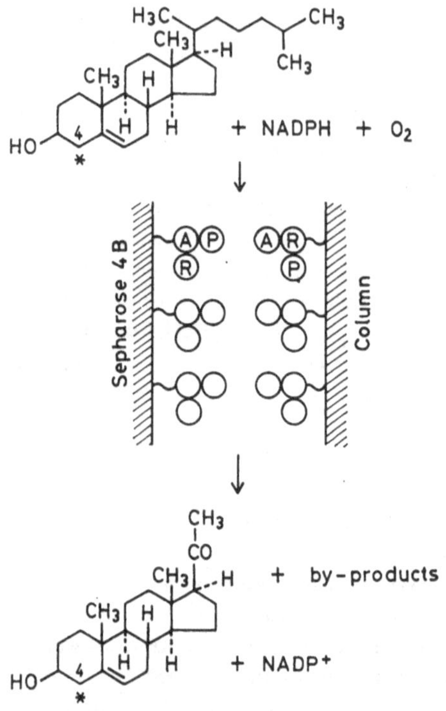

Fig. 8. Conversion of cholesterol into pregnenolone by immobilized P–450_{AC} (AKHREM, 1979) A = adrenoxin; R = adrenoxin reductase; P = cytochrome P–450

the enzyme system by adding adrenodoxin reductase and the terminal oxidase.

The reaction was carried out by passing $NADP^+$ and the NADPH generating system glucose-6-phosphate dehydrogenase and glucose-6-phosphate as well as [4-^{14}C]-cholesterol through a column filled with the immobilized P–450 (Fig. 8). Under these conditions 25—30% of the cholesterol were transformed into pregnenolone.

Some other substrate conversion processes catalyzed by P–450_{LM} could be carried out by SCHELLER et al. (1979) in an electro enzyme reactor (SCHELLER et al., 1976, 1979).

Liver microsomes from phenobarbital treated rabbits are capable to catalyze the demethylation of a variety of substrates by cathodically reduced molecular oxygen in the absence of pyridine nucleotides. Benzphetamine, aminopyrine and p-nitroanisole are demethylated by this system continuously as it is shown in

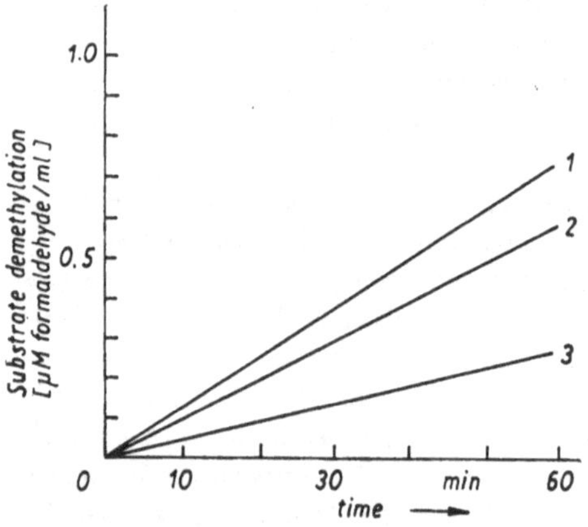

Fig. 9. P–450 catalyzed demethylation o benzphetamine (1), p-nitroanisole (2) and aminopyrine (3) by cathodically reduced oxygen. [P–450] = 5 μM (microsomes); [NaN_3] = 1 mM; 0.1 M phosphate buffer, pH 6.5; t = 15 °C

1 [benzphetamine] = 1.8 mM; I = 1 mA
2 [p-nitroanisole] = 2 mM; I = 4 mA
3 [aminopyrine] = 2 mM; I = 1 mA

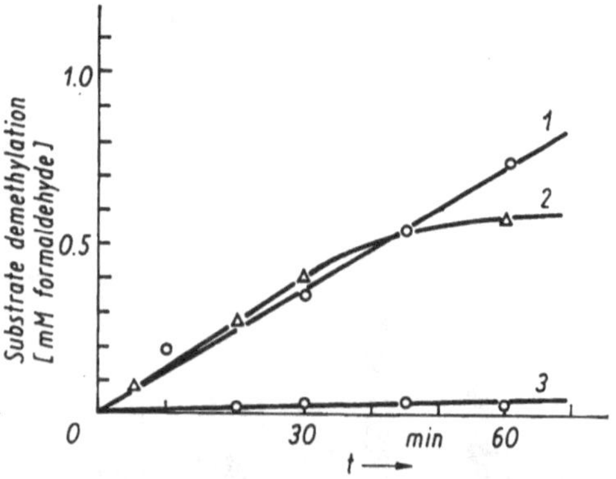

Fig. 10. P–450 catalyzed demethylation of benzphetamine by cathodically generated H_2O_2 (curve 1) and added H_2O_2 (curve 2). Curve 3: Inhibition of the reaction (conditions adequate to curve 1) by catalase.

[P–450] = 5 μ (microsomes); [benzphetamine] = 1.8 mM

0.1 M phosphate buffer, pH 6.5, t = 15 °C
curve 1: [NaN_3] = 1 mM, I = 2 mA
curve 2: [NaN_3] = 1 mM; [H_2O_2] = 2 mM
curve 3: [catalase] = 1000 units (according to BERGMAYER, 1962)

Fig. 9 (curves 1—3). Formaldehyde, 4-amino-antipyrine and p-nitrophenol have been identified as the respective reaction products.

The demethylation reaction of benzphetamine by cathodically generated and added H_2O_2 (Fig. 10, curves 1 and 2) is completely inhibited by addition of catalase evidencing the essential role of hydrogen peroxide in this system (Fig. 10, curve 3). On the other hand, superoxide dismutase (an O_2^-—scavenger) and mannitol (an OH· radical-scavenger) do not inhibit the reaction. The rate of the benzphetamine demethylation at 15 °C is constant at least for one hour (Fig. 10, curve 1). When the reaction is carried out in the presence of microsomes which have been heated up to 90 °C for 5 min before being used in the experiment no benzphetamine was demethylated giving evidence for the enzymatic nature of the reaction.

By addition of exogenous H_2O_2 to microsomal preparations benzphetamine is demethylated in the same way as by electrochemically generated H_2O_2.

Equal amounts of H_2O_2 generated either enzymatically (by the system glucose oxidase/glucose) or electrochemically form the same amounts of metabolites (Fig. 11). This identity shows that microsomal P-450 does not specifically interact with the cathodically reduced oxygen species and the surface of the electrode, respectively.

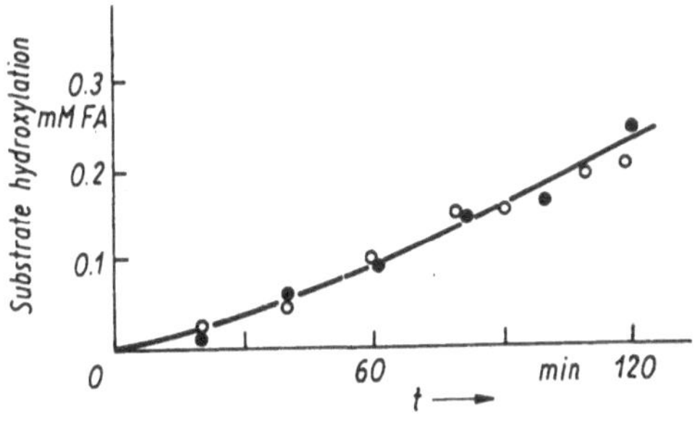

Fig. 11. Comparison of the benzphetamine demethylation by cathodically and enzymatically generated H_2O_2 in presence of liver microsomes.
[P-450] = 1 μM, [benzphetamine] = 4 mM, [NaN_3] = 1 mM, 0.1 M phosphate buffer pH 6.5; t = 37 °C
black points: I = 860 μA; empty circles: [glucose oxidase] = 0.27 units; [glucose] = 0.7 mM

Summarizing these results it has to be assumed that the P-450 catalyzed substrate conversion in the presence of electrochemically reduced oxygen is a H_2O_2-supported catalysis.

A model of an electro enzyme reactor (EER) based on P-450 as catalyst and utilized in some continuous substrate conversion processes under laboratory conditions is shown in Fig. 12 (Renneberg 1979). The molecular oxygen is reduced at the cathode to H_2O_2 and HO_2^-, respectively. Simultaneously the substrate passes the electrode and consequently relatively high concentrations of enzyme, cosubstrate (H_2O_2) and substrate are obtained at the cathode.

The flow of the substrate can be arranged in a closed circle. In a detoxification reactor the non-consumed H_2O_2 can be destroyed in a following step electronchemically at the counter-electrode or by catalase. This is an important advantage in comparison to organic peroxides which form alcohols as by-products difficult to be separated.

Benzphetamine has been demethylated in the laboratory model reactor (Figure 12) within 6 hours to 95% utilizing microsomal P-450 (Fig. 13, curve 1). Using solubilized P-450 aniline, benzphetamine and aminopyrine could be hydroxylated or demethylated to the respective products. Curves 2 and 3 in Fig. 13 show that in case of the benzphetamine demethylation the systems P-450/O_2/NADPH and P-450/H_2O_2 added are less effective than P-450 in the presence of electrochemically generated H_2O_2. By further research work it has to be clarified which factors are responsible for this behaviour and whether it is also valid for other substrates. Fig. 14

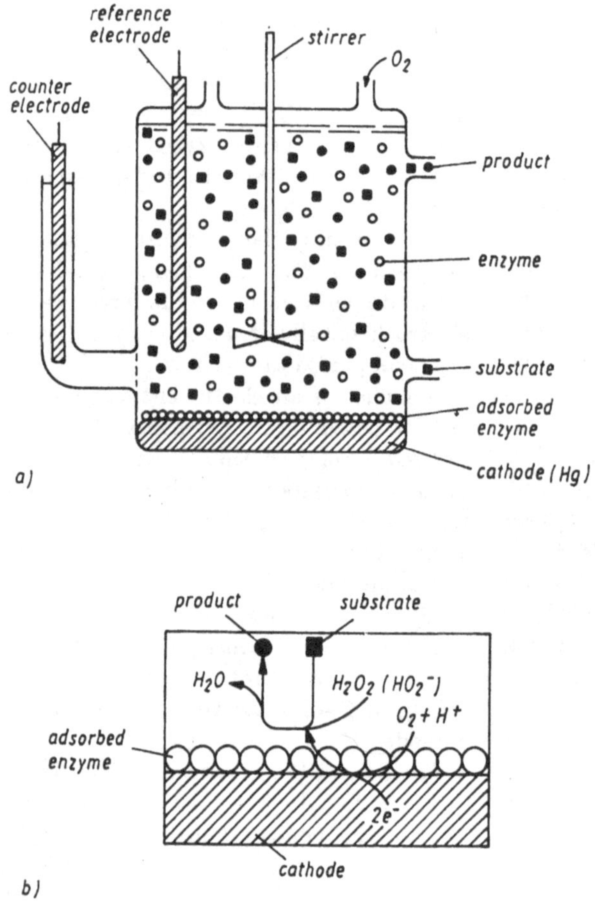

Fig. 12.
a) Schematic representation of the electro enzyme reactors
b) processes on the electrode surface

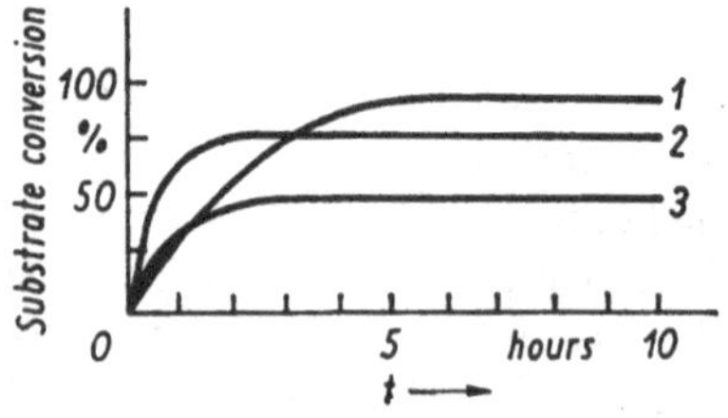

Fig. 13. Benzphetamine demethylation in the electro enzyme reactor (compared with the demethylation by added H_2O_2 and NADPH respectively) Ordinate: benzphetamine demethylation in %. (100% correspond to 0.8 mM formed (CH_2O) ·
Abszissa: time t in hours
[P–450] = 5 μM (microsomes); [benzphetamine] = 0.8 mM; [NaN_3] = 1 mM
0.1 M Na-phosphate buffer, pH
Curve 1: I = 430 μA;
curve 2: [H_2O_2] = 0.8mM;
curve 3: [NADPH] = 0.8 mM

shows the stability of microsomal P–450 in the model reactor. It is clear that the stability of P–450 is smaller when the reaction is started by addition of exogenous H_2O_2 compared with the reaction of P–450 with electrochemically reduced oxygen because in this case the reaction starts at zero H_2O_2 concentration. Solubilized P–450 is more rapidly destroyed during the catalysis in the electro enzyme reactor than the microsomal enzyme. Efforts to increase the stability by immobilization had no significant effect as yet.

First experiments on the development of an extracorporal detoxification reactor are in the very beginning. As catalytically active material intact microsomes or the immobilized enzyme system were used. First experiences in this direction were

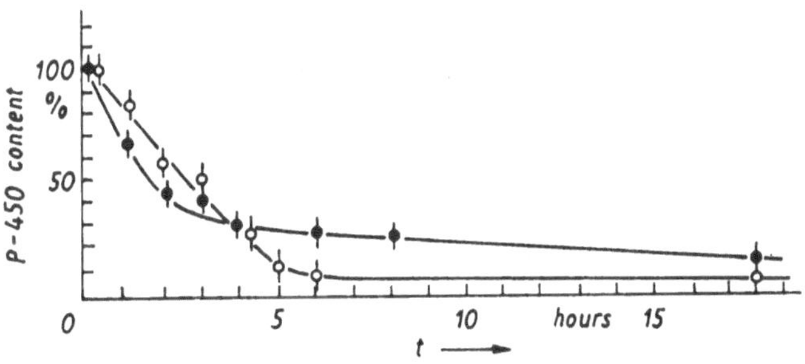

Fig. 14. Stability of P–450 in the electro enzyme reactor (compared with the stability of P–450 when H_2O_2 is added)
The P–450 content in % was determined by difference spectroscopy (100% correspond to 5 μM P–450)
[P–450] = 5 μM (microsomes); [benzphetamine] = 0.8 mM
[NaN_3] = 1 mM, 0,1 M phosphate buffer, pH
curve 1: addition of 0.8 mM H_2O_2
curve 2: I = 430 μA

achieved utilizing solubilized and purified rabbit P–450_{LM} immobilized on CNBr-activated Sepharose 4B (Brunner and Loesgen, 1977). The continuous regeneration of the reduced cofactor NADPH, however, was not accomplished.

Denti et al. (1976) immobilized rat liver microsomes by entrapment in polyvinylpyrrolidone gels performed by γ-ray crosslinking. These gels were used in a bioartificial column reactor system where aniline, p-nitroanisole, phenobarbital and diazepam (7-chloro-1.3-dihydro-1-methyl-5-phenyl-24-1.4 benzodiazepin-2-one) circulating in buffer solution could be converted (Table 2). In experiments with calf blood only aniline hydroxylase activity was found.

Rat liver microsomes were used by Cohen et al. (1977) and Kastl et al. (1978, 1979) in a hollow fiber reactor (Fig. 15, 16). Since the hollow fiber membrane is impermeable for all the components of the system undesired adverse immunological reactions can be avoided.

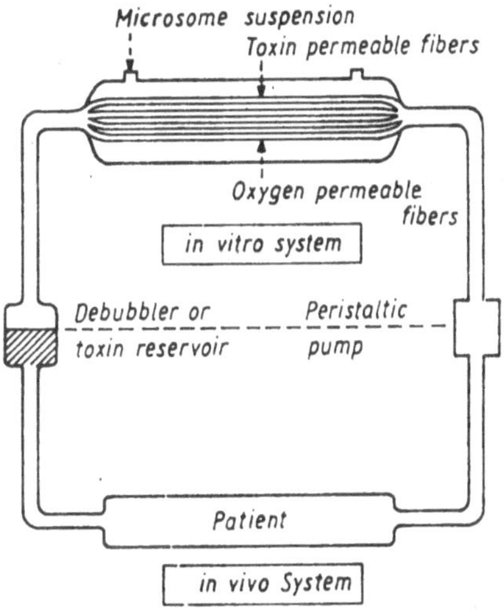

Fig. 15. Scheme of a detoxification reactor (Cohen et al., 1977).

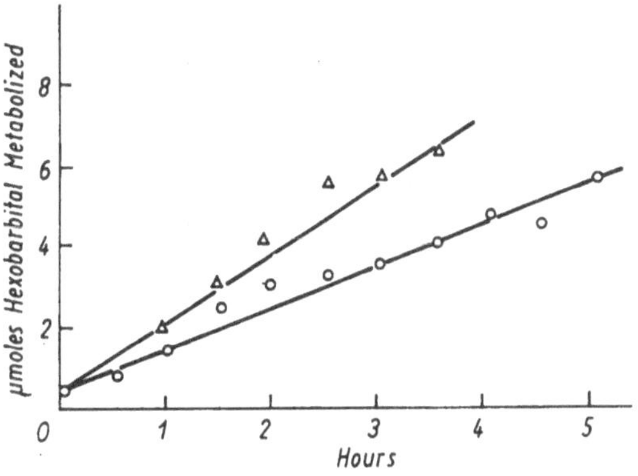

Fig. 16. Hexobarbital metabolism in a hollow fiber reactor (Cohen et al., 1977)
△ Oxygen bubbled through the circulating solution; ○ no oxygen bubbled through the circulating solution

The hollow fiber device contains two types of fibers: (a) cellulose fibers, being permeable for drugs and toxic substances and (b) silicone fibers, being permeable for oxygen. Therefore, oxygen and the toxic compounds can permeate from the blood into the microsomal suspension. The modified drug or toxin then can reenter the blood and will be excreted by the kidney.

Two drugs were applied by these authors: (a) p-nitroanisole (its product can be continuously monitored by spectrophotometry) and (b) hexobarbital as a representative for barbiturates which very frequently lead to intoxications. Though the NADPH-generating system glucose-6-phosphate dehydrogenase/glucose-6-phosphate was utilized in these experiments the authors recommended the use of the system glucose dehydrogenase/glucose for future investigations because in that case the ambient blood glucose could be utilized. Fig. 16 shows that the rate of the hexobarbital metabolism in the reactor is linear for 5 h. During this period approximately 25% of the total hexobarbital (5.5 μmoles) were metabolized.

From experiments with human plasma a theoretical clearance of 8.4 ml/min for hexobarbital was extrapolated for a clinical size hollow fiber reactor. This figure approaches the hemodialysis clearance of 15 ml/min reported for secobarbital, but it is significantly lower than 50 ml/min achieved using charcoal hemoperfusion. More permeable membranes, increased surface area, and more active microsomes could surpass at least the hemodialysis clearance.

Another potential application of immobilized P-450 is the preparation of selected P-450 metabolites of pharmaceutical importance. In this respect, LEHMAN et al. (1981) investigated the efficiency of a reconstituted P-450 system bound to CNBr-activated Sepharose in a fixed bed reactor for the conversion of model substrates (p-nitroanisole, ethylmorphine). A similar approach was decsribed by SCHUBERT et al. (1982) who used alginate entrapped microsomes and aminopyrine. However, the question of the economical usefulness of such processes remains to be solved.

2.5. Enzyme Tests, Electrodes and Sensors

In the last years several investigations were aimed on the analytical application of P-450 preparations in medicine and industry. In the following the main directions of the respective research work are characterized.

The Salmonella mammalian liver test: Damage to DNA by environmental mutagens is likely to be a major cause of cancer. A rapid, sensitive and economical test method for the identification of such chemicals was developed by AMES (1976a, b, 1979) using a special set of histidine-requiring mutants of *Salmonella typhimurium* and rodent or human liver microsomes (human microsomes can be obtained by autopsy). A simplified scheme demonstrating the principle of this test is given in Fig. 17. No doubt P-450 plays an essential role in this test by converting substrates into mutagenic or potential carcinogenic products.

The test procedure is done by combining on a petri disk the compound to be tested, about 1 billion *Salmonella bacteria* and the homogenized liver. After incubation at 37 °C for two days the number of bacterial revertant colonies is recorded. Each of these colonies originates from a bacterium whose defective histidine gene was reverted to a functional one. Normally, the dose response curves obtained from a test set are almost linear as it is shown in Fig. 18 on some selected examples. This method allows to detect mutagens already at low doses, in some cases nanogram amounts. There is a certain correlation

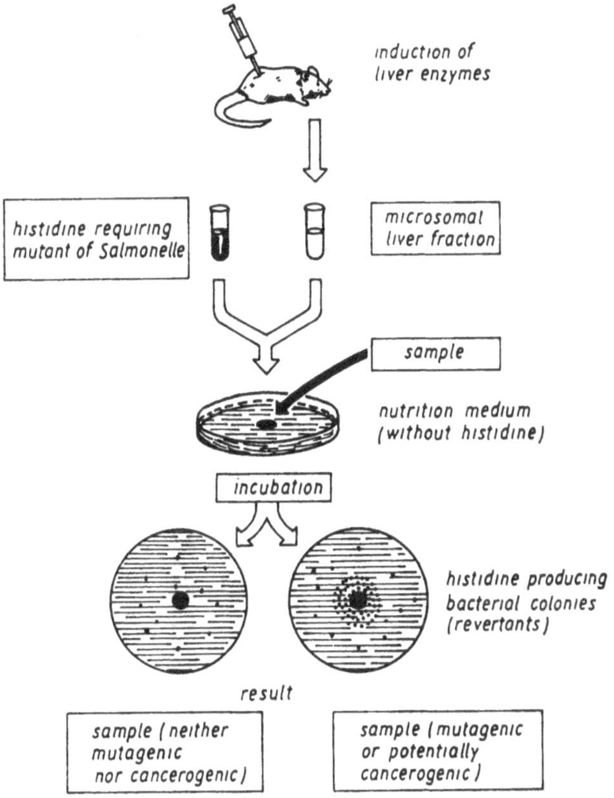

Fig. 17. Principle of the AMES-Test

between mutagenic and potential carcinogenic activities of chemical compounds. The authors assume that this test can play an important role in a long-term program of cancer preventation by identifying and minimizing human exposure to environmental carcinogens and mutagens.

Enzyme electrodes:

P-450-supported hydroxylations of organic substrates require molecular oxygen. In principle, therefore, a P-450 enzyme electrode should be constructed by coupling this enzyme with an electrode being able to monitor the oxygen consumption. This method can be successfully used with soluble or suspended P-450 when a closed measuring cell is used. The O_2-consumption results both from the NADPH oxidase activity and from the real substrate conversion. Therefore, the amount of O_2 consumed does not correlate directly with the substrate concentration in the solution.

Furthermore, an alteration of the O_2-solubility takes place after adding hydrophobic substrates, which can falsify the measurements. Therefore, the principle of a product indication seems to be more suitable as the basis of a P-450 electrode. Moreover, a high sensitivity can be attained hereby. P-aminophenol being formed by the P-450 supported hydroxylation of aniline, for example, can be oxidized to a well-defined quinoidic structure (Fig. 19) at +250 mV vs. S.C.E., where the limiting current is proportional to the p-aminophenol concentration. On the basis of this electrode reaction gelatin entrapped rat liver microsomes were attached directly on the top surface of a rotating disc carbon electrode which

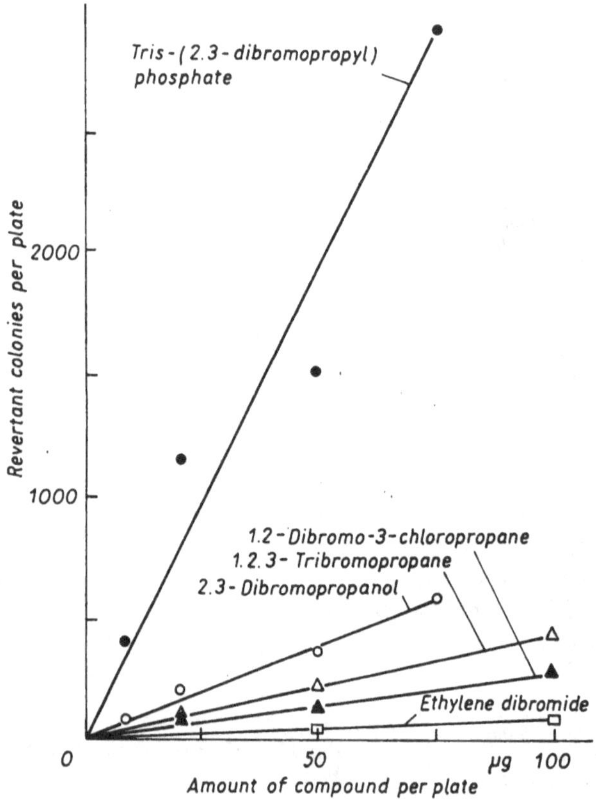

Fig. 18. Linear dose responses obtained with the Salmonella test (AMES, 1979). The flame retardant tris BP (1.2.3-tribromopropane), its metabolite dibromopropanol and the pesticide dibromochloropropane were tested in presence of rat liver homogenate. All compounds were tested on Salmonella strain TA 100. The amount of the industrial chemical ethylene dibromide added was ten times that indicated on the scale.

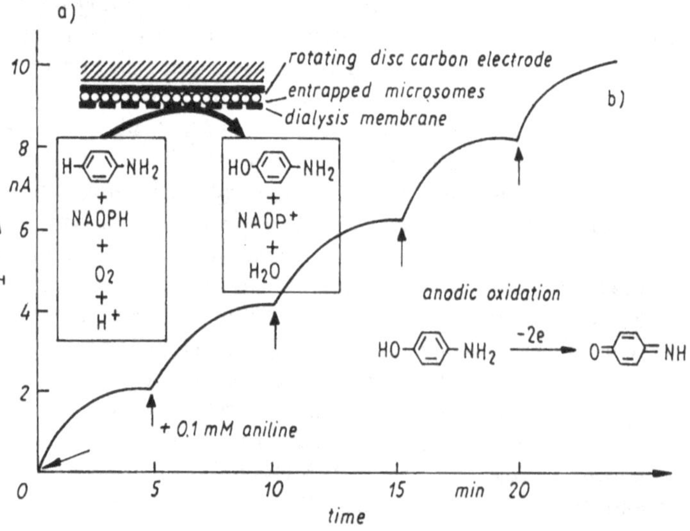

Fig. 19. Aniline specific P–450 enzyme electrode

a. Schematic representation of the membrane near section
b. Dependence of the limiting current on the aniline concentration. Each addition indicated by a black arrow enhances the aniline concentration by 0.1 mM

was polarized to +250 mV (SCHUBERT et al., 1980). To the measuring cell containing 3 ml of the background solution (0.1 M phosphate buffer, pH 7.0) NADPH was added to give a final concentration of 0.1 mM. When a constant current was attained repetitive additions of 50 µl-samples of 6.2 mM aniline hydrochloride solution were performed. The formation of p-aminophenol in the enzyme layer causes a stepwise increase of the current as it is demonstrated in Fig. 19. In the region between 0.1–0.5 mM aniline in the measuring cell the current depends linearly on the aniline concentration. Similar results have been achieved with coimmobilized glucose oxidase in the P-450 gelatin membrane and a modified background solution. In this case NADPH/O_2 is replaced by enzymatically generated H_2O_2. Though the described electrodes are based on the indication of a conversion product being formed only in the discussed reaction this example demonstrates in principle the suitability of P-450 supported mixed function oxidation for application in enzyme electrodes.

Table 4. Rates of hydroxylation of aniline by various reconstituted enzyme systems (MIEYAL et al., 1976)

System	Turnover number[1])
Rat P-450 (uninduced + P-450 reductase (0.05 units)[2]) + lipid	0.22
Human hemoglobin + P-450 reductase (0.005 units)	0.20
Bovine hemoglobin + P-450 reductase (0.01 units)	0.23
Whale myoglobin + P-450 reductase (0.009 units)	0.08

1) mol p-aminophenol/mol hemoprotein/min

2) units of reductase activity are defined as micromoles of cytochrome c reduced/min

Table 5. Dependence of p-aminophenol formation on the components of the reconstituted hemoglobin system (MIEYAL et al., 1976)

System	Activity[1])
Complete system[2]) (ferrihemoglobin)	100
heated hemoglobin	n.d.[3])
minus ferrihemoglobin	n.d.
minus reductase	38
minus NADPH	n.d.
minus O_2	n.d.
bubbled with O_2	174
plus dilaurolylphosphatidylcholin	100
Complete system (oxyhemoglobin)	67
minus NADPH	n.d.

[1]) Relatively to the complete system (activity = 100%) and p-aminophenol formed in 15 min.

[2]) [Human ferrihemoglobin] = 1 µM; [NADPH] = 1 mM; solubilized liver microsomal P-450 reductase = 12 µg; [aniline] = 40 mM; [K-phosphate] = 20 mM, pH 6.8; t = 38 °C

[3]) None detectable

3. Substitution of the Terminal Oxidase by Other Hemoproteins

The fact that the active component of P-450 systems is a hemoprotein raises the question to what extent the terminal oxidase can be exchanged by other known hemoproteins. MIEYAL et al. (1976) and MIEYAL and BLUMER (1976) succeeded in substituting the terminal oxidase by bovine hemoglobin or sperm whale myoglobin. These artificial enzyme complexes are able to catalyze the hydroxylation of aniline to p-aminophenol in presence of NADPH nearly with the same effectivity as the microsomal or solubilized P-450 complex as it is shown in Table 4. In case of the aniline hydroxylation this is supported by the turnover numbers as well as by the Michaelis-Menten constants.

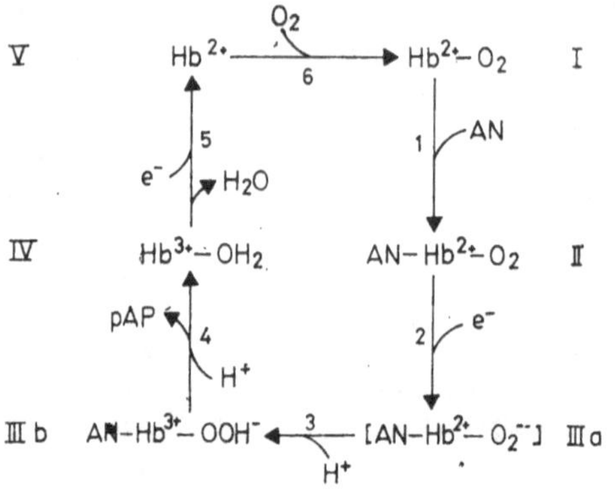

Fig. 20. Hypothetical scheme for the hydroxylation of aniline to p-aminophenol by the system metHb/P–450 reductase/NADPH/O_2 (Hb^{2+} = ferro-, Hb^{3+} = ferrihemoglobin; AN = aniline, pAP = p-aminophenol)

Furthermore benzphetamine can be demethylated under formation of formaldehyde in presence of thiosulfate instead of NADPH.

Native P–450 reductase, O_2 and NADPH are essential components for a complete effectivity of this system as it is shown in Table 5. Contrary to P–450 in the artificial complexes the lipid component, however, does not influence the catalytic behaviour.

The mechanism of the aniline hydroxylation catalyzed by the methemoglobin substituted system is discussed on the basis of the following reaction scheme (Fig. 20):
The catalysis is initiated by the reduction of ferrihemoglobin to the ferro form. In the next steps O_2 and aniline are bound which is followed by the insertion of an electron as well as of a proton. Different to the P–450 cycle it is proposed that aniline is bound to hemoglobin because ferrihemoglobin has only a small affinity to this substrate. The hypothetical peroxide intermediate then decomposes to p-aminophenol, water and the ferrihemoglobin complex. It is assumed that this reaction is the rate limiting step in the hydroxylation cycle.

Experiments using horseradish peroxidase instead of P–450 in the peroxide dependent catalysis provided negative results. It has been shown that H_2O_2-supported reactions of P–450 and horseradish peroxidase (HRP) follow different mechanisms (Renneberg et al., 1978).

The only known substrate converted by both horseradish peroxidase and P–450 under formation of the same products is aminopyrine. Dimethylaniline is demethylated by both enzymes, too. In the P–450 catalysis N-methylaniline and HCHO are formed. In the peroxidase-supported reaction, however, a tetramerization of the aromatic part takes place (beside a conversion of a N-methyl group to CH_2O) which indicates a radicalic mechanism.

Several other P–450 substrates (imipramine, ethylmorphine, methamphetamine, ketamine and mephentermine) show no significant conversion by H_2O_2 and HRP (Renneberg et al., 1981b).

Aniline was hydroxylated by H_2O_2 in presence of horseradish peroxidase to p-aminophenol only as a short living

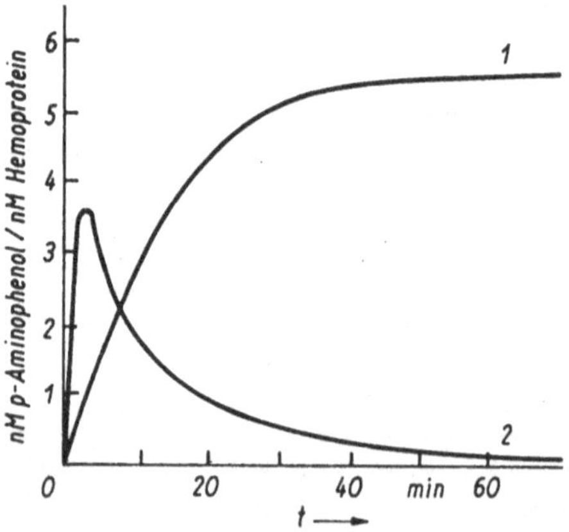

Fig. 21. Formation of p-aminophenol at the hydroxylation of aniline catalyzed by P–450 (curve 1) and HRP (curve 2)
[NaN_3] = 1 mM, 0.1 M phosphate buffer pH 6.5, z = 37 °C
curve 1: [P–450] = 1 µM (liver microsomes); [H_2O_2] = 8 mM [aniline · HCl] = 8 mM
curve 2: [HRP] = 1 µM, [H_2O_2] = 0.8 mM, [aniline · HCl] = 0.8 mM

intermediate contrary to the H_2O_2-supported hydroxylation of aniline by P–450 where p-aminophenol as a stable product is formed (Fig. 21) (RENNEBERG et al., 1978).

With regard to a practical application it has to be stated that experiments to substitute P–450 by other hemoproteins were not very successful up to now.

4. Simulation of the Catalytic Principle of Cytochrome P-450 by Iron Porphyrin Complexes

A lot of papers have been published dealing with the modeling of the terminal oxidase and active intermediates of P–450 (STERN and PEISACH, 1974; COLLMAN et al., 1975; CHANG and DOLPHIN, 1975; TANG et al., 1976; SAKURAI, 1976; RUF et al., 1979; TRAYLOR and MINCEY, 1979; KÜHN and BENES, 1980) as well as with investigations of the catalytic properties of such models (SAKURAI and OGAWA, 1979; GROVES et al., 1979; CHANG and KUO, 1979; MANSUY, 1980; KÜHN et al., 1981).

By various authors it was demonstrated that "peroxidatic" reactions can be also catalyzed by iron porphyrin complexes in aqueous solution but also in organic solvents (TOHJO et al., 1962; PORTSMOUTH and BEAL, 1971; PSHEZHETSKII and JAROSLAVOV, 1977; MOHR et al., 1979; MANSUY et al., 1980; KÜHN and BENES, 1980). The activity of iron porphyrin is very small, however, compared with horseradish peroxidase. After adding ligands which are capable to form iron porphyrin complexes an increase of the catalytic activity can occur.

Table 6 shows that the reaction of hemin complexes with H_2O_2 (7) (which can be compared with the formation of peroxidase compound I) (PORTSMOUTH and BEAL, 1971) is relatively slow.

$$\text{hemin} + H_2O_2 \xrightarrow{k_1} \text{hemin-}H_2O_2 \quad (7)$$

$$\text{hemin-}H_2O + \text{substrate} \xrightarrow{k_4} \text{hemin} + \text{product} \quad (8)$$

The following step (8) is more rapid as it is shown by the rate constant k_4. Moreover, Table 6 demonstrates that k_1 and k_4 depend strongly on the chemical structure of the ligand. An active hemin intermediate being comparable to compound II of horseradish peroxidase could not be observed spectrophotometrically till now. To clarify the question whether the corresponding mechanism is more related to the P–450 or to the peroxidase catalysis the kinetic constants for the formation of the typical peroxidase intermediates compound I and II were com-

Table 6. Rate constants k_1 (7) and k_4 (8) of iron porphyrin complexes in aqueous solution (TOHJO et al., 1962)

Ligand	pH	k_1	k_4
Without ligand	4.0	$(2.2 \pm 0.2) \cdot 10^{-1}$	$(9.6 \pm 1.4) \cdot 10$
11.9 M methylalcohol	5.3	4.8 ± 0.4	$(5.5 \pm 0.9) \cdot 10^3$
9.5 M formamide	6.5	$(5.0 \pm 1.0) \cdot 10$	$(6.9 \pm 0.5) \cdot 10^3$
1.2 M arginine	6.6	$(3.3 \pm 0.7) \cdot 10^2$	$(2.6 \pm 0.5) \cdot 10^4$
6.2 M acetate	4.5	$(2.2 \pm 0.2) \cdot 10^2$	$(3.2 \pm 0.5) \cdot 10^4$
0.9 M pyridine	5.4	$(9.8 \pm 0.2) \cdot 10^2$	$(3.6 \pm 0.9) \cdot 10^4$
0.26 M histidine	6.5	$(2.2 \pm 0.4) \cdot 10^2$	$(7.7 \pm 1.6) \cdot 10^4$
6.1 M guanidine	7.1	$(4.1 \pm 0.5) \cdot 10^2$	$(1.8 \pm 0.7) \cdot 10^5$

Table 7. Kinetic constants of peroxidatic and peroxide dependent reactions

Reaction	Kinetic constants ($k/M^{-1} \cdot s^{-1}$)			
	HRP	Fe^{3+} PP-dihistidine[1]) (TOHJO et al. 1962)	metHb[1]) (INADA et al. 1961)	P-450[3]) (COON, 1978)
Calatytic cpd I[2]) active → system	$0.9 \cdot 10^7$ (CHANCE 1951)	$(2.2 \pm 0.4) \cdot 10^2$	$6.7 \cdot 10^2$	$5 \cdot 10^3$
cpd I → cpd II	$10^6 - 10^5$ (YAMAZAKI et al. 1973)	$(7.7 \pm 1.6) \cdot 10^4$	$5.8 \cdot 10^4$	?
cpd II → catalytic active system	$10^3 - 10^4$ (YAMAZAKI et al. 1973)			

[1]) Substrate = leucomalachite green; [2]) cpd I = compound I; cpd II = compound II [3]) interaction with cumene hydroperoxide

pared with the analogous constants of the iron porphyrin complexes in which the histidine complex was selected as an example (Table 7). The constants listed in Table 7 show that the formation of compound I of peroxidase takes place very rapidly whereas the formation of compound II is the slower step. This behaviour of horseradish peroxidase deviates clearly from the kinetic behaviour of iron porphyrin complexes and ferrihemoglobin. Unfortunately the adequate constants of P–450 are unknown. Furthermore, it is not clear whether a compound I analogous form exists in the reaction cycle of P–450. On the basis of experiments carried out with cumene hydroperoxide it has to be assumed, however, that the proposed second step takes place more rapidly than the first one (COON, 1978). Therefore, it can be postulated that the mechanisms of the "peroxidatic" reaction being catalyzed by iron porphyrins and the mechanism of the peroxide dependent catalysis of P–450 have some common characteristics.

These results raise the question whether iron porphyrin complexes are able to catalyze substrate conversion reactions by H_2O_2 like P–450. A matrix bound Fe^{3+}

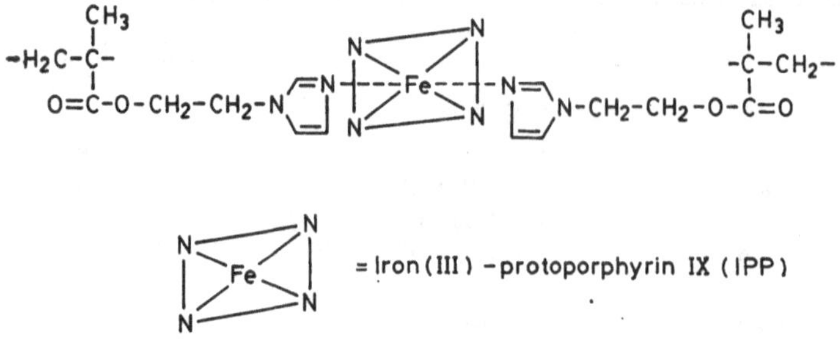

Fig. 22. Structure of the polymeric imidazole-Fe^{3+} protoporphyrin IX complex

protoporphyrin IX imidazole complex of the following structure (Fig. 22) was synthesized by KÜHN and BENES (1980) and used as catalytically active material. In this complex one iron porphyrin contacts 2 imidazole groups as it was indicated by optical spectroscopy and EPR-measurements (Fig. 22).

Analogously to P–450 the iron porphyrin polymer complex is able to catalyze demethylation as well as hydroxylation reactions of various organic substrates by H_2O_2 (Table 8). Against it typical peroxidase substrates (as leukomalachite green, guaiacol, benzidine) are not dehydrated. This also supports the hypothesis that the peroxide dependent catalysis of these iron porphyrin complexes is more related to the P–450 catalysis.

The activity of the iron porphyrin polymer complex is in the same order of magnitude as the activity of microsomal P–450 under identical conditions as it is proved by the turnover numbers as well as by V_{max} and K_M (the Michaelis-Menten constant) in case of the aniline hydroxylation and benzphetamine demethylation (Table 9).

Table 8. Turnover rates of H_2O_2-dependent hydroxylation and demethylation reactions catalyzed by a matrix bound Fe^{3+}-protoporphyrin IX imidazole complex (KÜHN and MOHR, 1981)

Substrate	Turnover number[1])	
Aniline	135	hydroxylation
p-Toluidine	117	
N.N-Dimethylaniline	170	
Acetanilide	45	
Substrate	Turnover number[2])	
Aminopyrine	180	N-demethylation
Ethylmorphine	156	
Benzphetamine	240	
N.N-Dimethylaniline	165	
p-Nitroanisole[3])		O-demethylation

[1]) n moles product/mg immobilized IPP/min
[2]) n moles formaldehyde/mg immobilized IPP/min
[3]) p-Nitrophenol determined as product

MANSUY et al. (1980) found that Fe^{3+}-mesotetraphenyl-porphyrin chloride catalyzes at 20 °C the oxidation of cyclohexane to cyclohexanol and cyclohexanon by cumene and tert. butylhydroperoxide in benzene as solvent (Table 10).

An enzyme model system consisting of [chloro-meso-tetraphenylporphynatoiron-(III)] and imidazole as ligand has been found to catalyze the H_2O_2-dependent S-oxidation and oxidative S-dealkylation of organic sulfides. In acetonitrile as reaction medium imidazole accelerates the oxidation extremely but it is not absolutely necessary. Against it, in chloroform imidazole is an essential part of this catalytic active system. The oxidation process is quite similar to the enzymatic oxidation by P–450 concerning the following points: (a) duality of the reaction path; (b) polar substituent effects on rate; (c) stereoselectivity of S-oxidation of unsymmetrical sulfides (OAE, 1982).

Molecular oxygen can be activated by iron porphyrins too (NAKAMURA et al., 1963; KAUFMANN and SCHILLER, 1971; MÜLLER et al., 1973; O'CARRA, 1975; ADAMS and BERMAN, 1980; KÜHN et al., 1981) but in most cases the mechanism is unknown or not comparable with the activation of O_2 by P–450. The hydroxylation and demethylation of several organic substrates in a mixed function oxidation corresponding to reaction (1) is also catalyzed by the polymeric iron porphyrin complexes shown in Figure 22. The turnover rates in Table 11, however, are considerably lower and amount to about 3% of the rates obtained with the same amount of P–450 under identical conditions.

With its ability to catalyze hydroxylation and demethylation reactions of organic substrates the polymeric complex differs from the corresponding low molecular iron porphyrin imidazole complex

Table 9. Kinetic constants of the H_2O_2 dependent aniline hydroxylation and benzphetamine demethylation, catalyzed by microsomal P–450 and by the polymeric diimidazole/iron (III)-protoporphyrin IX complex (MOHR et al., 1979)

Catalytically active system[2])	Turnover number[1]) (n mole product/n mole IPP · min)		K_M (mM)		V_{max} (n mole/ n mole · min)
	aniline hydroxylation	benzphetamine demethylation	related to aniline	related to H_2O_2	
Polymeric diimidazole/IPP complex	0.29	0.5	1.7	14	1.5
Microsomal P–450 rabbit	0.10	1.0	0.6	18	1.3

[1]) H_2O_2 = 800 μM; substrate = 800 μM; pH 7.0; [2]) conc. = 1 μM

Table 10. Oxidation of cyclohexane by organic hydroperoxides in presence of Fe^{3+}-mesotetraphenylporphyrin chloride (MANSUY et al., 1980)

Hydroperoxide	Product after 15 min (%)	
	$C_6H_{11}OH$	$C_6H_{10}O$
Cumene	40	20
tert. Butyl	20	12

Table 11. Turnover rates of the mixed function oxidation and demethylation, respectively, of various substrates catalyzed by a matrix bound Fe^{3+} protoporphyrin IX imidazole complex (KÜHN et al., 1981)

Substrate	Product	Turnover number[1])
Aniline	aminophenol[2])	10.5
Acetanilide	4-hydroxyacetanilid	16
N.N-Dimethylaniline	hydroxy-N.N-dimethyl-aniline[2/3])	26
Aminopyrine	formaldehyde	95
Ethylmorphine	formaldehyde	22
Benzphetamine	formaldehyde	46
N.N-Dimethylaniline	formaldehyde[3])	41

[1]) n moles product/mg immobilized IPP/min

[2]) mixture of isomers

[3]) the substrate is both hydroxylated and demethylated

which is only able to initiate a dehydration of organic substrates analogously to horseradish peroxidase (PSHEZHETSKII and JAROSLAVOV, 1977; TOHJO et al., 1962).

5. Concluding remarks

The research work concerning the practical utilization of P–450 and its enzymatic principle was considerably stimulated in the last period by an increased deepening of our insight into the structure of this enzyme system and the mechanism of its catalytic effectivity. Nevertheless the field of application of P–450 at present is in a pioneering state of development.

With the exception of the *Salmonella mammalian* liver test the enzymatic principle of P–450 is not utilized up to now. Essential preconditions to apply P–450 in continuously working enzyme reactors and sensors are (a) the improvement of the enzyme stability and (b) a selection of the most favourable way to supply the electrons for each device.

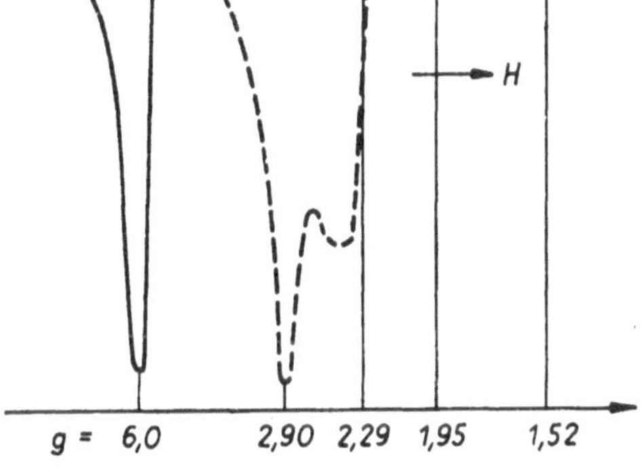

Fig. 23. EPR spectra (——) of iron (III)-protoporphyrin IX, (– – –) polymeric diimidazole/iron-(III)-protoporphyrin IX complex

The utilization of whole living cells in suspended or immobilized form offers a further real possibility in this direction e.g. in steroid metabilizing and drug biotransforming reactors (WISEMAN, 1977). A further essential progress can be expected by the application of stable iron porphyrin derivatives possessing P–450 like properties.

Summarizing it can be stated that the enzymatic principle of P–450 is of a considerable practical interest. However, it is impossible at present to forecast which directions of application will be successful in the next future.

References

ADAMS, P. A., and M. C. BERMAN (1980): Int. Conference on Xenobiochemistry, Bratislava, Abstracts, p. 1

AIZAWA, M., R. COUGHLIN, and M. CHARLES (1976a): Biochim. Biophys. Acta **440**, 233–239

AIZAWA, M., R. COUGHLIN, and M. CHARLES (1976b): Biotechnol. Bioengng. **18**, 209–215

AKHREM, A. A., D. I. METELITSA, S. M. BIELSKI, P. A. KISELEV, M. E. SKURKO, and S. A. USANOV (1977): Croat. Chim. Acta **49**, 223–235

AKHREM, A. A., V. N. LAPKO, A. G. LAPKO, V. M. SHKUMATOV, and V. I. CHASHCHIN (1979): Acta biol. med. germ. **38**, 257–273

AMES, B. N. (1976a): Science **191**, 241–244

AMES, B. N. (1976b): in: In vitro metabolic activation in mutagenesis testing (DESERIES, F. J., J. R. FOUTS, J. R. BEND, R. M. PHILPOT, ed.), North-Holland-Publishing Company, Amsterdam–New York–Oxford, pp. 57–62

AMES, B. N. (1979): Science **204**, 587–593

Anonymus (1974): Chem. Engng. News **17**, 19

ARCHAKOV, A. I., B. A. KUSNETSOV, M. V. ISOTOV and I. I. KURUZINA (1981): Dokl. AN SSSR **258**, 216–219

BACH, D., and J. MILLER (1970): Electrochim. Acta **15**, 533–540

BAESS, D., G.-R. JÄNIG, and K. RUCKPAUL (1975): Acta biol. med. germ. **34**, 1745–1754

BERG, A., J.-Å. GUSTAFSSON, and M. INGELMAN-SUNDBERG (1976): J. Biol. Chem. **251**, 2831–2838

BERGER, R. (1976): Z. Chem. **16**, 386–395

BERGMEYER, H.-U. (1962): Methoden der enzymatischen Analyse-Verlag, Chemie GmbH, Weinheim

BRUNNER, G., and H. LOESGEN (1977): in Artificial Organs (KENEDI, R., J. COURTNEY, J. GAYLOR, and T. GILCHRIST, eds.), Mcmillan Press, Ltd., London, pp. 388–394

BRUNNER, G., W. LOESGEN, B. GAWLIK and K. BELSNER (1980): in "Biochemistry, Biophysics and Regulation of Cytochrome P-450" (J.-Å. GUSTAFSSON, Edit.), Elsevier/North Holland Biomedical Press, Amsterdam–New York–Oxford 1980, pp. 573–576

CHANG, C. K., and D. DOLPHIN (1975): J. Amer. Chem. Soc. **97**, 5948–5950

CHANG, C. K., and M. S. KUO (1979): J. Amer. Chem. Soc. **101**, 3413–3415

CHEVALET, J., F. ROUELLE, L. GIERST, and J. LAMBERT (1972): J. Electroanal. Chem. Interfacial Electrochem. **39**, 201–216

COHEN, W., W. H. BARICOS, P. R. KASTL, and R. P. CHAMBERS (1977): in Biomedical applications of immobilized enzymes and proteins (CHANG, T. M. S., ed.) Plenum Press, New York–London, pp. 319–329

COLLMAN, J. P., T. N. SORRELL, and B. HOFMAN (1975): J. Amer. Chem. Soc. **97**, 913–914

COON, M. J. (1978): personal communication

COOPER, D. Y., H. SCHLEYER, and O. ROSENTHAL (1973): Drug. Metab. Dispos. **1**, 21–28

CUNNINGHAM, A., and L. UNDERWOOD (1966): Arch. Biochem. Biophys. **117**, 88–94

DAVIES, P., and K. MOSBACH (1974): Biochim. Biophys. Acta **370**, 329–338

DENTI, E., J. W. FRESTON, M. MARCHISIO, and W. J. KOLFF (1976): Trans. Amer. Soc. Artifical Internal. Organs **22**, 693–700

DIVISEK, I., and J. KASTENIG (1975): J. Electroanal. Chem. Interfacial Electrochem. **65**, 603–612

DUPPEL, W., J.-M. LEBEAULT, and M. J. COON (1973): Eur. J. Biochem. **36**, 583–592

ELVING, P. J. (1976): Topics in Bioelectrochem. **1**, 179–286

ERYOMIN, A. N., S. A. USANOV, and D. I. METELITSA (1979): Prikl. Biokhim. Mikrobiol. **15**, 861–867

ERYOMIN, A. N., D. I. METELITSA, S. A. USANOV, and A. A. AKHREM (1980): Prikl. Biokhim. Mikrobiol. **16**, 206–211

GROVES, J. T., T. E. NEMO, and R. S. MYERS (1979): J. Amer. Chem. Soc. **101**, 1032–1033

GUSTAFSSON, J.-Å., and J. BERGMAN (1976): FEBS-Lett. **70**, 276–280

GUSTAFSSON, J.-Å., E. G. HRYCAY, and L. ERNSTER (1976): Arch. Biochem. Biophys. **174**, 440–453

HRYCAY, E. G., and P. J. O'BRIEN (1972): Arch. Biochem. Biophys. **153**, 480–494

HRYCAY, E. G., and P. J. O'BRIEN (1974): Arch. Biochem. Biophys. **160**, 230–245

HRYCAY, E. G., J.-Å. GUSTAFSSON, M. INGELMAN-SUNDBERG, and L. ERNSTER (1975): FEBS-Lett. **56**, 161–165

HRYCAY, E. G., J.-Å. GUSTAFSSON, M. INGELMAN-SUNDBERG, and L. ERNSTER (1976): Eur. J. Biochem. **61**, 43–52

INADA, Y., T. KUROZUMI, and K. SHIBATA (1961): Arch. Biochem. Biophys. **93**, 30–36

JÄNIG, G.-R., D. PFEIL, H. HONECK, and K. RUCKPAUL (1977): Acta biol. med. germ. **36**, 35–44

JANIK, B., and P. ELVING (1968): Chem. Rev. **68**, 295–319

KADLUBAR, F. F., K. C. MORTON, and D. M. ZIEGLER (1973): Biochim. Biophys. Res. Commun. **54**, 1255–1261

KASTL, P. R., W. H. BARICOS, R. P. CHAMBERS, and W. COHEN (1978): Enzyme Engineering Vol. 4 (BROUN, G. B., G. MANECKE, and L. B. WINGARD, Jr., eds.), Plenum Press New York–London, pp. 199–206

KASTL, P. R., W. H. BARICOS, W. COHEN, and R. P. CHAMBERS (1979): in Immobilized Microbial Cells (K. Venkatsubramanian, ed.) ACS Symposium Series, Washington DC, **106**, pp. 237–247

KATAGIRI, M., B. N. GANGULI, and J. C. GUNSALUS (1968): J. Biol. Chem. **243**, 3543–3546

KAUFMANN, H. P., and H. SCHILLER (1971): Fette, Seifen, Anstrichmittel **73**, 209

KAWAI, M., and K. SCRIMGEOUR (1972): Can. J. Biochem. **50**, 1191–1196

KE, B. (1956): J. Amer. Chem. Soc. **78**, 3649–3651

KIRSTEIN, D., F. SCHELLER and F. SCHUBERT (1982): Priklad. Biokhim. Mikrobiol. (russ.) **18**, 471–480

KÜHN, M., and M. BENES (1980): Z. Chem. **20**, 351–352

KÜHN, M., O. RISTAU, and J. ČOUPEK (1981): Z. Chem. 21 231–232
KÜHN, M., and P. MOHR (1981): Pharmazie **36**, 382–383
KÜHN, M. (1982): Prikl. Biochim. Mikrobiol. (russ.) **18**, 489–498
KUZNETSOV, B., N. MESTECHKINA, M. IZOTOV, J. KARUZINA, A. KARYAKIN, and A. ARCHAKOV (1979): Biokhim. **44**, 1569–1575
LEHMAN, J. P., L. FERRIN, C. FENSELAU, and G. S. YOST (1981): Drug Metab. Dispos. **9**, 15–18
LICHTENBERGER, F., NASTAINZYK, W., and V. ULLRICH (1976): Biochim. Biophys. Res. Commun. **70**, 939–946
LU, A. Y. H., R. KUNTZMANN, S. WEST, M. JACOBSON, and A. H. CONNEY (1972): J. Biol. Chem. **247**, 1727–1734
MANSUY, D. (1980a): Pure Appl. Chem. **52**, 681–690
MANSUY, D., J.-F. BARTOLI, J.-C. CHOTTARD, and M. LANGE (1980b): Angew. Chem. **92**, 938–939
MANSUY, D., P. M. DANSETTO, F. PEOQUET, and J.-C. CHOTTARD (1980c): Biochem. Biophys. Res. Commun. **96**, 433–439
MANSUY, D. (1981): Rev. Biochem. Toxicol. **3**, 283–320
MIEYAL, J. J., R. S. ACKERMAN, J. L. BLUMER, and L. S. FREEMAN (1976): J. Biol. Chem. **251**, 3436–3441
MIEYAL, J. J., and J. L. BLUMER (1976): J. Biol. Chem. **251**, 3442–3446
MOHR, P., F. SCHELLER, R..RENNEBERG, M. KÜHN, and W. SCHELER (1978): Pharmazie **33**, 415–418
MOHR, P., M. KÜHN, E. WESULS, R. RENNEBERG, and F. SCHELLER (1979): Acta biol. med. germ. **38**, 495–501
MÜLLER, H.-G., P. MOHR, and W. SCHELER (1973): Abhandl. der AdW der DDR, 155 –158
MÜLLER, H.-G., W.-H. SCHUNCK, P. RIEGE, and H. HONECK (1979): Acta biol. med. germ. **38**, 345
NAKAMURA, Y., M. TOHJO, and K. SHIBATA (1963): Arch. Biochem. Biophys. **102**, 144
NORDBLOM, G. D., R. E. WHITE, and M. J. COON (1976): Arch. Biochem. Biophys. **175**, 524–533
OAE, SH., Y. WATANABE, and K. FUJIMORI (1982): Tetrahedron Letters **23**, 1189–1192
O'CARRA, P. (1975): in Porphyrins and Metalloporphyrins, (SMITH, K. M., ed.) Elsevier; Amsterdam, pp. 123–153
OMURA, T., E. SANDERS, R. W. ESTABROOK, P. Y. COOPER, and O. ROSENTHAL (1966): Arch. Biochem. Biophys. **117**, 660–673
PFEIL, D., G.-R. JÄNIG, J. BEHLKE, and K. RUCKPAUL (1977): Acta biol. med. germ. **36**, 1243–1251
POMMERENING, K., H. HONECK, and J. BEHLKE (1974): GDR-Patent 107703
PORTSMOUTH, D., and E. A. BEAL (1971): Eur. J. Biochem. **19**, 479–487
POWNING, R., and G. KRATZING (1957): Arch. Biochem. Biophys. **66**, 249–252
PSHEZHETSKII, V. S., and A. A. JAROSLAVOV (1977): Bioorgan. Khim. **3**, 1117–1125
RAHIMTULA, A. D., and P. J. O'BRIEN (1974): Biochem. Biophys. Res. Commun. **60**, 440–447
RAHIMTULA, A. D., and P. J. O'BRIEN (1975): Biochem. Biophys. Res. Commun. **62**, 268–275
RENNEBERG, R., F. SCHELLER, K. RUCKPAUL, J. PIRRWITZ, and P. MOHR (1978): FEBS-Lett. **96**, 349–353
RENNEBERG, R., J. CAPDEVILA, N. CHACOS, R. W. ESTABROOK, and R. A. PROUGH (1981a): Biochem. Pharmacol. **30**, 843–848
RENNEBERG, R., W. DAMERAU, CH. JUNG, and F. SCHELLER (1981b): unpublished results
RIEGE, P., W.-H. SCHUNCK, R. BLASIG, H. HONECK, and H.-G. MÜLLER (1977). 4. FEMS-Symposium, Vienna (Austria) 28 march–1. april 1977
RIEGE, P., W.-H. SCHUNCK, H. HONECK, and H.-G. MÜLLER (1981): Biochim. Biophys. Res. Commun. **98**, 527–534
RUF, H. H., P. WENDE, and V. ULLRICH (1979): J. Bioorg. Chem. **11**, 189–204
SAKURAI, H. (1976): Chem. Pharm. Bull. **24**, 1686–1688
SAKURAI, H., and S. OGAWA (1979): Chem. Pharm. Bull. **27**, 2171–2176
SCHELLER, F., R. RENNEBERG, P. MOHR, G.-R. JÄNIG, and K. RUCKPAUL (1976): FEBS-Lett. **71**, 309–312
SCHELLER, F., R. RENNEBERG, G. STRNAD, K. POMMERENING, and P. MOHR (1977b): Bioelectrochem. Bioenerg. **4**, 500–507
SCHELLER, F., R. RENNEBERG, W. SCHWARZE, G. STRNAD, K. POMMERENING, H.-J. PRÜMKE, and P. MOHR (1979): Acta biol. med. germ. **38**, 503–509

SCHELLER, F. (1980): Doctoral Thesis, Berlin, GDR

SCHELLER, F., P. MOHR, I. SEYER, G.-R. JÄNIG, and R. RENNEBERG (1977a): GDR-patent 126807

SCHMAKEL, C. O., K. S. V. SANTHANAM, and P. ELVING (1975): J. Am. Chem. Soc. **97**, 5083–5092

SCHUBERT, F., D. KIRSTEIN, F. SCHELLER, and P. MOHR (1980): Anal. Lett. **13**, 1167–1178

SCHUBERT, F., D. KIRSTEIN, and F. SCHELLER (1982): Acta Biotechnol. **2**, 187–191 .

SKINNER, K. J. (1975): Chem. Engng. News **18**, 22–24

SOFER, S. S., D. M. ZIEGLER, and R. P. POPOVICH (1974): Biochem. Biophys. Res. Commun. **57**, 183–189

SOFER, S. S., D. M. ZIEGLER, and R. P. POPOVICH (1975): Biotechnol. Bioengng. **17**, 107–117

STERN, J. O., and M. PEISACH (1974): J. Biol. Chem. **249**, 7495–7498

TANG, S. C., S. KOCH, G. C. PAPAEFTHYMION, S. FONER, R. B. FRANKEL, J.-A. IBERS, and R. H. HOLM (1976): J. Am. Chem. Soc. **98**, 2414–2434

THEVENOT, D., and G. H. HAMMOUYA (1971): Experientia Suppl. **18**, 631–645

THEVENOT, D., and R. BUVET (1972): J. Electroanal. Chem. Interfacial Electrochem. **39**, 429–446

TOHJO, M., Y. NAKAMURA, K. KURIHARA, T. SAMEJIMA, Y. GLACHIMORI, and K. SHIBATA (1962): Arch. Biochem. Biophys. **99**, 222–240

TRAYLOR, T. G., and T. MINCEY (1969): Acta biol. med. germ. **38**, 351–355

WELIKY, N. J. DAY, H. GALE, T. SEV, and P. SILVERMAN (1972): Trans. N.Y. Acad. Sci. **34**, 647–655

WHITE, R. W., and M. J. COON (1980): Ann. Rev. Biochem. **49**, 315–356

WISEMAN, A. (1977): in Topics in Enzyme and Fermentation Biotechnology (WISEMAN, A., ed.) Ellis Horwood Ltd. Publisher, Chichester, 172–184

YAO, T., and S. MUSHA (1979): Anal. Chim. Acta **110**, 203

YOST, G. S., L. FERRIN, J. P. LEHMANN, and C. FENSELAU (1980): Fed. Proc. **39**, 3941

List of Contributors

J. BLANCK, Department of Biocatalysis, Central Institute of Molecular Biology, Academy of Sciences of the GDR, Berlin-Buch

R. BERNHARDT, Department of Biocatalysis, Central Institute of Molecular Biology, Academy of Sciences of the GDR, Berlin-Buch

G. BUTSCHAK, Department of Immunology, Central Institute of Cancer Research, Academy of Sciences of the GDR, Berlin-Buch

J. FRIEDRICH, Department of Biocatalysis, Central Institute for Molecular Biology, Academy of Sciences of the GDR, Berlin-Buch

G. GRESCHNER, Department of Biocatalysis, Central Institute of Molecular Biology, Academy of Sciences of the GDR, Berlin-Buch

H. HONECK, Department of Applied Enzymology, Central Institute of Molecular Biology, Academy of Sciences of the GDR, Berlin-Buch

G.-R. JÄNIG, Department of Biocatalysis, Central Institute of Molecular Biology, Academy of Sciences of the GDR, Berlin-Buch

C. JUNG, Department of Hemecatalysis, Central Institute of Molecular Biology, Academy of Sciences of the GDR, Berlin-Buch

M. KÜHN, Department of Applied Enzymology, Central Institute of Molecular Biology, Academy of Sciences of the GDR, Berlin-Buch

J. LAMPE, Department of Hemecatalysis, Central Institute of Molecular Biology, Academy of Sciences of the GDR, Berlin-Buch

S. MARICIC, Zagreb, Proleterskih brigada 35, Yugoslavia

P. MOHR, Department of Applied Enzymology, Central Institute of Molecular Biology, Academy of Sciences of the GDR, Berlin-Buch

H.-G. MÜLLER, Department of Applied Enzymology, Central Institute of Molecular Biology, Academy of Sciences of the GDR, Berlin-Buch

D. PFEIL, Department of Biocatalysis, Central Institute of Molecular Biology, Academy of Sciences of the GDR, Berlin-Buch

K. POMMERENING, Department of Hemecatalysis, Central Institute of Molecular Biology, Academy of Sciences of the GDR, Berlin-Buch

H. REIN, Department of Hemecatalysis, Central Institute of Molecular Biology, Academy of Sciences of the GDR, Berlin-Buch

R. RENNEBERG, Department of Applied Enzymology, Central Institute of Molecular Biology, Academy of Sciences of the GDR, Berlin-Buch

P. RIEGE, Department of Applied Enzymology, Central Institute of Molecular Biology, Academy of Sciences of the GDR, Berlin-Buch

O. RISTAU, Department of Hemecatalysis, Central Institute of Molecular Biology, Academy of Sciences of the GDR, Berlin-Buch

K. RUCKPAUL, Department of Biocatalysis, Central Institute of Molecular Biology, Academy of Sciences of the GDR, Berlin-Buch

W. SCHELER, Department of Hemecatalysis, Central Institute of Molecular Biology, Academy of Sciences of the GDR, Berlin-Buch

F. SCHELLER, Department of Applied Enzymology, Central Institute of Molecular Biology, Academy of Sciences of the GDR, Berlin-Buch

F. SCHUBERT, Department of Applied Enzymology, Central Institute of Molecular Biology, Academy of Sciences of the GDR, Berlin-Buch

W.-H. SCHUNCK, Department of Applied Enzymology, Central Institute of Molecular Biology, Academy of Sciences of the GDR, Berlin-Buch

G. SMETTAN, Department of Biocatalysis, Central Institute of Molecular Biology, Academy of Sciences of the GDR, Berlin-Buch

Subject Index

FSC
www.fsc.org
MIX
Papier | Fördert
gute Waldnutzung
FSC® C083411

Kolibri 360 GmbH
Ferdinand-Jühlke-Straße 7
99095 Erfurt, Deutschland
produktsicherheit@kolibri360.de